Lecture Notes in Computer Science   14459

Founding Editors

Gerhard Goos
Juris Hartmanis

The series Lecture Notes in Computer Science (LNCS), including its subseries Lecture Notes in Artificial Intelligence (LNAI) and Lecture Notes in Bioinformatics (LNBI), has established itself as a medium for the publication of new developments in computer science and information technology research, teaching, and education.

LNCS enjoys close cooperation with the computer science R & D community, the series counts many renowned academics among its volume editors and paper authors, and collaborates with prestigious societies. Its mission is to serve this international community by providing an invaluable service, mainly focused on the publication of conference and workshop proceedings and postproceedings. LNCS commenced publication in 1973.

Anupam Chattopadhyay · Shivam Bhasin ·
Stjepan Picek · Chester Rebeiro
Editors

Progress in Cryptology – INDOCRYPT 2023

24th International Conference on Cryptology in India
Goa, India, December 10–13, 2023
Proceedings, Part I

 Springer

Editors
Anupam Chattopadhyay 🆔
Nanyang Technological University
Singapore, Singapore

Shivam Bhasin 🆔
Nanyang Technological University
Singapore, Singapore

Stjepan Picek 🆔
Radboud University
Nijmegen, The Netherlands

Chester Rebeiro 🆔
Indian Institute of Technology Madras
Chennai, India

ISSN 0302-9743   ISSN 1611-3349 (electronic)
Lecture Notes in Computer Science
ISBN 978-3-031-56231-0   ISBN 978-3-031-56232-7 (eBook)
https://doi.org/10.1007/978-3-031-56232-7

This Springer imprint is published by the registered company Springer Nature Switzerland AG
The registered company address is: Gewerbestrasse 11, 6330 Cham, Switzerland

Paper in this product is recyclable.

Foreword

We, at BITS Pilani, K K Birla Goa Campus, were immensely pleased to organize and host INDOCRYPT 2023, the 24th International Conference on Cryptology in India, jointly with Indian Statistical Institute, Kolkata, under the aegis of the Cryptology Research Society of India (CRSI). INDOCRYPT began in 2000 under the leadership of Prof. Bimal Roy at the Indian Statistical Institute, Kolkata, with an intention to target researchers and academicians in the domain of cryptology. Since its inception, this annual conference has not only been considered as the leading Indian venue on cryptology but also has gained recognition among the prestigious cryptology conferences in the world. This was the first time the conference has been held on one of the Indian campuses of BITS Pilani. I gather that INDOCRYPT 2023 received diverse and significant submissions from different countries. The conference also had exciting workshops lined up! I'm grateful to the General Co-chairs, Bimal Kumar Roy, Subhamoy Maitra (ISI Kolkata) and Indivar Gupta (DRDO), for coordinating all the issues related to the organization of the event. We would also like to take this opportunity to thank the Organizing Chair, Organizing Co-chairs, and all members of BITS Goa without whom this conference would not have taken shape.

INDOCRYPT targets researchers from Industry and academia in areas which include but are not limited to foundations, new primitives, cryptanalysis, provable security, cryptographic protocols, and (post)-quantum cryptography. Needless to say, in this era of fast-moving Information Technology, the focus areas of Indocrypt are relevant and impactful. At BITS Goa we are trying our best to take tiny steps in these emerging domains of IT to augment our own strength in related fields. We were excited to host INDOCRYPT while we were celebrating 20 years of our sojourn!

We welcome all delegates, guests and participants, approximately 250 in total, to the post-proceedings of INDOCRYPT 2023 and hope that they immersed themselves in thoughtful discussions and interaction during December 10–13, 2023, the dates of the conference, while enjoying our beautiful campus in the scenic state of Goa.

December 2023

Suman Kundu
Indivar Gupta
Bimal Kumar Roy

Preface

It is our utmost pleasure to welcome you to go through the following pages, which contain the proceedings of the 24th International Conference on Cryptology in India, commonly known as Indocrypt. The conference took place during 10th–13th December, 2023, at the beautiful campus of BITS Pilani, Goa.

We gathered a brilliant group of 62 researchers, who graciously agreed to serve on the Technical Program Committee. Based on a well-publicized Call for Papers, the conference received 86 full paper submissions. Out of these, 27 papers were finally accepted based on double-blind reviews and extensive discussions among the TPC members. Each paper received at least 3 reviews. For several papers, shepherding was done after the review process to ensure the quality of the final manuscript. The acceptance rate was 31.39%, which is slightly lower than that of the previous edition of this conference - 35.23%. The general theme of this edition of the conference was to look into quantum-resilient security. Matching this theme, we had a complete session organized by Bosch India Private Limited, who contributed to 2 invited papers. As part of the technical program, we also solicited tutorial speakers (from CDAC, Bengaluru, India), panel discussions and keynote speeches, which are not included in these proceedings.

We are very thankful to several people and organizations who played a huge supporting role behind the successful organization of this conference. The following list is our humble attempt to acknowledge their service and support. First and foremost, we would like to thank our sponsors, listed in no order of priority - CRSI, CDAC, DRDO, Bosch India Private Limited, Google India, Microsoft Research Lab - India, and BITS Pilani, Goa campus for being gracious hosts. We are very thankful for the service of the entire organization committee for their hard work through the last few months of event organization. Last but not least, we would like to thank our General Chairs, Bimal Roy, Indivar Gupta (SAG, DRDO) and Suman Kundu, for guidance and motivation.

To the scientifically curious reader, we thank you for your engagement in the Indocrypt conference, and hope to see you at future editions!

December 2023

Shivam Bhasin
Anupam Chattopadhyay
Stjepan Picek
Chester Rebeiro

Organization

General Chairs

Bimal Kumar Roy   Indian Statistical Institute, India
Suman Kundu   BITS Pilani, India
Indivar Gupta   DRDO, India

Technical Program Co-chairs

Anupam Chattopadhyay   NTU, Singapore
Shivam Bhasin   Temasek Labs, NTU, Singapore
Stjepan Picek   Radboud University, The Netherlands
Chester Rebeiro   IIT Madras, India

Organizing Chairs

Snehanshu Saha   BITS Pilani, India
Santonu Sarkar   BITS Pilani, India

Organizing Co-chairs

Hemant Rathore   BITS Pilani, India
Sravan Danda   BITS Pilani, India
Diptendu Chatterjee   BITS Pilani, India
Gargi Alavani   BITS Pilani, India
Prabal Paul   BITS Pilani, India

Publicity Chair

Shilpa Gondhali   BITS Pilani, India

Industry Chair

Kunal Korgaonkar   BITS Pilani, India

Web Co-chairs

Nilanjan Datta   IAI TCG CREST, India
Hemant Rathore   BITS Pilani, India
Shreenivas A. Naik   BITS Pilani, India

Publication Co-chairs

Debolina Ghatak   BITS Pilani, India
Saranya G. Nair   BITS Pilani, India

Program Committee

Avishek Adhakari   Presidency University, Kolkata, India
Subidh Ali   IIT Bhilai, India
Nalla Anandakumar   Continental Automotive, Singapore
Shi Bai   Florida Atlantic University, USA
Anubhab Baksi   Nanyang Technological University, Singapore
Shivam Bhasin   Temasek Lab, Nanyang Technological University,
                           Singapore
Rishiraj Bhattacharyya   University of Birmingham, UK
Christina Boura   University of Versailles, France
Suvradip Chakraborty   VISA Inc, USA
Anupam Chattopadhyay   Nanyang Technological University, Singapore
Sherman S. M. Chow   Chinese University of Hong Kong, China
Debayan Das   Purdue University, USA
Prem Laxman Das   SETS Chennai, India
Nilanjan Datta   IAI, TCG CREST, Kolkata, India
Avijit Dutta   IAI, TCG CREST, Kolkata, India
Ratna Dutta   IIT Kharagpur, India
Keita Emura   Kanazawa University, Japan
Andre Esser   Technology Innovation Institute, Abu Dhabi, UAE
Satrajit Ghosh   Indian Institute of Technology Kharagpur, India
Indivar Gupta   SAG DRDO, India
Takanori Isobe   University of Hyogo, Japan
Dirmanto Jap   Nanyang Technological University, Singapore
Mahavir Jhanwar   Ashoka University, India
Selçuk Kavut   Balikesir University, Turkey
Mustafa Khairallah   Seagate Technologies, Singapore

Jason LeGrow   Virginia Polytechnic Institute and State
                           University, USA
Chaoyun Li   COSIC, KU Leuven, Belgium
Fukang Liu   University of Hyogo, Japan
Arpita Maitra   IAI, TCG CREST, Kolkata, India
Monosij Maitra   Ruhr-Universität Bochum and Max Planck
                           Institute for Security and Privacy, Germany
Luca Mariot   University of Twente, The Netherlands
Willi Meier   FHNW, Windisch, Switzerland
Alfred Menezes   University of Waterloo, Canada
Nele Mentens   KU Leuven, Belgium
Sihem Mesnager   Universities of Paris VIII and XIII, LAGA Lab,
                           France
Marine Minier   LORIA, France
Pratyay Mukherjee   Swirlds Labs/Hedera, USA
Debdeep Mukhopadhyay   Indian Institute of Technology Kharagpur, India
Mridul Nandi   Indian Statistical Institute, Kolkata, India
Saibal Pal   DRDO, Delhi, India
Sumit Kumar Pandey   Indian Institute of Technology Jammu, India
Thomas Peyrin   Nanyang Technological University, Singapore
Stjepan Picek   Radboud University, The Netherlands
Prasanna Ravi   Nanyang Technological University, Singapore
Chester Rebeiro   IIT Madras, Chennai, India
Francesco Regazzoni   University of Amsterdam, The Netherlands
Raghavendra Rohit   Technology Innovation Institute, Abu Dhabi, UAE
Sushmita Ruj   University of New South Wales, Sydney, Australia
Somitra Sanadhya   IIT Jodhpur, India
Santanu Sarkar   IIT Madras, India
Sourav Sen Gupta   IMEC, Leuven, Belgium
Nicolas Sendrier   INRIA, France
Yixin Shen   Royal Holloway, University of London, UK
Bhupendra Singh   CAIR, DRDO, Bangalore, India
Sujoy Sinha Roy   IAIK, TU Graz, Austria
Pantelimon Stanica   Naval Postgraduate School, Monterey, USA
Atsushi Takayasu   University of Tokyo, Japan
Alexandre Wallet   INRIA, France
Jun Xu   Institute of Information Engineering, Chinese
                           Academy of Sciences, China
Haoyang Wang   Shanghai Jiao Tong University, China
Yuyu Wang   University of Electronic Science and Technology
                           of China, China

Invited Papers

Secure Boot in Post-Quantum Era

Megha Agrawal⑩, Kumar Duraisamy⑩, Karthikeyan Sabari Ganesan⑩,
Shivam Gupta⑩, Suyash Kandele⑩, Sai Sandilya Konduru⑩,
Harika Chowdary Maddipati⑩, K. Raghavendra⑩, Rajeev Anand Sahu⑩,
and Vishal Saraswat⑩

Bosch Global Software Technologies, Bangalore, India

Abstract. Secure boot is a standard feature for ensuring the authentication and integrity of software. For this purpose, secure boot leverages the advantage of Public Key Cryptography (PKC). However, the fast-developing quantum computers have posed serious threats to the existing PKC. The cryptography community is already preparing to thwart the expected quantum attacks. Moreover, the standardization of post-quantum cryptographic algorithms by NIST has advanced to 4th round, after selecting and announcing the post-quantum encryption and signature schemes for standardization. Hence, considering the recent developments, it is high time to realize a smooth transition from conventional PKC to post-quantum PKC. In this paper, we have implemented the PQ algorithms recently selected by NIST for standardization– CRYSTALS-Dilithium, FALCON and SPHINCS$^+$ as candidate schemes in the secure boot process. Furthermore, we have also proposed an idea of double signing the boot stages, for enhanced security, with signing a classical signature by a post-quantum signature. We have also provided efficiency analysis for various combinations of these double signatures.

Keywords: Secure Boot · Post-Quantum Cryptography · Dilithium · FALCON · SPHINCS$^+$

Patent Landscape in the field of Hash-Based Post-Quantum Signatures

Megha Agrawal⬤, Kumar Duraisamy⬤, Karthikeyan Sabari Ganesan⬤,
Shivam Gupta⬤, Suyash Kandele⬤, Sai Sandilya Konduru⬤,
Harika Chowdary Maddipati⬤, K. Raghavendra⬤, Rajeev Anand Sahu⬤,
and Vishal Saraswat⬤

Bosch Global Software Technologies, Bangalore, India

Abstract. Post-Quantum Cryptography (PQC) is one of the most fascinating topics of recent developments in cryptography. Following the ongoing standardization process of PQC by NIST, industry and academia both have been engaged in PQC research with great interest. One of the candidate algorithms finalized by NIST for the standardization of post-quantum digital signatures belongs to the family of Hash-based Signatures (HBS). In this paper, we thoroughly explore and analyze the state-of-the-art patents filed in the domain of post-quantum cryptography, with special attention to HBS. We present country-wise statistics of the patents filed on the topics of PQC. Further, we categorize and discuss the patents on HBS based on the special features of their construction and different objectives. This paper will provide scrutinized information and a ready reference in the area of patents on hash-based post-quantum signatures.

Keywords: Post-Quantum Cryptography · Patents · Hash-based signatures · XMSS · LMS · SPHINCS$^+$

Contents – Part I

Attacks

Contents – Part II

Symmetric-Key Cryptography, Hash Functions, Authenticated Encryption Modes

Multimixer-156: Universal Keyed Hashing Based on Integer Multiplication and Cyclic Shift

Koustabh Ghosh[(✉)], Parisa Amiri Eliasi, and Joan Daemen

Digital Security Group, Radboud University, Nijmegen, The Netherlands
{koustabh.ghosh,parisa.amirieliasi,joan.daemen}@ru.nl

Abstract. In this paper, we introduce Multimixer-156, a new keyed hash function based on 32-bit integer multiplication and cyclic shift operations. Presence of vector instructions for these operations on many CPUs make Multimixer-156 an efficient keyed hash function in software on many platforms.

We claim Multimixer-156 is 2^{-156}-universal and provide concrete arguments supporting our claim. This small universality is achieved at an expense of efficiency and we treat the interesting security/efficiency trade-off in the paper. We compare the efficiency of our implementation of Multimixer-156 with the NH hash function family and Multimixer-128, other keyed hash functions based on integer multiplication that are 2^{-128}-universal and 2^{-127}-universal respectively, and also with the two fastest NIST LWC candidates on ARMv7 architecture. While Multimixer-156 is slower than the keyed hash functions, it is significantly faster than the other candidates.

Keywords: Keyed Hashing · Parallel Construction · Multimixer-156

1 Introduction

Cryptographic functions such as Message Authentication Code (MAC) functions [21] and doubly extendable cryptographic keyed (Deck) functions [5] require the compression of variable-length inputs into a fixed-size state under a secret key. For that they can make use of keyed hash functions, functions that take an input of arbitrary number of blocks together with a secret key, and return a digest. The sole security requirement for these keyed hash functions is that a non-adaptive attacker that does not know the secret key shall not be able to find a state collision: different input strings leading to the same state value. Here non-adaptive means that the attacker is not given the state values, but only the information whether a collision was obtained. The probability of finding such a state collision can be measured in terms of its ε-universality and ε-Δuniversality [20].

There are various approaches to building secure keyed hash functions, like constructions based on cryptographic hash functions such as HMAC [2] and

A. Chattopadhyay et al. (Eds.): INDOCRYPT 2023, LNCS 14459, pp. 3–24, 2024.
https://doi.org/10.1007/978-3-031-56232-7_1

NMAC [2], or block ciphers like CBC-MAC [3], CMAC [1,7,17] and PMAC [8], or more efficiently by making use of finite field arithmetic such as GHASH [19], the function used to compute a MAC in GCM mode and Poly-1305 [4].

Further efficiency is achieved in software for MMH and NMH family of hash functions [9], which are based on multiplication in a finite field, but are implemented by means of integer multiplication and the modular reduction is deferred to the very end of the algorithm. The efficiency is improved on designs that are based on integer multiplications such as NH hash function family [6] used in UMAC [6], Adiantum [10], HS1-SIV [18], and Multimixer-128 [15]. These functions make use of 16 or 32-bit integer multiplication instructions available in software that results in a 32 or 64-bit output respectively; eliminating the need for modular reductions since modular reduction in this case is a simple truncation to 32 or 64-bits. As a consequence, these are the fastest keyed hash functions in software exhibiting good universality [10,15]. Concretely NH^T, that applies NH on a message 4 times in parallel with independent keys, and Multimixer-128 attain universalities $\varepsilon \approx 2^{-128}$ and $\varepsilon \approx 2^{-127}$ respectively in the ε-Δuniversality setting, where the digests have lengths 256 bits and 512 bits respectively.

Multimixer-128 follows the parallel construction [14], where a fixed length public permutation or function [16] is processed in parallel to form a keyed hash function. In this construction, a message and key are first parsed into blocks and then added block by block. Each masked message block is then an input to the underlying public permutation or function and the outputs of each block are added to form the digest. The security of this construction was first studied by Fuchs et al. [14] in the case that the underlying primitive is a permutation and then further generalized in [16] to include public functions. Apart from offering a high degree of parallelism, another advantage of this construction is that the universality of such a keyed hash function is determined solely by the propagation probabilities of the underlying public function/permutation. This leads to a much simpler security analysis while offering a high level of efficiency. To the best of our knowledge, Multimixer-128 is the fastest keyed hash function in software [15] and this efficiency is attained due to the usage of only integer multiplications and simple linear transformations in the underlying public function. The public function of Multimixer-128 has an input size of 256 bits and the digest size of Multimixer-128 is 512 bits.

Our Contribution: Most modern block ciphers, including AES, has a block-size of 128-bits. Due to the birthday bound, in many modes of operation the expected number of queries required for an adversary to perform a distinguishing attack is 2^{64}. There is a lot of interest in designing primitives that attain beyond birthday bound security. In our work, we look to build a keyed hash function, where the expected number of queries required to obtain a collision is significantly higher than 2^{64}.

We also follow the parallel construction and build a public function based on integer multiplication, linear transformation and cyclic shift. Our goal is to build a keyed hash function that is more secure than Multimixer-128 at a moderate expense in efficiency. We propose $\mathcal{F}$-156, a public function whose parallelization is called Multimixer-156. We provide strong arguments supporting our claim that Multimixer-156 is 2^{-156}-Δuniversal. This means that the expected number of queries required for an adversary to obtain a collision in Multimixer-156 is 2^{78}, significantly larger than 2^{64}.

Due to efficiency requirements, $\mathcal{F}$-156 makes use of operations that have fast vector instructions available and since the input size of 256-bits is very well suited to the Single Instruction Multiple Data (SIMD) architecture, $\mathcal{F}$-156, like Multimixer-128 also has this input size and Multimixer-156 outputs a digest of 512-bits similar to Multimixer-128.

2   Preliminaries and Notations

A public function is denoted as $f \colon G \to G'$, where G and G' are abelian groups $\langle G, + \rangle$ and $\langle G', + \rangle$. The elements of G are called *blocks*. The set containing ℓ-block string is denoted as G^ℓ, i.e., $G^\ell = \{(x_0, x_1, \ldots, x_{\ell-1}) \mid x_i \in G \text{ for each } i = 0, 1, \ldots, \ell - 1\}$. The set of strings of length 1 upto κ is denoted as $\mathrm{BS}(G, \kappa) = \cup_{\ell=1}^{\kappa} G^\ell$. We denote strings in bold uppercase letters, like $\mathbf{M}$, its blocks by M_i, where indexing starts from 0 and the length of that string by $|\mathbf{M}|$. Given any set S, the cardinality of that set is denoted by $\#S$.

2.1 ε and ε-Δuniversality

Let $F_\mathbf{K}$ denote a keyed hash function where the key $\mathbf{K}$ is sampled uniformly at random from the key space. The security of $F_\mathbf{K}$ is measured by the probability of generating a collision at the output of $F_\mathbf{K}$ or more strongly, by the probability of two distinct input strings exhibiting a specific output difference.

Definition 1 (ε-universality [20]). *A keyed hash function F is said to be ε-universal if for any distinct strings $\mathbf{M}, \mathbf{M}^*$*

$$\Pr[F_\mathbf{K}(\mathbf{M}) = F_\mathbf{K}(\mathbf{M}^*)] \leq \varepsilon.$$

Definition 2 (ε-Δuniversality [20]). *A keyed hash function F is said to be ε-Δuniversal if for any distinct strings $\mathbf{M}, \mathbf{M}^*$ and for all $\Delta \in G$*

$$\Pr[F_\mathbf{K}(\mathbf{M}) - F_\mathbf{K}(\mathbf{M}^*) = \Delta] \leq \varepsilon.$$

2.2   Key-then-Hash Functions

Key-then-hash functions [14] are a special type of keyed hash functions. They take as input elements of $\mathrm{BS}(G, \kappa)$ and return an element of G'. The keys are elements of G^κ. When processing an input, the key is first added to the input and then an unkeyed function is applied to the result. A key-then-hash function is defined as: $F \colon \mathrm{BS}(G, \kappa) \to G'$ with $F_\mathbf{K}(\mathbf{M}) := F(\mathbf{K} + \mathbf{M})$. The addition of two strings $\mathbf{M} = (M_0, M_1, \ldots, M_{|\mathbf{M}|-1})$ and $\mathbf{M}^* = (M_0^*, M_1^*, \ldots, M_{|\mathbf{M}^*|-1}^*)$ with $|\mathbf{M}| \leq |\mathbf{M}^*|$ is defined as $\mathbf{M}' := \mathbf{M} + \mathbf{M}^* = (M_0 + M_0^*, M_1 + M_1^*, \ldots, M_{|\mathbf{M}|-1} + M_{|\mathbf{M}|-1}^*)$ with $|\mathbf{M}'| = |\mathbf{M}|$. In Sect. 2.3 we demonstrate how to build such functions using a public function as the underlying primitive [16].

2.3   Parallel Universal Hashing

The parallelization of a public function to build key-then-hash function is described in Algorithm 1 and depicted in Fig. 1 [16]. The construction takes as parameters a public function $f\colon G \to G'$ and a maximum string length κ. The inputs to the construction are a key $\mathbf{K} \in G^\kappa$ and a string $\mathbf{M} \in \mathrm{BS}(G, \kappa)$. The construction returns a digest $h \in G'$. Given any f, its parallelization is the key-then-hash function denoted as Parallel $[f]$. The key space of Parallel $[f]$ is G^κ. So, in the construction the existence of long keys with independent key blocks is assumed, which is a commonly made assumption for keyed hash functions.

Algorithm 1: The parallelization Parallel $[f]$ [16]

Parameters: A public function $f\colon G \to G'$ and a maximum string length κ
Inputs   : A key $\mathbf{K} \in G^\kappa$ and a message $\mathbf{M} \in \mathrm{BS}(G, \kappa)$
Output   : A digest $h \in G'$

$\mathbf{X} \leftarrow \mathbf{M} + \mathbf{K}$
$h \leftarrow 0$
for $i \leftarrow 0$ **to** $|\mathbf{M}| - 1$ **do**
   | $h \leftarrow h + f(X_i)$
end
return h

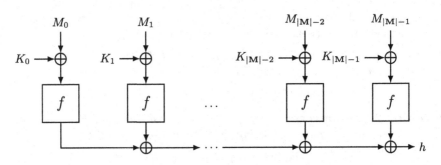

Fig. 1. The parallelization Parallel $[f]$ adapted from [14].

The universality of Parallel $[f]$ is upper-bounded by the propagation probabilities of the underlying fixed length function f, which are defined as follows.

A differential defined over f is the tuple (A, Δ), where $A \in G/\{0\}$ is called the input difference and $\Delta \in G'$ is called the output difference. We remind the reader of differential probability of a differential and image probability of an output over fixed-length public functions.

Definition 3 (Differential probability). *Let $f: G \to G'$ be a public function. The differential probability of a differential (A, Δ) of f, denoted as $\mathsf{DP}_f(A, \Delta)$, is:*

$$\mathsf{DP}_f(A, \Delta) = \frac{\#\{X \in G \mid f(X + A) - f(X) = \Delta\}}{\#G}.$$

We say that input difference A propagates to output difference Δ with probability $\mathsf{DP}_f(A, \Delta)$. We denote the set $\{X \in G \mid f(X + A) - f(X) = \delta\}$ as $\mathsf{S}_f(A, \delta)$ and call it the solution set of the differential.

Definition 4 (Differential weight). *Let $f: G \to G'$ be a public function. The differential weight of a differential (A, Δ) of f denoted as $\mathsf{w}_f(A, \Delta)$ is:*

$$\mathsf{w}_f(A, \Delta) = -\log_2(\mathsf{DP}_f(A, \Delta)) = \log_2(\#G) - \log_2(\#\mathsf{S}_f(A, \Delta)).$$

Definition 5 (Image probability [16]). *Let $f: G \to G'$ be a public function. The image probability of an output $Z \in G'$ of f, denoted as $\mathrm{IP}_f(Z)$, is the number of inputs that f maps to Z divided by the total number of possible inputs, namely,*

$$\mathrm{IP}_f(Z) = \frac{\#\{X \in G \mid f(X) = Z\}}{\#G}.$$

The maximum possible value of DP_f and IP_f over all differentials and outputs of the fixed-length public function f respectively are denoted as:

$$\mathrm{MDP}_f = \max_{(A, \Delta)} \mathsf{DP}_f(A, \Delta) \quad \text{and} \quad \mathrm{MIP}_f = \max_Z \mathrm{IP}_f(Z).$$

Theorem 1 (Theorem 1 [16]). *The parallelization of a public function f, $Parallel[f]$, is $\max\{\mathrm{MDP}_f, \mathrm{MIP}_f\}$-$\Delta universal$.*

Thus the problem of obtaining universality of $Parallel[f]$ is reduced to obtaining MDP_f and MIP_f of the underlying fixed length function f. The tightness of the ε-$\Delta universality$ bound in Theorem 1 depends solely on the tightness of the bounds for MDP_f and MIP_f of the underlying public function f.

2.4 Notations

In this paper, $\mathbb{Z}/2^w\mathbb{Z}$ denotes the ring of integer residues modulo 2^w with the addition and multiplication. $(\mathbb{Z}/2^w\mathbb{Z})^n$ denotes the Cartesian product of $\mathbb{Z}/2^w\mathbb{Z}$ n-times. For the public functions proposed in this paper, $G = (\mathbb{Z}/2^w\mathbb{Z})^8$ and $G' = (\mathbb{Z}/2^{2w}\mathbb{Z})^8$ with $w = 32$. So, the input and the output of our public function both consist of 8-tuples with elements from $\mathbb{Z}/2^w\mathbb{Z}$ and $\mathbb{Z}/2^{2w}\mathbb{Z}$ respectively.

We represent $X \in G - (\mathbb{Z}/2^w\mathbb{Z})^8$ as $X = (x_0, x_1, x_2, x_3, y_0, y_1, y_2, y_3)^\mathsf{T}$. For simplicity, we slightly abuse the notations to denote $X = (\mathbf{x}, \mathbf{y})$, where $\mathbf{x} = (x_0, x_1, x_2, x_3)^\mathsf{T} \in (\mathbb{Z}/2^w\mathbb{Z})^4$ and $\mathbf{y} = (y_0, y_1, y_2, y_3)^\mathsf{T} \in (\mathbb{Z}/2^w\mathbb{Z})^4$. Similarly input differences and key blocks are denoted as $A = (\mathbf{a}, \mathbf{b})$ and $K = (\mathbf{h}, \mathbf{k})$

respectively. An output $Z \in G' = (\mathbb{Z}/2^{2w}\mathbb{Z})^8$ is denoted as $Z = (z_0, z_1, \ldots, z_7)^{\mathsf{T}}$ and an output difference Δ is given by $\Delta = (\delta_0, \delta_1, \ldots, \delta_7)^{\mathsf{T}}$. This means that throughout this paper for $i \in \{0, 1, 2, 3\}$, each of $x_i, y_i, a_i, b_i, h_i, k_i \in \mathbb{Z}/2^w\mathbb{Z}$ and for $i \in \{0, 1, \ldots, 7\}$, each of $z_i, \delta_i \in \mathbb{Z}/2^{2w}\mathbb{Z}$.

For $n \geq 1$, the number of non-zero components of a vector $\mathbf{x} \in (\mathbb{Z}/2^w\mathbb{Z})^n$ is its Hamming Weight denoted as $w(\mathbf{x})$. We further denote $(0, 0, \ldots, 0) \in (\mathbb{Z}/2^w\mathbb{Z})^n$ as 0^n. $\mathbb{Z}_{\geq 0}$ is used to denote the set of positive integers including 0.

For two elements $x, y \in \mathbb{Z}/2^w\mathbb{Z}$, $x \boxplus y$, $x \boxminus y$ and $x \cdot y$ denote respectively $(x + y) \bmod 2^w$, $(x - y) \bmod 2^w$ and $(x \cdot y) \bmod 2^w$. For any element $x \in \mathbb{Z}/2^w\mathbb{Z}$, $\overline{x}$ denotes the additive inverse of x, i.e., $\overline{x} = 2^w \boxminus x$.

The integer multiplication of two elements of $\mathbb{Z}/2^w\mathbb{Z}$ is called the *w-bit multiplication* and is denoted as $M[w]$. This operation is defined as

$$M[w] \colon (\mathbb{Z}/2^w\mathbb{Z})^2 \to \mathbb{Z}/2^{2w}\mathbb{Z} \colon M[w](x, y) = x \times y. \tag{1}$$

To differentiate between w-bit multiplication and the ring multiplication in $\mathbb{Z}/2^w\mathbb{Z}$, $\times$ will be used to denote the w-bit multiplication and $\cdot$ will be used to denote the ring multiplication throughout this paper.

2.5 Differential Properties of Integer Multiplication

The differential properties of $M[w]$ have been studied in [15]. We highlight the key properties of $M[w]$ outlined in that work that are directly relevant to our research. For $M[w]$, the input difference is defined by the group operation of addition modulo 2^w and the output difference by the group operation of addition modulo 2^{2w}. Thus a differential to $M[w]$ has the form $((a, b), \delta)$, where $a, b \in \mathbb{Z}/2^w\mathbb{Z}$ and $\delta \in \mathbb{Z}/2^{2w}\mathbb{Z}$. Naturally the solution set of the differential is given by:

$$S_{M[w]}((a, b), \delta) = \{(h, k) \mid ((a \boxplus h) \times (b \boxplus k) - h \times k) \bmod 2^{2w} = \delta\}. \tag{2}$$

Throughout this paper we will use $N((a, b), \delta)$ to denote $\#S_{M[w]}((a, b), \delta)$.

Input differences of the form $a = 0$ or $b = 0$ and the corresponding differentials are referred to as *unilateral differences and unilateral differentials* respectively. All other input differences and their corresponding differentials are referred to as *bilateral*. Clearly from (2) we see that $N((a, b), \delta) = N((b, a), \delta)$ and as such for unilateral differentials, it suffices to only look at $DP_{M[w]}((a, 0), \delta)$.

Lemma 1. *[15] For a unilateral differential $((a, 0), \delta)$ to $M[w]$ with $\delta \neq 0$,*

$$\text{For } \delta < 2^w a \colon \quad DP((a, 0), \delta) = \begin{cases} \frac{\overline{a}}{2^{2w}}, & \text{if } a \mid \delta \\ 0, & \text{otherwise} \end{cases}$$

$$\text{For } \delta > 2^w a \colon \quad DP((a, 0), \delta) = \begin{cases} \frac{a}{2^{2w}}, & \text{if } \overline{a} \mid 2^{2w} - \delta \\ 0, & \text{otherwise} \end{cases}$$

$$\text{For } \delta = 2^w a \colon \quad DP((a, 0), \delta) = 0.$$

Lemma 2. *[15] For the differential $((a, 0), 0)$ to $M[w]$, $DP_{M[w]}((a, 0), 0) = \frac{1}{2^w}$.*

For a bilateral difference (a, b), we are primarily interested in upper-bounding the value of $\mathrm{DP}_{M[w]}((a, b), \delta)$ over all δ and another differential, namely $((a, b), 0)$ is of special interest since it corresponds to collision at the output of $M[w]$.

Lemma 3. *[15] Let $((a, b), \delta)$ be a bilateral differential to $M[w]$. Then*

$$\mathrm{N}((a,b),\delta) \leq \max\left(\left\lceil \gcd(a,b)\min\left(\tfrac{\overline{a}}{a},\tfrac{\overline{b}}{b}\right)\right\rceil + \left\lceil \gcd(\overline{a},\overline{b})\min\left(\tfrac{a}{\overline{a}},\tfrac{b}{\overline{b}}\right)\right\rceil, \left\lceil \gcd(a,\overline{b})\min\left(\tfrac{\overline{a}}{a},\tfrac{b}{\overline{b}}\right)\right\rceil + \left\lceil \gcd(\overline{a},b)\min\left(\tfrac{a}{\overline{a}},\tfrac{\overline{b}}{b}\right)\right\rceil\right).$$

Lemma 4. *[15] For any bilateral differential $((a, b), 0)$ to $M[w]$, we have*

$$\mathrm{N}((a,b),0) = \left\lceil \gcd\left(a, \overline{b}\right)\min\left(\frac{\overline{a}}{a}, \frac{b}{\overline{b}}\right)\right\rceil + \left\lceil \gcd\left(\overline{a}, b\right)\min\left(\frac{a}{\overline{a}}, \frac{\overline{b}}{b}\right)\right\rceil.$$

It follows from Lemmas 1, 2 and 3 that $\mathrm{DP}_{M[w]}((a, b), \delta) \leq 2^{-w}$ for any differential $((a, b), \delta)$. Now, $\mathrm{DP}_{M[w]}((a, b), 0) = 2^{-w}$ iff $a = 0$ or $b = 0$ or $b = \overline{a}$. We also see that for differences of the type (a, a) as well, the bound in Lemma 3 gives the maximum value of DP is 2^{-w}. Input differences of the type (a, a) and $(a, \overline{a})$ are called diagonal and counter-diagonal respectively. Thus the input differences that correspond to the highest possible value of 2^{-w} are the unilateral, diagonal and counter-diagonal differences. In fact for most input differences (a, b), $\max_\delta \mathrm{DP}_{M[w]}((a, b), \delta)$ is significantly smaller as is evident from Fig. 2.

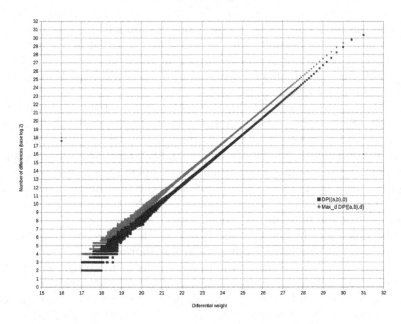

Fig. 2. Upper-bound of $\max_\delta \mathrm{DP}((a, b), \delta)$ and $\mathrm{DP}((a, b), 0)$ vs. Number of differences for $w = 16$ [15]

3   Multimixer-156

The authors of [15] have proposed a highly efficient public function $\mathcal{F}$-128, whose parallelization Multimixer-128 is 2^{-127}-Δuniversal. Our goal now is to design another public function based on integer multiplication, whose parallelization offers more security for a modest increase in computational cost while not increasing the digest size of 512-bits. We call the public function $\mathcal{F}$-156 and its parallelization is the key-then-hash function Multimixer-156. Before we look into the specifications of $\mathcal{F}$-156, we first motivate the design choices that we make. $\mathcal{F}$-156 makes use of 32-bit integer multiplication, but we prove the results in this section for a generic w unless otherwise specified.

3.1   Motivation and Design Rationale

Based on the security analysis of $\mathcal{F}$-128 in [15], to build a public function with a similar design, whose parallelization is ε-Δuniversal with $\varepsilon < 2^{-128}$ and which has the same digest size of 512-bits, we need to: (1) Decrease the DP of unilateral differentials, i.e., differentials with input difference of the form $((\mathbf{a}, 0^4), \Delta)$ or $((0^4, \mathbf{a}), \Delta)$, (2) Decrease the value of IP(0^8). We make use of bitwise circular shift operation and use matrices with better diffusion properties to achieve this.

The maximum value of DP of unilateral differentials to $\mathcal{F}$-128 is 2^{-128}. Furthermore there are several unilateral differentials to $\mathcal{F}$-128 that lead to only 4 active multiplications, where by active multiplication with respect to a differential we mean multiplications inside a public function that attain a non-zero input difference as a result of propagation of the corresponding input difference to the public function. So, first we avoid most of these differentials by adopting matrices with better diffusion properties.

As a matter of fact we chose $M_\alpha = \text{circ}(1, 1, 2, 3)$ and $M_\beta = \text{circ}(3, 1, 1, 2)$. M_α is similar to the MDS matrix used in the MixColumns step in the AES round function [12] that has branch number 5. However, $\mathbb{Z}/2^{32}\mathbb{Z}$ is just a ring, not a field and it turns out M_α has branch number 4. Still, there are only a few inputs that achieve this low branch number and those inputs have as non-zero entries multiples of 2^{29}. We refer to these inputs as corner case inputs and tabulate them in Table 2. Thus, $\mathcal{F}$-156 has 4 active multiplications only for unilateral differences with the non-zero component being a corner case input. We deal with these differences by adding some processing that turns them into bilateral differences. To that end, we make use of circular shift operation.

Elements of $\mathbb{Z}/2^w\mathbb{Z}$ are encoded in computers as w-bit strings. Many CPUs have dedicated (vector) shift instructions for these w-bit strings. Bitwise circular shift operations can be efficiently implemented by using these instructions, typically requiring very few cycles. We use this circular shift in a Feistel structure that we call FTR$_r$ with r being the rotation offset. FTR$_r$ being a Feistel is invertible. Both its input and output are elements of $(\mathbb{Z}/2^w\mathbb{Z})^2$.

FTR$_r$ converts unilateral differences of the form $(a, 0)$ or $(0, a)$, where a is a multiple of 2^{w-3} to differences that correspond to a low DP in the corresponding w-bit multiplication. FTR$_r$ effectively deals with other unilateral differentials as well. In fact since for all other input differences, at least 5 multiplications are active, our conjecture is that for all corresponding differentials MDP$_{\mathcal{F}\text{-}156} \leq 2^{-156}$. Due to FTR$_r$, we further prove that IP$_{\mathcal{F}\text{-}156}(Z) < 2^{-156}$ for any Z.

As $\mathcal{F}$-128, $\mathcal{F}$-156 has 8 multiplications, but for an input $(\mathbf{x}, \mathbf{y})$ the inputs to the 4 multiplications are instead FTR$_r(x_i, y_i)$ for $i \in \{0, 1, 2, 3\}$. The remaining 4 multiplications have (u_i, v_i) as inputs for $i \in \{0, 1, 2, 3\}$. Here $\mathbf{u}$ and $\mathbf{v}$ are defined as $M_\alpha \cdot \mathbf{x}$ and $M_\beta \cdot \mathbf{y}$ respectively. We now look at the bitwise right shift operation and the FTR$_r$ function that is employed in $\mathcal{F}$-156.

3.2 Analysis of Bitwise Cyclic Shift

We first look at the encoding of elements of $\mathbb{Z}/2^w\mathbb{Z}$ as w-bit words. Given $x \in \mathbb{Z}/2^w\mathbb{Z}$, we define $\langle x \rangle_w = (x_0, x_1, \ldots, x_{w-1})$ where for $i \in \{0, 1, \ldots, w-1\}$, $x_i \in \{0, 1\}$ and $x = \sum_{i=0}^{w-1} x_{w-1-i} 2^i$. Similarly, given a bitstring $(y_0, y_1, \ldots, y_{w-1})$, we denote its integer counterpart as $\mathrm{int}(y_0, y_1, \ldots, y_{w-1}) \in \mathbb{Z}/2^w\mathbb{Z}$, where naturally $\mathrm{int}(y_0, y_1, \ldots, y_{w-1}) = \sum_{i=0}^{w-1} y_{w-1-i} 2^i$.

Definition 6 (Left shift). *Left shift of $x \in \mathbb{Z}/2^w\mathbb{Z}$ with $\langle x \rangle_w = (x_0, x_1, \ldots, x_{w-1})$ over r positions denoted as $x \ll r$ is given by:*

$$x \ll r = \mathrm{int}(x_r, x_{r+1}, \ldots, x_{w-1}, 0, \ldots, 0).$$

The operation in $\mathbb{Z}/2^w\mathbb{Z}$ that corresponds with a left shift over 1 bit is the multiplication by 2. More generally, a left shift over r bits is the same as multiplication by 2^r modulo 2^w. The left shift is a linear operation in $\mathbb{Z}/2^w\mathbb{Z}$ as:

$$2^r \cdot (mx \boxplus ny) = m \cdot 2^r x \bmod 2^w \boxplus n \cdot 2^r y \bmod 2^w.$$

It is non-invertible as $2^r \cdot x$ is independent of the r most significant bits of x.

Given an input difference a, the output difference under the left shift operation is $a \ll r$. Indeed for any $x \in \mathbb{Z}/2^w\mathbb{Z}$,

$$(x \boxplus a) \ll r \boxminus x \ll r = 2^r \cdot (x \boxplus a) \boxminus 2^r \cdot x = 2^r \cdot a \bmod 2^w = a \ll r.$$

Definition 7 (Right shift). *The right shift of $x \in \mathbb{Z}/2^w\mathbb{Z}$, where $\langle x \rangle_w = (x_0, x_1, \ldots, x_{w-1})$, over r positions denoted as $x \gg r$ is given by:*

$$x \gg r = \mathrm{int}(0, \ldots, 0, x_0, x_1, \ldots, x_{w-r-1}).$$

12   K. Ghosh et al.

Intuitively, since the operation in $\mathbb{Z}/2^w\mathbb{Z}$ that corresponds with the bitwise left shift over r bits is ring multiplication by 2^r, we expect a right shift over r bits to be something like a division by 2^r.

For x and y positive integers, we will write $\lfloor \frac{x}{y} \rfloor$ to denote the integer floor division of x by y. Its output is the *quotient* Q, the largest integer such that $Qy \leq x$. Clearly, the remainder $R = x - Qy$ is equal to $R = x \bmod y$. Thus

$$Q = \left\lfloor \frac{x}{y} \right\rfloor = \frac{x - (x \bmod y)}{y}.$$

Lemma 5. *For $x \in \mathbb{Z}/2^w\mathbb{Z}$, we have: $x \gg r = \lfloor \frac{x}{2^r} \rfloor$.*

Proof. Due to the definition of the integer floor division, $\lfloor \frac{x}{2^r} \rfloor$ and $\lfloor \frac{x'}{2^r} \rfloor$ with $x' = x - (x \bmod 2^r)$ are equal. A right shift moves the $w - r$ most significant bits of x' to the right and so does division by 2^r. □

We now see that right shift is not a linear operation and an input difference to the right shift function can propagate to up to 4 possible output differences.

Lemma 6. *Let a and c be respectively an input and output difference to $\gg r$. Then $\mathrm{DP}_{\gg r}(a, c)$ is*

$$\mathrm{DP}_{\gg r}(a,c) = \begin{cases} \frac{(2^{w-r} - \lfloor \frac{a}{2^r} \rfloor)(2^r - (a \bmod 2^r))}{2^w}, & \text{for } c = a \gg r \\ \frac{(2^{w-r} - \lfloor \frac{a}{2^r} \rfloor - 1)(a \bmod 2^r)}{2^w}, & \text{for } c = a \gg r \boxplus 1 \\ \frac{\lfloor \frac{a}{2^r} \rfloor (2^r - (a \bmod 2^r))}{2^w}, & \text{for } c = a \gg r \boxminus 2^{w-r} \\ \frac{(\lfloor \frac{a}{2^r} \rfloor + 1)(a \bmod 2^r)}{2^w}, & \text{for } c = a \gg r \boxminus 2^{w-r} \boxplus 1 \\ 0, & \text{otherwise.} \end{cases}$$

Proof. For $x \in \mathbb{Z}/2^w\mathbb{Z}$,

$$c = (x \boxplus a) \gg r \boxminus x \gg r$$
$$= \left\lfloor \frac{(x \boxplus a)}{2^r} \right\rfloor \boxminus \left\lfloor \frac{x}{2^r} \right\rfloor$$
$$= \frac{(x \boxplus a) \boxminus ((x \boxplus a) \bmod 2^r)}{2^r} \boxminus \frac{x - (x \bmod 2^r)}{2^r}$$
$$= \frac{((x \boxplus a) \boxminus x) \boxminus (((x \boxplus a) \bmod 2^r) \boxminus (x \bmod 2^r))}{2^r}.$$

We note that

$$(x \boxplus a) \boxminus x = \begin{cases} a, & \text{if } x + a < 2^w \\ a - 2^w, & \text{otherwise.} \end{cases}$$

Furthermore,

$$(x \boxplus a) \bmod 2^r \boxminus x \bmod 2^r = \begin{cases} a \bmod 2^r, & \text{if } x \bmod 2^r + a \bmod 2^r < 2^r \\ a \bmod 2^r - 2^r, & \text{otherwise.} \end{cases}$$

Depending on whether $(x + a) < 2^w$ or not and $x \bmod 2^r + a \bmod 2^r < 2^r$ or not, we have

$$c = \begin{cases} \frac{a - (a \bmod 2^r)}{2^r}, & \text{if } x + a < 2^w \ \& \ x \bmod 2^r + a \bmod 2^r < 2^r \\ \frac{a - (a \bmod 2^r)}{2^r} \boxplus 1, & \text{if } x + a < 2^w \ \& \ x \bmod 2^r + a \bmod 2^r \geq 2^r \\ \frac{a - (a \bmod 2^r)}{2^r} \boxminus 2^{w-r}, & \text{if } x + a \geq 2^w \ \& \ x \bmod 2^r + a \bmod 2^r < 2^r \\ \frac{a - (a \bmod 2^r)}{2^r} \boxminus 2^{w-r} \boxplus 1, & \text{if } x + a \geq 2^w \ \& \ x \bmod 2^r + a \bmod 2^r \geq 2^r. \end{cases}$$

Computing sizes of the relevant ranges, we arrive at the lemma.   □

So, right shift is a non-linear operation; but its non-linearity is rather weak. It is non-invertible as $\lfloor \frac{a}{2^r} \rfloor$ is independent of the r least significant bits of a.

Definition 8 (Cyclic shift). *The cyclic shift of $x \in \mathbb{Z}/2^w\mathbb{Z}$ over r positions denoted as $x \ggg r$ is given by:*

$$x \ggg r = int(x_{w-r}, \ldots, x_{w-1}, x_0, x_1, \ldots, x_{w-r-1}).$$

A cyclic shift over r positions is the combination of a left shift over $w-r$ positions and a right shift over r positions.

Lemma 7. *For $a \in \mathbb{Z}/2^w\mathbb{Z}$, we have: $a \ggg r = 2^{w-r} \cdot a \boxplus \lfloor \frac{a}{2^r} \rfloor$.*

Proof. Integer division by 2^r shifts the $(w - r)$ left-most bits to the $(w - r)$ right-most positions and fills the freed bits positions with zeroes. Multiplication by 2^{w-r} shifts the r right-most bits to the left-most positions and fills the freed bits positions with zeroes. The addition of the two computes the cyclic shift. □

Lemma 8. *Let (a, c) be a differential to $\ggg r$. Then $\mathrm{DP}_{\ggg r}(a, c)$ is given by*

$$\mathrm{DP}_{\ggg r}(a, c) = \begin{cases} \frac{(2^{w-r} - \lfloor \frac{a}{2^r} \rfloor)(2^r - (a \bmod 2^r))}{2^w}, & \text{for } c = a \ggg r \\ \frac{(2^{w-r} - \lfloor \frac{a}{2^r} \rfloor - 1)(a \bmod 2^r)}{2^w}, & \text{for } c = a \ggg r \boxplus 1 \\ \frac{\lfloor \frac{a}{2^r} \rfloor (2^r - (a \bmod 2^r))}{2^w}, & \text{for } c = a \ggg r \boxminus 2^{w-r} \\ \frac{(\lfloor \frac{a}{2^r} \rfloor + 1)(a \bmod 2^r)}{2^w}, & \text{for } c = a \ggg r \boxminus 2^{w-r} \boxplus 1 \\ 0, & \text{otherwise.} \end{cases}$$

Proof. Since bitwise left shift is a linear operation, the proof follows from Lemmas 6 and 7.   □

The cyclic shift inherits the weak non-linearity of the right shift. As opposed to the left and right shifts, it is an invertible operation as the inverse operation is just a cyclic shift over $w - r$ bits. Corner case inputs that lead to 4 active multiplications in $\mathcal{F}$-156 have as non-zero entries multiples of 2^{29} and as such for our choice of $r = 29$, 2^r divides each such element.

Corollary 1. *Let a be such that $2^r \mid a$. Then $\mathrm{DP}_{\ggg r}(a, c)$ is given by:*

$$\mathrm{DP}_{\ggg r}(a, c) = \begin{cases} 1 - \frac{a}{2^w}, & \text{for } c = a \ggg r \\ \frac{a}{2^w}, & \text{for } c = a \ggg r \boxminus 2^{w-r} \\ 0, & \text{otherwise.} \end{cases}$$

Proof. The proof follows from Lemma 8 since $2^r \mid a$ implies that $a \bmod 2^r = 0$ and $\lfloor \frac{a}{2^r} \rfloor = \frac{a}{2^r}$.   □

We now look at the Feistel-with-rotation function and its composition with the w-bit multiplication, that we call the Rotate-then-multiply.

3.3   Feistel-with-Rotation and the Rotate-then-Multiply Functions

The Feistel-with-rotation denoted as FTR_r is a balanced 2-round Feistel network, where the non-linear function in the 2nd round is the cyclic shift by an offset r. FTR_r can be implemented with a single circular shift and 2 modular additions.

Definition 9 (FTR_r). *Given $w \geq 2$ and $1 \leq r < w$, FTR_r is defined as:*

$$FTR_r \colon (\mathbb{Z}/2^w\mathbb{Z})^2 \mapsto (\mathbb{Z}/2^w\mathbb{Z})^2 : FTR_r(x, y) = (x \boxplus y, (x \boxplus y) \ggg r \boxplus y).$$

Since FTR_r is a Feistel, it is necessarily invertible. We depict FTR_r in Fig. 3. By Lemma 8 an input difference to $\ggg r$ can propagate to upto 4 possible output differences. Unilateral input differences of the type $(a, 0)$ turn into bilateral

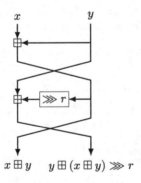

Fig. 3. Feistel with rotation FTR_r

differences under FTR_r. Indeed for input difference $(a, 0)$, the possible output differences in FTR_r are $\{(a, c_i) \mid i = 0, 1, 2, 3\}$, where for $i \in \{0, 1, 2, 3\}$, c_i denotes the possible output differences for the input difference a to $\ggg r$. Since cyclic shift is an invertible operation, by Lemma 8 each c_i is necessarily non-zero.

For input difference of the type $(0, a)$, the possible output differences are $\{(a, a \boxplus c_i) \mid i = 0, 1, 2, 3\}$, where for $i \in \{0, 1, 2, 3\}$, c_i denotes the possible output differences for the input difference a to $\ggg r$. Clearly unilateral differences of this type stay unilateral whenever $a \boxplus c_i = 0$. But, in such a case the unilateral difference $(a, 0)$ converts into $(0, a)$. This property of FTR_r is sufficient to deal with all unilateral differentials in our public function $\mathcal{F}$-156.

FTR_r converts a counter-diagonal difference $(a, \bar{a})$ into the unilateral difference $(0, \bar{a})$. We now define RTM_r that combines FTR_r with w-bit multiplication.

Definition 10 (RTM_r). *Given $w \geq 2$ and $1 \leq r < w$, the $RTM_r \colon (\mathbb{Z}/2^w\mathbb{Z})^2 \to \mathbb{Z}/2^{2w}\mathbb{Z}$ is defined as:*

$$RTM_r(x, y) = M[w] \circ FTR_r(x, y) = (x \boxplus y) \times ((x \boxplus y) \ggg r \boxplus y).$$

Lemma 9. *For any $w \geq 2$ and $1 \leq r < w$, $IP_{RTM_r}(0) = \frac{2^{w+1}-1}{2^{2w}}$. Furthermore, for every value of y, there are at most 2 values of x satisfying $RTM_r(x, y) = 0$.*

Proof. Since the FTR_r is invertible, for any $z \in \mathbb{Z}/2^{2w}\mathbb{Z}$, $IP_{RTM_r}(z) = IP_{M[w]}(z)$ and $IP_{M[w]}(0) = \frac{2^{w+1}-1}{2^{2w}}$ since $x \times y = 0$ implies $x = 0$ or $y = 0$.

Now, let $RTM_r(x, y) = 0$, i.e., $(x \boxplus y) \times ((x \boxplus y) \ggg r \boxplus y) = 0$. This implies that $(x \boxplus y) = 0$ or $(x \boxplus y) \ggg r \boxplus y = 0$. Clearly $x \boxplus y = 0$ implies $y = \bar{x}$.

When $(x \boxplus y) \ggg r \boxplus y = 0$, we have

$$(x \boxplus y) \ggg r = \bar{y}. \tag{3}$$

Since addition modulo 2^w and cyclic shift are both invertible operations, for every value of y, there is at most one value of x that satisfies (3). Thus for every value of y, there are at most 2 values of x, one corresponding to $(x \boxplus y) = 0$ and another corresponding to $(x \boxplus y) \ggg r \boxplus y = 0$, such that $RTM_r(x, y) = 0$. $\square$

Lemma 10. *Let (a, b) be any input difference to RTM_r. Then $DP_{RTM_r}((a, b), \delta)$ is upper-bounded by:*

$$DP_{RTM_r}((a, b), \delta) \leq DP_{M[w]}((x, y), \delta) + DP_{M[w]}((x, y \boxplus 1), \delta) +$$
$$DP_{M[w]}((x, y \boxminus 2^{w-r}), \delta) + DP_{M[w]}((x, y \boxminus 2^{w-r} \boxplus 1), \delta).$$

Here $(a \boxplus b) = x$ and $(a \boxplus b) \ggg r \boxplus b = y$.

Proof. To compute $DP_{RTM_r}((a, b), \delta)$, we look at the propagation of (a, b) under FTR_r. The input difference (a, b) to FTR_r can propagate to the following 4 values: (x, y), $(x, y \boxplus 1)$, $(x, y \boxminus 2^{w-r})$ $(x, y \boxminus 2^{w-r} \boxplus 1)$.

Clearly the solution set of the differential $((a, b), \delta)$ under RTM_r is

$$S_{RTM_r}((a, b), \delta) \subseteq S_{M[w]}((x, y), \delta) \cup S_{M[w]}((x, y \boxminus 2^{w-r}), \delta)$$
$$\cup S_{M[w]}((x, y \boxplus 1), \delta) \cup S_{M[w]}((x, y \boxminus 2^{w-r} \boxplus 1), \delta).$$

Hence, from the union-bound we obtain our desired result. $\square$

Corollary 2. *For a differential* $((a,0),\delta)$ *to* RTM_r *with* $2^r \mid a$, *we have*

$$\mathrm{DP}_{RTM_r}((a,0),\delta) \leq \mathrm{DP}_{M[w]}((a,a \ggg r),\delta) + \mathrm{DP}_{M[w]}((a,a \ggg r \boxminus 2^{w-r}),\delta).$$

For a differential $((0,b),\delta)$ *to* RTM_r *with* $2^r \mid b$, *we have*

$$\mathrm{DP}_{RTM_r}((0,b),\delta) \leq \mathrm{DP}_{M[w]}((b,b \boxplus b \ggg r),\delta) + \mathrm{DP}_{M[w]}((b,b \boxplus b \ggg r \boxminus 2^{w-r}),\delta).$$

Proof. The proof follows directly from Lemma 10 and Corollary 1.   □

Since $\mathrm{DP}_{M[w]}((a,b),\delta) \leq 2^{-w}$ for any differential $((a,b),\delta)$ to $M[w]$, we have a trivial bound of $\mathrm{DP}_{RTM_r}((a,b),\delta) \leq 2^{-w+2}$ from Lemma 10. But, we conjecture that actually $\mathrm{DP}_{RTM_r}((a,b),\delta) \leq 2^{-w+1}$. Indeed, for $\mathrm{DP}_{RTM_r}((a,b),\delta) > 2^{-w+1}$, (a,b) must propagate to multiple differences at the output of FTR_r that have a very high maximum value of $\mathrm{DP}_{M[w]}$. To be more precise, for $\mathrm{DP}_{RTM_r}((a,b),\delta) > 2^{-w+1}$, (a,b) must satisfy one of the following conditions:

1. (a,b) propagates to at least 3 differences at the output of FTR_r with a minimum of two of these differences satisfying $\mathrm{DP}_{M[w]} = 2^{-w}$.
2. (a,b) propagates to 4 differences such that, for one of them, $\mathrm{DP}_{M[w]} = 2^{-w}$, while for at least two others, it is 2^{-w-1}.

The maximum value of $\mathrm{DP}_{M[w]}$ is very low for most of the differentials [15]. In fact $\mathrm{DP}_{M[w]} = 2^{-w}$ only for unilateral, diagonal and counter-diagonal differences, and $\mathrm{DP}_{M[w]} = 2^{-w-1}$ also holds for very few (usually only 4) differentials.

Furthermore, in the bound of Lemma 10, we sum the sizes of the respective solution sets of the 4 differentials to $M[w]$. But, S_{RTM_r} is actually smaller since further conditions are imposed on the elements of S_{RTM_r} due to $\ggg r$, e.g., in Lemma 10, we assume that each $(h,k) \in \mathcal{S}_{M[w]}((x,y),\delta)$ also belongs to $\mathcal{S}_{RTM_r}((a,b),\delta)$. But, such (h,k) must also satisfy: $h \boxplus k \boxplus x \leq 2^w$ and $(h \boxplus k) \bmod 2^r + x \bmod 2^r < 2^r$. Such restrictions reduce the size of the solution set substantially. So we conjecture that $\mathrm{DP}_{RTM_r}((a,b),\delta) \leq 2^{-w+1}$.

For $w = 8, 10, 12, 16$ and for different corresponding values of r, we computed the upper-bound of $\mathrm{DP}_{RTM_r}((a,b),\delta)$ from Lemma 10 for all input differences. Then, for all the input differences (a,b) that led to an upper-bound greater than 2^{-w+1}, we computed the real value of $\max_\delta \mathrm{DP}_{RTM_r}((a,b),\delta)$ exhaustively and found out that for such differences $\mathrm{DP}_{RTM_r}((a,b),\delta) \leq 2^{-w}$ for all w, r.

3.4   Specifications of $\mathcal{F}$-156

We propose the public function $\mathcal{F}$-156 and describe it in Algorithm 2. The additions in the indexes for $\mathbf{x}$ and $\mathbf{y}$ in Algorithm 2 are computed modulo 4.

The input size of $\mathcal{F}$-156 is 256-bits. This means that Multimixer-156 can only process inputs whose lengths are multiples of 256 bits. To mitigate this, an input of arbitrary length can be padded via an injective padding to make its length in bits a multiple of 256, and thus suitable for processing via Multimixer-156.

$\mathcal{F}$-156 makes use of RTM_{29}. $\mathcal{F}$-156$(\mathbf{x},\mathbf{y}) = Z$ implies that for $i \in \{0,1,2,3\}$,

$$RTM_{29}(x_i,y_i) = z_i \quad \text{and} \quad u_i \times v_i = z_{i+4}.$$

Here $\mathrm{M}_\alpha \cdot \mathbf{x} = \mathbf{u}$ and $\mathrm{M}_\beta \cdot \mathbf{y} = \mathbf{v}$ with $\mathrm{M}_\alpha = \mathrm{circ}(1,1,2,3)$ and $\mathrm{M}_\beta = \mathrm{circ}(3,1,1,2)$.

Algorithm 2: The public function of Multimixer-156, $\mathcal{F}$-156

Inputs   : $X = (x_0, x_1, x_2, x_3, y_0, y_1, y_2, y_3)^\mathsf{T} \in (\mathbb{Z}/2^{32}\mathbb{Z})^8$
Output   : $Z = (z_0, z_1, z_2, z_3, z_4, z_5, z_6, z_7)^\mathsf{T} \in (\mathbb{Z}/2^{64}\mathbb{Z})^8$

for $i \leftarrow 0$ **to** 3 **do**   // Computation of the matrices
$\quad\mid\quad u_i \leftarrow x_i \boxplus x_{i+1} \boxplus 2x_{i+2} \boxplus 3x_{i+3}$
$\quad\mid\quad v_i \leftarrow 3y_i \boxplus y_{i+1} \boxplus y_{i+2} \boxplus 2y_{i+3}$
end
for $i \leftarrow 0$ **to** 3 **do**   // Computation of $\mathrm{FTR}_{29}(x_i, y_i)$
$\quad\mid\quad x_i^* \leftarrow x_i \boxplus y_i$
$\quad\mid\quad y_i^* \leftarrow (x_i^* \ggg 29) \boxplus y_i$
end
for $i \leftarrow 0$ **to** 3 **do**   // Computation of w-bit multiplications
$\quad\mid\quad z_i \leftarrow x_i^* \times y_i^*$
$\quad\mid\quad z_{i+4} \leftarrow u_i \times v_i$
end
$Z \leftarrow (z_0, z_1, z_2, z_3, z_4, z_5, z_6, z_7)$
return Z

3.5   Maximum Image Probability of $\mathcal{F}$-156

Lemma 11. *For the public function $\mathcal{F}$-156, $\mathrm{IP}_{\mathcal{F}\text{-}156}(0^8) < 2^{-157}$.*

Proof. $\mathcal{F}$-156$(\mathbf{x}, \mathbf{y}) = 0^8$ implies for $i \in \{0, 1, 2, 3\}$,

$$\mathrm{RTM}_{29}(x_i, y_i) = 0 \quad \text{and} \quad u_i \times v_i = 0.$$

Now, clearly for $i \in \{0, 1, 2, 3\}$, at least one of u_i and v_i must be 0.

When each $v_i = 0$, each $y_i = 0$ due to the invertibility of M_β. $y_i = 0$ together with $\mathrm{RTM}_{29}(x_i, y_i) = 0$ implies $x_i = 0$ for each i and thus in this case $(\mathbf{x}, \mathbf{y}) = 0^8$.

Now let three of the v_is be 0. Thus one of the u_is must be 0. Without loss of generality let us assume that $v_0 = 0, v_1 = 0, v_2 = 0$ and $u_3 = 0$. This implies

$$3 \cdot y_0 \boxplus y_1 \boxplus y_2 \boxplus 2 \cdot y_3 = 0$$
$$2 \cdot y_0 \boxplus 3 \cdot y_1 \boxplus y_2 \boxplus y_3 = 0$$
$$y_0 \boxplus 2 \cdot y_1 \boxplus 3 \cdot y_2 \boxplus y_3 = 0. \tag{4}$$

(4) is a system of three homogeneous linear equations in 4 variables y_i. So, for every choice of y_0, there is a unique choice of y_1, y_2 and y_3 such that (4) is satisfied. But by Lemma 9, for $i \in \{0, 1, 2\}$, every choice of y_i with $\mathrm{RTM}_{29}(x_i, y_i) = 0$ results in at most 2 solutions in each of the corresponding x_i. Each value of x_i for $i \in \{0, 1, 2\}$ together with $u_3 = 0$ results in a unique solution in x_3. So for every choice of y_0, there are at most 2 choices for x_i for $i \in \{0, 1, 2\}$ and a unique choice for the rest of the variables. Thus total no of such solution is $2^w \cdot 2^3 < 2^{w+3}$.

Similarly, the number of solutions when two of the v_is are 0 and when only one $v_i = 0$ are upper-bound by $2^2 \cdot 2^{2w} < 2^{2w+2}$ and $2 \cdot 2^{3w} < 2^{3w+1}$ respectively.

Each $v_i \neq 0$ implies that each $u_i = 0$ and that implies $\mathbf{x} = 0^4$. Now, we see that $\text{RTM}_{29}(0, y_i) = 0$ for each $i \in \{0, 1, 2, 3\}$ has at most 2 solutions in the corresponding variable y_i and this leads to at most $2^4 = 16$ solutions.

Thus $\text{IP}_{\mathcal{F}\text{-}156}(0^8) \leq \frac{2^{3w+1} + 2^{2w+2} + 2^{w+3} + 16}{2^{8w}} < \frac{1}{2^{5w-3}} = \frac{1}{2^{157}}$.   □

A number $n \in \mathbb{N}$ is called highly composite [15] if it has more divisors than any number smaller than n. $36, 802, 111, 876, 251, 321, 600$ is the largest highly composite number smaller than 2^{65} and has $207360 < 2^{18}$ divisors [15] . Thus, any non-zero output of a 32-bit multiplication has at most 2^{18} pre-images.

Lemma 12. *Let $Z \neq 0^8$. Then,* $\text{IP}_{\mathcal{F}\text{-}156}(Z) < 2^{-156}$.

Proof. $\mathcal{F}\text{-}156(\mathbf{x}, \mathbf{y}) = Z$ implies for $i \in \{0, 1, 2, 3\}$,

$$\text{RTM}_{29}(x_i, y_i) = z_i \quad \text{and} \quad u_i \times v_i = z_{i+4}. \tag{5}$$

(5) is a system of 8 equations. We can solve any 4 of these equations in their respective variable (x_i, y_i) or (u_i, v_i). Each of these solutions leads to at most a unique solution in $(\mathbf{x}, \mathbf{y})$ since $M_\alpha \cdot \mathbf{x} = \mathbf{u}$ and $M_\beta \cdot \mathbf{y} = \mathbf{v}$.

If $w(Z) \geq 3$, we solve the 3 equations corresponding to the non-zero components of Z and one of the remaining 5 equations. Each of the 3 equations with non-zero z_i has at most 2^{18} solutions and the other equation has at most $2^{w+1} - 1 < 2^{33}$ solutions. Thus the number of solutions to (5) for such a Z is upper-bounded by $(2^{18})^3 \cdot 2^{33} = 2^{87}$ and thus, $\text{IP}_{\mathcal{F}\text{-}156}(Z) < \frac{2^{87}}{2^{256}} = 2^{-169}$.

So, we only need to look at Zs with $w(Z) \leq 2$, i.e., when

1. Only one of the z_i is non-zero for some $i \in \{0, 1, 2, 3\}$.
2. Only one of the z_{i+4} is non-zero for some $i \in \{0, 1, 2, 3\}$.
3. Two of the z_i are non-zero for some $i \in \{0, 1, 2, 3\}$.
4. Two of the z_{i+4} are non-zero for some $i \in \{0, 1, 2, 3\}$.
5. One of $z_i = 0$ and one of z_{i+4} is non-zero for some $i \in \{0, 1, 2, 3\}$.

We look at the first case. We assume without loss of generality that $z_0 \neq 0$. Since each of $z_{i+4} = 0$, one of u_i and v_i must be 0 for each $i \in \{0, 1, 2, 3\}$.

Each $v_i = 0$ implies that each $y_i = 0$. Then for $i \in \{1, 2, 3\}$, $\text{RTM}_{29}(x_i, 0) = 0$, i.e., $x_i = 0$. The number of solutions in x_0 is upper-bounded by the number of factors of z_0, i.e., 2^{18}. So the total number of solutions is upper-bounded by 2^{18}.

When three of the $v_i = 0$ for $i \in \{0, 1, 2, 3\}$, we assume without loss of generality that $v_0 = v_1 = v_3 = 0$ and $u_3 = 0$. For every choice of y_0, there is a unique choice of y_1, y_2, y_3 satisfying $v_0 = v_1 = v_3 = 0$. For each such choice of $\mathbf{y}$, there are at most 2 values of x_1, x_2, x_3 that can lead to $\text{RTM}_{29}(x_i, y_i) = 0$ for $i \in \{1, 2, 3\}$, and for each choice of these variables, $u_3 = 0$ leads to a unique choice in x_0. Hence the total number of solutions in this case is at most $2^3 \cdot 2^w$.

Similarly when two, one or none of the $v_i = 0$, the number of solutions are upper-bounded by $2^2 \cdot 2^{2w}$, $2^3 \cdot 2^w$ and $2^3 \cdot 2^{18}$ respectively.

Thus, when only one of the z_i is non-zero for some $i \in \{0, 1, 2, 3\}$, $\text{IP}_{\mathcal{F}\text{-}156}(Z) \leq \frac{2^{18} + 2 \cdot 2^{w+3} + 2^{2w+2} + 2^{21}}{2^{8w}} < \frac{2^{2w+3}}{2^{8w}} = 2^{-(6w-3)} = 2^{-189}$. For all other cases for Z, we similarly upper-bound their IPs and tabulate the results in Table 1.   □

Table 1. Upper-bound of IP_Z with $w(Z) \leq 2$

Z	Upper-bound of $IP_{\mathcal{F}\text{-}156}(Z)$
$w(Z) = 1$, only one $z_{i+4} \neq 0$ for $i \in \{0,1,2,3\}$	$2^{-(6w-6)} = 2^{-186}$
$w(Z) = 2$, 2 of the $z_i \neq 0$ for $i \in \{0,1,2,3\}$	$2^{-(6w-3)} = 2^{-189}$
$w(Z) = 2$, 2 of the $z_{i+4} \neq 0$ for $i \in \{0,1,2,3\}$	$2^{-(5w-4)} = 2^{-156}$
$w(Z) = 2$, one of $z_i, z_{j+4} \neq 0$ for $i,j \in \{0,1,2,3\}$	$2^{-(6w-5)} = 2^{-187}$

Corollary 3. *For the public function* $\mathcal{F}$-156, $MIP_{\mathcal{F}\text{-}156} < 2^{-156}$.

Proof. The proof follows immediately from Lemmas 11 and 12. $\qquad\square$

3.6   Maximum Differential Probability of $\mathcal{F}$-156

In this section we make use of the following notation: $\mathbf{c} = M_\alpha \cdot \mathbf{a}$, $\mathbf{d} = M_\beta \cdot \mathbf{b}$, $\mathbf{p} = M_\alpha \cdot \mathbf{h}$ and $\mathbf{q} = M_\beta \cdot \mathbf{k}$.

We first look at the corner case inputs that lead to 4 active multiplications. We computed all such inputs and keeping rotational symmetries, we tabulate these inputs $\mathbf{a}$ along with the value of corresponding $\mathbf{c}$ in Table 2.

Table 2. Corner case inputs that lead to 4 active multiplications in $\mathcal{F}$-156

#	$\mathbf{a}^{\mathsf{T}}$	$\mathbf{c}^{\mathsf{T}}$	#	$\mathbf{a}^{\mathsf{T}}$	$\mathbf{c}^{\mathsf{T}}$
1	$(2^{w-1},0,0,0)$	$(2^{w-1},2^{w-1},0,2^{w-1})$	6	$(3 \cdot 2^{w-3},0,7 \cdot 2^{w-3},0)$	$(2^{w-3},0,5 \cdot 2^{w-3},0)$
2	$(2^{w-1},2^{w-1},0,0)$	$(0,0,2^{w-1},2^{w-1})$	7	$(2^{w-1},0,2^{w-1},0)$	$(2^{w-1},0,2^{w-1},0)$
3	$(2^{w-3},0,5 \cdot 2^{w-3},0)$	$(0,0,2^{w-1},2^{w-1})$	8	$(2^{w-1},2^{w-1},2^{w-1},0)$	$(0,2^{w-1},0,0)$
4	$(2^{w-2},0,2^{w-2},0)$	$(3 \cdot 2^{w-2},0,3 \cdot 2^{w-2},0)$	9	$(3 \cdot 2^{w-2},0,3 \cdot 2^{w-2},0)$	$(2^{w-2},0,2^{w-2},0)$
5	$(2^{w-2},2^{w-2},3 \cdot 2^{w-2},0)$	$(0,7.2^{w-2},0,0)$	10	$(3 \cdot 2^{w-2},3 \cdot 2^{w-2},2^{w-2},0)$	$(0,2^{w-2},0,0)$

Lemma 13. *Let* $((\mathbf{a},0),\Delta)$ *be a unilateral differential to* $\mathcal{F}$-156, *where* $\mathbf{a} = (2^{w-1},0,0,0)$. *Then we have*

$$DP_{\mathcal{F}\text{-}156}((\mathbf{a},0^4),\Delta) \leq \frac{DP_{RTM_{29}}((2^{w-1},0),\delta_0)}{2^{3w}}.$$

Proof. The input difference $((2^{w-1},0,0,0),0^4)$ propagates to the output difference Δ iff

$$\left(RTM_r(h_0 \boxplus 2^{w-1}, k_0) - RTM_r(h_0,k_0)\right) \bmod 2^{2w} = \delta_0 \tag{6.1}$$

$$\left((p_0 \boxplus 2^{w-1}) \times q_0 - p_0 \times q_0\right) \bmod 2^{2w} = \delta_4 \tag{6.2}$$

$$\left((p_1 \boxplus 2^{w-1}) \times q_1 - p_1 \times q_1\right) \bmod 2^{2w} = \delta_5 \tag{6.3}$$

$$\left((p_3 \boxplus 2^{w-1}) \times q_3 - p_3 \times q_3\right) \bmod 2^{2w} = \delta_7. \tag{6.4}$$

(6.2), (6.3) and (6.4) correspond to solution sets of unilateral differentials to $M[w]$ with variables (p_0, q_0), (p_1, q_1) and (p_3, q_3) respectively. As such, these equations are satisfied for at most one value of the corresponding q_i. Assuming all these equations to be consistent, for $i \in \{0, 1, 3\}$ we let these solutions be $q_i = \beta_{i+4}$.

(6.1) corresponds to the solution set of the differential $(2^{w-1}, 0)$ to RTM_{29}. Let $h_0 = \alpha_0, k_0 = \beta_0$ be a solution to (6.1). Thus from $q_i = \beta_{i+4}$, for $i \in \{0, 1, 3\}$,

$$k_1 \boxplus k_2 \boxplus 2k_3 = \beta_4 \boxminus 3\beta_0$$
$$3k_1 \boxplus k_2 \boxplus k_3 = \beta_5 \boxminus 2\beta_0$$
$$k_1 \boxplus 2k_2 \boxplus 3k_3 = \beta_7 \boxminus \beta_0.$$

These sets of equations lead to at most one solution in the variables k_i for $i \in \{1, 2, 3\}$. Thus, we conclude that the cardinality of the solution set of the differential $(((2^{w-1}, 0, 0, 0), 0^4), \Delta)$ is upper-bounded by the product of the number of solutions to (6.1), (6.2), (6.3) and (6.4). Thus for $\mathbf{a} = (2^{w-1}, 0, 0, 0)$, we have

$$\text{DP}_{\mathcal{F}\text{-156}}((\mathbf{a}, 0^4), \Delta) \leq \text{DP}_{\text{RTM}_{29}}((2^{w-1}, 0), \delta_0) \times \prod_{i \in \{4,5,7\}} \text{DP}_{M[w]}((2^{w-1}, 0), \delta_i)$$

$$\leq \frac{\text{DP}_{\text{RTM}_{29}}((2^{w-1}, 0), \delta_0)}{2^{3w}}.$$

$\square$

$\text{DP}_{\mathcal{F}\text{-156}}$ can be upper-bounded for all differentials $((\mathbf{a}, 0^4), \Delta)$ and $((0^4, \mathbf{a}), \Delta)$, where $\mathbf{a}$ is an element from Table 2 in a similar way. We also consider differentials of the form $((0^4, \mathbf{a}), \Delta)$ since RTM_{29} is not a commutative operation. By Lemma 13, to compute $\text{DP}_{\mathcal{F}\text{-156}}$ of such a differential, we need to compute the values of $\text{DP}_{\text{RTM}_{29}}$ for the corresponding differential to the active RTM_{29}. By Corollary 2, $\text{DP}_{\text{RTM}_{29}}$ is upper-bounded by the sum of DPs of two differentials to $M[w]$ and $\text{DP}_{M[w]}$ can be upper-bounded efficiently due to Lemma 3.

Applying this strategy, we computed the upper-bound for $\text{DP}_{\mathcal{F}\text{-156}}$ of all differentials $((\mathbf{a}, 0^4), \Delta)$ and $((0^4, \mathbf{a}), \Delta)$, where $\mathbf{a}$ is an element from Table 2. We also compute $\text{DP}_{\mathcal{F}\text{-156}}$ values for $\Delta = 0^8$ since that corresponds to a collision at the output of $\mathcal{F}\text{-156}$. We can compute these values thanks to Lemma 4. We tabulate these results in terms of minimum differential weight in Table 3. We omit $\mathcal{F}\text{-156}$ from the subscripts of differential weight for notational simplicity and $w_{\min}((\mathbf{a}, 0^4), \Delta)$ and $w_{\min}((0^4, \mathbf{a}), \Delta)$ refer to $\min_\Delta w((\mathbf{a}, 0^4), \Delta)$ and $\min_\Delta w((0^4, \mathbf{a}), \Delta)$ respectively. From Table 3 we can conclude that for differentials $((\mathbf{a}, \mathbf{b}), \Delta)$ leading to 4 active multiplications, $\text{DP}_{\mathcal{F}\text{-156}}((\mathbf{a}, \mathbf{b}), \Delta) \leq 2^{-156}$.

Table 3. Differential weights of differentials leading to 4 active multiplication

a	$w((\mathbf{a}, 0^4), 0^8)$	$w_{\min}((\mathbf{a}, 0^4), \Delta)$	$w((0^4, \mathbf{a}), 0^8)$	$w_{\min}((0^4, \mathbf{a}), \Delta)$
$(2^{w-1}, 0, 0, 0)$	156.67	156.67	156	156
$(2^{w-1}, 2^{w-1}, 0, 0)$	185.35	185.35	184	184
$(2^{w-3}, 0, 5 \cdot 2^{w-3}, 0)$	186.35	184.67	187.67	184.54
$(2^{w-2}, 0, 2^{w-2}, 0)$	185.66	184.38	187.35	184.18
$(2^{w-2}, 2^{w-2}, 3 \cdot 2^{w-2}, 0)$	214.49	212.57	217.03	212.27
$(3 \cdot 2^{w-3}, 0, 7 \cdot 2^{w-3}, 0)$	186.35	184.67	187.67	184.54
$(2^{w-1}, 0, 2^{w-1}, 0)$	185.35	185.35	184	184
$(2^{w-1}, 2^{w-1}, 2^{w-1}, 0)$	214.03	214.03	212	212
$(3 \cdot 2^{w-2}, 0, 3 \cdot 2^{w-2}, 0)$	185.66	184.38	187.35	184.18
$(3 \cdot 2^{w-2}, 3 \cdot 2^{w-2}, 2^{w-2}, 0)$	214.49	212.57	217.03	212.27

We claim that for differentials $((\mathbf{a}, \mathbf{b}), \Delta)$ that lead to at least 5 active multiplications inside $\mathcal{F}$-156, $\mathrm{DP}_{\mathcal{F}\text{-}156}((\mathbf{a}, \mathbf{b}), \Delta) \leq 2^{-156}$. Indeed, now we need to solve at least 5 equations involving either RTM_{29} or $M[32]$. We first choose to solve 3 of these equations. $\mathbf{h}$, $\mathbf{p}$ and $\mathbf{k}$, $\mathbf{q}$ are related linearly via M_α and M_β respectively. Thus, corresponding to each solution to these 3 equations, all the other remaining equations involving RTM_{29} or $M[32]$ can be considered to have the same variable. For example, if we solve the RTM_{29}s involving (h_0, k_0), (h_1, k_1) and $M[32]$ involving (p_0, q_0) such that one solution is given by $h_0 = \alpha_0$, $k_0 = \beta_0$, $h_1 = \alpha_1$, $k_1 = \beta_1$ and $p_0 = \alpha_4$, $q_4 = \beta_4$. Then clearly we can substitute $2h_2 = \alpha_4 \boxminus \alpha_0 \boxminus \alpha_1 \boxminus 3h_3$ and $k_2 = \beta_4 \boxminus 3\beta_0 \boxminus \beta_1 \boxminus 2k_3$ in the remaining equations, each of which then turn into an equation involving h_3, k_3.

For each solution to the 3 equations, the remaining equations now turn into solution sets for differentials to RTM_{29} or $M[32]$ with the same variable and we need to obtain the intersection of all these solution sets. The differentials to the remaining equations all depend on the solutions to the 3 equations and for most of the solutions, they would lead to different differentials. Due to Fig. 2, we know that for most of the differentials to $M[32]$ and consequently RTM_{29}, the value of DP is very low. Furthermore, we are required to find the intersection of at least 2 such solution sets. Thus, we claim that for most of the solutions, there can be at most 1 or 2 solutions of the remaining equations, if any. But, it is also possible that for some solutions to the 3 equations, all the remaining equations become identical and as such lead to significantly more solutions. However, the number of such solutions would be very low and thus such cases do not impact the total size of the solution sets.

More concretely, the 3 equations that we choose to solve can have at most $(2^{33})^3 = 2^{99}$ solutions due to our conjecture that $\mathrm{DP}_{\mathrm{RTM}_r}$ is upper-bounded by 2^{-w+1}. Now, if for n of these solutions, the remaining equations all become identical, then the size of the solution set to the differential to $\mathcal{F}$-156 is upper-bounded by $(2^{99} - n)2 + n2^{32}$. Since, n is expected to be very small for any differential, we claim that the size of the solution set is upper-bound by 2^{100} and consequently for all such differentials, $\text{DP}_{\mathcal{F}\text{-}156} \leq 2^{-156}$.

Also, the DP of any unilateral differential to the public function of Multimixer-128 is upper-bound by 2^{-128} since their solution sets involve either only $\mathbf{h}$ or only $\mathbf{k}$ [15]. But, due to the presence of FTR_{29} in $\mathcal{F}\text{-}156$, even when the input differences in all the active multiplications become unilateral, their solution sets involve variables from both $\mathbf{h}$ and $\mathbf{k}$. Thus, for such differences as well, via similar arguments, $\text{DP}_{\mathcal{F}\text{-}156} \leq 2^{-156}$.

Hence, we claim that $\text{MDP}_{\mathcal{F}\text{-}156} \leq 2^{-156}$ and thus thanks to Corollary 3 and Theorem 1, we claim that Multimixer-156 is 2^{-156}-Δuniversal.

4   Implementation and Benchmarking Results

We take advantage of the inherent parallelism present in Multimixer-156 by leveraging NEON instructions in our implementation, and write an optimized code for Multimixer-156 on a 32-bit ARMv7 Cortex-A processor. The NEON vector operations used in our implementation work on data elements of size 32-bits. The inputs to $\mathcal{F}\text{-}156$ are 8 key and 8 message words that can be stored in 4 vector registers. These message words are loaded from memory and then mapped as $\{x_0, x_2, y_1, y_3\}$ and $\{x_1, x_3, y_2, y_0\}$. This re-mapping makes the implementation of the circulant matrices more efficient without impacting the security.

The circulant matrix symmetry allows an efficient implementation using only vector additions, a word shuffle called *vector reverse in double-words*, and a vector swap on double-words. The circular shift operation is implemented using two instructions, *vector shift right* and *vector shift left and insert*. All in all, we implemented both matrix multiplications in $\mathcal{F}\text{-}156$ using only one `vswp` instruction, two `vrev64` instructions, and four `vadd` instructions. The circular shift is done with one `vshr` and one `vshr` instructions. Two `vadd` instructions add the key to the message in the beginning, and 4 `vmlal` instructions operate the integer multiplication and add the result to the register that is used to keep the output. We end up with a total of 20 instructions per call to $\mathcal{F}\text{-}156$, while Multimixer-128 and NH^{T} are using 11 and 16 instructions to process same amount of data respectively. We compare the size features and number of arithmetic operations in Multimixer-156 with that of Multimixer-128 and NH^{T} in Table 4. We benchmarked our Multimixer-156 code[1] and the code of Multimixer-128[2] and NH^{T}[3] on an ARM Cortex-A7 processor in the Broadcom BCM2836 chipset used in the Raspberry Pi 2 model B single board computer and report the results in Table 5. We also compare the performance of Multimixer-156 with those of Xoodyak[4] [11] and ASCON [13]. For ASCON we used the reported performance per authenticated/encrypted byte on an Armv7a architecture[5]. Xoodyak's performance is

---

[1] https://github.com/Parisaa/Multimixer-156/.

[2] https://github.com/Parisaa/Multimixer.

[3] https://github.com/google/adiantum/tree/master/benchmark/src/arm.

[4] https://github.com/XKCP/XKCP/tree/master/lib/low/Xoodoo/ARMv7A-NEON.

[5] https://ascon.iaik.tugraz.at/implementations.html.

Table 4. Feature comparison between NH^T, Multimixer-128, and Multimixer-156.

Algorithm	length in bits			# ops. per 32-bit word			
	digest	block	key	$\ggg$	$\times$	$+ \bmod 2^{32}$	$+ \bmod 2^{64}$
NH^T	256	64	64(# blocks) +192	–	2	4	2
Multimixer-128	512	256	256(# blocks)	–	1	2.5	1
Multimixer-156	512	256	256(# blocks)	0.5	1	5	1

around 24.5 cycles per authenticated byte for processing 1400 bytes, and the ASCON team reports performance of 30.7 cycles per authenticated/encrypted byte on an Armv7a architecture, while for Multimixer-156, it is at most close to 3 cycles per byte.

Table 5. Performance on ARM Cortex-A7 in cycles per byte.

Algorithm	ε-Δ universality	Input length in bytes		
		512	4096	32768
Multimixer-156	$\varepsilon = 2^{-156}$	3.051	2.518	2.662
Multimixer-128	$\varepsilon = 2^{-127}$	1.830	1.233	1.396
NH^T	$\varepsilon = 2^{-128}$	2.033	1.500	1.558

Acknowledgements. The authors would like to thank the anonymous reviewers of Indocrypt 2023 for their valuable feedback. Koustabh Ghosh is supported by the Netherlands Organisation for Scientific Research (NWO) under TOP grant TOP1.18.002 SCALAR, Parisa Amiri Eliasi is supported by the Cryptography Research Center of the Technology Innovation Institute (TII), Abu Dhabi (UAE), under the TII-Radboud project with title Evaluation and Implementation of Lightweight Cryptographic Primitives and Protocols and Joan Daemen is supported by the European Research Council under the ERC advanced grant agreement under grant ERC-2017-ADG Nr. 788980 ESCADA.

References

1. NNSP 800-38B: Recommendation for Block Cipher Modes of Operation: The CMAC Mode for Authentication (2005). https://csrc.nist.gov/pubs/sp/800/38/b/upd1/final
2. Bellare, M., Canetti, R., Krawczyk, H.: Keying hash functions for message authentication. In: Koblitz, N. (ed.) CRYPTO 1996. LNCS, vol. 1109, pp. 1–15. Springer, Heidelberg (1996). https://doi.org/10.1007/3-540-68697-5_1
3. Bellare, M., Kilian, J., Rogaway, P.: The security of cipher block chaining. In: Desmedt, Y.G. (ed.) CRYPTO 1994. LNCS, vol. 839, pp. 341–358. Springer, Heidelberg (1994). https://doi.org/10.1007/3-540-48658-5_32

4. Bernstein, D.J.: The Poly1305-AES message-authentication code. In: Gilbert, H., Handschuh, H. (eds.) FSE 2005. LNCS, vol. 3557, pp. 32–49. Springer, Heidelberg (2005). https://doi.org/10.1007/11502760_3
5. Bertoni, G., Daemen, J., Hoffert, S., Peeters, M., Assche, G.V., Keer, R.V.: Farfalle: parallel permutation-based cryptography. IACR Trans. Symmetric Cryptol. **2017**(4), 1–38 (2017)
6. Black, J., Halevi, S., Krawczyk, H., Krovetz, T., Rogaway, P.: UMAC: fast and secure message authentication. In: Wiener, M. (ed.) CRYPTO 1999. LNCS, vol. 1666, pp. 216–233. Springer, Heidelberg (1999). https://doi.org/10.1007/3-540-48405-1_14
7. Black, J., Rogaway, P.: CBC MACs for arbitrary-length messages: the three-key constructions. In: Bellare, M. (ed.) CRYPTO 2000. LNCS, vol. 1880, pp. 197–215. Springer, Heidelberg (2000). https://doi.org/10.1007/3-540-44598-6_12
8. Black, J., Rogaway, P.: A block-cipher mode of operation for parallelizable message authentication. In: Knudsen, L.R. (ed.) EUROCRYPT 2002. LNCS, vol. 2332, pp. 384–397. Springer, Heidelberg (2002). https://doi.org/10.1007/3-540-46035-7_25
9. Carter, L., Wegman, M.N.: Universal classes of hash functions. J. Comput. Syst. Sci. **18**(2), 143–154 (1979)
10. Crowley, P., Biggers, E.: Adiantum: length-preserving encryption for entry-level processors. IACR Trans. Symmetric Cryptol. **2018**(4), 39–61 (2018)
11. Daemen, J., Hoffert, S., Peeters, M., Assche, G.V., Keer, R.V.: Xoodyak, a lightweight cryptographic scheme. IACR Trans. Symmetric Cryptol. **2020**(S1), 60–87 (2020)
12. Daemen, J., Rijmen, V.: The Design of Rijndael - The Advanced Encryption Standard (AES). ISC, 2nd edn. Springer, Cham (2020). https://doi.org/10.1007/978-3-662-60769-5
13. Dobraunig, C., Eichlseder, M., Mendel, F., Schläffer, M.: ASCON v1.2: lightweight authenticated encryption and hashing. J. Cryptol. **34**(3), 33 (2021)
14. Fuchs, J., Rotella, Y., Daemen, J.: On the security of keyed hashing based on public permutations. In: Handschuh, H., Lysyanskaya, A. (eds.) CRYPTO 2023. LNCS, vol. 14083, pp. 607–627. Springer, Cham (2023). https://doi.org/10.1007/978-3-031-38548-3_20
15. Ghosh, K., Eliasi, P., Daemen, J.: Multimixer-128: universal keyed hashing based on integer multiplication. IACR Trans. Symmetric Cryptol. **2023**(3), 1–24 (2023)
16. Ghosh, K., Fuchs, J., Eliasi, P.A., Daemen, J.: Universal hashing based on field multiplication and (near-)MDS matrices. In: El Mrabet, N., De Feo, L., Duquesne, S. (eds.) AFRICACRYPT 2023. LNCS, vol. 14064, pp. 129–150. Springer, Cham (2023). https://doi.org/10.1007/978-3-031-37679-5_6
17. Iwata, T., Kurosawa, K.: OMAC: one-key CBC MAC. In: Johansson, T. (ed.) FSE 2003. LNCS, vol. 2887, pp. 129–153. Springer, Heidelberg (2003). https://doi.org/10.1007/978-3-540-39887-5_11
18. Krovetz, T.: HS1-SIV (v2) (2015). Submission to CAESAR competition
19. McGrew, D.A., Viega, J.: The security and performance of the Galois/Counter mode (GCM) of operation. In: Canteaut, A., Viswanathan, K. (eds.) INDOCRYPT 2004. LNCS, vol. 3348, pp. 343–355. Springer, Heidelberg (2004). https://doi.org/10.1007/978-3-540-30556-9_27
20. Stinson, D.R.: On the Connections Between Universal Hashing, Combinatorial Designs and Error-Correcting Codes. Electron. Colloquium Comput. Complex. TR95-052 (1995)
21. Wegman, M.N., Carter, L.: New hash functions and their use in authentication and set equality. J. Comput. Syst. Sci. **22**(3), 265–279 (1981)

On the Security of Triplex- and Multiplex-Type Constructions with Smaller Tweaks

Nilanjan Datta[1], Avijit Dutta[1], Eik List[2], and Sougata Mandal[1,3(✉)]

[1] Institute for Advancing Intelligence, TCG CREST, Kolkata, India
{nilanjan.datta,avijit.dutta}@tcgcrest.org
[2] Nanyang Technological University, Singapore, Singapore
eik.list@ntu.edu.sg
[3] Ramakrishna Mission Vivekananda Educational and Research Institute, Howrah, India
sougata.mandal@tcgcrest.com

Abstract. In TCHES'22, Shen et al. proposed Triplex, a single-pass leakage-resistant authenticated encryption scheme based on Tweakable Block Ciphers (TBCs) with $2n$-bit tweaks. Triplex enjoys beyond-birthday-bound ciphertext integrity in the CIML2 setting and birthday-bound confidentiality in the CCAmL1 notion. Despite its strengths, Triplex's operational efficiency was hindered by its sequential nature, coupled with a rate limit of $2/3$. In an endeavor to surmount these efficiency challenges, Peters et al. proposed Multiplex, a variant of Triplex with increased parallelism and a flexible rate of $d/(d+1)$ that retains similar security guarantees. However, the innovation came at the price of requiring TBCs with dn-bit tweaks, which are unusual and potentially costly for $d > 3$. In this paper, we investigate the limits of generalized Triplex- and Multiplex-type constructions for single-pass leakage-resilient authenticated encryption. Our contributions are threefold. First, we show that such constructions cannot provide CIML2 integrity for any tweak lengths below $dn/2$ bits. Second, we provide a birthday-bound attack for constructions with TBCs of tweak lengths between $dn/2$ and $(d-1)n+n/2$ bits. Finally, on the constructive side, we propose a family of single-pass leakage-resilient authenticated ciphers, dubbed Tweplex, that uses tweaks of $dn/2$ bits and provides a rate of $d/(d+1)$ while providing $n/2$-bit CIML2 integrity and CCAmL1 confidentiality.

1 Introduction

The design and analysis of schemes for authenticated encryption (with associated data) has been a highly active research area since it had been postulated to be a primitive of its kind [38] that shall protect both the confidentiality of the message and the integrity of the ciphertext. Throughout the decades, variants like online schemes [4,29], nonce-based, or deterministic authenticated encryption [40] arose. With them, a vast number of designs have been proposed that tried to optimize various needs between efficiency and security, most recently as the

A. Chattopadhyay et al. (Eds.): INDOCRYPT 2023, LNCS 14459, pp. 25–47, 2024.
https://doi.org/10.1007/978-3-031-56232-7_2

outcomes of the CAESAR [7] and NIST Lightweight competitions [42]. Beyond those general aspects, a series of research has been addressing robustness aspects, such as security under nonce-misuse [40], accidental nonce repetitions [19], or settings where unverified would-be plaintexts could be released [1,11]. This led to the strongest theoretical security notions for AE in Robust [28] and Subtle AE [3].

1.1 Leakage-Resilient Authenticated Encryption

The usual theoretical security notions of AE treat the primitives as black boxes whereas real-world adversaries are free to exploit additional information from side channels. Such leakage can include but is not limited to information about timing, power consumption, or electromagnetic radiation, which can leak significant amounts of information about the internal state of the mode including its keys. One differentiates between two main attack vectors in the context of power consumption: Simple Power Analysis (SPA) and Differential Power Analysis (DPA) [30]. In SPA attacks, an adversary observes leakages from encrypting a single input message under potentially multiple measurements to remove noise. DPA attacks study leakage from encrypting multiple inputs which provides new information about the internals of a cipher. Thus, the privacy of internal states can be reduced at a rate exponential in the number of distinct inputs.

Widespread AEAD schemes such as OCB [31,39], GCM [21,33], or CCM [20], which invoke a block cipher multiple times with a single key, are typically susceptible to DPA attacks. However, the protection of the underlying block cipher against leakage is usually left to the implementors and engineers who implement the components of a scheme so to minimize leakages as much as possible. For example, on a hardware level, the usual approaches for preventing leakage are to blur the signal with noise or special circuits. In contrast, on the implementation level, countermeasures include masking [12,22], where the internal state of the device is split into several shares, which are then used in individual computations, or shuffling [26,43]. Strong protection often adds considerable amounts of additional area, power, or efficiency penalties. Protection against DPAs lowers the performance of both software and hardware implementations of the algorithm by several orders of magnitude compared to the unprotected implementations in standard metrics (e.g. [23]). Consequently, a line on research of developing more efficient schemes that provide trade-offs has emerged. For an in-depth survey, we refer the interested reader to [6].

Instead of using a strongly protected block-cipher implementation for all invocations in an AEAD scheme, leakage-resistant modes of operations [3,9,10,17] have shifted the paradigm to support dedicated leveled implementations. In this scenario, a crytographic primitive is called multiple times in an AEAD mode, but only certain calls to the primitive are strongly protected against DPA attacks while the remaining calls that usually perform the majority of computation are allowed to leak a certain amount of information to the adversary every time they are used. In summary, leakage-resilient AE schemes ensure security despite leakage at the cost of requiring strong protection for a few primitive calls. This allows for reasonable efficiency with sufficient protection in practice.

1.2 Security Models for Leakage-Resilient Authenticated Encryption

We consider the notions for leakage-resilient authenticated encryption by Guo et al. [9,10,24]. In [24], they proposed a comprehensive framework and the relations between them. As the strongest notions for AE, they identified (1) Ciphertext Integrity with Misuse-resistance and Leakage in encryption and decryption, or CIML2 [9,10]; (2) chosen-ciphertext security with misuse-resilience and Leakage in encryption and decryption oracle, called CCAmL2 [24]; (3) moreover, schemes that process the messages in multiple passes could furthermore achieve the usual nonce-misuse resistance in the black-box setting without leakage [40].

In general, authenticated encryption schemes can be categorized into one- or multipass schemes. The latter can achieve both CIML2 and CCAmL2 security. However, in the context of lightweight cryptography single-pass modes are often preferred over two-pass modes. While nonce-based single-pass schemes can also achieve CIML2 security, CCAmL2 is out of range, but they can achieve CCA security with misuse-resilience and leakage in encryption, which was formulated as CCAmL1 [24].

A portfolio of leakage-resilient schemes for leveled implementations has been developed in the past few years. Misuse-resistant two-pass schemes include ISAP [17], ISAPv2 [16], or TEDT [8]. One-pass schemes include AEDT [10], or Triplex [41], and Multiplex [36]. All these constructions share a common design structure that consists of three independent modules [6,15]: (i) the first module is called the *key-derivation function* (KDF) that employs a protected primitive to derive a session state from the nonce and the long-term master secret key; (ii) the second module is called the *message-processing function* (MPF), in which the plaintext (or the ciphertext) is encrypted (or decrypted) with a less protected primitive. In security treatments, the MPF module is often assumed to leak continuously. However, it adopts the idea of frequent rekeying to ensure that the security does not degrade badly. (iii) The third module is called the *tag-generation function* (TGF). It also employs a heavily protected primitive to derive the authentication tag. For nonce-based one-pass AE schemes, the MPF module will finally output a state as the input data to the TGF module, whereas for two-pass AE schemes [8,15,24,32] such a state is produced by hashing the ciphertext, nonce, and associated data with a less protected primitive. Following [24], Bellizia et al. [6] referred to leveled designs achieving both CIML2 and CCAmL1 security as *Grade-2-protected*. Recent lightweight one-pass leakage-resilient schemes, such as Ascon [18], Spook [5], or Triplex [41] are Grade-2 designs.

1.3 Revisiting Triplex and Multiplex

Leakage-resilient AE schemes are built primarily upon public permutations and tweakable block ciphers (TBCs). TBC-based leakage-resilient AE schemes require at least two calls to a primitive with n-bit block size for encrypting an n-bit message. While TEDT and TEDT2, which are two-pass AE schemes, achieve strong security guarantees (i.e., CIML2 and CCAmL2), they inherently

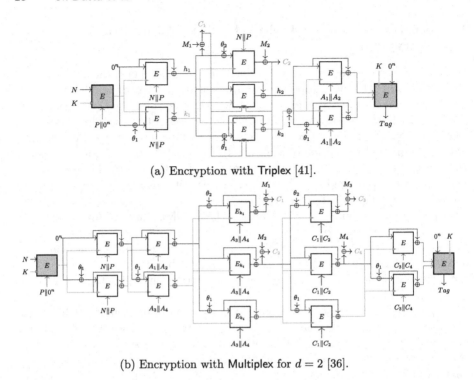

(a) Encryption with **Triplex** [41].

(b) Encryption with **Multiplex** for $d = 2$ [36].

Fig. 1. Triplex and Multiplex. The darkened TBC calls in the KDF and TGF modules need strong protection whereas the white calls to E do not. The blue, green, and red lines represent block, key, and tweak input to the TBCs, respectively.

offer a lower performance compared to single-pass modes. In this respect, Shen et al. [41] introduced a nonce-based single-pass AE scheme, called Triplex, which offered a rate of $2/3$ while providing $n - \log_2(n)$-bit CCAmL1 and CIML2 security. Triplex is shown schematically in Fig. 1a.

Despite its efficiency under leakage, its design limits its throughput. In particular, one cannot make parallel TBC calls for en- or decryption. To address this shortcoming, Peters et al. proposed Multiplex [36], that allows a higher degree of parallelization at every round of the algorithm and offers a flexible rate of $d/(d+1)$, where d denotes the degree of parallelization, with $O(n - \log_2(dn))$-bits of security. In particular, at each round, it processes dn bits message using $d+1$ TBC calls such that each one of them requires dn bits tweak. Despite achieving a higher throughput, its primary disadvantage is the use of large tweaks. Although, a few long-tweak variants of TBCs have been proposed [13,34,35], their security is far less understood compared to established designs and demands more cryptanalysis [25,37] to be stable. Moreover, instantiating Multiplex with TBCs of tweak lengths between $2n$ and $3n$ bits does not add anything extra over Triplex. Therefore, it is an interesting question how the security of Triplex and Multiplex is affected when they are instantiated with smaller tweak size TBCs. In particular, our study is narrowed down to ask the following two questions:

1. *Can we instantiate* Triplex *using a TBC with* $< 2n$-bit tweaks?
2. *In the light of the importance of using TBCs with established tweak lengths, can we process d message blocks in* Multiplex *with* $(d+1)$ *calls to a TBC with* $< dn$-bit tweaks?

1.4   Our Contribution

Answering the question above is the central theme of the paper. More precisely, we tackle it in three steps as follows:

1. We show that for any choice of linear functions $f, g : \{0,1\} \rightarrow \{0,1\}^l$ that take $2n$-bit ciphertexts and produce l-bit outputs, if $l < n$, one can mount a forging attack with probability one on Triplex with a small constant number of queries. Furthermore, we show for tweak lengths of $l \in [n, 3n/2]$ bits, one can mount a forging attack on the construction with probability one in approximately $2^{n/2}$ queries. Finally, we show for tweak lengths of $l \in (3n/2, 2n)$ bits, an adversary can mount a forgery attack on the construction with success probability 2^{3n-2l} by making at least 2^{2n-l} queries.
2. For a given fixed parameter d, we show that for any choice of a pair of linear functions (f, g) that takes dn-bit ciphertexts and produces l-bits output, if $l < dn/2$, then one can mount a forging attack with probability 1 on Multiplex with a constant number of queries. We further show for tweak lengths of $l \in [dn/2, dn-n/2]$ bits, then one can mount a forging attack on the construction with probability 1 by making at least $2^{n/2}$ queries. Finally, we show for tweak lengths of $l \in (dn-n/2, dn)$ bits, an adversary can mount a forgery attack on the construction with success probability 2^{3n-2l} by making at least 2^{2n-l} queries. This provides an answer to our second question.
3. On a constructive side and to transform our theoretical results into practice, we propose an efficient Multiplex-type construction, dubbed Tweplex, which employs a TBC with $dn/2$-bits tweak and a rate of $d/(d+1)$. We show that Tweplex achieves the maximally possible $O(n/2)$-bit CIML2 and birthday-bound CCAmL1 security in the multi-user setting. Our construction maintains the rate of $d/(d+1)$ while using only half of the tweak size of Triplex and Multiplex. Hence, our construction provides higher throughput over Triplex and Multiplex while allowing still a rate of $4/5$ for $d = 4$ with established ciphers such as Deoxys-BC-128-384 or Skinny-128-384.

2   Preliminaries

NOTATIONS: For a finite set $\mathcal{X}$, we write $X \xleftarrow{\$} \mathcal{X}$ to denote that X is uniformly sampled from $\mathcal{X}$. We write $(X_1, X_2, \ldots, X_q) \xleftarrow{\$} \mathcal{X}$ to denote that each X_i is sampled uniformly at random from $\mathcal{X}$. For a set $\mathcal{X}$, we write $\mathcal{X} \xleftarrow{\cup} X$ to denote that $\mathcal{X} \leftarrow \mathcal{X} \cup \{X\}$. For a fixed n, we write the set of all n-bit binary strings as $\{0,1\}^n$, and $\{0,1\}^*$ denotes the set of all binary strings of arbitrary length. ε is used to denote the empty string. $|x|$ denotes the length of the bit string x. $\mathsf{msb}_c(Z)$ and $\mathsf{lsb}_c(Z)$ return the c most and least significant bits of a bit string Z, respectively. $x[i,j]$ denotes the substring from i-th bit to j-th bit of x. The concatenation of two strings x and y is denoted as $x\|y$. We also often write it as (x,y). If $\mathcal{A}$ is an algorithm, then $y \leftarrow \mathcal{A}(x_1, x_2, \ldots, ; r)$ denotes running the algorithm $\mathcal{A}$ with randomness r on inputs $x_1, \ldots,$ and assigning the output to y. Equivalently, we can express the notation above as follows: let $y \xleftarrow{\$} \mathcal{A}(x_1, \ldots)$ be the result of picking r uniformly at random and then compute $\mathcal{A}(x_1, \ldots; r)$ and assign the result to the variable y. For an algorithm $\mathcal{A}$ and an oracle $\mathcal{O}$, we write $\mathcal{A}^{\mathcal{O}}$ to denote the output of $\mathcal{A}$ at the end of its interaction with $\mathcal{O}$.

2.1 Security Notions

A distinguisher $\mathcal{A}$ is an algorithm that tries to distinguish between two oracles $\mathcal{O}_0$ and $\mathcal{O}_1$ via black-box interaction with one of them. At the end of its interaction, it returns a bit $b \in \{0,1\}$. The distinguishing advantage of $\mathcal{A}$ against $\mathcal{O}_0$ and $\mathcal{O}_1$ is defined as $\Delta_{\mathcal{A}}[\mathcal{O}_0; \mathcal{O}_1] \triangleq \left| \Pr[\mathcal{A}^{\mathcal{O}_0} = 1] - \Pr[\mathcal{A}^{\mathcal{O}_1} = 1] \right|$, where the probabilities depend on the random coins of $\mathcal{O}_0$ and $\mathcal{O}_1$ and the random coins of the distinsguisher $\mathcal{A}$. The time complexity of the adversary is defined over the usual RAM model of computations. We call $\mathcal{A}$ a (q,t)-adversary if it asks at most q queries and runs in time at most t. We augment this notation by parameters e.g. in settings, where queries consist of en- and decryption oracles, we consider (q_e, q_d, t)-adversaries, assuming it asks at most q_e en- and q_d decryption queries, respectively. When queries consist of multiple blocks or bits, we augment it by σ for the number of blocks an adversary asks, and by p if an additional oracle to a primitive is given.

2.2 Tweakable Block Cipher

A tweakable block cipher with key space $\{0,1\}^\kappa$, tweak space $\{0,1\}^t$ and domain $\{0,1\}^n$ is a function $\widetilde{E} : \{0,1\}^\kappa \times \{0,1\}^t \times \{0,1\}^n \to \{0,1\}^n$ such that for each key $k \in \{0,1\}^\kappa$ and each tweak $\mathsf{t} \in \{0,1\}^t$, the function $\widetilde{E}(k, \mathsf{t}, \cdot)$ is a permutation over $\{0,1\}^n$. We call such TBCs (κ, t, n)-TBCs and define $\mathsf{TBC}(\kappa, t, n)$ for the set of all (κ, t, n)-TBCs. We call a function $\mathsf{IC} : \{0,1\}^\kappa \times \{0,1\}^t \times \{0,1\}^n$ an ideal TBC if $\mathsf{IC} \xleftarrow{\$} \mathsf{TBC}(\kappa, t, n)$. In this case, IC_k^t is a random independent permutation over $\{0,1\}^n$, for each $(k, \mathsf{t}) \in \{0,1\}^\kappa \times \{0,1\}^t$, even if k is public. We write $\widetilde{E}$ for TBCs, and in our ideal TBC-based security proofs, we use the notation IC. $\mathsf{TP}(t, n)$ denotes the set of all functions $\widetilde{\pi} : \{0,1\}^t \times \{0,1\}^n \to \{0,1\}^n$ such that for all $\mathsf{t} \in \{0,1\}^t$, $\widetilde{\pi}(\mathsf{t}, \cdot)$ is a permutation over $\{0,1\}^n$. We define the strong tweakable pseudorandom permutation (stprp) advantage of $\mathcal{A}$ against $\widetilde{E}$ as

$$\mathbf{Adv}_{\widetilde{E}}^{\mathsf{stprp}}(\mathcal{A}) \triangleq \Delta_{\mathcal{A}} \left[(\widetilde{E}_k, \widetilde{E}_k^{-1}); (\widetilde{\pi}, \widetilde{\pi}^{-1}) \right],$$

where $k \xleftarrow{\$} \{0,1\}^\kappa$ and $\widetilde{\pi} \xleftarrow{\$} \mathsf{TP}(t, n)$.

2.3   Nonce-Based Single-Pass Authenticated Encryption

Let $\mathcal{K}, \mathcal{N}, \mathcal{A}, \mathcal{M}, \mathcal{C}, \mathcal{T}$ be non-empty sets for keys, nonces, associated data, messages, ciphertext, and authentication tags, respectively. A nonce-based authenticated encryption scheme (nAE) consists of a pair of deterministic algorithms, called the encryption algorithm $\mathcal{E} : \mathcal{K} \times \mathcal{N} \times \mathcal{A} \times \mathcal{M} \rightarrow \mathcal{C} \times \mathcal{T}$ and the decryption algorithm $\mathcal{D} : \mathcal{K} \times \mathcal{N} \times \mathcal{A} \times \mathcal{C} \times \mathcal{T} \rightarrow \mathcal{M} \cup \{\perp\}$. The correctness condition of a nAE scheme states that for every $K \in \mathcal{K}, N \in \mathcal{N}, A \in \mathcal{A}$, and $M \in \mathcal{M}$, we have $\mathcal{D}(K, N, A, \mathcal{E}(K, N, A, M)) = M$ and the tidiness condition of the nAE scheme states that for all $(K, N, A, C, T) \in \mathcal{K} \times \mathcal{N} \times \mathcal{A} \times \mathcal{C} \times \mathcal{T}$, where $\exists M \in \mathcal{M}$ such that $\mathcal{E}(K, N, A, M) = (C, T)$, it holds that $\mathcal{E}(K, N, A, \mathcal{D}(K, N, A, (C, T))) = (C, T)$. Let $\Pi = (\mathcal{E}, \mathcal{D})$ be a nAE scheme.

In this work, we focus on various security models of single-pass AEAD schemes under leakage; in particular, we limit our interest to notions for AEAD with nonce-misuse-resistant integrity and nonce-misuse-resilient confidentiality under potential leakage in encryption queries. Note that nonce-misuse-resistant confidentiality in the context of both en- and decryption-oracle leakage is impossible to achieve for a single-pass AE mode. We refer to the interested reader to Appendix A of [8] for a detailed discussion.

Leakage depending on the implementation of an AEAD scheme can be viewed as two functions: leakage during encryption queries and leakage during decryption queries. In [9], Berti et al. have defined the leakage integrity notion with nonce-misuse-resistant by allowing only encryption leakage, which is referred to as CIML [9] notion, in which an adversary makes encryption and decryption queries, and obtains the corresponding responses. Along with that, the adversary also obtains leakages corresponding to the encryption queries. The final goal of the adversary in this model is to forge the construction with a valid tuple. However, this security notion has been extended from CIML [9] to CIML2 [10], where the latter allows leakage from not only encryption but also from decryption queries. Berti et al. [8] defined muCIML2 as a multi-user distinguishing version of CIML2. We focus on the strong multi-user notions (in their non-distinguishing variants) muCIML2 for integrity and muCCAmL1 for confidentiality both under leakage.

2.4   (Multi-user) Ciphertext Integrity Under Misuse Leakage

We consider Ciphertext Integrity under Misuse Leakage (CIML2) under leakage in both en- and decryption queries. In this security notion, an adversary $\mathscr{A}$ is allowed to make queries to the encryption $L\mathcal{E}_{K_i}$ for any user i, and the decryption oracle $L\mathcal{D}_{K_i}$, and obtains the corresponding responses along with the leakages corresponding to the encryption and the decryption. Finally, the adversary submits a forging tuple (i, N, A, M, C, Tag) such that it is fresh. The forging advantage of the muCIML2 notion is then defined as the probability that (i, N, A, M, C, Tag) is valid. Formally, we define the muCIML2 notion as follows:

Definition 1. *Let $\mathscr{A}$ be a (q,t)-muCIML2 adversary on an AEAD scheme $\Pi :=$ $(\mathcal{E}, \mathcal{D})$. Then, the advantage $\mathscr{A}$ is defined as*

$$\mathbf{Adv}_{\Pi}^{mu\,CIML2}(\mathscr{A}) \triangleq \Pr\left[\mathscr{A}^{L\mathcal{E}_\mathbf{K}, L\mathcal{D}_\mathbf{K}, \widetilde{E}, \widetilde{E}^{-1}} \; forges\right],$$

where the probability is taken over the u many user keys $\mathbf{K} = (K_1, \ldots, K_u)$, randomness of $\mathscr{A}$, and the ideal TBC $\widetilde{E}$.

The algorithmic description of muCIML2 is given in the full version [14].

2.5 (Multi-user) Chosen-Ciphertext Indistinguishability Under Nonce Misuse and Leakage

For privacy under nonce repetitions, we follow [2]. In this context, an adversary $\mathscr{A}$ is allowed to make encryption queries and ideal TBC queries. We split the encryption oracle into two categories: $\mathcal{E}_1$ and $\mathcal{E}_2$, where $\mathscr{A}$ is allowed to repeat the nonce for the same user during $\mathcal{E}_1$ queries but has to use a fresh nonce for every query to $\mathcal{E}_2$. Ultimately, the adversary has to distinguish between $\mathcal{E}_2$ and a random function. We define the advantages as follows:

$$\mathbf{Adv}_{\Pi}^{\mathrm{conf}}(\mathscr{A}) \triangleq \Delta_{\mathscr{A}}\left[(\mathcal{E}_\mathbf{K}^1, \mathcal{E}_\mathbf{K}^2, \widetilde{E}, \widetilde{E}^{-1}); (\mathcal{E}_\mathbf{K}^1, \$, \widetilde{E}, \widetilde{E}^{-1})\right].$$

We extend the notion above to incorporate leakage. Then, the adversary interacts with the en- and decryption oracles with possibly repeating nonces. It obtains the corresponding en- or decryption responses plus potential leakage during the encryption queries, i.e., the *decryption oracle does not leak*. Finally, the adversary submits a challenge encryption query corresponding for some user under a fresh nonce N. The challenger either encrypts that query or encrypts a randomly chosen message and responds with the corresponding ciphertext-tag pair. The security advantage of muCCAmL1 is then defined as the distinguishing advantage. We define muCCAmL1 security of an authenticated encryption scheme Π with respect to leakage and nonce-misuse resilience. An adversary $\mathscr{A}$ can query four oracles: a primitive oracle, the decryption oracle without leakage, the encryption oracle $L\mathcal{E}^1$ with a leakage function $\mathfrak{L}^{\mathcal{E}}$, and another encryption oracle $L\mathcal{E}^2$ with a leakage function $\mathfrak{L}^{\mathcal{E}}$. Using these queries, $\mathscr{A}$ has to distinguish between $L\mathcal{E}^2$ and a random function. We define the muCCAmL1 advantage as

$$\mathbf{Adv}_{\Pi}^{mu\mathsf{CCAmL1}}(\mathscr{A}) \triangleq \Delta_{\mathscr{A}}\left[(L\mathcal{E}_\mathbf{K}^1, L\mathcal{E}_\mathbf{K}^2, \mathcal{D}, \widetilde{E}, \widetilde{E}^{-1}); (L\mathcal{E}_\mathbf{K}^1, L\$, \mathcal{D}, \widetilde{E}, \widetilde{E}^{-1})\right],$$

where the probability is taken over the u user keys $\mathbf{K} = (K_1, \ldots, K_u)$, the randomness of $\mathscr{A}$, and the ideal TBC $\widetilde{E}$. Note that we have considered leakage from only the encryption oracle. This is a weaker notion as compared to muCCAmL2, where both en- and decryption oracles leak. However, it is well-known that the notion muCCAmL2 is impossible to achieve for single-pass authenticated encryption schemes. Thus, we restrict our interest to muCCAmL1. An alternative game-based description of this notion is given in full version [14].

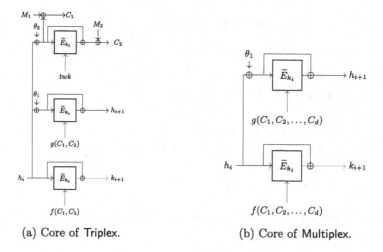

(a) Core of Triplex.   (b) Core of Multiplex.

Fig. 2. Core of the Triplex and Multiplex constructions. where $C_1, C_2, \ldots$ stem from the preceding iteration.

3 Forgery Complexity on Triplex- and Multiplex-Type Constructions

In this section, we show forging attacks on Triplex- and Multiplex-type constructions based on TBCs with varying tweak lengths. Since our attack algorithms and the corresponding analysis depend on a well-known combinatorial result, we briefly recall it here.

Theorem 1. *Let A and B be linear spaces and $f : A \to B$ be a linear map. Then, if $dim(A)$ is finite, we have $dim(A) = rank(f) + nullity(f)$, where $rank(f) = dim(f(A))$ and $nullity(f) = dim\{x : x \in A \land f(x) = 0\}$.*

3.1 Forging Attack on Triplex with Smaller Tweak

We start with the attack algorithm on Triplex. In each iteration of its message-processing module, Triplex encrypts two n-bit message blocks M_{2i}, M_{2i+1} and produces two ciphertext blocks C_{2i}, C_{2i+1}. Moreover, it generates the successive key and chaining-value pair (h, k) using two tweakable block ciphers with $2n$-bit tweak using C_{2i}, C_{2i+1} as the tweak. It has been shown in [41] that Triplex achieves n-bit CIML2 security. Moreover, it can be easily seen from the design of Triplex that it is impossible to process more than $2n$-bit message material in one iteration.

In the following, let those ciphertext blocks be denoted as C_1 and C_2. Let further t_1 be an l-bit tweak to the bottom TBC and t_2 be the other l-bit tweak to the middle TBC call in a round of Triplex with $l < 2n$. Those tweaks are used for creating the successive pair of key and chaining value. Assume that these two tweaks t_1, t_2 are created using two linear transformations $f, g : \{0, 1\}^{2n} \to \{0, 1\}^l$

CASE A: $\exists\ x, y \in \mathsf{Ker}(f) : g(x) = g(y)$ | CASE B: $\exists\ x, y \in \mathsf{Ker}(g) : f(x) = f(y)$

1: **Choose** $z \in \{0,1\}^{2n} : f(z) \neq 0^l$; | **Choose** $z \in \{0,1\}^{2n} : g(z) \neq 0^l$;

2: **Make an Encryption Query** $(N, A, M = (M_1 \| M_2))$;

3: **Let the response be** $(C = C_1 \| C_2,\ T)$;

4: **Compute** $e_1 = C_1 \oplus M_1,\ e_2 = C_2 \oplus M_2$;

5: **Compute** $M_1' \| M_2' = (x \oplus z) \oplus (e_1 \| e_2)$;

6: **Make an Encryption Query** $(N, A, M' = (M_1' \| M_2'))$;

7: **Let the response be** $(C',\ T')$;

8: **Forge** $(N, A, C'' = (y \oplus z),\ T')$;

Fig. 3. Forgery algorithm in Cases A (without the boxed statements) and B (with the boxed statements) for Triplex.

such that $t_1 = f(C_1, C_2)$ and $t_2 = g(C_1, C_2)$. Let $\mathsf{Ker}(f) = \{x \in \{0,1\}^{2n} : f(x) = 0^l\}$ be the set of all preimages of f which are mapped to 0^l. $\mathsf{Ker}(g)$ is defined similarly. There are three distinct cases in which we attempt forgery on Triplex with l-bit tweak. The cases are as follows:

- Case A: There exist $x, y \in \mathsf{Ker}(f)$ such that $g(x) = g(y)$.
- Case B: There exist $x, y \in \mathsf{Ker}(g)$ such that $f(x) = f(y)$.
- Case C: None of the two conditions above holds.

We describe forgery-attack algorithms for each case in the following.

☐ *Forgery Algorithm for Case A:* The attack algorithm for Case A is depicted in Fig. 3. Let us briefly justify how the attack works. Since $x, y \in \mathsf{Ker}(f)$ such that $g(x) = g(y)$, we have $f(x \oplus z) = f(y \oplus z)$ and $g(x \oplus z) = g(y \oplus z)$ due to the linearity of f and g. We will use this fact to mount the forgery attack. Note that we have used the intermediate values (e_1 and e_2) of the first encryption query to set up the second encryption query such that the tweak used at the core component of the same round of the construction for generating the ciphertext blocks (C_1', C_2') is $f(C_1' \| C_2') = f(M_1' \oplus e_1 \| M_2' \oplus e_2) = f(x \oplus z)$. A similar argument holds for the tweak used at the middle TBC call $g(x \oplus z)$. Next, we have explicitly used $y \oplus z$ as the two respective ciphertext blocks in the forging attempt that ensure the tweaks used to generate the successive chaining and updated key values are $f(y \oplus z)$ and $g(y \oplus z)$, respectively. The property that $f(x \oplus z) = f(y \oplus z)$ and $g(x \oplus z) = g(y \oplus z)$ ensures the forgery.

☐ *Forging Algorithm for Case B:* The attack algorithm for Case B is similar to that of Case A, but concentrates on $f(x) = f(y)$ instead of $g(x) = g(y)$. When the boxed statements are included, Fig. 3 also describes Case B. The analysis of this case is then naturally almost identical to that of Case A. Note that similarly to Case A, the success probability depends on the proper choice of x, y, and z such that $f(x \oplus z) = f(y \oplus z)$, and $g(x \oplus z) = g(y \oplus z)$.

CASE C: CASES A AND B DO NOT HOLD

1 :   **Make an Encryption Query** $(N, A, M = (M_1 \| M_2))$;

2 :   **Let the resonse be** $(C = (C_1 \| C_2), T)$;

3 :   **Note internal round pair** (h_α, k_α), **used to create** $C_1 \| C_2$;

4 :   **Choose** $z \in \{0, 1\}^{2n} : g(z) \neq 0$;

5 :   **Let** $a = \min\{2^{2n-l}, 2^{\frac{n}{2}}\}$;

6 :   **Choose distinct** $x_1, x_2, \ldots, x_a \leftarrow \mathsf{Ker}(g)$;

7 :   **Compute** $k_i \leftarrow \widetilde{E}(k_\alpha, t_1^i, h_\alpha)$, **where** $t_1^i \leftarrow f(x_i \oplus z)$, **for** $i = 1, \ldots, a$;

8 :   **Find** (i, j) **such that** $k_i = k_j$;

9 :   **Compute** $e_1 := C_1 \oplus M_1$, $e_2 := C_2 \oplus M_2$;

10 :   **Compute** $M_1' \| M_2' = (x_i \oplus z) \oplus (e_1 \| e_2)$;

11 :   **Make an Encryption Query** $(N, A, M' = (M_1' \| M_2'))$;

12 :   **Let the response be** (C', T');

13 :   **Forge with** $(N, A, C'' = (x_j \oplus z), T')$;

Fig. 4. Forgery algorithm in Case C for Triplex.

□ *Forging Algorithm for Case C:* In this case, we also try to achieve an internal two-block (h, k) collision. While Cases A and B achieved this by a collision in the tweak values, Case C constructs a collision in the tweak for one TBC call (that generates the chaining value). For the other calls, we make several queries with different tweaks and expect to get a collision. Thus, our attack will succeed probabilistically depending on the tweak length l. The algorithm for Case C is given in Fig. 4.

Let us briefly discuss how the attack works. Since $x_1, \ldots, x_a \in \mathsf{Ker}(g)$, we have $g(x_1 \oplus z) = \cdots = g(x_a \oplus z) = g(z)$ due to the linearity of g. Now we try to see if we could find (i, j) such that $k_i = k_j$, then we would be done and could set up the forgery in a similar manner that we have done for Case A. Now let us look at the probability of matching two keys k_i and k_j. Note that the k_i values are generated from the invocation of the TBC with the same key and input but with different tweaks. Hence,

$$\Pr[\exists\, i, j \in \{1, 2, \ldots, a\} : k_i = k_j] \leq a^2 \cdot 2^{-n}.$$

It is easy to see that a successful forgery happens in this case whenever we obtain a pair (i, j) with $k_i = k_j$. Hence, the probability of getting a successful forgery is bounded by $a^2 \cdot 2^{-n}$ (Table 1).

Now let us summarize the different settings depending on the used tweak lengths:

(i) $\boxed{l < n.}$ From the rank-nullity theorem, we have $\dim(\mathsf{Ker}(f)) \geq 2n - l > n$ and the same follows for $\mathrm{ker}(g)$. Let $\mathcal{B}_f$ and $\mathcal{B}_g$ be two bases for $\mathrm{ker}(f)$ and $\mathsf{Ker}(g)$, respectively. Then $|\mathcal{B}_f| \geq n + 1$ and $|\mathcal{B}_g| \geq n + 1$. Moreover,

Table 1. Forgery-attack properties for instantiations of Triplex and Multiplex with TBCs of varying tweak lengths.

(a) For Triplex.			(b) For Multiplex.		
Tweak length l	#Queries q	Success prob.	Tweak length l	#Queries q	Success prob.
$l < n$	2	1	$l < dn/2$	2	1
$n \leq l \leq 3n/2$	$2^{n/2}$	1	$dn/2 \leq l \leq dn - n/2$	$2^{n/2}$	1
$3n/2 < l < 2n$	2^{2n-l}	2^{3n-2l}	$dn - n/2 < l < dn$	2^{2n-l}	2^{3n-2l}

the dimension of $\{0,1\}^{2n}$ ensures that one of following properties will be satisfied: $\exists\ x, y \in \mathcal{B}_f$ which are linearly dependent to $\mathcal{B}_g$ which satisfy Case A or $\exists\ x, y \in \mathcal{B}_g$ which are linearly dependent on $\mathcal{B}_f$ which satisfies Case B. Thus, for $l < n$, at least one of Case A or B will happen. Hence, for tweaks of less than n bits, we can forge successfully with a constant number of queries.

(ii) $\boxed{n \leq l \leq 3n/2.}$ Here, we can have multiple cases: if any of the first two cases is satisfied for f and g, we will have successful forgery with only three queries. Otherwise, we consider Case C. Since $a = \min\{2^{2n-l}, 2^{\frac{n}{2}}\} = 2^{\frac{n}{2}}$, from the analysis of Case C, the adversary will be successful with probability one using at most $2^{\frac{n}{2}}$ queries.

(iii) $\boxed{l > 3n/2.}$ Again, if any of the first two cases are satisfied, we will have successful forgery with only three queries. Otherwise, as the analysis of Case C in Fig. 4 states, the success probability of the forgery is at least $\frac{a^2}{2^n}$ for the value of a as in Line 5 of Fig. 4. Clearly, the value of a decreases as l increases. Thus, the security increases as the tweak length increases.

3.2   Forgery Attacks on Multiplex with $< dn$-bit TBCs

Multiplex encrypts d n-bit message blocks using $d+1$ TBC calls in each iteration. It processes these d blocks in the next iteration using them as the tweak of the TBC calls. Multiplex uses the same Hirose's compression function as Triplex for generating the subsequent key and chaining value. Unlike Triplex, Multiplex allows different values of d while achieving almost n-bit CIML2 security. In this section, we will analyze the security of Multiplex when instantiated with a TBC with a tweak length of $< dn$ bits.

Similarly, as for Triplex, the core component of Multiplex (see Fig. 2b) processes d ciphertext blocks (from the preceding iteration). Let t_1 and t_2 be the tweaks to two TBC calls in an iteration of the compression function to create the pair of key and chaining value. These two tweaks t_1, t_2 are created using two linear transformations $f, g : \{0,1\}^{dn} \rightarrow \{0,1\}^l$ such that $t_1 = f(C_1, C_2, \ldots, C_d)$ and $t_2 = g(C_1, C_2, \ldots, C_d)$. Let $\mathsf{Ker}(f) = \{x \in \{0,1\}^{dn} : f(x) = 0^l\}$ be the set of all preimages of f which are mapped to 0^l. $\mathsf{Ker}(g)$ is defined similarly. Similarly, as for Triplex, there are three distinct cases in which we attempt a forgery on Multiplex when instantiated with an l-bit TBC. Now we will analyze the security for three cases depending on f and g as follows:

- Case A: There exist $x, y \in \mathsf{Ker}(f)$ such that $g(x) = g(y)$.
- Case B: There exist $x, y \in \mathsf{Ker}(g)$ such that $f(x) = f(y)$.
- Case C: None of the two conditions above holds.

Similarly, as for Triplex, we can mount forgery attacks for the three cases above. The detailed attacks and analysis can be found in the full version [14].

According to the cases mentioned above, one can mount the following generic attacks based on the length of the tweaks used:

(i) $\boxed{l < dn/2:}$ In this case, one can show that one of Case A or Case B gets satisfied, and hence, we can forge successfully with only two encryption queries.

(ii) $\boxed{dn/2 \le l \le dn - n/2:}$ In this case, if one of the first two cases gets satisfied for f and g, we will have a successful forgery with only three queries. Otherwise, we will mount an attack following Case C with $a = \min\{2^{dn-l}, 2^{n/2}\} = 2^{n/2}$. From the analysis of Case C, the adversary will be successful with probability one using at most $2^{n/2}$ queries.

(iii) $\boxed{l > dn - n/2:}$ Similarly as in the previous settings, it holds that if any of the first two cases is satisfied, we will have a successful forgery with three queries. Otherwise, the success probability of the forgery is at least $a^2 \cdot 2^{-n}$ for the value of a. Clearly, the value of a decreases as l increases. Thus, the security will increase as the tweak length increases.

4 The Tweplex Authenticated Cipher

Tweplex follows the three-step model suggested in [6] for designing leakage-resilient AEAD schemes. It uses a TBC with dn-bit tweaks as the underlying primitive. Like Triplex and Multiplex, Tweplex consists of three distinct modules called key-derivation (KDF), message-processing (MPF), and tag-generation function (TGF), respectively. In a broader sense, the KDF module consists of a call to a TBC implementation that is strongly protected against DPA. It uses the secret key, the public key of the user, and the nonce as inputs and produces a pair of an initial key and a chaining-value (h_1, k_1). The MPF module takes (h_1, k_1) as its input and processes the associated data using two tweakable block-cipher calls in each iteration. This module is structurally very similar to Hirose's compression function [27]. Thereafter, it computes a multi hash based on the MBL compression function, which is used in the design of the multi hash function of Multiplex. Recall that the MBL compression function used in Multiplex consists of $(d + 1)$ calls to a tweakable block cipher to encrypt a dn-bit part of the message such that each of the $(d + 1)$ TBC calls uses a dn-bit tweak. Our design does the same but uses a TBC with only $dn/2$-bit tweaks. The final two

Algorithm 1. Encryption Algorithm of Tweplex

11: **function** $\mathcal{E}(i, K_i, P_i, N, A, M)$	31: **function** $\text{MPF}(h_1, k_1, A, M)$
12: $d' \leftarrow dn/2$	32: $A_1\|A_2\|\cdots\|A_{da} \leftarrow PAD(A)$
13: $h_1\|k_1 \leftarrow \text{KDF}(i, K_i, P_i, N)$	33: $M_1\|M_2\|\cdots\|M_{dm} \leftarrow PAD(M)$
14: $(C, twk) \leftarrow \text{MPF}(h_1, k_1, A, M)$	34: **for** $i \leftarrow 2 \ldots a$ **do**
15: $\text{T} \leftarrow \text{TGF}(i, K_i, twk, 0^n)$	35: $t_{i,1} \leftarrow \text{msb}_{d'}(A_{d(i-2)+1}\|A_{d(i-2)+2}\|\cdots\|A_{d(i-2)+d})$
16: Return (C, T)	36: $t_{i,2} \leftarrow \text{lsb}_{d'}(A_{d(i-2)+1}\|A_{d(i-2)+2}\|\cdots\|A_{d(i-2)+d})$
	37: $k_i \leftarrow \widetilde{E}(k_{i-1}, t_{i,1}, h_{i-1}) \oplus h_{i-1}$
21: **function** $\text{KDF}(i, K_i, P_i, N)$	38: $h_i \leftarrow \widetilde{E}(k_{i-1}, t_{i,2}, h_{i-1} \oplus \theta_1) \oplus h_{i-1} \oplus \theta_1$
22: $k_0 \leftarrow \widetilde{E}(K_i, P_i\|0^{d'-n}, N)$	39: $X_1 \leftarrow \text{msb}_{d'}(A_{d(a-1)+1}\|A_{d(a-1)+2}\|\cdots\|A_{d(a-1)+d})$
23: $a \leftarrow \widetilde{E}(k_0, N\|P_i\|0^{d'-2n}, 0^n)$	40: $X_2 \leftarrow \text{lsb}_{d'}(A_{d(a-1)+1}\|A_{d(a-1)+2}\|\cdots\|A_{d(a-1)+d})$
24: $b \leftarrow \widetilde{E}(k_0, N\|P_i\|0^{d'-2n}, \theta_1) \oplus \theta_1$	41: $X_3 \leftarrow X_1 \oplus X_2$
25: **return** (a, b)	42: **for** $i \leftarrow 1 \ldots m$ **do**
	43: $k_{a+i} \leftarrow \widetilde{E}(k_{a+i-1}, X_1, h_{a+i-1}) \oplus h_{a+i-1}$
41: **function** $\text{TGF}(i, K_i, twk, X)$	44: $h_{a+i} \leftarrow \widetilde{E}(k_{a+i-1}, X_2, h_{a+i-1} \oplus \theta_1) \oplus h_{a+i-1} \oplus \theta_1$
42: **return** $\widetilde{E}(K_i, twk\|0^{d'-2n}, X)$	45: $C_{d(i-1)+1} \leftarrow M_{d(i-1)+1} + h_{a+i}$
	46: **for** $j \leftarrow 2 \ldots d$ **do**
	47: $e_j \leftarrow \widetilde{E}(k_{a+i-1}, X_3, h_{a+i-1} \oplus \theta_j) \oplus h_{a+i-1} \oplus \theta_j$
	48: $C_{d(i-1)+j} \leftarrow M_{d(i-1)+j} + e_j$
	49: $X_1 \leftarrow \text{msb}_{d'}(C_{d(i-1)+1}\|C_{d(i-1)+2}\|\cdots\|C_{d(i-1)+d})$
	50: $X_2 \leftarrow \text{lsb}_{d'}(C_{d(i-1)+1}\|C_{d(i-1)+2}\|\cdots\|C_{d(i-1)+d})$
	51: $X_3 \leftarrow X_1 \oplus X_2$
	52: $twk_1 \leftarrow \widetilde{E}(k_{a+m}, X_1, h_{a+m}) \oplus h_{a+m}$
	53: $twk_2 \leftarrow \widetilde{E}(k_{a+m}, X_2, h_{a+m} \oplus \theta_1) \oplus h_{a+m+1} \oplus \theta_1$
	54: $twk \leftarrow twk_1\|twk_2$
	55: $C = C_1\|C_2\|C_3\|\cdots\|C_{dm}$
	56: **return** (C, twk)

TBC calls produce the pair of successive key and chaining value (h, k) for the next iteration. Finally, the TGF module consists again of a strongly protected call to the TBC, taking the fixed input 0^n, the secret key, and the output of the multi hash concatenated with the required number of zeroes as the tweak and outputs T.

An algorithmic description of the construction is given in Algorithm 1. An illustration for $d = 4$, with $8n$-bit message and associated data each and tweak sizes of $2n$ bit is shown in Fig. 5.

5   Authentication Security of Tweplex

In this section, we show that Tweplex achieves ciphertext integrity in the muCIML2 setting for up to $2^{n/2}$ queries.

Theorem 2. *Consider an adversary $\mathscr{A}$ that tries to break the muCIML2 security of Tweplex. Assuming that $\mathscr{A}$ makes at most q_e encryption queries, q_d decryption queries with at most σ blocks in total, q_p primitive queries, over at most u users. Then, we have*

$$\mathbf{Adv}^{\text{muCIML2}}_{\text{Tweplex}}(\mathscr{A}) \leq \frac{u^2}{2^{2n+1}} + \frac{4(q^2 + q)}{2^n} + \frac{q^2}{2^{2n}}.$$

where $q_c = q_e + q_d$, $q = \max\{q_p, q_c\}$ and $u \leq q$.

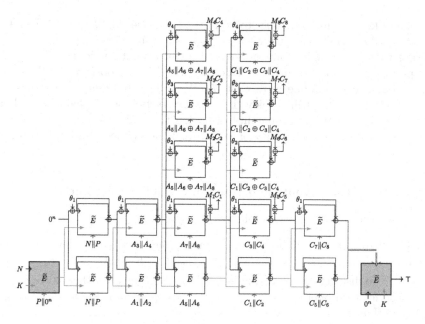

Fig. 5. Tweplex for $d = 4$.

5.1 Query Types and Responses

The adversary $\mathscr{A}$ can make three types of queries: primitive, encryption, and decryption (or forging) queries. Here, we describe how the challenger responds to each query:

☐ **Primitive Query:** For a primitive (ideal TBC) query $(J, t, x, \rightarrow)$, $\mathscr{A}$ will obtain $y = \widetilde{E}_J(t, x)$ and for query of the form $(J, t, y, \leftarrow)$, $\mathscr{A}$ gets $x = \widetilde{E}_J^{-1}(t, y)$. For both query types, the transcript will store an entry $(\text{Prim}, J, t, x, y, dir)$, where $dir \in \{\leftarrow, \rightarrow\}$. Hereafter, we will use $\star$ to denote that the direction is arbitrary.

☐ **Encryption Query:** For an encryption query of the form (i, N, A, M), $\mathscr{A}$ will obtain a ciphertext-tag pair $C\|T$ such that $C\|T \leftarrow \mathcal{E}(i, N, A, M)$. Let $\mathbf{h}$ be the set of all h_i's generated during the production of $C\|T$ and $\mathbf{k}$ be the set of all k_i's generated during the production of $C\|T$. Due to the unbounded leakage assumption $\mathscr{A}$ will have full access to $\mathbf{h}$ and $\mathbf{k}$. An encryption query is stored in the form $(\text{Enc}, i, N, A, M, C\|T, \mathbf{h}, \mathbf{k})$. In addition, all internal primitive in- and outputs are stored in τ_p.

☐ **Decryption Query:** For an decryption query of the form $(i, N, A, C\|T)$, $\mathscr{A}$ will get M such that $M \leftarrow \mathcal{D}(i, N, A, C\|T)$. Note that M can be a message or $\perp$ depending on whether authentication is successful or not, respectively. Let $\mathbf{h}$ be the set of all h_i's generated during decryption and $\mathbf{k}$ be the set of all k_i's generated during the same. Due to the unbounded leakage assumption $\mathscr{A}$ will have $\mathbf{h}$ and $\mathbf{k}$. This query will be stored in the form $(\mathsf{Dec}, i, N, A, M, C\|\mathsf{T}, \mathbf{h}, \mathbf{k})$. Moreover, all internal primitive in- and outputs are stored in τ_p.

Note that we consider only non-trivial adversaries that do not make any decryption query of the form $(i, N, M, C\|\mathsf{T})$, after having observed $C\|\mathsf{T}$ as the output of an encryption query (i, N, A, M). We will try to bound $\mathscr{A}$'s successful forging probability. We will employ the standard technique: (i) define some bad events and show that the probability of having the bad events is low, and (ii) show that the forging probability of $\mathscr{A}$ is low if the defined bad events do not occur.

5.2 Defining Bad Events and Bounding Their Probabilities

We define a set of bad events as follows.

- <u>Bad1</u>: There are two different users sharing the same secret key and same public constant, i.e., $\exists\, i \neq j$, $i, j \in \{1, 2, \dots, u\}$: $K_i = K_j$, $P_i = P_j$.
- <u>Bad2</u> : Two distinct primitive queries lead to a collision after their corresponding feed-forward operations, i.e. $\exists\, (\mathsf{Prim}, J, t, x, y, \star)$ and $(\mathsf{Prim}, J', t', x', y', \star)$ such that $x \oplus y = x' \oplus y'$.
- <u>Bad3</u>: There is a collision in one of the KDF calls with a primitive query, i.e., $\exists\, (\mathsf{Prim}, J, t, x, y, \star)$: $J = K_i$, $t = P_i\|0^{(dn/2-n)}$, for some user i.
- <u>Bad4</u>: There is a collision in the output of a protected block of two different users having the same public constant and the same nonce, i.e., $\exists\, i \neq j$, $(\mathcal{D}, i, N_i^a, A_i^a, M_i^a, C_i^a\|\mathsf{T}_i^a, \mathbf{h}_i^a, \mathbf{k}_i^a)$, $(\mathcal{E}, j, N_j^b, A_j^b, M_j^b, C_j^b\|\mathsf{T}_j^b, \mathbf{h}_j^b, \mathbf{k}_j^b)$ such that $k_{i,0}^a = k_{j,0}^b$, $N_i^a\|P_i = N_j^b\|P_j$. This condition considers the case of having the same input for hash for two different users.
- <u>Bad5</u>: There is a collision between tweak of two TGF calls, i.e., $\exists\, (i, N_i^a, A_i^a, M_i^a, C_i^a\|\mathsf{T}_i^a, \mathbf{h}_i^a, \mathbf{k}_i^a) \in \tau_d$, $(j, N_j^b, A_j^b, M_j^b, C_j^b\|\mathsf{T}_j^b, \mathbf{h}_j^b, \mathbf{k}_j^b) \in \tau_{e/d}$ such that $h_{l_a+v_a+1}\|k_{l_a+v_a+1} = h_{l_b+v_b+1}\|k_{l_b+v_b+1}$, where $|A_i^a| = 4l_a$, $|C_i^a| = 4v_a$, $|A_j^b| = 4l_b$, $|C_j^b| = 4v_b$.
- <u>Bad6</u>: There is key-tweak pair collision among a KDF and TGF query, i.e., $\exists\, (i, N_i^a, A_i^a, M_i^a, C_i^a\|\mathsf{T}_i^a, \mathbf{h}_i^a, \mathbf{k}_i^a) \in \tau_d$, $(j, N_j^b, A_j^b, M_j^b, C_j^b\|\mathsf{T}_j^b, \mathbf{h}_j^b, \mathbf{k}_j^b)) \in \tau_c$ such that $K_i = K_j$, $h_{l_a+v_a+1}\|k_{l_a+v_a+1} = P_j\|0^n$, for some $j \in \{1, 2, \dots, u\}$.
- <u>Bad7</u> : There is a collision between a TGF call and a primitive query, i.e., $\exists\, (i, N_i^a, A_i^a, M_i^a, C_i^a\|Tag_i^a, \mathbf{h}_i^a, \mathbf{k}_i^a) \in \tau_c$, $(J, t, x, y, \star) \in \tau_p$ such that $J = k_i$, $t = h_{l_a+v_a+1}\|k_{l_a+v_a+1}\|0^{(dn/2-2n)}$, for some user i.

We define an event Bad that is true if and only if any of the conditions above are satisfied. The following lemma bounds the probability of the event Bad:

Lemma 1. *It holds that*

$$\Pr[\mathsf{Bad}] \leq \frac{u^2}{2^{2n+1}} + \frac{4q^2}{2^n} + \frac{2q}{2^n} + \frac{q^2}{2^{2n}}.$$

Lemma 2 upper bounds the probability of forging when Bad did not occur:

Lemma 2. *It holds that*

$$\Pr[\mathscr{A} \ forges \wedge \overline{\mathsf{Bad}}] \leq \frac{2q}{2^n}.$$

Proof of both the lemmas can be found in full version [14]. The result in Theorem 2 follows directly from Lemma 1 and 2.

6 Confidentiality Analysis of **Tweplex**

For privacy, we first define nonce-misuse-resilient security under the black-box assumption. Then, we will show muCCAmL1 security in the presence of leakage under the bounded-leakage assumption following a standard way as it does not have much technical novelty.

Theorem 3. *Suppose, an adversary $\mathscr{A}$ makes at most q_e encryption queries, that contain a total number of primitive calls of at most q_p, over at most u users. Then, for $q = \max\{q_p, q_e\}$, we have*

$$\mathbf{Adv}^{conf}_{\mathsf{Tweplex}}(\mathscr{A}) \leq \frac{u^2}{2^{2n+1}} + \frac{4q^2 + 2q}{2^n} + \frac{q^2}{2^{2n}}.$$

IMPLICATION OF THE BOUND: This bound shows that we can achieve confidentiality for an adversary under the black-box assumption and up to $2^{n/2}$ queries. We will not consider security beyond the birthday bound since it is out of reach anyway.

Theorem 4. *Let $\widetilde{E} \in TBC(\{0,1\}^n, \mathcal{T}, \{0,1\}^n)$. Let $\mathscr{A}$ be a muCCAmL1 adversary on $\Pi[\widetilde{E}]_K = \mathsf{Tweplex}[\widetilde{E}]_K$ that is allowed to ask at most q_e encryption queries of at most σ dn-bit blocks and q primitive queries in total. Let $\mathcal{F}[\widetilde{E}]$ denote an iteration of $\mathsf{Tweplex}$ for message encryption, and $\mathcal{L}^{in}$, $\mathcal{L}^{out}$, and $\mathcal{L}^{\oplus}$ be leakage functions. Then*

$$\mathbf{Adv}^{muCCAmL1}_{\Pi[\widetilde{E}]}(\mathscr{A}) \leq \mathbf{Adv}^{conf}_{\Pi[\widetilde{E}]}(q, \sigma) + 2\sigma \cdot \mathbf{Adv}^{LUP\text{-}d}_{\mathcal{F}[\widetilde{E}], \mathcal{L}^{in}, \mathcal{L}^{out}}(p, q) +$$

$$\sigma \cdot \mathbf{Adv}^{XOR\$}_{\mathcal{F}[\widetilde{E}], \mathcal{L}^{out}, \mathcal{L}^{\oplus}}(q) + \mathbf{Adv}^{muCIML2}_{\Pi[\widetilde{E}]}(q, \sigma),$$

where LUP-d is similar to LOR2 leakage assumption as defined for TEDT [8] and the XOR\$ assumption is similar as in TEDT2 [32]. The concrete definitions can be found in the full version [14].

The proof follows similar steps as the qCPAmL2 proof by [32] on TEDT2 and the muCCAmL2 proof on TEDT [8]. However, we consider a multi-user version and assume that the adversary can query its decryption oracle only with some previous outputs from the encryption oracle due to the muCIML2 assumption. Moreover, we assume that the decryption oracle does not give any leakage. For the proof of the black-box privacy game conf, we first define some Bad events in the nonce-misuse setting. Assuming those events did not happen, we will study the distinguishing probability under bounded leakage. Let $\mathscr{A}$ be a muCCAmL1 adversary against Tweplex and it is allowed to make three kinds of queries.

6.1 Query Types and Responses

☐ **Primitive Query:** For a primitive query $(J, t, x, \rightarrow)$, $\mathscr{A}$ will get $y = \widetilde{E}_J(t, x)$ and for a query of the form $(J, t, y, \leftarrow)$, $\mathscr{A}$ will get $x = \widetilde{E}_J^{-1}(t, y)$. For both kinds of queries, the challenger will store an entry $(\mathsf{Prim}, J, t, x, y, dir)$.

☐ **Encryption Query:** For an encryption query of the form (i, N, A, M), $\mathscr{A}$ will get $C\|\mathsf{T}$ such that $C\|\mathsf{T} \leftarrow \mathcal{E}(i, N, A, M)$. $\mathscr{A}$ will also get the leakage corresponding to all internal primitive and other computations. This query will be stored in the form $(\mathcal{E}, i, N, A, M, C\|\mathsf{T}, \mathfrak{L}, \mathbf{hk})$, where $\mathbf{hk}$ is the set of all input-key pairs used during encryption in internal primitive calls and $\mathfrak{L}$ is the set of all leakage results during this computation. We divide all encryption queries into two types: $\mathcal{E}_1$, where the adversary can repeat the nonce for the same user, and $\mathcal{E}_2$, where the adversary has to make queries for the same user under a fresh nonce. The $\mathcal{E}_2$ queries act as the *challenge query* for muCCAmL1 security.

☐ **Decryption Query:** For a decryption query of the form $(i, N, A, C\|\mathsf{T})$, $\mathscr{A}$ will get M such that $M \leftarrow \mathcal{D}(i, N, A, C\|\mathsf{T})$. Note that M can be a message or $\perp$ depending on whether the authentication has been successful or not, respectively. $\mathscr{A}$ will not receive any leakage during the decryption process.

Note that muCCAmL1 security is defined as the distinguishing advantage between a real and an ideal world. Without loss of generality, we can assume that the muCIML2 security of Tweplex as the leakage assumptions are weaker in the muCCAmL1 model. Then, there will be no valid decryption query. So, it remains to prove the muCPAmL1 security of Tweplex. For that, we will first upper bound the probability in the nonce-misuse scenario by defining Bad events. Assuming $\overline{\mathsf{Bad}}$, we show then that the output of Tweplex is indistinguishable from a randomly chosen message in the black-box setting. From this, we will argue muCCAmL1 security in the presence of (bounded) leakage. Let us define two games as follows:

- Game $\mathscr{G}_1$: This is the real-world encryption of a given message M, $\mathsf{Real}(M)$.
- Game $\mathscr{G}_4$: This is equivalent to the real-world $\mathsf{Real}(\$)$, that encrypts not the given M but a randomly chosen message M^* of the same length as M with the real encryption algorithm.

It holds that $\mathbf{Adv}_{\mathcal{E}_k, \mathcal{D}_k, \mathcal{L}_{\mathcal{E}}}^{muCCAmL1}(\mathscr{A}) = \Delta\mathscr{G}_{14}$.

6.2 Confidentiality Under Nonce Misuse and Bounded Leakage

Here, we define Bad events similar to those defined in our CIML2 proof.

- <u>Bad1</u>: There are two different users sharing the same secret key and the same public constant, i.e., $\exists\, i \neq j$, $i, j \in \{1, 2, \dots, u\}$: $K_i = K_j$, $P_i = P_j$.
- <u>Bad2</u> : Two distinct primitive queries lead to a collision after their corresponding feed-forward operations, i.e. $\exists\, (\mathsf{Prim}, J, t, x, y, \star)$ and $(\mathsf{Prim}, J', t', x', y', \star)$ such that $x \oplus y = x' \oplus y'$.

- **Bad3**: There is a collision between one of the KDF calls and a primitive query, i.e., $\exists$ $(\mathtt{Prim}, J, t, x, y, \star)$: $J = K_i$, $t = P_i \| 0^{(dn/2-n)}$, for some user i.
- **Bad4**: There is a collision in the output of a protected block of a $\mathcal{E}_2$ queries and another encryption query, along with the same public constant and the same nonce, i.e., $\exists$ $(\mathcal{E}_2, i, N_i^a, A_i^a, M_i^a, C_i^a \| \mathsf{T}_i^a, \mathfrak{L}, \mathbf{hk}_i^a)$ the a-th query to a user i and another query $(\star, j, N_j^b, A_j^b, M_j^b, C_j^b \| \mathsf{T}_j^b, \mathfrak{L}, \mathbf{hk}_j^b)$ such that $k_{i,0}^a = k_{j,0}^b$, $N_i^a \| P_i = N_j^b \| P_j$.
- **Bad5** : There is double block collision between an internal (h, k) pair, i.e., $\exists$ $(\mathcal{E}_2, i, N_i^a, A_i^a, M_i^a, C_i^a \| \mathsf{T}_i^a, \mathfrak{L}, \mathbf{hk}_i^a)$ the a-th query to user i and another query $((\star, j, N_j^b, A_j^b, M_j^b, C_j^b \| Tag_j^b, \mathbf{hk}_j^b))$ such that $(h', k') = (h^\star, k^\star)$, where $(h', k') \in \mathbf{hk}_i^a$ and $(h^\star, k^\star) \in \mathbf{hk}_j^b$.
- **Bad6**: There is a key-tweak pair collision among a KDF and TGF query, i.e., $\exists$ $(\mathcal{E}_2, i, N_i^a, A_i^a, M_i^a, C_i^a \| \mathsf{T}_i^a, \mathfrak{L}, \mathbf{hk}_i^a)$ and another query $\exists$ $(\star, j, N_j^b, A_j^b, M_j^b, C_j^b \| \mathsf{T}_j^b, \mathfrak{L}, \mathbf{hk}_j^b)$ such that $(K_i = K_j) \wedge (P_i \| 0^n = h_{l_a+v_a+1} \| k_{l_b+v_b+1})$, for some $i, j \in \{1, 2, \ldots, u\}$.
- **Bad7** : There is a collision between TGF call and primitive call, i.e., $\exists$ $(\mathcal{E}_2, i, N_i^a, A_i^a, M_i^a, C_i^a \| \mathsf{T}_i^a, \mathfrak{L}, \mathbf{hk}_i^a)$ and $(\mathtt{Prim}, J, t, x, y, \star)$ such that $t = h_{l_a+v_a+1} \| k_{l_a+v_a+1} \| 0^{(dn/2-2n)} \wedge J = K_i$.

We define the event **Bad** that is true if and only if any one of the conditions above is satisfied. One can see that these **Bad** conditions are similar to those of CIML2. Moreover, note that the adversary can not repeat a nonce for $\mathcal{E}_2$ queries which restrains it from obtaining information compared to a CIML2 adversary. The probability of these **Bad** conditions is upper bounded by the probability for a CIML2 adversary to forge successfully, i.e., $\Pr[\mathtt{Bad}] \leq \frac{u^2}{2^{2n+1}} + \frac{4q^2}{2^n} + \frac{2q}{2^n} + \frac{q^2}{2^{2n}}$.

If none of these events happen, all outputs from the KDF will be fresh for each $\mathcal{E}_2$ query. Moreover, from the absence of $\mathtt{Bad}_2$, it follows that all internal primitive calls are pairwise unique. The tweaks used in the TGF module for every $\mathcal{E}_2$ query are also fresh. Then, there is no difference between $Real(M)$ and $Real(\$)$ in the black-box setting under the assumption of nonce-misuse resilience.

6.3 Proof Idea of muCCAmL1 Security

A detailed treatment and the game definitions are in full version [14], where we show that

$$\Delta \mathcal{G}_{14} \leq \mathbf{Adv}_{\Pi[\widetilde{E}]}^{\mathsf{conf}}(q, \sigma) + 2\sigma \cdot \mathbf{Adv}_{\mathcal{F}[\widetilde{E}], \mathcal{L}^{in}, \mathcal{L}^{out}}^{\mathsf{LUP\text{-}d}}(p, q) +$$
$$\sigma \cdot \mathbf{Adv}_{\mathcal{F}[\widetilde{E}], \mathcal{L}^{out}, \mathcal{L}^{\oplus}}^{\mathsf{XOR\$}}(q) + \mathbf{Adv}_{\Pi[\widetilde{E}]}^{\mathsf{muCIML2}}(q, \sigma).$$

Though, the reductions are standard, and for the sake of space limitations, here, we provide only the core ideas.

Without leakage and in the absence of **Bad**, we can ensure indistinguishability between $\mathcal{G}_1$ and $\mathcal{G}_4$. In the presence of leakage, we show that the security of the scheme reduces to the SPA security of a single block-cipher call under the security assumption in [6]. Following the definition of multi-user CCAmL1 security, this is equivalent to the difference between $\mathscr{G}_1$ and $\mathscr{G}_4$. Now, we will argue the muCCAmL1 security as follows:

From the definition of muCCAmL1, we consider leakage only in the encryption direction. The weaker (bounded-)leakage assumption in muCCAmL1 compared to CIML2 adversary allows us to reduce security to a CIML2 adversary with equal resources and assume CIML2 security in the following. We assume the bound on the leakage function using the hard-to-invert property introduced in [44] and also used in TEDT [8], TEDT2 [32].

For challenge queries (to $\mathcal{E}_2$), the adversary has to submit a fresh nonce. Then, k_0 will be fresh up to the birthday bound as it is the output of a block-cipher call with the secret key. This k_0 will be used as the key only in two block-cipher calls with different inputs for computing (h_1, k_1). For processing the associated data, some key k_i can be used twice with input h_i and $h_i \oplus \theta_1$. Following a similar argument, two such invocations pass randomness about h_i, k_i to h_{i+1}, k_{i+1}. Similarly, while processing messages, the key and chaining values are used only a few times and are updated in each iteration. Each key k_s is used in $(d+1)$ block-cipher calls with input $h_s, h_s \oplus \theta_1, h_s \oplus \theta_2, \ldots, h_s \oplus \theta_d$. Therefore, the secrecy and randomness will be maintained in the next iteration. Moreover, at the end in the TGF, there is a protected block-cipher call with a long-term secret key that is assumed to not leak any significant information. The only remaining leakage is due to the XORs for creating the ciphertext as $C_i^j \leftarrow M_i^j \oplus e_i^1$ where $e_i^j \leftarrow E(k_{a+i-1}, t, h_{a+i}^j \oplus \theta_j) \oplus h_{a+i}^j \oplus \theta_j$ for $j = 2, 3, \ldots, d$ and $e_i^1 = h_{a+i}$, where subscript i denoted number of block and superscript j denote the number of sub-block of block i. For these, we use the similar LOR2 leakage assumption as defined for TEDT [8] and the XOR\$ assumption in TEDT2 [32]. Moreover, those computations involve the hidden internal value h_{a+i-1}. Thus, from the low probability of the Bad events and the mentioned leakage assumptions above, we achieve muCCAmL1 security up to the birthday bound.

7   Conclusion

We have shown that Multiplex-type constructions that use TBCs with shorter tweaks than proposed cannot achieve beyond-birthday-bound CIML2 security but can increase the throughput. We presented Tweplex, a birthday-bound-secure scheme that uses the minimal tweak length under a higher throughput. It remains an interesting research problem to find another rekeying or message processing function that will give a Grade-2 AE scheme with a higher rate and $> n/2$-bit security.

Acknowledgments. This research was partially supported by Nanyang Technological University in Singapore under Start-up Grant 04INS000397C230 and Ministry of Education in Singapore under Grants RG91/20 and MOE2019-T2-1-060.

References

1. Andreeva, E., Bogdanov, A., Luykx, A., Mennink, B., Mouha, N., Yasuda, K.: How to securely release unverified plaintext in authenticated encryption. In: Sarkar, P., Iwata, T. (eds.) ASIACRYPT 2014. LNCS, vol. 8873, pp. 105–125. Springer, Heidelberg (2014). https://doi.org/10.1007/978-3-662-45611-8_6

2. Ashur, T., Dunkelman, O., Luykx, A.: Boosting authenticated encryption robustness with minimal modifications. In: Katz, J., Shacham, H. (eds.) CRYPTO 2017. LNCS, vol. 10403, pp. 3–33. Springer, Cham (2017). https://doi.org/10.1007/978-3-319-63697-9_1

3. Barwell, G., Martin, D.P., Oswald, E., Stam, M.: Authenticated encryption in the face of protocol and side channel leakage. In: Takagi, T., Peyrin, T. (eds.) ASIACRYPT 2017. LNCS, vol. 10624, pp. 693–723. Springer, Cham (2017). https://doi.org/10.1007/978-3-319-70694-8_24

4. Bellare, M., Boldyreva, A., Knudsen, L., Namprempre, C.: Online ciphers and the hash-CBC construction. In: Kilian, J. (ed.) CRYPTO 2001. LNCS, vol. 2139, pp. 292–309. Springer, Heidelberg (2001). https://doi.org/10.1007/3-540-44647-8_18

5. Bellizia, D., et al.: Spook: sponge-based leakage-resistant authenticated encryption with a masked tweakable block cipher. IACR Trans. Symmetric Cryptol. **2020**(S1), 295–349 (2020)

6. Bellizia, D., et al.: Mode-level vs. implementation-level physical security in symmetric cryptography. In: Micciancio, D., Ristenpart, T. (eds.) CRYPTO 2020. LNCS, vol. 12170, pp. 369–400. Springer, Cham (2020). https://doi.org/10.1007/978-3-030-56784-2_13

7. Bernstein, D.J.: CAESAR: Competition for Authenticated Encryption: Security, Applicability, and Robustness (2014). https://competitions.cr.yp.to/caesar.html. Update 20 Feb 2019. Accessed 18 July 2023

8. Berti, F., Guo, C., Pereira, O., Peters, T., Standaert, F.-X.: TEDT, a leakage-resistant AEAD mode for high physical security applications. IACR Trans. Cryptogr. Hardw. Embed. Syst. **2020**(1), 256–320 (2020)

9. Berti, F., Koeune, F., Pereira, O., Peters, T., Standaert, F.-X.: Ciphertext integrity with misuse and leakage: definition and efficient constructions with symmetric primitives. In: Kim, J., Ahn, G.-J., Kim, S., Kim, Y., López, J., Kim, T. (eds.) AsiaCCS, pp. 37–50. ACM (2018)

10. Berti, F., Pereira, O., Peters, T., Standaert, F.-X.: On leakage-resilient authenticated encryption with decryption leakages. IACR Trans. Symmetric Cryptol. **2017**(3), 271–293 (2017)

11. Chang, D., et al.: Release of unverified plaintext: tight unified model and application to ANYDAE. IACR Trans. Symmetric Cryptol. **2019**(4), 119–146 (2019)

12. Chari, S., Jutla, C.S., Rao, J.R., Rohatgi, P.: Towards sound approaches to counteract power-analysis attacks. In: Wiener, M. (ed.) CRYPTO 1999. LNCS, vol. 1666, pp. 398–412. Springer, Heidelberg (1999). https://doi.org/10.1007/3-540-48405-1_26

13. Cogliati, B., Jean, J., Peyrin, T., Seurin, Y.: A long tweak goes a long way: high multi-user security authenticated encryption from tweakable block ciphers. IACR Cryptology ePrint Archive, p. 846 (2022)

14. Datta, N., Dutta, A., List, E., Mandal, S.: On the security of triplex- and multiplex-type constructions with smaller tweaks. Cryptology ePrint Archive, Paper 2023/1658 (2023). https://eprint.iacr.org/2023/1658

15. Degabriele, J.P., Janson, C., Struck, P.: Sponges resist leakage: the case of authenticated encryption. In: Galbraith, S.D., Moriai, S. (eds.) ASIACRYPT 2019. LNCS,

vol. 11922, pp. 209–240. Springer, Cham (2019). https://doi.org/10.1007/978-3-030-34621-8_8

16. Dobraunig, C., et al.: ISAP v2.0. IACR Trans. Symmetric Cryptol. **2020**(S1), 390–416 (2020)

17. Dobraunig, C., Eichlseder, M., Mangard, S., Mendel, F., Unterluggauer, T.: ISAP - towards side-channel secure authenticated encryption. IACR Trans. Symmetric Cryptol. **2017**(1), 80–105 (2017)

18. Dobraunig, C., Eichlseder, M., Mendel, F., Schläffer, M.: ASCON v1.2 Submission to the CAESAR Competition (2016). https://competitions.cr.yp.to/round3/asconv12.pdf. Submission to the CAESAR competition

19. Dutta, A., Nandi, M., Talnikar, S.: Beyond birthday bound secure MAC in faulty nonce model. In: Ishai, Y., Rijmen, V. (eds.) EUROCRYPT 2019. LNCS, vol. 11476, pp. 437–466. Springer, Cham (2019). https://doi.org/10.1007/978-3-030-17653-2_15

20. Dworkin, M.: NIST Special Publication 800-38C – Recommendation for Block Cipher Modes of Operation: The CCM Mode for Authentication and Confidentiality [including updates through 7/20/2007]. Technical report, U.S. National Institute of Standards and Technology (2004)

21. Dworkin, M.: NIST Special Publication 800-38D – Recommendation for block cipher modes of operation: Galois/Counter Mode (GCM) and GMAC. Technical report, U.S. National Institute of Standards and Technology (2007)

22. Goubin, L., Patarin, J.: DES and differential power analysis the "Duplication" method. In: Koç, Ç.K., Paar, C. (eds.) CHES 1999. LNCS, vol. 1717, pp. 158–172. Springer, Heidelberg (1999). https://doi.org/10.1007/3-540-48059-5_15

23. Grosso, V., Standaert, F.-X., Faust, S.: Masking vs. multiparty computation: how large is the gap for AES? In: Bertoni, G., Coron, J.-S. (eds.) CHES 2013. LNCS, vol. 8086, pp. 400–416. Springer, Heidelberg (2013). https://doi.org/10.1007/978-3-642-40349-1_23

24. Guo, C., Pereira, O., Peters, T., Standaert, F.-X.: Authenticated encryption with nonce misuse and physical leakage: definitions, separation results and first construction. In: Schwabe, P., Thériault, N. (eds.) LATINCRYPT 2019. LNCS, vol. 11774, pp. 150–172. Springer, Cham (2019). https://doi.org/10.1007/978-3-030-30530-7_8

25. Hadipour, H., Sadeghi, S., Eichlseder, M.: Finding the impossible: automated search for full impossible-differential, zero-correlation, and integral attacks. In: Hazay, C., Stam, M. (eds.) EUROCRYPT 2023. LNCS, vol. 14007, pp. 128–157. Springer, Cham (2023). https://doi.org/10.1007/978-3-031-30634-1_5

26. Herbst, C., Oswald, E., Mangard, S.: An AES smart card implementation resistant to power analysis attacks. In: Zhou, J., Yung, M., Bao, F. (eds.) ACNS 2006. LNCS, vol. 3989, pp. 239–252. Springer, Heidelberg (2006). https://doi.org/10.1007/11767480_16

27. Hirose, S.: Some plausible constructions of double-block-length hash functions. In: Robshaw, M. (ed.) FSE 2006. LNCS, vol. 4047, pp. 210–225. Springer, Heidelberg (2006). https://doi.org/10.1007/11799313_14

28. Hoang, V.T., Krovetz, T., Rogaway, P.: Robust authenticated-encryption AEZ and the problem that it solves. In: Oswald, E., Fischlin, M. (eds.) EUROCRYPT 2015. LNCS, vol. 9056, pp. 15–44. Springer, Heidelberg (2015). https://doi.org/10.1007/978-3-662-46800-5_2

29. Hoang, V.T., Reyhanitabar, R., Rogaway, P., Vizár, D.: Online authenticated-encryption and its nonce-reuse misuse-resistance. In: Gennaro, R., Robshaw, M.

(eds.) CRYPTO 2015. LNCS, vol. 9215, pp. 493–517. Springer, Heidelberg (2015). https://doi.org/10.1007/978-3-662-47989-6_24

30. Kocher, P., Jaffe, J., Jun, B.: Differential power analysis. In: Wiener, M. (ed.) CRYPTO 1999. LNCS, vol. 1666, pp. 388–397. Springer, Heidelberg (1999). https://doi.org/10.1007/3-540-48405-1_25

31. Krovetz, T., Rogaway, P.: OCB (v1.1) (2016). https://competitions.cr.yp.to/round3/ocbv11.pdf

32. List, E.: TEDT2 – highly secure leakage-resilient TBC-based authenticated encryption. In: Longa, P., Ràfols, C. (eds.) LATINCRYPT 2021. LNCS, vol. 12912, pp. 275–295. Springer, Cham (2021). https://doi.org/10.1007/978-3-030-88238-9_14

33. McGrew, D.A., Viega, J.: The security and performance of the Galois/Counter mode (GCM) of operation. In: Canteaut, A., Viswanathan, K. (eds.) INDOCRYPT 2004. LNCS, vol. 3348, pp. 343–355. Springer, Heidelberg (2004). https://doi.org/10.1007/978-3-540-30556-9_27

34. Naito, Y., Sasaki, Yu., Sugawara, T.: Lightweight authenticated encryption mode suitable for threshold implementation. In: Canteaut, A., Ishai, Y. (eds.) EUROCRYPT 2020. LNCS, vol. 12106, pp. 705–735. Springer, Cham (2020). https://doi.org/10.1007/978-3-030-45724-2_24

35. Naito, Y., Sasaki, Y., Sugawara, T.: Secret can be public: low-memory AEAD mode for high-order masking. In: Dodis, Y., Shrimpton, T. (eds.) CRYPTO 2022. LNCS, vol. 13509, pp. 315–345. Springer, Cham (2022). https://doi.org/10.1007/978-3-031-15982-4_11

36. Peters, T., Shen, Y., Standaert, F.-X.: Multiplex: TBC-based Authenticated Encryption with Sponge-Like Rate (2023). https://dial.uclouvain.be/pr/boreal/object/boreal%3A273131/datastream/PDF_01/view

37. Qin, L., Dong, X., Wang, A., Hua, J., Wang, X.: Mind the TWEAKEY schedule: cryptanalysis on SKINNYe-64-256. In: Agrawal, S., Lin, D. (eds.) ASIACRYPT 2022. LNCS, vol. 13791, pp. 287–317. Springer, Cham (2022). https://doi.org/10.1007/978-3-031-22963-3_10

38. Rogaway, P.: Authenticated-encryption with associated-data. In: Atluri, V. (ed.) ACM CCS, pp. 98–107. ACM (2002)

39. Rogaway, P., Bellare, M., Black, J., Krovetz, T.: OCB: a block-cipher mode of operation for efficient authenticated encryption. In: Reiter, M.K., Samarati, P. (eds.) CCS, pp. 196–205. ACM (2001)

40. Rogaway, P., Shrimpton, T.: A provable-security treatment of the key-wrap problem. In: Vaudenay, S. (ed.) EUROCRYPT 2006. LNCS, vol. 4004, pp. 373–390. Springer, Heidelberg (2006). https://doi.org/10.1007/11761679_23

41. Shen, Y., Peters, T., Standaert, F.-X., Cassiers, G., Verhamme, C.: Triplex: an efficient and one-pass leakage-resistant mode of operation. IACR Trans. Cryptogr. Hardw. Embed. Syst. 2022(4), 135–162 (2022)

42. Turan, M.S., et al.: NIST IR 8454 - Status Report on the Final Round of the NIST Lightweight Cryptography Standardization Process. Technical report, US National Institute of Standards and Technology (2023)

43. Veyrat-Charvillon, N., Medwed, M., Kerckhof, S., Standaert, F.-X.: Shuffling against side-channel attacks: a comprehensive study with cautionary note. In: Wang, X., Sako, K. (eds.) ASIACRYPT 2012. LNCS, vol. 7658, pp. 740–757. Springer, Heidelberg (2012). https://doi.org/10.1007/978-3-642-34961-4_44

44. Yu, Yu., Standaert, F.-X.: Practical leakage-resilient pseudorandom objects with minimum public randomness. In: Dawson, E. (ed.) CT-RSA 2013. LNCS, vol. 7779, pp. 223–238. Springer, Heidelberg (2013). https://doi.org/10.1007/978-3-642-36095-4_15

From Substitution Box to Threshold

Anubhab Baksi[1(✉)], Sylvain Guilley[2], Ritu-Ranjan Shrivastwa[2],
and Sofiane Takarabt[2]

[1] Nanyang Technological University, Singapore, Singapore
anubhab001@e.ntu.edu.sg
[2] Secure-IC, Rennes, France
{sylvain.guilley,ritu-ranjan.shrivastwa,sofiane.takarabt}@secure-ic.com

Abstract. With the escalating demand for lightweight ciphers as well as side channel protected implementation of those ciphers in recent times, this work focuses on two related aspects. First, we present a tool for automating the task of finding a Threshold Implementation (TI) of a given Substitution Box (SBox). Our tool returns 'with decomposition' and 'without decomposition' based TI. The 'with decomposition' based implementation returns a combinational SBox; whereas we get a sequential SBox from the 'without decomposition' based implementation. Despite being high in demand, it appears that this kind of tool has been missing so far. In the process, we report new decomposition for the PRESENT SBox (improving from Poschmann et al.'s JoC'11 paper) and that of the GIFT SBox (improving from Jati et al.'s TIFS'20 paper). Second, we show an algorithmic approach where a given cipher implementation can be tweaked (without altering the cipher specification) so that its TI cost can be significantly reduced. We take the PRESENT cipher as our case study (our methodology can be applied to other ciphers as well). Indeed, we show over 31% reduction in area and over 52% reduction in depth compared to the basic threshold implementation.

Keywords: Lightweight Cryptography · Block Cipher · SBox · Side Channel Countermeasure · Threshold Implementation · PRESENT

1 Introduction

Over the last few years, we have observed a surge of research works dedicated to finding new lightweight ciphers and/or low-cost implementation of those ciphers. Recently we have also seen side channel protected implementation of the lightweight ciphers gaining traction [14,21]. While the community is proactive in advocating side channel protected implementation, surprisingly, a systematic and generic study on how to do that appears to be missing from the literature. There is an overall theory, but for the most part, a detailed study is required

The first author would like to thank Robert Hines, Bijoy Das (IIT Bhilai, India), Gan Peizhou (NTU, Singapore) and Aneesh Kandi (IIT Madras, India). An extended version of the paper is available as [5].

to be undertaken for better understanding of the context. The authors seem to come up with some ad-hoc approach (see, e.g., [21]) for the implementation. This further hinders the overdue tasks such as, finding proper algorithms, easy-to-use and publicly available tools, and various optimisations that can be employed to reduce the cost.

This work makes a humble attempt to look into the problem of finding a *Threshold Implementation* (TI) of a given SBox. The TI is a well-known concept that aims at protecting against the most common type of side channel attack which relies on power or electromagnetic leakage [10,21,24]. Such SBoxes are probably the most common choice for the non-linear component in the modern lightweight ciphers, thus an automated tool to find TI of those SBoxes is of prime importance.

Further, we observe that with a slight modification in the implementation, it is possible to reduce the cost for a TI of cipher. This does not alter the actual description of the cipher, hence there is no need to redo the security analysis. We show how to adopt a systematic approach with the lightweight cipher PRESENT [11], and how it successfully reduces the TI cost, by more than 31% in terms of area or by more of 52% in terms of depth. This methodology is generic, hence it can be applied to other ciphers that use a similar structure.

Our Contributions

The side channel attack is considered among the major threats, though, the field of studying the protected implementations seems less than developed at multiple levels. While working on this general area, we find ourselves in a situation where we need/want to find threshold implementation of a fairly large pool of SBoxes. Effectively, we look for automated tools to do the batch processing, but to our dismay, cannot find anything suitable (except for the tool by Petkova-Nikova mentioned in Sect. 4). Parallel to this, the problem of finding some optimisations at the algorithmic level is another interesting direction, which seems underdeveloped too.

Ultimately, we decide to make our own tool for this purpose. As we go along, we discover a proper algorithm is missing. For the most part, the previous authors such as [21, Section III] or [10, Chapter 3] convey the idea behind threshold through examples, instead of a well-formed algorithm. Naturally, several pertinent issues remain unexplored, such as automating the process or dealing with the corner cases. We therefore look further down with adequate clarity to come up with algorithms (as well as release our codes as open-source[1]). Our tool has two segments, one segment returns 'without decomposition' based threshold implementation (Sect. 3), and the other segment returns 'with decomposition' based threshold implementation (Sect. 4). The idea for decomposition presented here, while being similar to that one used in [21,35], is simplified. On top of that, we show decomposition for the PRESENT and GIFT SBoxes, which improve

---

[1] Our codes can be accessed at https://github.com/anubhab001/sbox-threshold-public.

in terms of ASIC cost compared to the currently best-known results from [35] (for PRESENT) and [21] (for GIFT).

On the other hand, better understanding of TI motivates us to find other avenues for cost reduction. With the lightweight block cipher PRESENT [11], we show an optimisation strategy (Sect. 5) that uses less resources. The basic idea stems from the concept of *affine equivalence* of SBoxes, i.e., by opting for another SBox than specified by the designers. This SBox is chosen in such as a way that the netlist optimisation tool can leverage on the new SBox's algebraic properties. However, our approach does not change the overall cipher specification (so one does not need to carry out the usual security analysis). Thus, our analysis reveals, some alternative representations are indeed more efficient to implement than the naïve implementation (which is used as a baseline). Further, this strategy can be applied to any other cipher with a similar structure.

2   Background

2.1   Side Channel Attack and Countermeasure

Side channel attacks, particularly those relying on information from power consumption or electromagnetic emanation, are of prominent concern while dealing with the physical security of the ciphers [6,10,23,25,29,34]. It has been systematically shown that a cipher with sufficient classical security claims falls short against an adversary equipped with a side channel attack set-up. It therefore goes without saying, understanding the attacks and finding low-cost countermeasures are among the top research priorities by/for the community.

Side-channel attacks are based on the connection between a (learned) model and any intermediate variable in the implementation that might be leaking. Therefore, the countermeasures attempt to destroy the linkage of the model and the intermediate variables. *Masking* [29, Chapter 9] is considered a prominent countermeasure. A masking scheme randomly distributes each intermediate to introduce randomness in a way that the overall algorithmic flow in the cipher is unchanged, but the randomised operations makes the side channel leakage free from the intermediate variables. Depending on the strength of the attacker, various *degrees* of masking can be adopted.

Threshold Implementation

The threshold implementation (TI) is a form of masking, and is among the top recommendations against the side channel attacks [10,31] specially for protecting the hardware implementation.

Typically, the TI of an affine function is considered straightforward, while that of a non-linear (in most block ciphers, the only non-linear component is the SBox) function is considered a strenuous task to accomplish. The TI of a given SBox can be realised either as *without decomposition* (the SBox is implemented as a combinational circuit) or as *with decomposition* (the SBox is implemented as a sequential circuit, typically it allows for smaller and more shallow netlists at the expense of pipelining) [21].

3 Threshold Without Decomposition (Combinational SBox)

In essence, the without decomposition based threshold implementation takes the coordinate function (in ANF) form of an SBox, then converts into a suitable implementation satisfying a set of conditions. We consider the usual notation here, i.e., for an $n \times n$ SBox S we denote the input variables as $x_0, x_1, \ldots, x_{n-1}$ and the output variables as $y_0, y_1, \ldots, y_{n-1}$. Then it substitutes each x_i and each y_j by $d + 1$ shares (where $d \geq$ the algebraic degree of the SBox). In the process, each the x_i and y_i variables are respectively replaced by the new variables $x_{i,j}$ and $y_{i,j}$; where $i \in \{0, 1, \ldots, n - 1\}$ and $j \in \{0, 1, \ldots, d\}$.

Based on the existing literature [10, 15, 21, 24, 32], the 'without decomposition' based TI of S satisfies the following conditions:

(α) **Sharing of input variables.** $\bigoplus_{j=0}^{d} x_{i,j} = x_i$, for $i = 0, 1, \ldots, n - 1$.

(β) **Sharing of output variables/Correctness.** $\bigoplus_{j=0}^{d} y_{i,j} = y_i$ for $i = 0, 1, \ldots, n - 1$.

(γ) **Non-completeness.** At least one variable from $\{x_{i,0}, x_{i,1}, \ldots, x_{i,d}\}$ is missing in each of $y_{j,0}, y_{j,1}, \ldots, y_{j,d}$ for all $i, j \in \{0, 1, \ldots, n - 1\}$.

(δ) **Uniformity.** All non-zero entries in the x_i ($\forall i$) versus $y_{j,k}$ ($\forall j, k$) frequency distribution table are equal.

After this, $d + 1$ separate SBoxes $S_0, S_1, \ldots, S_d$ are implemented in parallel, where these SBoxes are given by the following arrangement of the coordinate functions:

$$
S_0 : \begin{cases} y_0 & = y_{0,0} \\ y_1 & = y_{1,0} \\ \quad \vdots \\ y_{n-1} & = y_{n-1,0} \end{cases}
$$

$$
\vdots
$$

$$
S_d : \begin{cases} y_0 & = y_{0,d} \\ y_1 & = y_{1,d} \\ \quad \vdots \\ y_{n-1} & = y_{n-1,d} \end{cases}
$$

Thanks to the ingenious arrangement, each of the component SBoxes $(S_0, \ldots, S_d)$ computes some share of the original SBox (S). When combined, the coordinate functions of S (namely, $y_0, \ldots, y_{n-1}$) can be realised even if one $x_{i,j}$ variable in each $S_0, \ldots, S_d$ is randomised. Since one input variable $x_{i,j}$ is randomised, it makes the corresponding side channel leakage random, making it hard for the attacker to exploit secret information from the leakage (this outlines the philosophy of the side channel protection). At the beginning of the (protected) cipher execution, some input variables are fed random inputs and at the end all the shares are combined to get the intended cipher output. In other words the fundamental concept can be described as running multiple randomised modules (each is a bit different from the others) among which the inputs to the cipher are distributed. One module does not give exploitable information to the attacker, still the combined output from the modules results in the desired cipher output.

Example 1 (Uniformity). For the sake of clarity, here we present an example of a threshold sharing that does not have the uniformity property (Condition (δ)), which is adopted from [10, Chapter 3.3.1]. Consider the function, $y = x_0 x_1$ with the sharing (where $x_i = x_{i,0} \oplus x_{i,1} \oplus x_{i,2}$ for $i = 0, 1$; and $y = y_0 \oplus y_1 \oplus y_2$)[2]:

$$y_0 = x_{0,1}x_{1,1} \oplus x_{0,1}x_{1,2} \oplus x_{0,2}x_{1,1},$$
$$y_1 = x_{0,2}x_{1,2} \oplus x_{0,0}x_{1,2} \oplus x_{0,2}x_{1,0},$$
$$y_2 = x_{0,0}x_{1,0} \oplus x_{0,0}x_{1,1} \oplus x_{0,1}x_{1,0}.$$

Table 1. Non-conformity to uniformity property (example with $y = x_0 x_1$)

(a) Frequency distribution table for (x_0, x_1) versus (y_0, y_1, y_2)

x_0	x_1				(y_0, y_1, y_2)				
		000	011	101	110	001	010	100	111
0	0	7	3	3	3	0	0	0	0
0	1	7	3	3	3	0	0	0	0
1	0	7	3	3	3	0	0	0	0
1	1	0	0	0	0	5	5	5	1

(b) Computation for $(x_0, x_1) = (1,1)$

		$(x_{1,0}, x_{1,1}, x_{1,2})$			
		$(0,0,1)$	$(0,1,0)$	$(1,0,0)$	$(1,1,1)$
.333	$(0,0,1)$	$y_0 = 0$ $y_1 = 1$ $y_2 = 0$	$y_0 = 1$ $y_1 = 0$ $y_2 = 0$	$y_0 = 0$ $y_1 = 1$ $y_2 = 0$	$y_0 = 1$ $y_1 = 0$ $y_2 = 0$
$(x_{0,0}, x_{0,1}, x_{0,2})$	$(0,1,0)$	$y_0 = 1$ $y_1 = 0$ $y_2 = 0$	$y_0 = 1$ $y_1 = 0$ $y_2 = 0$	$y_0 = 0$ $y_1 = 0$ $y_2 = 1$	$y_0 = 0$ $y_1 = 0$ $y_2 = 1$
	$(1,0,0)$	$y_0 = 0$ $y_1 = 1$ $y_2 = 0$	$y_0 = 0$ $y_1 = 0$ $y_2 = 1$	$y_0 = 0$ $y_1 = 0$ $y_2 = 1$	$y_0 = 0$ $y_1 = 1$ $y_2 = 0$
	$(1,1,1)$	$y_0 = 1$ $y_1 = 0$ $y_2 = 0$	$y_0 = 0$ $y_1 = 0$ $y_2 = 1$	$y_0 = 0$ $y_1 = 1$ $y_2 = 0$	$y_0 = 1$ $y_1 = 1$ $y_2 = 1$

The example is given through Table 1. In particular, the corresponding frequency distribution table is shown in Table 1(a). An instance on how Table 1(a) is computed is shown in Table 1(b); where the computation corresponding to the last row (i.e., $(x_0, x_1) = (1, 1)$; and consequently, $y = 1$) is shown. One may note from Table 1(b) that (y_0, y_1, y_2) equals $(0, 0, 1), (0, 1, 0), (1, 0, 0))$ and $(1, 1, 1)$ with respective frequency of 5, 5, 5 and 1.

Notice from Table 1(a) that, the non-zero entries are not equal. Hence, this sharing does not conform to the uniformity property.   □

---

[2] Notice that all $9 = 3 \times 3$ cross-terms (i.e., quadratic monomials) appear in the expressions. Further, notice that y_j does not contain any $x_{.,j}$, this satisfies non-completeness (Condition (γ)).

Remark 1 (Note on uniformity). As stated in [10, Chapter 3.3], the absence of the uniformity property (Condition (δ)) in a threshold circuit does not leak side channel information by itself; but if it is used to drive another circuit, this second circuit may leak side channel information. However, we could not find any experimental evaluation in the literature, and this could be an interesting problem for future study. $\quad\Box$

Remark 2 (Operational uniformity (relaxed) condition). In this article, we consider a more natural uniformity constraint. We aim at specifying a condition whereby the template attacks on the SBoxes are not possible. For all coordinate $0 \le i < n$ and share $0 \le j \le d$, each distribution of $Y_{i,j}|X = x$ must be the same, irrespective the value of $x \in \mathbb{F}_2^n$.

As each $Y_{i,j}$ is a Boolean variable, this means that for all fixed value of $x \in \mathbb{F}_2^n$, we check that $\Pr(Y_{i,j}) \in [0,1]$ is the same. This reflects the first-order probing resistance in the *Strong Non-Interference* (SNI, see [8]) setting. Now, note that $\Pr(Y_{i,j} = 1) = E(Y_{i,j})$. For the PRESENT SBox, we get for all x (see Sect. 5) the same distributions that follow:

$$\begin{cases} \Pr(Y_{0,0}|X = x) = \frac{1}{2^5}, \Pr(Y_{0,1}|X = x) = \frac{1}{2^5}, \Pr(Y_{0,2}|X = x) = \frac{3}{2^7}, \Pr(Y_{0,3}|X = x) = 0 \quad ; \\ \Pr(Y_{1,0}|X = x) = \frac{1}{2^5}, \Pr(Y_{1,1}|X = x) = \frac{1}{2^5}, \Pr(Y_{1,2}|X = x) = \frac{27}{2^{10}}, \Pr(Y_{1,3}|X = x) = \frac{21}{2^{10}}; \\ \Pr(Y_{2,0}|X = x) = \frac{1}{2^5}, \Pr(Y_{2,1}|X = x) = \frac{1}{2^5}, \Pr(Y_{2,2}|X = x) = \frac{27}{2^{10}}, \Pr(Y_{2,3}|X = x) = \frac{21}{2^{10}}; \\ \Pr(Y_{3,0}|X = x) = \frac{1}{2^5}, \Pr(Y_{3,1}|X = x) = \frac{1}{2^5}, \Pr(Y_{3,2}|X = x) = \frac{27}{2^{10}}, \Pr(Y_{3,3}|X = x) = \frac{21}{2^{10}}. \end{cases}$$

For instance, the fact that '$\Pr(Y_{0,3}) = 0$' is not a weakness. This results from the fact that the first coordinate of the SBox could be split in simply in $d = 3$ shares while obeying TI precepts. From what we can tell, this is not a vulnerability. Otherwise, plunging any licit TI implementation enjoying a sharing into d into a sharing into $d + 1$ would no longer be considered a valid, as the extra added share would be consistently equal to 0 (as in our case of the first coordinate function of the PRESENT SBox). $\quad\Box$

3.1 Need for a Well-Developed Algorithm

As noted (Sect. 1), the corresponding engineering methods to apply TI does not appear to be developed enough. No specific algorithm is available to the best of our understanding/finding, rather the concept is described mostly through some cherry-picked examples [10,21]. This is somewhat surprising, specially given the state-of-the-art explores advanced side channel attack (e.g., [26,27]). The problems that arise from explaining a concept through examples instead of a developing a proper algorithm can be manifested in a number of ways, as listed next.

Problem 1 (Ambiguous sharing). Since the concept is mostly communicated through examples, it is in general hard to come up with a proper algorithm. To illustrate the point, we look into the example given in [21, Section III.A]. The 3-variable quadratic function $f(x, y, z) = xy \oplus z$ is shown with the following 3-shares: $f_1 = z_2 \oplus x_2y_2 \oplus x_2y_3 \oplus x_3y_2, f_2 = z_3 \oplus x_1y_3 \oplus x_3y_1 \oplus x_3y_3, f_3 = z_1 \oplus$

$x_1y_1 \oplus x_1y_2 \oplus x_2y_1$. This sharing is not unique. For instance, the following is also valid sharing of f_1 and f_2 (f_3 is unchanged): $f_1 = \mathbf{z_3} \oplus x_2y_2 \oplus x_2y_3 \oplus x_3y_2$, $f_2 = \mathbf{z_2} \oplus x_1y_3 \oplus x_3y_1 \oplus x_3y_3$. □

Problem 2 (Corner case: Low algebraic degree of a coordinate function). Consider the 3-bit SBox, 03214756, which is given by the following coordinate functions: $y_0 = x_0 \oplus x_1x_2, y_1 = x_0 \oplus x_1x_2 \oplus x_1, y_2 = x_2$. Note that the coordinate function y_2 is of algebraic degree 1 whereas the rest are of algebraic degree 2 (i.e., not every coordinate function has the maximum algebraic degree). Thus, it is possible to consider 3-share masking for y_0 and y_1, but only 2-share is sufficient for y_2. Based on the existing literature (such as, the threshold implementation of PRESENT in [35]), it is not clear if this $(3, 3, 2)$-sharing is not considered valid (the number of shares for an SBox is taken as a constant for all its coordinate functions, regardless of the algebraic degree of the individual coordinate functions), even though it may be possible that this unequal sharing for individual coordinate functions does not leak any exploitable side channel information. Further research may be conducted on this. □

Problem 3 (No automation). There does not appear to be any publicly available tool to server the purpose. For instance, the without decomposition based threshold implementation for the GIFT SBox (1A4C6F392DB7508E) [7] is computed manually. It is a laborious task, as it requires to compute the entire expression consisting of a few hundred to a few thousands (if not more) of monomials. In general, this task is tedious to carry out manually beyond 3-bit SBoxes and borderline impossible starting from 5-bit SBoxes. □

Problem 4 (Lack of further optimisation). In certain cases, there could be opportunity to further optimise the threshold expressions. For example, one may look for factorisation so that the netlist can be further optimised (without compromising the side channel protection). Other logic gates than {AND, XOR} can also be used to reduce the hardware cost. □

3.2 Our Approach

In this work, we take each coordinate function of the given SBox (which is essentially a vectorial Boolean function) one at a time, before moving on the next one. A bird's eye view of our approach can be found in Algorithm 1.

First we introduce the input and output sharing variables $(x_{i,j}, y_{i,j})$ to compute the Boolean function of the form: $\bigoplus_j y_{i,j} = $ function of $x_{i,j}$ where i is fixed[3]. After this, each monomial from the RHS is taken one-by-one and assigned to

---

[3] Note that, the naïve AND count in the coordinate functions of the SBox affects the total number of monomials in the shares (more naïve AND count means more monomials in the coordinate functions of the SBox, this in turn leads to more monomials in the shares). The exact relationship in which the naïve AND count affects the total number of monomials is complicated, but it seems the relationship is similar to exponential.

the LHS variables (i.e., $y_{i,j}$ for a fixed i) so that the Condition (β)) (i.e., use all the monomials from the RHS) and non-completeness) is satisfied. Next, in order to satisfy Condition (γ) (i.e., not all input sharing variables are used in any $y_{i,j}$), we determine a conflict list. For instance, if this list looks like this: $(\{x_{i,0}\}, \{x_{i,1}\}, \ldots)$; it means that, no $y_{i,0}$ can contain $x_{i,0}$, no $y_{i,1}$ can contain $x_{i,1}$, and so on. Then each monomial (which is picked up from the RHS of the expression before sharing) is checked for intersection with the element of conflict list corresponding to each $y_{i,j}$. If the intersection is null, then that monomial is assigned to $x_{i,j}$. The remaining property, Condition (δ), is hard to ensure in this way; therefore we simply check if it holds after the threshold implementation is completed, if it does not hold then we randomise and try again.

In total, we offer 3 options to randomise. In the first option, we shuffle the RHS of the combined expression. In the second option, we allow the ordering in which the $y_{i,j}$ variables are picked (for a fixed i). In the third, we allow to randomise the conflict list (e.g., allowing $y_{i,0}$ to have conflict with $x_{i,1}$ instead of $x_{i,0}$).

Warning for Zero-Sharing. We display a warning message if some $y_{i,j}$ is a constant (happens if some coordinate function has less algebraic degree). As an illustration, reconsider the SBox described in Problem 2 (03214756). By exactly following Algorithm 1, the sharing is given as (notice that a warning is shown for $y_{2,2}$):$y_{2,0} = x_{2,1} \oplus x_{2,2}, y_{2,1} = x_{2,0}, y_{2,2} = 0$. It can be stated that, the presence of a zero sharing does not inherently contradict uniformity (Condition (δ)). Further, one may note that, the case for zero sharing can happen for a larger ($\geq$ 4-bit) SBox. For instance, consider a high degree SBox but some of its coordinate functions are affine (this kind of SBox is used in practice, such as that one used in DEFAULT-LAYER [3]).

Algorithm 1. Sharing for threshold (without decomposition) for an SBox

---

Input: An $n \times n$ SBox given as coordinate functions $(y_0, \ldots, y_{n-1}$ with input variables $x_0, \ldots, x_{n-1})$
Output: Assignment of $d + 1$ SBoxes given as coordinate functions
1: **for** $i \leftarrow 0$ to $n - 1$ **do**   ▷ Iterate over each coordinate function
2:     **for** $j \leftarrow 0$ to $n - 1$ **do**   ▷ Iterate over all input variables
3:       | $X_j \leftarrow x_{j,0} \oplus x_{j,1} \oplus \cdots \oplus x_{j,d}$ ▷ Share input variables
4:     $Y \leftarrow$ Substitute $x_j = X_j$ in y_i for all j ▷ Compute RHS of output sharing
5:     **for** $j \leftarrow 0$ to d **do**
6:       | $y_{i,j} \leftarrow 0$ ▷ Shares for y_i
7:     **for each** monomial $m \in Y$ **do**   ▷ Ordering can be randomised
8:       **for** $j \leftarrow 0$ to d **do**   ▷ Ordering can be randomised
9:         **if** m is not conflicting with y_j **then**   ▷ Variables can be randomised
10:         | $y_{i,j} \leftarrow y_{i,j} \oplus m$ ▷ Non-completeness is respected
11:     **if** Uniformity not satisfied **then**
12:       Go back to Step 5 with randomised options   ▷ Try again
13:     **return** Sharing of y_i as $(y_{i,0}, y_{i,1}, \ldots, y_{i,d})$ ▷ $\bigoplus_j y_{i,j} = y_i$

3.3   Results

In the following results, we do not enforce uniformity (Condition (δ)). This is in concordance with the existing literature (e.g., Jati et al. [21]). Further, although not conclusive yet, our experiment hints that it may be hard to ensure this property (as our tool sometimes has to iterate few thousand times before finding a uniform sharing) in general. It is even possible that the this property does not hold for some Boolean functions[4].

Minimal Order (Algebraic Degree + 1) Sharing. Our approach can deal with 3×3 and 4×4 SBoxes without any difficulty. Further, 5×5 SBoxes, such as that of ASCON's [18] works well (one may note from [33] that ASCON is recently selected as the primary choice in the LWC project run by NIST). Results with some SBoxes are summarised in Table 2, where we show the number of monomials along with STM 130 nm ASIC cost (HCMOS9GP, Cadence v14.20 2016) in terms of gate equivalent (rounded off to nearest integer). Each of the SBoxes is used in a cipher, save for 048AFC691EBD7532 which is presented in [19, Section 3.4]. The number of shares are taken as the algebraic degree of the SBox plus 1.

Table 2. Without decomposition cost of some SBoxes (uniformity not enforced)

		Shares	# Monomials	Hardware[☆]
GIFT [7]	1A4C6F392DB7508E	4	265[★]	526
PRESENT [11]	C56B90AD3EF84712	4	666	1132
PRINCE [12]	BF32AC916780E5D4	4	731	991
PICCOLO [36]	E4B238091A7F6C5D	4	399	645
SKINNY-64 [9]	C6901A2B385D4E7F	4	398	723
TWINE [37]	C0FA2B9583D71E64	4	626	723
PYJAMASK-128 [20]	2D397BA6E0F4851C	4	373	640
QARMA [1]	ADE6F735980CB124	4	562	826
NOEKEON Gamma [16]	7A2C48F0591E3DB6	4	387	697
DEFAULT [2,3] LS	037ED4A9CF18B265	3	102	60
Non-LS	196F7C82AED043B5	4	213	412
Gao-Roy-Oswald [19]	048AFC691EBD7532	4	988	1658
ASCON[★] [18]		3	160	85

☆: Gate equivalent in STM 130nm ASIC library (HCMOS9GP)

★: Used in [21, Appendix B]

★: 4B1F141A15921B58121D361C1E137E0D111810C11916AF17

Higher Order ($\geq$ Algebraic Degree + 2) Sharing. Our tool can work with higher number of shares as well, without any actual change in the algorithm flow. We show some examples in Table 3, where the number of shares is taken as the algebraic degree of the SBox plus 2. It is worth noting that ASCON has recently been selected as the NIST LWC winner [33].

---

[4] To the best of our knowledge, no proof about the existence of the uniformity property is presented in the literature.

Table 3. Higher order without decomposition cost of some SBoxes (uniformity not enforced)

		Shares	# Monomials	Hardware[☆]
GIFT [7]	1A4C6F392DB7508E	5	451	432
SKINNY-64 [9]	C6901A2B385D4E7F	5	682	681
QARMA [1]	ADE6F735980CB124	5	967	983
ASCON[★] [18]		4	257	142

☆: Gate equivalent in STM 130nm ASIC library (HCMOS9GP)

★: 4B1F141A15921B58121D361C1E137E0D111810C11916AF17

4 Threshold with Decomposition (Sequential SBox)

The theory for a decomposition based TI requires finding two other SBoxes such that the composition of these SBoxes is the target SBox. Given an $n \times n$ SBox S, we want to find two $n \times n$ (bijective) SBoxes F_0 and F_1 such that

(i) $F_1 \circ F_0 \equiv S$ (i.e., $F_1(F_0(x)) = S(x) \; \forall x$);
(ii) the algebraic degrees of both F_0 and F_1 are less than the algebraic degree of S .

Example 2 (PRESENT SBox decomposition from [35]). The authors in [35] find the following decomposition for the PRESENT SBox (C56B90AD3EF84712, cubic): $F_1 = $ 1A4F3C69D2875E0B (quadratic) and $F_0 = $ 5C6F7E184D3A2B09 (quadratic). The coordinate functions are:

$$F_1 : \begin{cases} y_0 = x_0 \oplus x_1 \oplus 1, \\ y_1 = x_0 \oplus x_2 x_3 \oplus x_2, \\ y_2 = x_0 x_2 \oplus x_0 x_3 \oplus x_1 \oplus x_3, \\ y_3 = x_0 \oplus x_2 x_3 \oplus x_3. \end{cases} \qquad F_0 : \begin{cases} y_0 = x_0 \oplus x_1 x_2 \oplus x_1 \oplus x_3 \oplus 1, \\ y_1 = x_1 \oplus x_2, \\ y_2 = x_1 x_2 \oplus x_1 x_3 \oplus x_2 x_3 \oplus 1, \\ y_3 = x_0. \end{cases}$$

Note that, when constructing the coordinate functions of the target SBox; y_i of F_0 becomes the corresponding x_i of F_1, $\forall i$. In other words, we first start with the coordinate functions of F_1, the replace each x_i with the RHS of the ith coordinate function of F_0. For instance, to get y_0 of the PRESENT SBox (= $x_0 \oplus x_1 x_2 \oplus x_2 \oplus x_3$); we start with y_0 of F_1 (= $\mathbf{x_0} \oplus \mathbf{x_1} \oplus 1$); then we replace $\mathbf{x_0}$ with y_0 of F_0 (= $x_0 \oplus x_1 x_2 \oplus x_1 \oplus x_3 \oplus 1$), and $\mathbf{x_1}$ with y_1 of F_0 (= $x_1 \oplus x_2$). □

Remark 3. It is not possible to decompose a quadratic SBox S in this way. In order to decompose a quadratic SBox, one needs that F_0 and F_1 both have lower algebraic degree than that of S, implying that both F_0 and F_1 are affine. However, composition of two affine SBoxes does not produce a quadratic SBox. Hence, the quadratic SBoxes such as those used in ASCON [18] or BAKSHEESH [4] cannot be decomposed in this way. □

Despite being quite popular [21,30,35], it appears that there exists only one (publicly available) tool which is a courtesy of Petkova-Nikova[5]. This tool, avail-

---

[5] Hosted at Svetla Petkova-Nikova's official web-page: https://homes.esat.kuleuven.be/~snikova/ti_tools.html.

able exclusively as executable files, can find decomposition based threshold for a given 3×3 or 4×4 SBox.

Here we describe a simple idea to find such decomposition in Algorithm 2. In short, we generate F_1 and F_0 in a way that $F_1 \circ F_0 \equiv S$ is always satisfied (by randomly constructing F_0 in a way so that the algebraic degree requirement is already satisfied, then adjusting F_1 according to F_0 and S); then we decide whether to keep or discard depending on the algebraic degree of F_1.

Algorithm 2. Threshold (with decomposition) for an SBox

Input: An $n \times n$ SBox S
Output: Two $n \times n$ SBoxes F_1 and F_0 such that $F_1 \circ F_0 \equiv S$
1: Pick F_0 randomly which is affine equivalent to a given lower algebraic degree SBox
2: **for** $i \leftarrow 0$ to $2^n - 1$ **do**   ▷ Iterate over all SBox inputs
3: | $r \leftarrow F_0(i)$
4: | $F_1(r) \leftarrow S(i)$ ▷ $F_0 : i \mapsto r,\ F_1 : r \mapsto S(i)$
5: **if** algebraic degree of F_1 that of S **then**
6: | Discard F_1, and go back to Step 1
7: **return** (F_1, F_0)

In Step 1, we specify that the algorithm be given candidate SBoxes from which it can generate F_0 (F_0 is affine equivalent to a given candidate SBox). For instance, suppose, we want to find the decomposition of a given cubic 4×4 SBox (which is typically the case for the common ciphers like PRESENT [11], SKINNY-64 [9] or GIFT [7]). In that case, we need to supply quadratic SBoxes to Algorithm 2. Following [10, Chapter 5], we choose the quadratic affine classes, $\{4, 12, 293, 294, 299, 300\}$.

Overall, Algorithm 2 simplifies the problem (cf. the complication in [35, Section 3.2] or [21, Section III.A]). In summary, the previous approach constructs an exhaustive pool of SBoxes first in a deterministic way, which is extensively time and space consuming.

Some examples of decomposition with respect to three high-profile ciphers can be found in Table 4 (Table 4(a) for PRESENT [11], 4(b) for SKINNY-64 [9], and 4(c) for GIFT [7]). In order to estimate the implementation cost in ASIC (TSMC 65nm and UMC 180nm logic libraries), we use the LIGHTER tool [22] (the backward-compatible code from [17] is used for this purpose, as the URL for the LIGHTER code does not seem accessible). From the estimates, one can see that the our (F_1, F_0) take lower total cost compared to the PRESENT (F_1 = 1A4F3C69D2875E0B F_0 = 5C6F7E184D3A2B09) and GIFT (F_1 = 7DB58E02CA4639F1, F_0 = F9A8BECD71203645) SBox decomposition given in [35] and [21], respectively. This observation is further supported when we check the total ASIC cost under the STM 130nm library (HCMOS9GP, Cadence v14.20 2016), as shown in Table 5. Thus, we infer that our research improves the state-of-the-art results for SBox decomposition, thereby opening the possibility for further reducing the threshold cost for PRESENT [35] and GIFT [21].

Table 4. ASIC costs (in GE) for SBox decomposition using LIGHTER

(a) PRESENT (C56B90AD3EF84712)

F_1	F_0	TSMC 65nm	UMC 180nm
08B7A31C46F9ED52*	7E92B04D5CA1836F*	33.50	27.67
547C0129BDA6E8F3	30B874A9FCED1256	36.50	30.67
47095C12E3DA8FB6	54FE32BA98DC0167	36.00	30.67
C5DAF706849BE312	017BA632DC4895EF	38.50	31.67
1A4F3C69D2875E0B	5C6F7E184D3A2B09	29.00	24.67
9B5C7F12DA483E06	32F10E98CD5BA467	37.50	32.67

✳: Taken from [35, Table 2]

(b) SKINNY-64 (C6901A2B385D4E7F)

F_1	F_0	TSMC 65nm	UMC 180nm
863C72FBA41E50D9	31FDA85720CE9B46	26.00	22.34
B72F0D95843E1CA6	DF64CE20A8759B13	26.50	22.67
2B7FA1E43C68D095	9AED54018BFC7623	28.00	24.34
863CFB72D950A41E	319BEC7520A8DF64	25.00	22.33
C46E18A37DF529B0	02DF46CE75B9138A	24.50	21.34
A3186EC45FD70B92	64EC20FD138A75B9	26.50	23.00
C368950D1EA47BF2	02468AFD1357B9CE	21.50	19.00
4AE1F27BD05968C3	ECB93157FDA80264	27.50	22.34
20138A9BCEDF7546	8F61250734DAE9CB	25.00	22.34

(c) GIFT (1A4C6F392DB7508E)

F_1	F_0	TSMC 65nm	UMC 180nm
5638127C9EF0DAB4★	4DF71A285CE60B39★	26.00	22.34
AC14D72B39F68E05	2031BA896475FECD	26.00	21.34
FB2673D90415C8AE	AE9C30572614B8DF	29.00	25.33
7DB58E02CA4639F1	F9A8BECD71203645	23.50	21.34
CA7DB1428E39F506	5160FCAB7342DE89	28.00	24.34
E596308FD2A1CB74	BAFC374298DE1560	26.50	22.67
C1A472DBF9630E58	1230A8B95674ECFD	27.50	23.00
754AF982BDC6310E	D32AB4C579801E6F	28.00	24.33
D5A6192E803F4CB7	42CD3BA560EF1987	28.00	23.67

★: Taken from [21, Table II]

Remark 4 (Third order decomposition). It is possible to find third order decomposition (i.e., F_2, F_1, F_0 such that $F_2 \circ F_1 \circ F_0 \equiv S$; and the algebraic degree of each of F_2, F_1, F_0 is less than that of S) using our tool (see also [13, Section 4.1] for an example of higher order decomposition). For instance, the SKINNY-64 SBox (C6901A2B385D4E7F, cubic) can be decomposed as $F_2 = $ 329EAD7645BC8F01 (quadratic), $F_1 = $ 5B941CE0A37D826F (quadratic) and $F_0 = $ 1AD6F3487520C9EB (quadratic). □

5   Further Optimisation Based on Affine Equivalence

In this part, we attempt to reduce the cost of an existing TI by optimising its SBox. In essence, we redesign the cipher with an affine equivalent SBox (so that

Table 5. Decomposition of PRESENT and GIFT SBoxes with ASIC costs

F_1		Area		F_0		Area		Total area	
		μm^2	GE			μm^2	GE	μm^2	GE
PRESENT	1A4F3C69D2875E0B	104.894	17.34	5C6F7E184D3A2B09		62.533	10.34	167.427	27.68
	08B7A31C46F9ED52*	90.774	15.00	7E92B04D5CA1836F*		88.757	14.67	179.531	29.67
GIFT	7DB58E02CA4639F1	94.808	15.67	F9A8BECD71203645		60.516	10.00	155.324	25.67
	5638127C9EF0DAB4★	86.740	14.34	4DF71A285CE60B39★		82.705	13.67	169.445	28.01

❋: Taken from [35, Table 2]

★: Taken from [21, Table II]

the cipher specification is unchanged) which reduces the TI cost. The SBox is implemented as a combinational only circuit (i.e., no register) per coordinate functions. We show our method on the lightweight block cipher PRESENT [11], though it can be applied to any other cipher with similar construction.

We study in this section the impact of affine equivalence on the SBox to reduce the cost of TI. Notice that the concept of affine equivalence we propose here is with respect to the representation of the cipher, not about changing the specification of the cipher. In other words, only the implementation changes, but the cipher description remains unchanged.

5.1 Motivation and Basic Observation

The complexity of threshold implementations directly relates to the number of monomials in the coordinate function of the SBox. Let us consider the PRESENT [11] (which is also a standard, ISO/IEC 29192-2:2012[6]) SBox, C56B90AD3EF84712. From its coordinate functions, we notice the following properties:

○ 8 monomials of degree 3,
○ 7 monomials of degree 2,
○ 10 monomials of degree 1, and
○ 2 monomials of degree 0 (constant 1).

Upon a closer inspection, however, we observe some monomials are duplicates. The following factors contribute to the total cost:

1. **Computation:** Number of unique monomials of given degree (i.e., each monomial of a given degree is counted only once even if its multiplicity is higher).
2. **Reduction:** Number of XORs in the coordinate functions.

About the PRESENT SBox, the contributing components are:

○ 3 unique monomials of degree 3 (namely, $x_0x_1x_2$, $x_0x_1x_3$, and $x_0x_2x_3$),
○ 5 unique monomials of degree 2, (the only absent one being x_0x_2),

---

[6] https://webstore.ansi.org/Standards/ISO/ISOIEC291922012-1383736.

○ 4 monomials of degree 1, and 1 monomial of degree 0 (constant 1),
○ 23 XORs.

Since the cost for TI for a higher degree monomial is much higher than a lower degree monomial, our aim here is to minimise the number of unique monomials of degree 3, then of degree 2, etc.

5.2   Improving Efficiency with Affine Equivalent SBox

From an implementation cost standpoint, it can be beneficial to represent the PRESENT block cipher in an equivalent notation where the SBox S is traded for one of its affine equivalent SBox, $S' = A^{-1} \circ S \circ A$, where A is an invertible affine mapping operating in $\mathbb{F}_2^4$. We reiterate that we do not aim at altering the cipher functionality, simply its representation. The usage of the affine mapping is compensated before and after the SBox application.

The use of the affine equivalent SBox is illustrated in Fig. 1. In the part (a) of the figure, a simplified block diagram of PRESENT is depicted, with the main components:

- Ports are plaintext (denoted as, "ptx"), round key (denoted as "K"), a selection signal indicating whether the encryption starts (denoted as, "round = 0?") or not, and an output ciphertext denoted as "ctx").
- Building blocks are the affine equivalent SBox S, the permutation layer P, a multiplexer allowing to input the plaintext or to iterate, and a DFF barrier storing the result computed till that particular round.

This can be implemented with a shorter critical path by pushing the conversion to/from affine representation outside of the main path. This is represented in the Fig. 1(b): Assuming that the plaintext and the round keys are applied the affine transformation, then a regular datapath can be used, provided finally the ciphertext is applied to inverse affine mapping. This equivalent representation leverages the fact that all elements in the PRESENT cipher (permutation layer and multiplexer) are linear. Notice that the scheme in Fig. 1(b) is correct only if A is linear. If it is affine, then the constant of the transformation shall be adapted for each operation, in particular the outer ones. The critical path, highlighted as the green box in Fig. 1(b), is no different than the original critical path. The architecture shown in Fig. 1 is thus suitable for masked TI with combinational SBox (i.e., without decomposition).

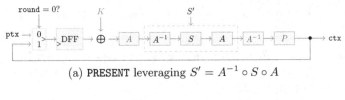

(a) PRESENT leveraging $S' = A^{-1} \circ S \circ A$

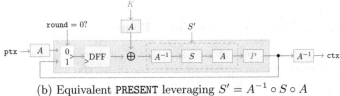

(b) Equivalent PRESENT leveraging $S' = A^{-1} \circ S \circ A$

Fig. 1. PRESENT leveraging affine equivalent SBox

5.3   Results

Efficiency Based on Algebraic Property. For 1000 random choices of A, we count the number of monomials in each degree, and we get statistics as depicted in Fig. 2. More specifically, Fig. 2(a) shows the relative frequency distribution for number of unique monomials for each individual degrees. For instance, there exists a unique monomial for degree 0 (constant 1) 100% of the cases. Figure 2(b) shows the probability distribution of XOR count.

In our case, the transformed SBox is found as follows. The binary matrix multiplied to obtain the linear part, and the constant binary vector are given respectively by: $\begin{pmatrix} 1\,0\,0\,0 \\ 1\,1\,1\,0 \\ 1\,1\,0\,0 \\ 1\,0\,1\,1 \end{pmatrix}$, $\begin{pmatrix} 0 \\ 1 \\ 1 \\ 1 \end{pmatrix}$. The transformed SBox, 4EC20B1A5F3D9867,

has the coordinate functions: $y_0 = x_0 x_2 \oplus x_1 x_2 \oplus x_3$, $y_1 = x_0 x_2 x_3 \oplus x_0 \oplus x_1 x_3$, $y_2 = x_0 x_1 x_2 \oplus x_0 x_1 \oplus x_1 x_3 \oplus x_2 \oplus 1$, $y_3 = x_0 x_2 x_3 \oplus x_0 \oplus x_1 x_2 \oplus x_1 x_3 \oplus x_1 \oplus x_2 x_3$. Therefore, we manage to get a transformed SBox with the following properties in its coordinate functions:

o 2 unique monomials of degree 3 ($x_0 x_2 x_3$ and $x_0 x_1 x_2$),
o 5 unique monomials of degree 2 ($x_0 x_2, x_1 x_2, x_1 x_3, x_0 x_1$ and $x_2 x_3$),
o 4 monomials of degree 1, and 1 monomial of degree 0 (constant 1),
o only 13 XORs.

Statistics on Netlist Parameters. Now we study the real impact of affine transformation to the property of the netlist; namely, we are interested in the algebraic features that drive the netlist properties. We study the netlist area (number of monomials vs. GE) and its logical depth (number of monomials vs. critical path). For the sake of simplicity, every gate is attributed a unitary area and propagation time.

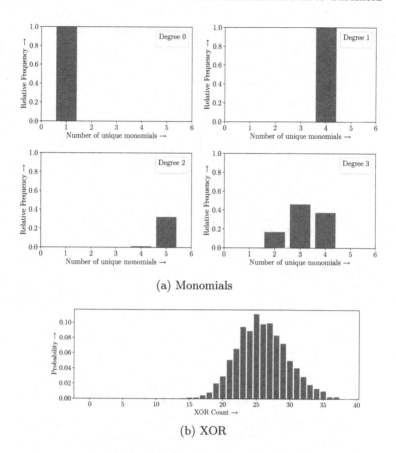

(a) Monomials

(b) XOR

Fig. 2. Statistics for affine equivalent SBox search for PRESENT

The goal is to figure out which property of the algebraic expression determines most the area and/or depth in the netlist. The following features are considered:

- The number of unique monomials of various degrees (3, 2, 1, and 0) – mostly, the number of unique monomials of degree 3 is to be minimised.
- XOR count is to be reduced to make area usage low.

The results are summarised in Fig. 3 (Fig. 3(a) for area, 3(b) for logical depth; and finally Fig. 3(c) for area versus logical depth); based on 10000 random selections of affine transformation $A : \mathbb{F}_2^4 \to \mathbb{F}_2^4$; where we group based on the number of unique monomials for better readability. It can be seen that, as expected, reducing the unique number of monomials of degree 3 is the main parameter to reduce the gate count and the depth of the netlists. The smallest netlist has 589 gates (of depth 18), and the most shallow one has depth 11 (and 728 gates). For comparison, the reference netlist (i.e., when A is the identity matrix) has 863 gates and a depth of 23. Hence an area reduction of more than 31% or a depth reduction of more than 52% is observed by applying our methodology. Further, as it can be seen from Fig. 3(c), larger netlists also have (slightly) longer critical path, on average.

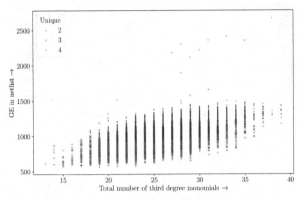

(a) Area (GE) vs. third degree monomials

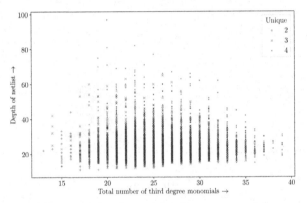

(b) Logical depth vs. third degree monomials

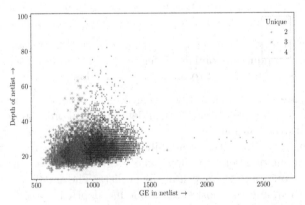

(c) Area (GE) vs. logical depth for third degree monomials

Fig. 3. SBoxes AE to PRESENT SBox

6   Conclusion

This work takes a deeper looks into the problem of finding threshold implementation of SBoxes. The first main contribution of this work is to present an open-source tool for automating the task for threshold implementation for a large pool of SBoxes. Our tool returns 'without decomposition' (Sect. 3) and 'with decomposition' (Sect. 4) based implementations. Despite being quite popular, such a tool seems overdue. Among other results, we show improvement over [35] and [21]. The second main contribution (Sect. 5) comes from an alternate representation of the PRESENT SBox [11] so that the TI cost can be reduced. The idea is to replace the original SBox by one of its affine equivalent SBox (so that the cipher description remains unchanged), but this new SBox has lower threshold cost. Overall, we show over 31% improved area and over 52% improved depth compared to the naïve implementation.

One interesting follow-up work could be to find SBoxes with lower AND count (but with other desirable cryptographic properties) so that the cipher is more suitable for adopting TI. Besides, as noted in Remark 1, it would be interesting to evaluate the amount of side channel leakage from the circuit which takes input from another circuit not obeying the uniformity property. As the main objective in Sect. 5 is to find another SBox, works like [28,38] can be incorporated in the search procedure in the future.

References

1. Avanzi, R.: The QARMA block cipher family - almost MDS matrices over rings with zero divisors, nearly symmetric even-mansour constructions with non-involutory central rounds, and search heuristics for low-latency s-boxes. Cryptology ePrint Archive, Report 2016/444 (2016). https://eprint.iacr.org/2016/444
2. Baksi, A.: Classical and physical security of symmetric key cryptographic algorithms. Ph.D. thesis, School of Computer Science & Engineering, Nanyang Technological University, Singapore (2021). https://dr.ntu.edu.sg/handle/10356/152003
3. Baksi, A.: DEFAULT: cipher-level resistance against differential fault attack. In: Tibouchi, M., Wang, H. (eds.) ASIACRYPT 2021. LNCS, vol. 13091, pp. 124–156. Springer, Singapore (2022). https://doi.org/10.1007/978-3-030-92075-3_5
4. Baksi, A., et al.: Baksheesh: similar yet different from gift. Cryptology ePrint Archive, Paper 2023/750 (2023). https://eprint.iacr.org/2023/750
5. Baksi, A., Guilley, S., Shrivastwa, R.R., Takarabt, S.: From substitution box to threshold. IACR Cryptol. ePrint Arch. 633 (2023)
6. Baksi, A., Kumar, S., Sarkar, S.: A new approach for side channel analysis on stream ciphers and related constructions. IEEE Trans. Comput. **71**(10), 2527–2537 (2021)
7. Banik, S., Pandey, S.K., Peyrin, T., Sasaki, Y., Sim, S.M., Todo, Y.: Gift: a small present. Cryptology ePrint Archive, Report 2017/622 (2017). https://eprint.iacr.org/2017/622
8. Barthe, G., et al.: Strong non-interference and type-directed higher-order masking. In: Proceedings of the 2016 ACM SIGSAC Conference on Computer and Communications Security, Vienna, Austria, 24–28 October 2016, pp. 116–129 (2016)

9. Beierle, C., et al.: The SKINNY family of block ciphers and its low-latency variant MANTIS. IACR Cryptology ePrint Archive **2016**, 660 (2016)
10. Bilgin, B.: Threshold implementations as countermeasure against higher-order differential power analysis. Ph.D. thesis, Katholieke Universiteit Leuven and University of Twente (2015). https://www.esat.kuleuven.be/cosic/publications/thesis-256.pdf
11. Bogdanov, A., et al.: PRESENT: an ultra-lightweight block cipher. In: Paillier, P., Verbauwhede, I. (eds.) CHES 2007. LNCS, vol. 4727, pp. 450–466. Springer, Heidelberg (2007). https://doi.org/10.1007/978-3-540-74735-2_31
12. Borghoff, J., et al.: Prince - a low-latency block cipher for pervasive computing applications (full version). Cryptology ePrint Archive, Report 2012/529 (2012). https://ia.cr/2012/529
13. Božilov, D., Knežević, M., Nikov, V.: Optimized threshold implementations: securing cryptographic accelerators for low-energy and low-latency applications. Cryptology ePrint Archive, Paper 2018/922 (2018). https://eprint.iacr.org/2018/922
14. Caforio, A., Collins, D., Glamocanin, O., Banik, S.: Improving first-order threshold implementations of skinny. Cryptology ePrint Archive, Report 2021/1425 (2021). https://ia.cr/2021/1425
15. Daemen, J.: Changing of the guards: a simple and efficient method for achieving uniformity in threshold sharing. Cryptology ePrint Archive, Report 2016/1061 (2016). https://ia.cr/2016/1061
16. Daemen, J., Peeters, M., Assche, G.V., Rijmen, V.: Nessie Proposal: NOEKEON (2000). http://gro.noekeon.org/Noekeon-spec.pdf
17. Dasu, V.A., Baksi, A., Sarkar, S., Chattopadhyay, A.: LIGHTER-R: optimized reversible circuit implementation for sboxes. In: 32nd IEEE International System-on-Chip Conference, SOCC 2019, Singapore, 3–6 September 2019, pp. 260–265 (2019)
18. Dobraunig, C., Eichlseder, M., Mendel, F., Schläffer, M.: Ascon v1.2. Submission to NIST (2019). https://csrc.nist.gov/CSRC/media/Projects/lightweight-cryptography/documents/round-2/spec-doc-rnd2/ascon-spec-round2.pdf
19. Gao, S., Roy, A., Oswald, E.: Constructing TI-friendly substitution boxes using shift-invariant permutations. In: Matsui, M. (ed.) CT-RSA 2019. LNCS, vol. 11405, pp. 433–452. Springer, Cham (2019). https://doi.org/10.1007/978-3-030-12612-4_22
20. Goudarzi, D., et al.: Pyjamask v1.0 (2019)
21. Jati, A., Gupta, N., Chattopadhyay, A., Sanadhya, S.K., Chang, D.: Threshold implementations of gift: a trade-off analysis. IEEE Trans. Inf. Forensics Secur. **15**, 2110–2120 (2020)
22. Jean, J., Peyrin, T., Sim, S.M., Tourteaux, J.: Optimizing implementations of lightweight building blocks. IACR Trans. Symmetric Cryptol. **2017**(4), 130–168 (2017)
23. Kumar, S., et al.: Side channel attack on stream ciphers: a three-step approach to state/key recovery. IACR Trans. Cryptogr. Hardw. Embed. Syst. **2022**(2), 166–191 (2022)
24. Kutzner, S., Nguyen, P.H., Poschmann, A.: Enabling 3-share threshold implementations for any 4-bit s-box. Cryptology ePrint Archive, Report 2012/510 (2012). https://eprint.iacr.org/2012/510
25. Lomné, V.: Power and electro-magnetic side-channel attacks: threats and countermeasures. Ph.D. thesis, Docteur de l'Université Montpellier II (2010). https://sites.google.com/site/victorlomne/research

26. Lomné, V., Prouff, E., Rivain, M., Roche, T., Thillard, A.: How to estimate the success rate of higher-order side-channel attacks. In: Batina, L., Robshaw, M. (eds.) CHES 2014. LNCS, vol. 8731, pp. 35–54. Springer, Heidelberg (2014). https://doi.org/10.1007/978-3-662-44709-3_3

27. Lomné, V., Prouff, E., Roche, T.: Behind the scene of side channel attacks. Cryptology ePrint Archive, Report 2013/794 (2013). https://eprint.iacr.org/2013/794

28. Lu, Z., Mesnager, S., Cui, T., Fan, Y., Wang, M.: An STP-based model toward designing s-boxes with good cryptographic properties. Des. Codes Cryptogr. 90(5), 1179–1202 (2022)

29. Mangard, S., Oswald, E., Popp, T.: Power Analysis Attacks - Revealing the Secrets of Smart Cards. Springer, New York (2007). https://doi.org/10.1007/978-0-387-38162-6

30. Müller, N., Moos, T., Moradi, A.: Low-latency hardware masking of PRINCE. In: Bhasin, S., Santis, F.D. (eds.) COSADE 2021. LNCS, vol. 12910, pp. 148–167. Springer, Cham (2021). https://doi.org/10.1007/978-3-030-89915-8_7

31. Nikova, S., Rechberger, C., Rijmen, V.: Threshold implementations against side-channel attacks and glitches. In: Ning, P., Qing, S., Li, N. (eds.) ICICS 2006. LNCS, vol. 4307, pp. 529–545. Springer, Heidelberg (2006). https://doi.org/10.1007/11935308_38

32. Nikova, S., Rijmen, V., Schläffer, M.: Secure hardware implementation of nonlinear functions in the presence of glitches. J. Cryptol. 24(2), 292–321 (2011)

33. NIST: Lightweight Cryptography Standardization Process: NIST Selects Ascon (2023). https://csrc.nist.gov/News/2023/lightweight-cryptography-nist-selects-ascon

34. Peeters, E.: Advanced DPA Theory and Practice: Towards the Security Limits of Secure Embedded Circuits, 1st edn. Springer, New York (2013). https://doi.org/10.1007/978-1-4614-6783-0

35. Poschmann, A., Moradi, A., Khoo, K., Lim, C., Wang, H., Ling, S.: Side-channel resistant crypto for less than 2,300 GE. J. Cryptol. 24(2), 322–345 (2011)

36. Shibutani, K., Isobe, T., Hiwatari, H., Mitsuda, A., Akishita, T., Shirai, T.: Piccolo: an ultra-lightweight blockcipher. In: Preneel, B., Takagi, T. (eds.) CHES 2011. LNCS, vol. 6917, pp. 342–357. Springer, Heidelberg (2011). https://doi.org/10.1007/978-3-642-23951-9_23

37. Suzaki, T., Minematsu, K., Morioka, S., Kobayashi, E.: Twine: a lightweight, versatile block cipher. ECRYPT (2011). https://www.nec.com/en/global/rd/tg/code/symenc/pdf/twine_LC11.pdf

38. Wadhwa, M., Baksi, A., Hu, K., Chattopadhyay, A., Isobe, T., Saha, D.: Finding desirable substitution box with SASQUATCH. IACR Cryptol. ePrint Arch. 742 (2023)

Tight Security Bound of 2k-LightMAC_Plus

Nilanjan Datta[1], Avijit Dutta[1], and Samir Kundu[2(✉)]

[1] TCG Centres for Research and Education in Science and Technology,
Kolkata, India
{nilanjan.datta,avijit.dutta}@tcgcrest.org
[2] Indian Statistical Institute, Kolkata, Kolkata, India
samirkundu3@gmail.com

Abstract. In ASIACRYPT'17, Naito proposed a beyond-birthday-bound variant of the LightMAC construction, called LightMAC_Plus, which is built on three independently keyed n-bit block ciphers, and showed that the construction achieves $2n/3$-bits PRF security. Later, Kim et al. claimed (without giving any formal proof) its security bound to $2^{3n/4}$. In FSE'18, Datta et al. have proposed a two-keyed variant of the LightMAC_Plus construction, called 2k-LightMAC_Plus, which is built on two independently keyed n-bit block ciphers, and showed that the construction achieves $2n/3$-bits PRF security. In this paper, we show a tight security bound on the 2k-LightMAC_Plus construction. In particular, we show that it provably achieves security up to $2^{3n/4}$ queries. We also exhibit a matching attack on the construction with the same query complexity and hence establishing the tightness of the security bound. To the best of our knowledge, this is the first work that provably shows a message length independent $3n/4$-bit tight security bound on a block cipher based variable input length PRF with two block cipher keys.

Keywords: LightMAC_Plus · H-Coefficient technique · Beyond Birthday Bound · Double Block Hash-then-Sum · 2k-LightMAC_Plus

1 Introduction

In FSE'16 [8], Luykx et al. have proposed LightMAC, which has been standardized by ISO/IEC standardization process. LightMAC is a block cipher based PRF that operates in parallel mode, i.e., for an n-bit block cipher E instantiated with two independently sampled keys K_1, K_2, and with a global counter size s, the LightMAC function is defined as follows:

$$\mathsf{LightMAC}_{\mathsf{E}_{K_1,K_2}}(M) = \mathsf{E}_{K_2}\left(\sum_{i=1}^{\ell-1} \mathsf{E}_{K_1}(\langle i \rangle_s \| M[i]) \oplus \mathsf{pad}_n(M[\ell])\right),$$

where $\langle i \rangle_s$ denotes the s bit encoding of the integer i and $(M[1], \ldots, M[\ell])$ denotes the $n - s$ bit parsing of message M, where each $M[i]$ is an $n - s$ bit string, and pad_n is an injective function that takes a message and appends to it a suitable number of 10^* to make the length of the padded string to be exactly

© The Author(s), under exclusive license to Springer Nature Switzerland AG 2024
A. Chattopadhyay et al. (Eds.): INDOCRYPT 2023, LNCS 14459, pp. 68–88, 2024.
https://doi.org/10.1007/978-3-031-56232-7_4

n. However, this design comes at the cost of a reduced rate of construction, where the rate of a construction is determined by the ratio of the total number of n-bit message blocks in a message M to the total number of primitive calls with block size n required to process the message M. Despite having a reduced rate, the design of LightMAC is simple in the sense that it minimizes all auxiliary operations other than having the block cipher calls, which allows to have a low overhead cost, and hence obtains a more compact implementation than PMAC [1]. Moreover, due to the inherent parallelism in the design of the scheme, LightMAC outperforms all the other popular sequential MAC constructions in terms of throughput in the parallel computing infrastructure.

1.1   Beyond Birthday Bound Secure Variants of LightMAC

Over the years, there have been many proposals of variants of LightMAC construction achieving beyond the birthday bound security. In 2017, Naito [9] proposed LightMAC_Plus construction based on three block cipher keys and showed that the construction is secure against all adversaries that make roughly $2^{2n/3}$ queries. In fact, LightMAC_Plus is the first beyond the birthday bound-secure PRF which is proven to have a message length independent security bound. In the same paper, the author has also proposed LightMAC_Plus2 [9] that provides a higher security bound than LightMAC_Plus or LightMAC, but it comes at the increased number of block cipher calls. In CT-RSA'18 [10], Naito has improved the bound of the LightMAC_Plus construction from $q^3/2^{2n}$ to $q_t^2 q_v/2^{2n}$, where q_t is the number of tagging queries and q_v is the number of verification queries. This security bound implies that LightMAC_Plus is secure up to 2^n tagging queries if the number of verification queries is 1. Later, in [7], Leurent et al. have shown a forging attack on the construction that achieves a constant success probability when the number of tagging queries is $2^{3n/4}$ and the number of verification queries is 1, which in turn invalidates the security claim of Naito [10] on LightMAC_Plus. In EUROCRYPT'20, Kim et al. [6] have claimed an improved security bound (but did not supply any formal proof to back up the claim) of LightMAC_Plus construction from $2n/3$-bits to $3n/4$-bits (ignoring the maximum message length), and due to the result of [7], the improved bound of LightMAC_Plus turns out to be the tight one.

In FSE'19, Datta et al. [4] proposed a two-keyed variant of LightMAC_Plus, called 2K-LightMAC_Plus, where the sum function used in the finalization phase uses the same block cipher key that is independent to the block cipher key used in the internal hash computation of 2K-LightMAC_Plus. Authors have shown that 2K-LightMAC_Plus achieves $2n/3$-bits security bound. In [10], Naito has proposed a single-keyed variant of LightMAC_Plus, dubbed as LightMAC_Plus-1k, in which a single block cipher key is used in the entire construction. However, the $2n$-bits output (Σ, Θ) of the internal hash computation is domain separated by setting their two most significant bits to 10 and 11, respectively. Moreover, the checksum of the message blocks after padded with the string 0^{n-s} is masked with the Σ value. Author has shown that LightMAC_Plus-1k achieves $2n/3$-bits security. Recently, Song [12] proposed another variant of the single-keyed LightMAC_Plus construction dubbed as 1k-LightMAC_Plus, in which a single block cipher is used throughout the construction and the $2n$-bit hash value is domain separated by setting their most significant bit to 0 and 1 respectively. It has been shown in [12] that 1k-LightMAC_Plus also achieves $2n/3$-bits security bound.

Therefore, to summarize, only the LightMAC_Plus construction has been claimed to achieve a tight $3n/4$-bit security bound [6], and all its existing reduced-keyed variants achieve only $2n/3$-bits security. Therefore, the motivation for this paper stems from asking the question

> *Can we prove a tight $3n/4$-bit security bound on any reduced-keyed variants of the LightMAC_Plus construction ?*

1.2 Our Contribution

In this paper, we answer the above question affirmatively and show that the construction achieves a tight security bound up to $2^{3n/4}$ queries (ignoring the maximum message length). In particular, we have shown an upper bound on the PRF advantage of 2k-LightMAC_Plus in roughly of the order of $2^{3n/4}$ queries, provided the maximum number of message blocks in a query is at most $\min\{2^{n-2}-1, 2^s\}$, and the total number of distinct message blocks across all queries is at most 2^n, where n denotes the block size of the block cipher and s denotes the size of the block counter. Moreover, we have also shown a matching PRF attack on the construction with query complexity in roughly of the order of $2^{3n/4}$ queries. The schematic diagram of 2k-LightMAC_Plus is shown in Fig. 1. However, to prove the security bound of the construction, we deeply rely on the result of the mirror theory, where we lower the bound on the number of solutions of a given system of equations. The following result establishes an upper bound on the PRF advantage of 2k-LightMAC_Plus against all information-theoretic adversaries.

Theorem 1. *Let $\mathcal{K}$ be a finite and non-empty set. Let $\mathsf{E} : \mathcal{K} \times \{0,1\}^n \to \{0,1\}^n$ be a block cipher. Then, the PRF advantage for any (q, ℓ, σ, t) adversary against 2k-LightMAC_Plus[E] is given by,*

$$\mathbf{Adv}^{\mathrm{PRF}}_{\text{2k-LightMAC\_Plus[E]}}(q, \ell, \sigma, t) \leq 2\mathbf{Adv}^{\mathrm{PRP}}_{\mathsf{E}}(\sigma + 2q, t') + \frac{96q^4}{2^{3n}} + \frac{8\sqrt{2}q^2}{2^{3n/2}} + \frac{7q^{4/3}}{2^n}$$

$$+ \frac{39q^{8/3}}{2^{2n}} + \frac{244q^2}{2^{2n}} + \frac{32q^3}{2^{3n}} + \frac{6\sigma}{2^n} + \frac{q}{2^n} + \frac{8}{2^n},$$

where $\ell \leq \min\{2^{n-2}-1, 2^s\}$, is the maximum number of message blocks in a query, $\sigma \leq 2^n$, is the total number of distinct message blocks queried, and $t' = O((\sigma + 2q)t)$.

2 Preliminaries

GENERAL NOTATIONS: For $q \in \mathbb{N}$, we write $[q]$ to denote the set $\{1, \ldots, q\}$. For a natural number n, $\{0,1\}^n$ denotes the set of all binary strings of length

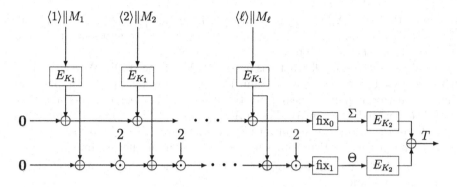

Fig. 1. Pictorial description of the 2k-LightMAC_Plus [4].

n and $\{0,1\}^*$ denotes the set of all binary strings of arbitrary length. For a natural number n, we call the elements of $\{0,1\}^n$ *blocks*. For any binary string $x \in \{0,1\}^*$, $|x|$ denotes the length i.e., the number of bits in x. For $x, y \in \{0,1\}^n$, we write $z = x \oplus y$ to denote the bitwise xor of x and y. For two binary strings $x, y \in \{0,1\}^*$, we write $x\|y$ to denote the concatenation of x followed by y. For a natural number n and $x \in \{0,1\}^*$, we write $(x_1, x_2, \ldots, x_{l-1}, x_l) \xleftarrow{n} x$ to denote the n-bit parsing of x, where $|x_i| = n$ for all $i \in [l-1]$ and $0 < |x_l| \le n-1$. For any $n \in \mathbb{N}$, we define an injective function pad_n that takes an arbitrary string $x \in \{0,1\}^*$ and returns $y \in (\{0,1\}^n)^*$, defined as follows:

$$\mathsf{pad}_n(x) \triangleq x\|10^d,$$

where d is the smallest integer such that $|\mathsf{pad}_n(x)|$ is a multiple of n. For two positive integers i, s such that $i < 2^s$, we write $\langle i \rangle_s$ to denote the s-bit representation of integer i. For $b \in \{0,1\}$, we consider the function fix_b that takes an n-bit binary string x and returns x except its least significant bit is changed to bit b. For $b \in \{10, 11\}$, we consider the function fix_b that takes an n-bit binary string x and returns x except its two most significant bits are changed to b. For a pair of positive integers $(i, j), (i', j') \in \mathbb{Z}^+ \times \mathbb{Z}^+$, we write $(i, j) \preceq (i', j')$ to denote that either $i < i'$ or $(i = i'$ and $j < j')$.

We write a q-tuple $\widetilde{x} = (x_1, \ldots, x_q)$ as $(x_i)_{i \in [q]}$. When all the elements of a tuple $\widetilde{x} = (x_1, \ldots, x_q)$ are distinct, then by abusing of notation, we often write $\widetilde{x}$ as the set $\widetilde{x} = \{x_i : i \in [q]\}$. We write $\mathcal{X}^{(q)}$ to denote the set of all q tuples whose all elements are distinct, i.e.,

$$\mathcal{X}^{(q)} = \{(x_1, \ldots, x_q) : x_i \ne x_j, \forall i \ne j\}.$$

We write $x \leftarrow y$ to denote the assignment of the variable y into x. For a set $\mathcal{X}$, $\mathsf{X} \leftarrow_\$ \{0,1\}^n$ denotes that X is sampled uniformly at random from $\{0,1\}^n$ and independent to all random variables defined so far. For a tuple of random variables $(X_1, \ldots, X_q)$, we write $(X_1, \ldots, X_q) \leftarrow_\$ \{0,1\}^n$ to denote that each X_i is sampled uniformly from $\{0,1\}^n$ and independent to all other previously sampled random variables. Similarly, we write $(X_1, \ldots, X_q) \xleftarrow{\mathsf{wor}} \{0,1\}^n$

to denote that each X_i is sampled uniformly from $\{0,1\}^n \setminus \{X_1, \ldots, X_{i-1}\}$, i.e., $X_i \xleftarrow{\$} \{0,1\}^n \setminus \{X_1, \ldots, X_{i-1}\}$. For integers $1 \leq b \leq a$, we write $(a)_b$ to denote $a(a-1)\ldots(a-b+1)$, where $(a)_0 = 1$ by convention.

We denote the set of all functions from $\mathcal{X}$ to $\mathcal{Y}$ as $\mathsf{Func}(\mathcal{X}, \mathcal{Y})$. We write $\mathsf{Func}_{\mathcal{X}}$ when $\mathcal{Y} = \{0,1\}^n$. Sometimes, we omit the set $\mathcal{X}$ from $\mathsf{Func}_{\mathcal{X}}$ and simply write Func when the domain is clear from the context. The set of all permutations over $\mathcal{X}$ is denoted as $\mathsf{Perm}(\mathcal{X})$. When $\mathcal{X} = \{0,1\}^n$, then we omit $\mathcal{X}$ and simply write Perm to denote the set of all permutations over $\{0,1\}^n$. We say that an n-bit permutation $\mathsf{P} \in \mathsf{Perm}$ maps a q-tuple $\widetilde{x}$ to an another q-tuple $\widetilde{y}$, denoted as $\widetilde{x} \overset{\mathsf{P}}{\mapsto} \widetilde{y}$, where each element of the $\widetilde{x}$ tuple and the $\widetilde{y}$ tuple is an n-bit string, if the following holds:

$$\forall i \in [q], \quad \mathsf{P}(x_i) = y_i.$$

We say that a q-tuple $\widetilde{x}$ is permutation compatible with an another q-tuple $\widetilde{y}$, where each element of both the tuples is an n-bit string, if there exists at least one permutation $\mathsf{P} \in \mathsf{Perm}$ such that $\widetilde{x} \overset{\mathsf{P}}{\mapsto} \widetilde{y}$.

2.1 Psuedorandom Function and Pseudorandom Permutation

Let $\mathsf{F} : \{0,1\}^k \times \mathcal{X} \to \{0,1\}^n$ be a family of keyed functions from $\mathcal{X}$ to $\{0,1\}^n$. We define the pseudorandom function (prf) advantage of F with respect to a distinguisher $\mathscr{A}$ as follows:

$$\mathbf{Adv}_{\mathsf{F}}^{\mathrm{prf}}(\mathscr{A}) \triangleq \Delta_{\mathscr{A}}\,[\mathsf{F}_K; \mathsf{R}] = \left| \Pr[K \leftarrow \{0,1\}^k : \mathscr{A}^{\mathsf{F}_K} = 1] - \Pr[\mathsf{R} \leftarrow \mathsf{Func} : \mathscr{A}^{\mathsf{R}} = 1] \right|.$$

When $\mathcal{X} = \{0,1\}^n$ such that for every $K \in \{0,1\}^k$, the function $\mathsf{E}_K : \{0,1\}^n \to \{0,1\}^n$ is bijective, then we call F to be a family of pseudorandom permutation. We say that F is $(q, \ell, \sigma, \mathsf{t}, \epsilon)$ secure if the maximum pesudorandom function (permutation) advantage of F is ϵ where the maximum is taken over all distinguishers $\mathscr{A}$ that makes q queries to its oracle such that the total number of message blocks queried across all q queries is σ, ℓ being the maximum number of message blocks among all q queries, and the adversary runs for time at most t, i.e.,

$$\mathbf{Adv}_{\mathsf{F}}^{W}(q, \ell, \sigma, \mathsf{t}) \triangleq \max_{\mathscr{A} \in \mathcal{C}} \mathbf{Adv}_{\mathsf{F}}^{W}(\mathscr{A}),$$

where W is either prf or prp, $\mathcal{C}$ is the class of all distinguishers $\mathscr{A}$ that makes at most q queries such that the total number of message blocks queried across all q queries is σ, and ℓ being the maximum number of message blocks among all q queries with run time at most t.

The following result from linear algebra will be very useful in establishing the security bound of our construction. Proof of this result can be found in Proposition 1 of [5].

Lemma 1. *Let* $(Z_1, \ldots, Z_q) \xleftarrow{\text{wor}} \mathcal{X} \subseteq \{0,1\}^n$ *with* $|\mathcal{X}| = N > q$. *Let A be a* $k \times q$ *binary matrix with rank r. We denote the column vector* $(Z_1, \ldots, Z_q)^{\text{tr}}$ *as* $\widetilde{Z}$. *Then, for any* $\widetilde{c} \in (\{0,1\}^n)^k$, *we have*

$$\Pr[A \cdot \widetilde{Z} = \widetilde{c}] \leq \frac{1}{(N - q + r)_r}.$$

2.2   Mirror Theory

Consider an undirected edge-labelled acylic graph $\mathsf{G} = (\mathcal{V}, \mathcal{E}, \mathcal{L})$ with edge labelling function $\mathcal{L} : \mathcal{E} \to \{0,1\}^n$, where $\mathcal{V} = \{P_1, \ldots, P_\alpha\}$ be the set of vertices of the graph. For an edge $\{P_i, P_j\} \in \mathcal{E}$, we write $\mathcal{L}(\{P_i, P_j\}) = \lambda_{ij}$. For a path $\mathcal{P}$ and a cycle $\mathcal{C}$ in the graph G, we define the label of the path and the label of the cycle as

$$\mathcal{L}(\mathcal{P}) \triangleq \sum_{e \in \mathcal{P}} \mathcal{L}(e), \ \mathcal{L}(\mathcal{C}) \triangleq \sum_{e \in \mathcal{C}} \mathcal{L}(e).$$

We say the graph G is **good** if the graph is acylic and for all paths $\mathcal{P}$ of arbitrary length in the graph G, one has $\mathcal{L}(\mathcal{P}) \neq \mathbf{0}$. For such a good graph G, we associate a system of bivariate affine equations as follows:

$$\mathcal{E}_{\mathsf{G}} = Y_i \oplus Z_j = \lambda_{ij} \ \forall \ \{Y_i, Z_j\} \in \mathcal{E}.$$

Note that, in the above system of bivariate affine equations, the variables are the vertices of the associated graph. We say that two variables are involved in an equation if the corresponding vertices are connected by an edge in the graph. The constants of the equations are the labels of the corresponding edges. Therefore, for the system of affine equations $\mathcal{E}_{\mathsf{G}}$, the variables are Y_i's and Z_i's. Now, we define an equivalence relation $\sim$ over $\mathcal{V}$ such that $u \sim v$ if and only if $(u, v) \in \mathcal{E}$. This equivalence relation induces a partition on $\mathcal{V}$ and each partition is called a component. The size of a component refers to the number of elements (i.e., the number of vertices) in the partition. The set of components in G is denoted by $\mathsf{comp}(\mathsf{G}) = (\mathsf{C}_1 \sqcup \ldots \sqcup \mathsf{C}_\alpha \sqcup \mathsf{D}_1 \sqcup \ldots \sqcup \mathsf{D}_\beta)$ where we assume that there are α many components of G (i.e., $\mathsf{C}_1, \ldots, \mathsf{C}_\alpha$) with component size greater than 2 and β many components of G (i.e., $\mathsf{D}_1, \ldots, \mathsf{D}_\beta$) having component size exactly 2. We write C to denote $\mathsf{C}_1 \sqcup \ldots \sqcup \mathsf{C}_\alpha$ and D to denote $\mathsf{D}_1 \sqcup \ldots \sqcup \mathsf{D}_\beta$. We write q_c to denote the total number of edges in C and q denotes the total number of edges in the graph G. Then, it is easy to see that $q = q_c + \beta$.

Notations: For the i-th component of C, i.e., C_i, which is acyclic and edge-labelled graph, let $\mathcal{V}_{\mathsf{C}_i}$ be the set of vertices of the component C_i and w_i denotes the cardinality of the set $\mathcal{V}_{\mathsf{C}_i}$. Let $\mathcal{V}_{\mathsf{C}}$ denotes the set of vertices of C. For $1 \leq i \leq \alpha$, we write $\sigma_i = w_1 + w_2 + \ldots + w_i$, with the convention that $\sigma_0 = 0$. Note that $q_c = \sigma_\alpha - \alpha$ as each component C_i is a tree. Let $h(\mathsf{G})$ denote the number of solutions to the graph G. Let $h_c(i)$ denote the number of solutions for the subgraph $\mathsf{C}_1 \sqcup \mathsf{C}_2 \sqcup \ldots \sqcup \mathsf{C}_i$ and $h_d(i)$ denote the number of solutions for the subgraph $\mathsf{C} \sqcup \mathsf{D}^i$ where $\mathsf{D}^i \triangleq \mathsf{D}_1 \sqcup \mathsf{D}_2 \sqcup \ldots \sqcup \mathsf{D}_i$. Therefore, $h_d(0) = h_c(\alpha)$ and $h_d(\beta) = h(\mathsf{G})$.

Definition 1. *Let $\mathcal{E}_G$ be a system of equations corresponding to a good acyclic edge-labeled graph G (as defined above). An injective function $\Phi : \mathcal{V} \to \{0,1\}^n$, is said to be an injective solution to $\mathcal{E}_G$ if $\Phi(P_i) \oplus \Phi(P_j) = \lambda_{ij}$ for all $\{P_i, P_j\} \in \mathcal{E}$ such that $\mathcal{L}(\{P_i, P_j\}) = \lambda_{ij}$.*

In [2], authors have proved that if G is a good acyclic edge-labeled graph such it is decomposed into finitely many components of size greater than 2 and exactly 2, then the number of injective solutions chosen from $\{0,1\}^n$, to $\mathcal{E}_G$, is very close to the average number of solutions until the number of edges in $\mathcal{E}$ is roughly $2^{3n/4}$. Formally, the result is as follows:

Theorem 2. *Let $G = (\mathcal{V}, \mathcal{E}, \mathcal{L})$ be a good acylic edge-labelled graph with $|\mathcal{E}| = q$ edges and s vertices such that G is decomposed into α many components $C_1 \sqcup \ldots \sqcup C_\alpha$ of size at least 3 and β many components $D_1 \sqcup \ldots \sqcup D_\beta$ of size exactly 2. For $1 \leq i \leq \alpha$, let w_i be the total number of vertices of $C_1 \sqcup \ldots \sqcup C_i$ and q_c be the total number of edges in $C_1 \sqcup \ldots \sqcup C_\alpha$. Let $\sigma_\alpha = w_1 + w_2 + \ldots + w_\alpha$ be the total number of vertices of $C_1 \sqcup C_2 \sqcup \ldots \sqcup C_\alpha$. Then the total number of injective solutions to $\mathcal{E}_G$ which are chosen from $\{0,1\}^n$ is at least:*

$$\frac{(2^n)_s}{2^{nq}} \left(1 - \frac{9q_c^2}{4 \cdot 2^n} - \frac{9q_c^2 q}{2^{2n}} - \frac{24q^2 q_c}{2^{2n}} - \frac{6qq_c}{2^{2n}} - \frac{40q^2}{2^{2n}} - \frac{16q^4}{2^{3n}} \right).$$

We refer the interested reader to [2] for proof of the result.

3   Proof of Theorem 1

As the first step of the proof, we replace the underlying block ciphers E_{K_1} and E_{K_2} of the construction with a pair of uniformly sampled n-bit random permutations P_1 and P_2 at the cost of the prp advantage of E and denote the resulting construction as 2k-LightMAC_Plus*[P_1, P_2], i.e.,

$$\mathbf{Adv}_{\text{2k-LightMAC\_Plus}[E]}^{\text{PRF}}(q, \sigma, t) \leq 2\mathbf{Adv}_E^{\text{PRP}}(\sigma, t') + \mathbf{Adv}_{\text{2k-LightMAC\_Plus}^*[P_1, P_2]}^{\text{PRF}}(q, \sigma).$$

We write 2k-LightMAC_Plus or 2k-LightMAC_Plus* instead of 2k-LightMAC_Plus[E] or 2k-LightMAC_Plus*[P_1, P_2] whenever the primitives are understood from the context. Now, our goal is to upper bound the information-theoretic PRF security of 2k-LightMAC_Plus*. For doing this, we bound the PRF security of 2k-LightMAC_Plus* in terms of the distinguishing advantage of an information-theoretic distinguisher D in distinguishing the output of 2k-LightMAC_Plus* from the output of an ideal world that consists of a random function RF which outputs a random n-bit tag T on every input $M \in \mathcal{M}$. We assume that the distinguisher D makes q queries to the oracle in either of the two worlds and at the end of the interaction, the oracle releases some additional information to D. If D interacts with the oracle in the real world, then it releases $\widetilde{\Sigma} = (\Sigma_1, \Sigma_2, \ldots, \Sigma_q)$ and $\widetilde{\Theta} = (\Theta_1, \Theta_2, \ldots, \Theta_q)$. However, if D interacts with the oracle in the ideal world, then the oracle also releases $\widetilde{\Sigma}, \widetilde{\Theta}$ tuple, where the tuple $\widetilde{\Sigma}$, and $\widetilde{\Theta}$ are computed in the ideal world as described in the following section.

3.1   Description of the Ideal World

The ideal oracle consists of two phases: (i) online phase in which for each queried message M^i, the oracle samples the response T_i uniformly at random from $\{0,1\}^n$ and returns it to the distinguisher D. If it happens that any of the sampled responses are all zero strings, then we set the bad flag Bad-Tag to 1 and abort the game, i.e.,

$$\text{Bad-Tag} \leftarrow 1 : \exists i \in [q] : T_i = 0^n.$$

When all the queries and responses are over, the offline phase of the ideal world begins. In this phase, we consider a function $\mathcal{L}_1$, which is initially undefined at every point of its domain. The oracle of the ideal world computes $X_j^i = \langle j \rangle_s \| M_j^i$ values for all $i \in [q], j \in [\ell_i]$ and samples Y_j^i as follows: (a) if X_j^i is fresh in $\widetilde{X}$, then Y_j^i is uniformly sampled from outside of the set $\text{Ran}(\mathcal{L}_1)$ followed by including it to the set $\text{Ran}(\mathcal{L}_1)$; (ii) on the other hand, if X_j^i collides with some previous $X_{j'}^{i'}$ value, where $(i',j') \preceq (i,j)$, then Y_j^i is set to the value $Y_{j'}^{i'}$. When all the Y_j^i, for $i \in [q], j \in [\ell_i]$ are determined, the oracle computes the tuple (Σ_i, Θ_i) for all $i \in [q]$ as

$$\Sigma_i = \text{fix}_0(Y_1^i \oplus Y_2^i \oplus \ldots \oplus Y_{\ell_i}^i), \Theta_i = \text{fix}_1(2^{\ell_i} Y_1^i \oplus 2^{\ell_i-1} Y_2^i \oplus \ldots \oplus 2Y_{\ell_i}^i).$$

After the computation of the tuple $(\widetilde{\Sigma}, \widetilde{\Theta})$ is over, we set the bad flag Bad1 to 1, if there exists two pairs (Σ_i, Θ_i) and (Σ_j, Θ_j) such that $(\Sigma_i, \Theta_i) = (\Sigma_j, \Theta_j)$ holds, i.e.,

$$\text{Bad1} \leftarrow 1 : \exists i \neq j \in [q] : (\Sigma_i, \Theta_i) = (\Sigma_j, \Theta_j).$$

Moreover, we set the bad flag Bad2 to 1, if there exists two pairs (Σ_i, T_i) and (Σ_j, T_j) such that $(\Sigma_i, T_i) = (\Sigma_j, T_j)$ holds, i.e.,

$$\text{Bad2} \leftarrow 1 : \exists i \neq j \in [q] : (\Sigma_i, T_i) = (\Sigma_j, T_j).$$

Similarly, we set the bad flag Bad3 to 1, if there exists two pairs (Θ_i, T_i) and (Θ_j, T_j) such that $(\Theta_i, T_i) = (\Theta_j, T_j)$ holds, i.e.,

$$\text{Bad3} \leftarrow 1 : \exists i \neq j \in [q] : (\Theta_i, T_i) = (\Theta_j, T_j).$$

We set the bad flag Bad4 to 1 if there exists three distinct indices $i_1, i_2, i_3 \in [q]$ such that $\Sigma_{i_1} = \Sigma_{i_2}, \Theta_{i_2} = \Theta_{i_3}, T_{i_1} \oplus T_{i_2} \oplus T_{i_3} = 0^n$ holds, i.e.,

$$\text{Bad4} \leftarrow 1 : \exists i_1, i_2, i_3 \in [q] : \Sigma_{i_1} = \Sigma_{i_2}, \Theta_{i_2} = \Theta_{i_3}, T_{i_1} \oplus T_{i_2} \oplus T_{i_3} = 0^n.$$

We set the bad flag Bad5 to 1 if there exists four distinct indices $i_1, i_2, i_3, i_4 \in [q]$ such that $\Sigma_{i_1} = \Sigma_{i_2}, \Theta_{i_2} = \Theta_{i_3}, \Sigma_{i_3} = \Sigma_{i_4}$ holds, i.e.,

$$\text{Bad5} \leftarrow 1 : \exists i_1, i_2, i_3, i_4 \in [q] : \Sigma_{i_1} = \Sigma_{i_2}, \Theta_{i_2} = \Theta_{i_3}, \Sigma_{i_3} = \Sigma_{i_4}.$$

We set the bad flag Bad6 to 1 if there exists four distinct indices $i_1, i_2, i_3, i_4 \in [q]$ such that $\Theta_{i_1} = \Theta_{i_2}, \Sigma_{i_2} = \Sigma_{i_3}, \Theta_{i_3} = \Theta_{i_4}, T_{i_1} \oplus T_{i_2} \oplus T_{i_3} \oplus T_{i_4} = 0^n$ holds, i.e.,

$$\text{Bad6} \leftarrow 1 : \exists i_1, i_2, i_3, i_4 \in [q] : \Theta_{i_1} = \Theta_{i_2}, \Sigma_{i_2} = \Sigma_{i_3}, \Theta_{i_3} = \Theta_{i_4}, T_{i_1} \oplus T_{i_2} \oplus T_{i_3} \oplus T_{i_4} = 0^n.$$

Finally, we set the bad flag Bad7 to 1 if the number of colliding pairs for Σ or Θ values is at least $q^{2/3}$, i.e.,

$$\mathsf{Bad7} \leftarrow 1 : \begin{cases} |\{(i,j) : i \neq j \in [q], \Sigma_i = \Sigma_j\}| \geq q^{2/3} \text{ or} \\ |\{(i,j) : i \neq j \in [q], \Theta_i = \Theta_j\}| \geq q^{2/3}. \end{cases}$$

The offline phase of the ideal world is depicted in Fig. 2.

Therefore, we summarize the interaction of D with the oracle in the following attack transcript

$$\tau = \{(M_1, T_1, \Sigma_1, \Theta_1), (M_2, T_2, \Sigma_2, \Theta_2), \ldots, (M_q, T_q, \Sigma_q, \Theta_q)\}.$$

Let T_{re} denote the random variable that takes a transcript τ realized in the real world. Similarly, T_{id} denotes the random variable that takes a transcript τ realized in the ideal world. The probability of realizing a transcript τ in the ideal (resp. real) world is called the *ideal (resp. real) interpolation probability*. A transcript τ is said to be attainable with respect to D if its ideal interpolation probability is non-zero, and Θ denotes the set of all such attainable transcripts. Following these notations, we now state the main theorem of the H-Coefficient technique [11]:

Theorem 3 (H-Coefficident Technique). *Let* $\Theta = \mathsf{GoodT} \sqcup \mathsf{BadT}$ *be a partition of the set of attainable transcripts. Suppose there exists* $\epsilon_{\mathrm{ratio}} \geq 0$ *such that for any* $\tau \in \mathsf{GoodT}$,

$$\frac{\mathsf{p}_{\mathrm{re}}(\tau)}{\mathsf{p}_{\mathrm{id}}(\tau)} \triangleq \frac{\Pr[\mathsf{T}_{\mathrm{re}} = \tau]}{\Pr[\mathsf{T}_{\mathrm{id}} = \tau]} \geq 1 - \epsilon_{\mathrm{ratio}},$$

and there exists $\epsilon_{\mathrm{bad}} \geq 0$ *such that* $\Pr[\mathsf{T}_{\mathrm{id}} \in \mathsf{BadT}] \leq \epsilon_{\mathrm{bad}}$. *Then*

$$\mathbf{Adv}_{\text{2k-LightMAC\_Plus*}}^{\mathrm{PRF}}(\mathsf{D}) \leq \epsilon_{\mathrm{ratio}} + \epsilon_{\mathrm{bad}}. \tag{1}$$

Therefore, to prove the security of the construction using the H-Coefficient technique, we need to identify the set of bad transcripts and compute an upper bound for their probability in the ideal world. Then we need to lower bound the ratio of the real to ideal interpolation probability for a good transcript.

Remark 1. Note that, $\overline{\mathsf{Bad}_4}$ allows the good graph to have a path of length three (an N-type graph) such that the sum of the labels of the edges of the path is non-zero. On the other hand, Bad_5 is stronger than Bad_6. Note that, $\overline{\mathsf{Bad}_5}$ allows the graph to have path length at most three (an N-type graph), whereas $\overline{\mathsf{Bad}_6}$ allows the graph to have path length at most four (an W-type graph) such that the sum of the labels of the edges of the path is non-zero. The asymmetry between Bad_5 and Bad_6 arises because it is easy to bound Bad_6 with the condition $T_{i_1} \oplus T_{i_2} \oplus T_{i_3} \oplus T_{i_4} = 0^n$

OFFLINE PHASE OF $\mathcal{O}_{\text{ideal}}$, INITIALIZE $\mathcal{L}_1 = \emptyset$

1 : $\forall i \in [q]$: compute $(\Sigma_i, \Theta_i) \leftarrow$ ┌ $\textsf{Internal}^{\mathcal{L}_1}(M^i)$

1 : $\forall j \in [\ell_i]$: $X_j^i \leftarrow \langle j \rangle_s \| M_j^i$;

2 :   if $\mathcal{L}_1(X_j^i) = \top$, then

3 : $\mathcal{L}_1(X_j^i) \leftarrow Y_j^i \xleftarrow{\$} \overline{\textsf{Ran}(\mathcal{L}_1)}$;

4 :   else $Y_j^i \leftarrow \mathcal{L}_1(X_j^i)$;

5 : $\Sigma_i := \textsf{fix}_0(Y_1^i \oplus \cdots \oplus Y_{\ell_i}^i)$;

6 : $\Theta_i := \textsf{fix}_1(2^{\ell_i} Y_1^i \oplus \cdots \oplus 2^2 Y_{\ell_i-1}^i \oplus 2 Y_{\ell_i}^i)$;

return (Σ_i, Θ_i);

2 :   Let $\widetilde{\Sigma} = (\Sigma_1, \ldots, \Sigma_q), \widetilde{\Theta} = (\Theta_1, \ldots, \Theta_q)$;

3 :   if $\exists i \neq j \in [q] : (\Sigma_i, \Theta_i) = (\Sigma_j, \Theta_j)$, then $\boxed{\textsf{Bad1} \leftarrow 1}$, $\perp$;

4 :   if $\exists i \neq j \in [q] : (\Sigma_i, T_i) = (\Sigma_j, T_j)$, then $\boxed{\textsf{Bad2} \leftarrow 1}$, $\perp$;

5 :   if $\exists i \neq j \in [q] : (\Theta_i, T_i) = (\Theta_j, T_j)$, then $\boxed{\textsf{Bad3} \leftarrow 1}$, $\perp$;

6 :   if $\exists i_1, i_2, i_3 \in [q] : \Sigma_{i_1} = \Sigma_{i_2}, \Theta_{i_2} = \Theta_{i_3}, T_{i_1} \oplus T_{i_2} \oplus T_{i_3} = 0^n$, then $\boxed{\textsf{Bad4} \leftarrow 1}$, $\perp$;

7 :   if $\exists i_1, i_2, i_3, i_4 \in [q] : \Sigma_{i_1} = \Sigma_{i_2}, \Theta_{i_2} = \Theta_{i_3}, \Sigma_{i_3} = \Sigma_{i_4}$, then $\boxed{\textsf{Bad5} \leftarrow 1}$, $\perp$;

8 :   if $\exists i_1, i_2, i_3, i_4 \in [q] : \Theta_{i_1} = \Theta_{i_2}, \Sigma_{i_2} = \Sigma_{i_3}, \Theta_{i_3} = \Theta_{i_4}, T_{i_1} \oplus T_{i_2} \oplus T_{i_3} \oplus T_{i_4} = 0^n$,

9 :   then $\boxed{\textsf{Bad6} \leftarrow 1}$, $\perp$;

10 : $\mathcal{F}_\Sigma \leftarrow \{(i,j) \in [q]^2 : \exists i \neq j, \ \Sigma^i = \Sigma^j\}$; $\mathcal{F}_\Theta \leftarrow \{(i,j) \in [q]^2 : \exists i \neq j, \ \Theta^i = \Theta^j\}$;

11 :   if $|\mathcal{F}_\Sigma| \geq q^{2/3} \vee |\mathcal{F}_\Theta| \geq q^{2/3}$, then $\boxed{\textsf{Bad7} \leftarrow 1}$, $\perp$;

12 :   return $\left((\widetilde{X}_i, \widetilde{Y}_i)_{i \in [q]}, (\widetilde{\Sigma}, \widetilde{\Theta}) \right)$;

Fig. 2. Offline phase of the Ideal oracle $\mathcal{O}_{\text{ideal}}$: Boxed statements denote bad events. Whenever a bad event is set to 1, the oracle immediately aborts (denoted as $\perp$) and returns the remaining values of the transcript in any arbitrary manner. So, if we proceed further we can surely assume that the event $\perp$ (and so any bad event so far) does not hold. We write $\top$ when the value of a variable is not defined.

3.2 Definition and Probability of Bad Transcripts

In this section, we define and bound the probability of bad transcripts in the ideal world. We say that an attainable transcript τ is a **bad** transcript if anyone of the bad flags, defined in the offline phase of the ideal world as shown in Fig. 2, is set to 1. Recall that $\textsf{BadT} \subseteq \Theta$ be the set of all attainable bad transcripts and $\textsf{GoodT} = \Theta \setminus \textsf{BadT}$ be the set of all attainable good transcripts. We bound the probability of bad transcripts in the ideal world as follows. Before we proceed to bound the above events in the ideal world, we state the following two lemmas that upper bounds the collision probability between two Σ (or Θ) values for two distinct queries. We emphasize that the following result will be frequently used in upper bounding the probability of the above bad events.

Lemma 2. *For distinct two messages M_α and M_β, we have*

$$(i)\ \Pr[\Sigma_\alpha = \Sigma_\beta] \leq \frac{4}{2^n}, \quad (ii)\ \Pr[\Theta_\alpha = \Theta_\beta] \leq \frac{4}{2^n}.$$

Proof. We prove only (i) as the proof of (ii) is exactly similar to (i). Suppose the number of blocks of M_α and M_β be ℓ_α and ℓ_β respectively. Without loss of generality, we assume that $\ell_\alpha \leq \ell_\beta$. Now,

$$\Sigma_\alpha = \Sigma_\beta \Rightarrow \mathsf{msb}_{n-1}\left(\underbrace{\bigoplus_{i=1}^{\ell_\alpha} Y_\alpha[i] \oplus \bigoplus_{i=1}^{\ell_\beta} Y_\beta[i]}_{\mathfrak{F}}\right) = 0^{n-1}. \tag{2}$$

For computing the probability of the above event, we consider the following three cases.

1. $(\ell_\alpha = \ell_\beta) \wedge (\exists a \in [\ell_\alpha] : X_\alpha[a] \neq X_\beta[a]) \wedge (\forall i \in [\ell_\alpha] \setminus \{a\} : X_\alpha[i] = X_\beta[i])$
2. $(\ell_\alpha = \ell_\beta) \wedge (\exists a, b \in [\ell_\alpha] : X_\alpha[a] \neq X_\beta[a] \wedge X_\alpha[b] \neq X_\beta[b])$
3. $(\ell_\alpha \neq \ell_\beta)$.

Case 1: Since $X_\alpha[a] \neq X_\beta[a] \Rightarrow Y_\alpha[a] \neq Y_\beta[a]$ and $X_\alpha[i] = X_\beta[i] \Rightarrow Y_\alpha[i] = Y_\beta[i]$, for $i \in [\ell_\alpha] \setminus \{a\}$, $\mathfrak{F} \neq 0^n$. So, the probability of $\Sigma_\alpha = \Sigma_\beta$ is $1/2^{n-1}$.

Case 2: Suppose $\exists a_1, a_2, \ldots, a_j \in [\ell_\alpha]$, $j \geq 2$ such that, for all $i \in [j]$, $X_\alpha[a_i] \neq X_\beta[a_i]$. After eliminating all the same outputs between $\{Y_\alpha[i] : 1 \leq i \leq \ell_\alpha\}$ and $\{Y_\beta[i] : 1 \leq i \leq \ell_\beta\}$, we have

$$\mathfrak{F} = \bigoplus_{i=1}^{j} (Y_\alpha[a_i] \oplus Y_\beta[a_i]).$$

Since $\mathfrak{F}$ has at most $\ell_\alpha + \ell_\beta$ outputs, the probability of $\mathfrak{F} = 0^n$ is $1/(2^n - \ell_\alpha - \ell_\beta - 1)$.

Case 3: Without loss of generality, we assume that $\ell_\alpha < \ell_\beta$. Similarly from the previous case, after eliminating the same outputs between $\{Y_\alpha[i] : 1 \leq i \leq \ell_\alpha\}$ and $\{Y_\beta[i] : 1 \leq i \leq \ell_\beta\}$, we have

$$\mathfrak{F} = \bigoplus_{i=1}^{j} Y_\alpha[a_i] \oplus \bigoplus_{i=1}^{k} Y_\beta[a_i],$$

where $a_1, \ldots, a_j \in [\ell_\alpha]$ and $b_1, \ldots, b_k \in [\ell_\beta]$. Also, by the similar argument of case 2, we have the probability of $\mathfrak{F} = 0^n$ is at most $1/(2^n - \ell_\alpha - \ell_\beta - 1)$. Hence,

$$\Pr[\Sigma_\alpha = \Sigma_\beta] \leq \frac{2}{(2^n - \ell_\alpha - \ell_\beta - 1)}$$

$$\leq \frac{4}{2^n}, \text{ assuming } \ell_\alpha + \ell_\beta \leq 2^{n-1}.$$

$\square$

Now, we are ready to bound the probability of the above bad events and hence, we bound the probability of realizing a bad transcript in the ideal world as follows:

Lemma 3 (Bad Lemma). *Let us define the event* $\mathsf{BadT} := Bad\text{-}Tag \vee Bad1 \vee Bad2 \vee Bad3 \vee Bad4 \vee Bad5 \vee Bad6 \vee Bad7_a \vee Bad7_b$. *Let* τ' *be any attainable transcript and* X_{id} *be defined as above. Then*

$$\Pr[X_{id} \in \mathsf{BadT}] \leq \frac{204q^2}{2^{2n}} + \frac{80q^4}{2^{3n}} + \frac{8\sqrt{2}q^2}{2^{3n/2}} + \frac{8}{2^n} + \frac{q}{2^n} + \frac{6\sigma}{2^n} + \frac{4q^{4/3}}{2^n} + \frac{32q^3}{2^{3n}}.$$

Proof. We upper bound the probability of individual bad events in the ideal world and then by the virtue of the union bound, we sum up the bounds to obtain the overall bound on the probability of bad transcripts in the ideal world.

1. Bound for Bad-Tag: For a fixed $i \in [q]$, the probability that $T_i = 0^n$ is exactly 2^{-n}, which follows from the uniform sampling of the output for the i-th query in the ideal world. Therefore, by varying over all possible choices for i, we have

$$\Pr[\text{Bad-Tag}] = \Pr[\exists i \in [q] : T_i = 0^n] \leq \frac{q}{2^n}. \tag{3}$$

2. Bound for Bad1: For a fixed $i \neq j \in [q]$, $(\Sigma_i, \Theta_i) = (\Sigma_j, \Theta_j)$ implies the following two equations:

$$\mathcal{E} = \begin{cases} \underbrace{\mathsf{msb}_{n-1}\Big((Y_i[1] \oplus \ldots \oplus Y_i[\ell_i]) \oplus (Y_j[1] \oplus \ldots \oplus Y_j[\ell_j]) \Big)}_{S_1} = 0^{n-1} \\ \underbrace{\mathsf{msb}_{n-1}\Big((2^{\ell_i}Y_i[1] \oplus \ldots \oplus 2Y_i[\ell_i]) \oplus (2^{\ell_j}Y_j[1] \oplus \ldots \oplus 2Y_j[\ell_j]) \Big)}_{S_2} = 0^{n-1}, \end{cases}$$

where ℓ_i and ℓ_j denotes the number of blocks of message M_i and M_j. We bound the probability of the above equation holds in the three disjoint cases as follows:

1. $(\ell_i = \ell_j) \wedge (\exists a \in [\ell_i] : X_i[a] \neq X_j[a]) \wedge (\forall \alpha \in [\ell_i] \setminus \{a\} : X_i[\alpha] = X_j[\alpha])$
2. $(\ell_i = \ell_j) \wedge (\exists a, b \in [\ell_i] : X_i[a] \neq X_j[a] \wedge X_i[b] \neq X_j[b])$
3. $(\ell_i \neq \ell_j)$.

Case 1: Since $X_i[a] \neq X_j[a] \Rightarrow Y_i[a] \neq Y_j[a]$ and $X_i[\alpha] = X_j[\alpha] \Rightarrow Y_i[\alpha] = Y_j[\alpha]$, for $\alpha \in [\ell_i] \setminus \{a\}$, $\bigoplus_{t=1}^{\ell_i} Y_i[t] \oplus \bigoplus_{t=1}^{\ell_j} Y_j[t] \neq 0^{n-1}$. So, the probability of $S_1 = 0^{n-1}$ is $1/2^n$ and also the probability of $S_2 = 0^{n-1}$ is $1/2^{n-1}$. Thus, the probability that satisfies equation $\mathcal{E}$ is $1/2^{2n-2}$.

Case 2: Suppose $\exists a_1, a_2, \ldots, a_p \in [\ell_i]$, $p \geq 2$ such that, for all $t \in [p]$, $X_i[a_t] \neq X_j[a_t]$. After eliminating all the same outputs between $\{Y_i[\alpha] : 1 \leq \alpha \leq \ell_i\}$ and $\{Y_j[\alpha] : 1 \leq \alpha \leq \ell_j\}$, we have

$$S_1 = \mathsf{msb}_{n-1}\Big(\bigoplus_{t=1}^{p} (Y_i[a_t] \oplus Y_j[a_t]) \Big), \quad S_2 = \mathsf{msb}_{n-1}\Big(\bigoplus_{t=1}^{p} 2^{\ell_i - a_t + 1} (Y_i[a_t] \oplus Y_j[a_t]) \Big).$$

$$\tag{4}$$

Note that, there are at most $\ell_i + \ell_j$ outputs in S_1 and S_2. Therefore, the numbers of possibilities for $Y_i[a_1]$ and $Y_i[a_2]$ are at least $2^n - (\ell_i + \ell_j - 2)$ and $2^n - (\ell_i + \ell_j - 1)$ respectively. Therefore, by fixing the values to the other output variables of equations in $\mathcal{E}$, the equations in $\mathcal{E}$ provide a unique solution for $Y_i[a_1]$ and $Y_i[a_2]$. As a result, the probability that equation $\mathcal{E}$ is satisfied is at most $4/(2^n - (\ell_i + \ell_j - 2))(2^n - (\ell_i + \ell_j - 1))$.

Case 3: Without loss of generality, we assume that $\ell_i < \ell_j$. Similar to the previous case, after eliminating the same outputs between $\{Y_i[\alpha] : 1 \le \alpha \le \ell_i\}$ and $\{Y_j[\alpha] : 1 \le \alpha \le \ell_j\}$, we have

$$
\begin{aligned}
S_1 &= \mathsf{msb}_{n-1}\left(\bigoplus_{t=1}^{p_1} Y_i[a_t] \oplus \bigoplus_{t=1}^{p_2} Y_j[a_t] \right), \\
S_2 &= \mathsf{msb}_{n-1}\left(\bigoplus_{t=1}^{p_1} 2^{\ell_i - a_t + 1} Y_i[a_t] \oplus \bigoplus_{t=1}^{p_2} 2^{\ell_j - a_t + 1} Y_j[a_t] \right),
\end{aligned}
\tag{5}
$$

where $a_1, \ldots, a_{p_1} \in [\ell_i]$ and $b_1, \ldots, b_{p_2} \in [\ell_j]$. By $\ell_i < \ell_j$, we have $\ell_j \in \{b_1, \ldots, b_{p_2}\}$ and $\ell_j \ne 1$. Since, there are at most $\ell_i + \ell_j$ outputs in S_1 and in S_2, the number of possibilities for $Y_j[b_1]$ and $Y_j[\ell_j]$ is at least $(2^n - (\ell_i + \ell_j - 2))(2^n - (\ell_i + \ell_j - 1))$. By fixing the values to the other output variables of equations in $\mathcal{E}$, the equations in $\mathcal{E}$ provide a unique solution for $Y_j[b_1]$ and $Y_j[\ell_j]$. As a result, the probability that equation $\mathcal{E}$ is satisfied is at most $4/(2^n - (\ell_i + \ell_j - 2))(2^n - (\ell_i + \ell_j - 1))$.

Therefore, we see that for each of the above case, equations in $\mathcal{E}$ holds with probability at most $4/(2^n - (\ell_i + \ell_j - 2))(2^n - (\ell_i + \ell_j - 1))$. Therefore, we have

$$
\Pr[\mathsf{Bad1}] \le \frac{4\binom{q}{2}}{(2^n - (\ell_i + \ell_j - 2))(2^n - (\ell_i + \ell_j - 1))} \le \frac{8q^2}{2^{2n}},
\tag{6}
$$

where the second last inequality follows due to the fact that $\ell_i + \ell_j - 1 \le 2^{n-1}$.

3. Bound for Bad2: To bound the probability of the event Bad2, for a fixed choice of indices $i \ne j \in [q]$,

$$
\Pr[\Sigma_i = \Sigma_j, T_i = T_j] \overset{(1)}{=} \Pr[\Sigma_i = \Sigma_j] \cdot \Pr[T_i = T_j] \overset{(2)}{=} \frac{4}{2^n} \times \frac{1}{2^n} = \frac{4}{2^{2n}},
$$

where (1) follows due to the fact that the distribution of T_i is independent over the distribution of Σ_i in the ideal world and (2) follows from Lemma 2 and from the event that $T_i = T_j$ holds with probability 2^{-n}. Therefore, by varying over all possible choices of indices, we have

$$
\Pr[\mathsf{Bad2}] = \Pr[\exists i \ne j \in [q] : (\Sigma_i, T_i) = (\Sigma_j, T_j)] \le \frac{2q^2}{2^{2n}}
\tag{7}
$$

4. Bound for Bad3: We bound the probability of the event Bad3 in a similar way as we have bounded the probability of the event Bad2. Using the exact argument as used in bounding the probability of the event Bad2, we similarly bound the probability of the event Bad3 and hence, we have

$$\Pr[\mathsf{Bad3}] \leq \frac{2q^2}{2^{2n}}. \tag{8}$$

5. Bound for Bad4: To obtain the bound for Bad4, we first define an auxiliary bad event

$$\mathsf{Aux\text{-}Bad} := Y_i[j] \in \{0^n, 0^{n-1}1\}.$$

It is easy to see that $\Pr[\mathsf{Aux\text{-}Bad}] \leq \frac{2\sigma}{2^n}$, if σ is the total number of blocks over all the q queries. Now we will obtain the bound for Bad4 assuming that the auxiliary bad doesn't occur. Suppose ℓ be the maximum number of message blocks among all the q queries. After fixing a triplet (i_1, i_2, i_3), $\Sigma_{i_1} = \Sigma_{i_2}, \Theta_{i_2} = \Theta_{i_3}$ can be represented by a system of three linear equations as follows:

$$\Sigma'_{i_1} = \Sigma'_{i_2} \oplus 0^{n-1}b_1 \Leftrightarrow \overset{t}{\underset{j=1}{\bigoplus}} A_{1,j} \cdot Y[j] = 0^{n-1}b_1,$$

$$\Theta'_{i_2} = \Theta'_{i_3} \oplus 0^{n-1}b_2 \Leftrightarrow \overset{t}{\underset{j=1}{\bigoplus}} A_{2,j} \cdot Y[j] = 0^{n-1}b_2, \tag{9}$$

for some $A_{\alpha,\beta}$, b_i, where $i \in [2]$ and $t \leq 3\ell$. The i-th row of the augmented matrix $(A|B)$ is denoted as $(A|B)_i$ and we denote the i-th row of the coefficient matrix A as A_i for $i = 1, 2$. Now, we assume that $b_1 = b_2 = 0$. If Aux-Bad doesn't occur then (i) A_1 contains at least three 1's, and (ii) A_2 contains at least two distinct entries and at most two 2^α for each α. Thus, A_2 is not a multiple of A_1, and hence rank of A is at least 2. For other choices of b_1, b_2 we can also show that the rank of A is at least 2. Thus, for a fixed choice of indices $i_1, i_2, i_3 \in [q]$ as follows:

$$\Pr[\Sigma_{i_1} = \Sigma_{i_2}, \Theta_{i_2} = \Theta_{i_3}, T_{i_1} \oplus T_{i_2} \oplus T_{i_3} = 0^n \wedge \overline{\mathsf{Aux\text{-}Bad}}]$$
$$= \Pr[\Sigma_{i_1} = \Sigma_{i_2}, \Theta_{i_2} = \Theta_{i_3} \wedge \overline{\mathsf{Aux\text{-}Bad}}] \cdot \Pr[T_{i_1} \oplus T_{i_2} \oplus T_{i_3} = 0^n]$$
$$= \frac{4}{(2^n - 3\ell)(2^n - 3\ell - 1)} \times \frac{1}{2^n - 2}$$
$$\leq \frac{32}{2^{3n}},$$

assuming $\ell \leq 2^{n-2} - 1$. Here we have used the facts that the distribution of $T_{i_1}, T_{i_2}, T_{i_3}$ are chosen uniformly at random and they are independent over the distribution of Σ_i in the ideal world. Therefore, by varying over all possible choices of indices, we have

$$\Pr[\mathsf{Bad4}] \leq \Pr[\mathsf{Aux\text{-}Bad}] + \Pr[\mathsf{Bad4} \wedge \overline{\mathsf{Aux\text{-}Bad}}] \leq \frac{2\sigma}{2^n} + \frac{32q^3}{2^{3n}}. \tag{10}$$

6. Bound for Bad5: To obtain the bound for Bad5, we first define an auxiliary bad event

$$\mathsf{Aux\text{-}Bad} := Y_i[j] \in \{0^n, 0^{n-1}1\}.$$

It is easy to see that $\Pr[\text{Aux-Bad}] \leq \frac{2\sigma}{2^n}$, if σ is the total number of blocks over all the q queries. Now we will obtain the bound for Bad5 conditioned on the auxiliary bad doesn't happen:

Lemma 4. *Let* Bad5 *and* Aux − Bad *be as defined above. Then,*

$$\Pr[\text{Bad5} \mid \overline{\text{Aux − Bad}}] \leq \frac{64q^4}{2^{3n}} + \frac{8\sqrt{2}q^2}{2^{3n/2}} + \frac{8}{2^n} + \frac{192q^2}{2^{2n}}.$$

For the proof of Lemma 4, see the full version of the paper [3]. We bound the bad event by applying Lemma 4 as follows:

$$\Pr[\text{Bad5}] \leq \Pr[\text{Bad5} \mid \overline{\text{Aux-Bad}}] + \Pr[\text{Aux-Bad}] \leq \frac{64q^4}{2^{3n}} + \frac{8\sqrt{2}q^2}{2^{3n/2}} + \frac{2\sigma + 8}{2^n} + \frac{192q^2}{2^{2n}}. \tag{11}$$

7. Bound for Bad6: To obtain the bound for Bad6, we first define an auxiliary bad event

$$\text{Aux-Bad} := Y_i[j] \in \{0^n, 0^{n-1}1\}.$$

It is easy to see that $\Pr[\text{Aux-Bad}] \leq \frac{2\sigma}{2^n}$, if σ is the total number of blocks over all the q queries. Now we will obtain the bound for Bad6 assuming that the auxiliary bad doesn't occur. Suppose ℓ be the maximum number of message blocks among all the q queries. After fixing a quadruple (i_1, i_2, i_3, i_4), $\Theta_{i_1} = \Theta_{i_2}, \Sigma_{i_2} = \Sigma_{i_3}, \Theta_{i_3} = \Theta_{i_4}$ can be represented by a system of three linear equations as follows:

$$\Theta'_{i_1} = \Theta'_{i_2} \oplus 0^{n-1}b_1 \Leftrightarrow \bigoplus_{j=1}^{t} A_{1,j} \cdot Y[j] = 0^{n-1}b_1,$$

$$\Sigma'_{i_2} = \Sigma'_{i_3} \oplus 0^{n-1}b_2 \Leftrightarrow \bigoplus_{j=1}^{t} A_{2,j} \cdot Y[j] = 0^{n-1}b_2, \tag{12}$$

$$\Theta'_{i_3} = \Theta'_{i_4} \oplus 0^{n-1}b_3 \Leftrightarrow \bigoplus_{j=1}^{t} A_{3,j} \cdot Y[j] = 0^{n-1}b_3,$$

for some $A_{\alpha,\beta}, b_\alpha$, where $\alpha \in [3]$ and $t \leq 4\ell$. The i-th row of the augmented matrix $(A|B)$ is denoted as $(A|B)_i$ and we denote the i-th row of the coefficient matrix A as A_i for $i = 1, 2$. Now we claim that if Bad-Aux doesn't occur, then rank of A is at least 2, for any choice of (b_1, b_2). Thus, for a fixed choice of indices $i_1, i_2, i_3, i_4 \in [q]$ as follows:

$$\Pr[\Theta_{i_1} = \Theta_{i_2}, \Sigma_{i_2} = \Sigma_{i_3}, \Theta_{i_3} = \Theta_{i_4}, T_{i_1} \oplus T_{i_2} \oplus T_{i_3} \oplus T_{i_4} = 0^n \wedge \overline{\text{Aux-Bad}}]$$
$$= \Pr[\Theta_{i_1} = \Theta_{i_2}, \Sigma_{i_2} = \Sigma_{i_3}, \Theta_{i_3} = \Theta_{i_4} \wedge \overline{\text{Aux-Bad}}] \cdot \Pr[T_{i_1} \oplus T_{i_2} \oplus T_{i_3} \oplus T_{i_4} = 0^n]$$
$$= \frac{4}{(2^n - 4\ell)(2^n - 4\ell - 1)} \times \frac{1}{2^n - 3} = \frac{16}{2^{3n}},$$

assuming $\ell \leq 2^{n-2} - 1$. Note that, we have used the facts that the distribution of $T_{i_1}, T_{i_2}, T_{i_3}, T_{i_4}$ are chosen uniformly at random and they are independent over the distribution of Y_i values in the ideal world. Therefore, by varying over all possible choices of indices, we have

$$\Pr[\mathsf{Bad6}] \leq \Pr[\mathsf{Aux\text{-}Bad}] + \Pr[\mathsf{Bad6} \wedge \overline{\mathsf{Aux\text{-}Bad}}] \leq \frac{2\sigma}{2^n} + \frac{16q^4}{2^{3n}}. \tag{13}$$

8. Bound for $\mathsf{Bad7}_a$ and $\mathsf{Bad7}_b$: We bound only the probability of the event $\mathsf{Bad7}_a$ as the analysis of bounding the probability of the event $\mathsf{Bad7}_b$ is exactly similar to that of bounding the probability of the event $\mathsf{Bad7}_a$. To bound the probability of the event $\mathsf{Bad7}_a$, we define an indicator random variable. For each $i \neq j \in [q]$, we define $\mathbb{X}_{i,j}$ which is defined as follows:

$$\mathbb{X}_{i,j} = \begin{cases} 1, & \text{if } \Sigma_i = \Sigma_j \\ 0, & \text{otherwise} \end{cases} \qquad \mathbb{X} := \sum_{i,j} \mathbb{X}_{i,j}$$

Due to Lemma 2, $\Pr[\mathbb{X}_{i,j} = 1] = \Pr[\Sigma_i = \Sigma_j] = 4/2^n$ and therefore, we have

$$\Pr[\mathsf{Bad7}_a] = \Pr[\mathbb{X} > q^{2/3}] \leq \frac{\mathbf{E}[\mathbb{X}]}{q^{2/3}} \leq \frac{4\binom{q}{2}}{2^n \cdot q^{2/3}} \leq \frac{2q^{4/3}}{2^n}. \tag{14}$$

Using the exact argument as used in bounding the probability of the event $\mathsf{Bad7}_a$, we similarly bound the probability of the event $\mathsf{Bad7}_b$ and hence, we have $\Pr[\mathsf{Bad7}_b] \leq \frac{2q^{4/3}}{2^n}$. Finally, the result follows as sum the probabilities of all these bad events. $\qquad \square$

3.3 Analysis of Good Transcript

In this section, we lower bound the ratio of the probability of realizing a good transcript τ in the real and the ideal world. Let τ be a good transcript, where

$$\tau = \{(M_1, T_1, \widetilde{X}_1, \widetilde{Y}_1, \Sigma_1, \Theta_1), (M_2, T_2, \widetilde{X}_2, \widetilde{Y}_2, \Sigma_2, \Theta_2), \dots, (M_q, T_q, \widetilde{X}_q, \widetilde{Y}_q, \Sigma_q, \Theta_q)\}.$$

In order to compute the real or ideal interpolation probability, let σ denote the distinct number of message blocks among all q queries. As a result of that, the ideal interpolation probability becomes $2^{-nq}/(2^n)_\sigma$.

Now, to compute the real interpolation probability, we first note that the permutation P_1 is invoked on a total of σ distinct input-output pairs and P_2 is invoked on at most $2q$ input-output pairs. Therefore, we have

$$\Pr[\mathsf{T}_{\mathrm{re}} = \tau] = \Pr[\mathsf{P}_1(X_j^i) = Y_j^i, \forall i \in [q], j \in [\ell_i], \mathsf{P}_2(\Sigma_i) \oplus \mathsf{P}_2(\Theta_i) = T_i, \forall i \in [q]]$$

$$= \Pr[\mathsf{P}_1(X_j^i) = Y_j^i, \forall i \in [q], j \in [\ell_i]] \cdot \underbrace{\Pr[\mathsf{P}_2(\Sigma_i) \oplus \mathsf{P}_2(\Theta_i) = T_i, \forall i \in [q]]}_{\mathsf{E}}$$

$$= \frac{1}{(2^n)_\sigma} \cdot \Pr[\mathsf{E}] \tag{15}$$

Therefore, it now boils down to compute a lower bound on the probability of the event E. To do this, we first consider that τ is a good transcript. As a result of it, none of the bad flags defined in the offline phase of the ideal world have been set to 1. Now, we consider the tuple $\widetilde{\Sigma} = (\Sigma_1, \Sigma_2, \ldots, \Sigma_q)$, $\widetilde{\Theta} = (\Theta_1, \Theta_2, \ldots, \Theta_q)$ corresponding to the good transcript τ. From the two tuples $\widetilde{\Sigma}$ and $\widetilde{\Theta}$, we construct an edge labeled graph G as follows: for each $i \in [q]$, Σ_i and Θ_i represents the vertices of the graph and for each $i \in [q]$, we put an edge between the vertices Σ_i and Θ_i with the label of the edge being T_i. Moreover, for any $i \neq j$, if $\Sigma_i = \Sigma_j$, then we merge the corresponding two vertices into one. Similarly, for any $i \neq j$, if $\Theta_i = \Theta_j$, then we merge the corresponding two vertices into one. This will end up with an edge-labeled graph having the following properties:

1. The graph does not have any cycle of length 2, otherwise the bad event Bad1 would have been hold true.
2. The label of an edge of any path is non-zero, otherwise bad event Bad-Tag would have been hold true.
3. For a path of length two in the graph, the xor of the label of the edges of the path is non-zero, otherwise, the bad event Bad2 or the bad event Bad3 would have been hold true.
4. The graph does not have any odd length cycle.
5. The graph contains path of length three, which we call N path, such that the xor of the label of the edges of the path is non-zero, otherwise bad event Bad4 would have been hold true.
6. The graph does not have any M-path, otherwise bad event Bad5 would have been hold true. A pictorial description of the M path is shown in (b) of Fig. 3
7. The graph contains a W path such that the xor of the label of the edges of the path is non-zero, otherwise bad event Bad6 would have been hold true. A pictorial description of the W path is shown in (a) of Fig. 3
8. The last three properties ensure that the graph does not have any cycle of length 4 or above and it does not have any path of length more than 4. Hence, the graph G becomes acyclic. Therefore, G is a collection of some disjoint components.
9. Finally, due to $\overline{\mathsf{Bad7}_a}$ and $\overline{\mathsf{Bad7}_b}$, each component is of size at most $q^{2/3}$.

Therefore, computing a lower bound on the probability of the event E is equivalent to computing a lower bound on the number of injective solutions which are chosen from $\{0,1\}^n$ to $\mathcal{E}_\mathsf{G}$. Therefore, by applying Theorem 2, we have

$$\Pr[\mathsf{E}] \geq \frac{1}{2^{nq}} \left(1 - \epsilon_{\mathrm{ratio}}\right). \tag{16}$$

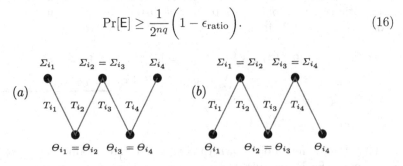

Fig. 3. (a) represents a W-path and (b) represents a M-path.

Therefore, from Eq. (15) and Eq. (16), we have

$$\Pr[\mathsf{T_{re}} = \tau] \geq \frac{1}{(2^n)_\sigma} \cdot \frac{1}{2^{nq}} \cdot \left(1 - \epsilon_{\text{ratio}}\right) \tag{17}$$

where ϵ_{ratio} is defined as follows:

$$\epsilon_{\text{ratio}} \triangleq \frac{9q_c^2}{4 \cdot 2^n} + \frac{9q_c^2 q}{2^{2n}} + \frac{24q^2 q_c}{2^{2n}} + \frac{6qq_c}{2^{2n}} + \frac{40q^2}{2^{2n}} + \frac{16q^4}{2^{3n}}. \tag{18}$$

where q_c denotes the total number of edges in the components having size greater than two. Since $q_c \leq q^{2/3} \leq q$, we have

$$\epsilon_{\text{ratio}} \leq \frac{9q^{4/3}}{4 \cdot 2^n} + \frac{9q^{7/3}}{2^{2n}} + \frac{24q^{8/3}}{2^{2n}} + \frac{6q^{5/3}}{2^{2n}} + \frac{40q^2}{2^{2n}} + \frac{16q^4}{2^{3n}} \tag{19}$$

Finally, the result follows by taking the ratio of real to ideal interpolation probability, and by combining Lemma 3 and Eq. (19). □

4   Matching Attack on 2k-LightMAC_Plus

In this section, we show an information-theoretic distinguishing attack on the construction 2k-LightMAC_Plus based on random permutations $\mathsf{P}_1, \mathsf{P}_2$ with $2^{3n/4}$ query complexity which establishes the proven information-theoretic security bound of 2k-LightMAC_Plus is tight. The distinguishing attack essentially follows a similar technique as described in [7]. Broadly speaking, we consider a computationally unbounded adversary $\mathcal{A}$ that makes a sufficient number of queries to the construction so that it satisfies a given relation $\mathcal{R}$. Once $\mathcal{A}$ gets a quadruple that satisfies the relation $\mathcal{R}$, it tries to distinguish. Note that it has been assumed that $s \leq n/4$ for the attack. Details of the attack are given as follows:

---

1. $\mathcal{A}$ performs the following for different choices of $x \leq 2^{3n/4}$:

   (a) $\mathcal{A}$ makes queries to the construction 2k-LightMAC_Plus on the following three inputs: (i) $0\|x$, (ii) $1\|x$, (iii) $2\|x$.

   (b) $L[x] \triangleq \|_{i=0}^2 \left(\text{2k-LightMAC\_Plus}(i\|x) \right)$.

2. For each (x_1, x_2, x_3, x_4) such that $L[x_1] \oplus L[x_2] \oplus L[x_3] \oplus L[x_4] = 0^{3n}$, $\mathcal{A}$ does the following:

   (a) $\mathcal{A}$ makes four additional queries to the construction 2k-LightMAC_Plus with the following inputs: (i) $3\|x_1$, (ii) $3\|x_2$, (iii) $3\|x_3$, (iv) $3\|x_4$.

   (b) If $\bigoplus_{i=1}^4 \text{2k-LightMAC\_Plus}(3\|x_i) = 0^n$ output 1.

3. Output 0.

4.1 Attack Idea

Due to the presence of collisions in the fix functions in the finalization process, we can construct a matching attack by utilizing differences in Σ' and/or Θ' that are absorbed by the fix functions. Our approach involves finding a quadruple of messages $(M_1 := u\|x_1, M_2 := u\|x_2, M_3 := u\|x_4, M_4 := u\|x_4)$ such that two values collide within half of the state. Specifically, we search for quadruples that satisfy a relation $\mathcal{R}(M_1, M_2, M_3, M_4)$ defined as:

$$\mathcal{R}(M_1, M_2, M_3, M_4) \triangleq \begin{cases} \Sigma'(M_1) = \Sigma'(M_2) \oplus 0^{n-1}1 \\ \Theta'(M_2) = \Theta'(M_3) \oplus 0^{n-1}1 \\ \Sigma'(M_3) = \Sigma'(M_4) \oplus 0^{n-1}1 \\ \Theta'(M_4) = \Theta'(M_2) \oplus 0^{n-1}1 \end{cases}$$

Note that, a quadruple (M_1, M_2, M_3, M_4) satisfies the relation $\mathcal{R}$, we must have

$$\bigoplus_{i=1}^{4} \text{2k-LightMAC\_Plus}(M_i) = 0^n.$$

Now, it is easy to see that our choice of messages, as shown in the attack algorithm, ensures the following:

$$\mathcal{R}(M_1, M_2, M_3, M_4) \Leftrightarrow \begin{cases} \mathsf{E}_{K_1}(\langle 2 \rangle \| x_1) = \mathsf{E}_{K_1}(\langle 2 \rangle \| x_2) \oplus 0^{n-1}1 \\ 2\mathsf{E}_{K_1}(\langle 2 \rangle \| x_2) = 2\mathsf{E}_{K_1}(\langle 2 \rangle \| x_3) \oplus 0^{n-1}1 \\ \mathsf{E}_{K_1}(\langle 2 \rangle \| x_3) = \mathsf{E}_{K_1}(\langle 2 \rangle \| x_4) \oplus 0^{n-1}1 \\ 2\mathsf{E}_{K_1}(\langle 2 \rangle \| x_4) = 2\mathsf{E}_{K_1}(\langle 2 \rangle \| x_1) \oplus 0^{n-1}1 \end{cases}$$

$$\Leftrightarrow \begin{cases} \bigoplus_{i=1}^{4} \mathsf{E}_{K_1}(\langle 2 \rangle \| x_i) = 0^n \\ \mathsf{E}_{K_1}(\langle 2 \rangle \| x_1) = \mathsf{E}_{K_1}(\langle 2 \rangle \| x_2) \oplus 0^{n-1}1 \\ \mathsf{E}_{K_1}(\langle 2 \rangle \| x_1) = \mathsf{E}_{K_1}(\langle 2 \rangle \| x_4) \oplus 0^{n-1}1 \end{cases}$$

Therefore, $\mathcal{R}$ defines a $3n$-bit relation which is independent of u, so that several quadruples can be made easily that satisfy $\mathcal{R}$. Now we consider a list:

$$L = \{\text{2k-LightMAC\_Plus}(0\|x)\|\text{2k-LightMAC\_Plus}(1\|x)\|\text{2k-LightMAC\_Plus}(2\|x)\},$$

where $x \in [2^{3n/4}]$ and looking for a quadruples (x_1, x_2, x_3, x_4) such that $L(x_1) \oplus L(x_2) \oplus L(x_3) \oplus L(x_4) = 0^{3n}$. This leads to an attack: we look for a quadruple (x_1, x_2, x_3, x_4) such that

$$\forall u \in \{0, 1, 2\}, \quad \bigoplus_{i=1}^{4} \text{2k-LightMAC\_Plus}(u\|x_i) = 0^n.$$

We expect, on average, one random quadruple (with 2^{3n} potential quadruples and a $3n$-bit filtering) and one quadruple satisfying $\mathcal{R}$ (also a $3n$-bit condition). The correct quadruple is checked with 4 extra queries (as given in line 2(a) of the algorithm). It is easy to see that the distinguisher succeeds with probability $(1 - \frac{1}{2^n})$. This is due to the fact that the probability that line 2(b) gets executed for (i) the real construction is 1, and for (ii) a random function is $\frac{1}{2^n}$.

4.2 Attack Complexity

It is easy to see that the number of queries made by the adversary is $\tilde{\mathcal{O}}(2^{3n/4})$. The searching required for step (iii) is done with at most $\tilde{\mathcal{O}}(2^{3n})$ operations, and using $\mathcal{O}(2^{3n/4})$ memory size (to store all the lists). We would like to point out that one can improve on the time complexity of the attack following the technique used in [7], which can report a quadruple used in line 2(a) in $\tilde{\mathcal{O}}(2^{3n/2})$ operations.

5 Conclusion

To the best of our knowledge, this is the first work that provably shows a message length-independent $3n/4$-bit tight security bound for a block cipher-based variable input length PRF with two block cipher keys. Proving a similar security bound for 1k-LightMAC_Plus is an interesting research problem. To prove the security of the 1k-LightMAC_Plus construction, we require to solve a combinatorial problem, called *mirror theory over a restricted set*, a variant of the mirror theory result, that considers establishing a lower bound on the solutions of a given system of bivariate affine equations which are chosen from a non-empty finite subset of $\{0,1\}^n$.

References

1. Black, J., Rogaway, P.: A block-cipher mode of operation for parallelizable message authentication. In: Knudsen, L.R. (ed.) EUROCRYPT 2002. LNCS, vol. 2332, pp. 384–397. Springer, Heidelberg (2002). https://doi.org/10.1007/3-540-46035-7_25
2. Datta, N., Dutta, A., Dutta, K.: Improved security bound of (E/D)WCDM. IACR Trans. Symmetric Cryptol. **2021**(4), 138–176 (2021)
3. Datta, N., Dutta, A., Kundu, S.: Tight security bound of 2k-LightMAC plus. Cryptology ePrint Archive, Paper 2023/1422 (2023). https://eprint.iacr.org/2023/1422
4. Datta, N., Dutta, A., Nandi, M., Paul, G.: Double-block hash-then-sum: a paradigm for constructing BBB secure PRF. IACR Trans. Symmetric Cryptol. **2018**(3), 36–92 (2018)
5. Datta, N., Dutta, A., Nandi, M., Paul, G., Zhang, L.: Single key variant of PMAC_Plus. IACR Trans. Symmetric Cryptol. **2017**(4), 268–305 (2017)
6. Kim, S., Lee, B., Lee, J.: Tight security bounds for double-block hash-then-sum MACs. In: Canteaut, A., Ishai, Y. (eds.) EUROCRYPT 2020, Part I. LNCS, vol. 12105, pp. 435–465. Springer, Cham (2020). https://doi.org/10.1007/978-3-030-45721-1_16
7. Leurent, G., Nandi, M., Sibleyras, F.: Generic attacks against beyond-birthday-bound MACs. In: Shacham, H., Boldyreva, A. (eds.) CRYPTO 2018, Part I. LNCS, vol. 10991, pp. 306–336. Springer, Cham (2018). https://doi.org/10.1007/978-3-319-96884-1_11
8. Luykx, A., Preneel, B., Tischhauser, E., Yasuda, K.: A MAC mode for lightweight block ciphers. In: Peyrin, T. (ed.) FSE 2016. LNCS, vol. 9783, pp. 43–59. Springer, Heidelberg (2016). https://doi.org/10.1007/978-3-662-52993-5_3

9. Naito, Y.: Blockcipher-based MACs: beyond the birthday bound without message length. In: Takagi, T., Peyrin, T. (eds.) ASIACRYPT 2017, Part III. LNCS, vol. 10626, pp. 446–470. Springer, Cham (2017). https://doi.org/10.1007/978-3-319-70700-6_16

10. Naito, Y.: Improved security bound of LightMAC_Plus and its single-key variant. In: Smart, N.P. (ed.) CT-RSA 2018. LNCS, vol. 10808, pp. 300–318. Springer, Cham (2018). https://doi.org/10.1007/978-3-319-76953-0_16

11. Patarin, J.: The "coefficients H" technique. In: Avanzi, R.M., Keliher, L., Sica, F. (eds.) SAC 2008. LNCS, vol. 5381, pp. 328–345. Springer, Heidelberg (2009). https://doi.org/10.1007/978-3-642-04159-4_21

12. Song, H.: A single-key variant of LightmAC_Plus. Symmetry **13**(10), 1818 (2021)

Designing Full-Rate **Sponge** Based AEAD Modes

Bishwajit Chakraborty[1,2], Nilanjan Datta[3(✉)], and Mridul Nandi[1,3]

[1] Indian Statistical Institute, Kolkata, India
mridul.nandi@gmail.com
[2] Nanyang Technological University, Singapore, Singapore
bishwajit.chakrabort@ntu.edu.sg
[3] Institute for Advancing Intelligence, TCG CREST, Kolkata, India
nilanjan.datta@tcgcrest.org

Abstract. Sponge based constructions have gained significant popularity for designing lightweight authenticated encryption modes. Most of the authenticated ciphers following the Sponge paradigm can be viewed as variations of the Transform-then-permute construction. It is known that a construction following the Transform-then-permute paradigm provides security against any adversary having data complexity D and time complexity T as long as $DT \ll 2^{b-r}$. Here, b represents the size of the underlying permutation, while r pertains to the rate at which the message is injected. The above result demonstrates that an increase in the rate leads to a degradation in the security of the constructions, with no security guaranteed to constructions operating at the full rate, where $r = b$. This present study delves into the exploration of whether adding some auxiliary states could potentially improve the security of the Transform-then-permute construction.

Our investigation yields an affirmative response, demonstrating that a special class of full rate Transform-then-permute with additional states, dubbed frTtP+, can indeed attain security when operated under a suitable feedback function and properly initialized additional state. To be precise, we prove that frTtP+ provides security as long as $D \ll 2^{s/2}$ and $T \ll 2^{s}$, where s denotes the size of the auxiliary state in terms of bits. To demonstrate the applicability of this result, we show that the construction ORANGE-ZEST$_{\mathrm{mod}}$ belongs to this class, thereby obtaining the desired security. In addition, we propose a family of full rate Transform-then-permute construction with Beetle like feedback function, dubbed fr-Beetle, which also achieves the same level of security.

1 Introduction

Since the inception of the Sponge function [2] as a mode of operation for variable output length hash functions, it has received major attention in the symmetric

Bishwajit Chakraborty was supported by the NRF-ANR project SELECT ("NRF2020-NRF-ANR072").

A. Chattopadhyay et al. (Eds.): INDOCRYPT 2023, LNCS 14459, pp. 89–110, 2024.
https://doi.org/10.1007/978-3-031-56232-7_5

key cryptography paradigm. With time, the Sponge mode found its application
in a variety of cryptographic protocols such as message authentication [2,6],
pseudorandom sequence generation [4], and the duplex mode [5] for authenticated encryption. This popularity of the Sponge mode is evident from the number of Sponge-based designs submitted in the CAESAR competition and the
recently concluded NIST lightweight cryptography (LwC) standardization process. A Sponge duplex type scheme ASCON [14] turned out to be the winner of
the NIST lightweight competition and one of the joint winners in the category of
lightweight applications (resource-constrained environments) in CAESAR competition.

At a high level, Sponge-type constructions consist of a b bit state, which is
split into a c bit inner state, called the capacity, and an r bit outer state, called
the rate, where $b = c + r$. Traditionally, in Sponge-like modes, r bits of data
absorption and squeezing are done via the rate part at a time. However, there
are a few exceptions, e.g., SpoC [1], where the absorption is done via the capacity
part while the squeezing is done from the rate part. In [3], Bertoni et al. proved
that the Sponge construction is indifferentiable from a random oracle with a
birthday-type bound in the capacity. While it is well-known that this bound is
tight for hashing, for keyed applications of the Sponge, especially authenticated
encryption schemes, such as duplex mode, the security could be significantly
higher.

1.1 Existing Security Bounds for Sponge-Type AEAD Schemes

Sponge-type authenticated encryption is mostly done via the duplex construction
[5]. The duplex mode is a stateful construction that consists of an initialization
interface and a duplexing interface. Initialization creates an initial state using
the underlying permutation π, and each duplexing call to π absorbs and squeezes
r bits of data. The security of Sponge-type AEAD modes can be represented and
understood in terms of two parameters, namely the data complexity D (total
number of initialization and duplexing calls to π), and the time complexity T
(total number of direct calls to π).

Initially, Bertoni et al. [5] proved that duplex is as strong as Sponge and
achieves security up to $DT \ll 2^c$. At Asiacrypt'14, Jovanovic et al. [16] proved
that sponge duplex achieves beyond the birthday bound of the capacity. To be
precise, they have shown that it achieves privacy up to $D \ll \min\{2^{b/2}, 2^\kappa\}$, $T \ll$
$\min\{2^{b/2}, 2^{c-\log_2 r}, 2^\kappa\}$, and integrity up to $DT \ll 2^c$, $D \ll \min\{2^{c/2}, 2^\kappa, 2^\tau\}$,
$T \ll \min\{2^{b/2}, 2^{c-\log_2 r}, 2^\kappa\}$, where τ denotes the tag size. Later, a tight privacy
analysis [17] was also provided. At Asiacrypt'15, Mennink et al. [20] introduced
the full-state duplex and proved that this variant is secure up to $DT \ll 2^\kappa$,
$D \ll 2^{c/2}$, where κ is the key size. In CHES'18 [8], Chakraborti et al. came
up with a variant of duplex mode, dubbed Beetle, that achieves privacy up to
$DT \ll 2^b$, $D \ll 2^{b/2}$, $T \ll 2^c$, and integrity up to $D \ll \min\{2^{b/2}, 2^{c-\log_2 r}, 2^\tau\}$,
$T \ll \min\{2^{c-\log_2 r}, 2^\tau, 2^{b/2}\}$, when set with $\kappa = c$ and $\tau = r$. Recently,
Chakraborty et al. [11] introduced the Transform-then-Permute construction
which encompasses most of the popular Sponge-type constructions and showed
that popular designs like Beetle can achieve security upto $D, T \ll 2^{b-r}$. All

the existing analysis show that increasing r degrades the security of a Sponge type design. If $2^r \geq \min\{2^{b-\log T}, 2^{b-\log D}\}$ then the security of all these existing constructions becomes void.

Chakraborty et al. designed a new Sponge-based authenticated encryption ORANGE-ZEST [12] where the designers introduced some extra-state in the protocol to construct a full-rate Sponge type AEAD scheme. This construction was a round 2 submission to the recently concluded NIST LwC standardization process. However, Dobraunig et al. [15] and Khairallah et al. [18] mounted forgery attacks on the original and a modified variant of ORANGE-ZEST. It seems interesting to investigate whether these attacks can be avoided with some minor changes in the design or if there is some inherent flaw in the overall design strategy.

1.2   Our Contributions

In this paper, we revisit the Transform-then-Permute construction introduced by Chakraborty et al. [11] and investigate the security dependency on the capacity in Sponge-type modes. Our contribution is two-fold.

1. **Full Rate Transform-then-Permute Mode with Extra State:** We show that at the cost of some additional state, suitable initialization of the extra state, one can indeed achieve a full rate Transform-then-Permute type authenticated encryption mode with security up to $D \ll 2^{s/2}, T \ll 2^s$, where s denotes the bit-size of this extra state. To do that, we first introduce a generic class of full rate Transform-then-Permute authenticated encryption constructions inspired by the Transform-then-Permute construction with the extra state in Sect. 3.2. First, we describe the general structure of such construction. Then, we consider a special class of Transform-then-Permute construction, dub it frTtP, by imposing several restrictions in the underlying feedback function. In Sect. 3.3, we provide the necessary justification for the choice of these restrictions to achieve the desired security. Roughly, the restrictions take care of all the necessary conditions required for the correctness and desired security of such constructions along with simplifying the proof. We prove the generic security of the class of frTtP construction in Sect. 4 (see Proposition 4). In addition, we also consider a special sub-class of frTtP constructions, called frTtP+, with some additional restrictions that obtain a much-simplified security bound (see Theorem 2).
2. **Concrete Instantiations:** Finally, we demonstrate the applicability of our results. First, in Sect. 4.1, we show that the modified ORANGE-ZEST [12] belongs to the frTtP+ class, and hence obtains the desired security. This essentially shows that the weakness in the original ORANGE-ZEST [12] was only due to improper initialization, not a flaw in the underlying design strategy. Next, in Sect. 4.2, we demonstrate that simple duplex sponge-type designs, even when extended to full rate using some extra state, do not satisfy one of the necessary conditions for security, and hence, are inherently insecure. Next, in Sect. 4.3, we consider a family of Transform-then-Permute constructions following Beetle like feedback, dub fr-Beetle that belong to the frTtP+

class and hence, achieve the desired security. As a concrete instantiation from the class of fr-Beetle, we demonstrate the example of fr-COFB that uses combined feedback, as used in CoFB [9].

1.3 Significance of the Result

In this subsection, we highlight the significance of our result. We provide comparative results among the proposed construction fr-COFB, Orange with existing constructions such as Sponge-Duplex, Beetle in terms of rate, state, security, and linear operations. As depicted in Table 1, consider Beetle with $b = 256$, $r = 128$ and fr-COFB with $s = 128$. They both achieve similar security (upto $D \ll 2^{64}$, $T \ll 2^{128}$). However, at the cost of the additional 128 bit additional state (and necessary additional xor operations), fr-COFB achieves double throughput as compared to Beetle.

Table 1. A Comparative Study of Sponge-based constructions. b and r denote the permutation size and message injection rate, respectively. By state, we mean the additional state required. The security bound only considers the major terms.

Mode	Rate	State	Linear operations/block	Security Bound
Sponge-Duplex [5]	r/b	0	r bit xor	$\mathcal{O}(\frac{q_p^2 + \sigma^2}{2^{b-r}})$
Beetle [8]	r/b	0	r bit xor, r bit shift	$\mathcal{O}(\frac{q_p}{2^{b-r}} + \frac{q_p \sigma}{2^b})$ [11]
ORANGE-ZEST$_{\mathrm{mod}}$ [12]	1	s	$2b$ bit xor, $(b-s)$ bit shift	$\mathcal{O}(\frac{\sigma^2 + q_p}{2^s} + \frac{q_p \sigma}{2^b})$ [Sect. 4.1]
fr-COFB [This paper]	1	s	$(2b+s)$ bit xor	$\mathcal{O}(\frac{\sigma^2 + q_p}{2^s} + \frac{q_p \sigma}{2^b})$ [Sect. 4.3]

We believe our result is significant in designing high throughput, lightweight authenticated encryption designs as it provides a general guideline for constructing full-rate sponge-based constructions.

2 Preliminaries

In this paper, for any $n \in \mathbb{N}$, $(n]$ (res. $[n]$) signifies the set $\{1, 2, \ldots, n\}$ (res. $\{0, 1, \ldots, n\}$). $\{0,1\}^n$ denotes the set of bit strings of length n, $\{0,1\}^* := \bigcup_{n \geq 0} \{0,1\}^n$, and $\mathsf{Perm}(n)$ signifies the set of all permutations over $\{0,1\}^n$. We say that the two distinct strings $a = a_1 \ldots a_m$ and $b = b_1 \ldots b_{m'}$ have a common prefix of length $n \leq \min\{m, m'\}$ if $a_i = b_i$ for all $i \in (n]$, and $a_{n+1} \neq b_{n+1}$. $\lceil x \rceil_n$ (res. $\lfloor x \rfloor_n$) designates the most (res. least) significant n bits of any bit string x with $|x| \geq n$. We use the notation $\langle N \rangle_x$ to denote the binary representation of N represented in x bits. We define the falling factorial $(n)_k := n(n-1) \cdots (n-k+1)$. For any finite set $\mathcal{X}$, $(\mathcal{X})_q$ signifies the set of all q-tuples with distinct elements from $\mathcal{X}$. $\mathsf{X} \xleftarrow{\$} \mathcal{X}$ signifies the uniform sampling of X from $\mathcal{X}$, which is independent of all other previously sampled random variables. An uniform sampling of t random variables $\mathsf{X}_1, \ldots, \mathsf{X}_t$ from $\mathcal{X}$ without replacement is denoted by $(\mathsf{X}_1, \ldots, \mathsf{X}_t) \xleftarrow{\text{wor}} \mathcal{X}$. We use the symbol $\star$ to denote that it can take any possible values.

2.1 Authenticated Encryption: Definition and Security Model

Given any *key space* $\mathcal{K}$, *nonce space* $\mathcal{N}$, *associated data space* $\mathcal{A}$, *message space* $\mathcal{M}$, *ciphertext space* $\mathcal{C}$, and *tag space* $\mathcal{T}$ an authenticated encryption scheme with associated data functionality (or AEAD in short), is a tuple of algorithms $\mathsf{AE} = (\mathsf{E} : \mathcal{K} \times \mathcal{N} \times \mathcal{A} \times \mathcal{M} \to \mathcal{C} \times \mathcal{T}, \mathsf{D} : \mathcal{K} \times \mathcal{N} \times \mathcal{A} \times \mathcal{C} \times \mathcal{T} \to \mathcal{M} \cup \{\perp\})$ such that for all $(K, N, A, M) \in \mathcal{K} \times \mathcal{N} \times \mathcal{A} \times \mathcal{M}$ and $(C, T) \in \mathcal{C} \times \mathcal{T}$, $\mathsf{D}(K, N, A, C, T) = M$ if and only if $\mathsf{E}(K, N, A, M) = (C, T)$. We call E (res. D) the encryption (res. decryption) algorithm of AE. For any key $K \in \mathcal{K}$, let $\mathsf{E}_K(\cdot)$ (res. $\mathsf{D}_K(\cdot)$) denotes $\mathsf{E}(K, \cdot)$ (res. $\mathsf{D}(K, \cdot)$). In this paper, we assume $\mathcal{K}, \mathcal{N}, \mathcal{A}, \mathcal{M}, \mathcal{T} \subseteq \{0,1\}^*$ and $\mathcal{C} = \mathcal{M}$.

For $b \in \mathbb{N}$, let $\Pi \leftarrow_\$ \mathsf{Perm}(b)$, and $\Gamma \leftarrow_\$ \mathsf{Func}(\mathcal{N} \times \mathcal{A} \times \mathcal{M}, \mathcal{M} \times \mathcal{T})$. Let $\perp$ denote the degenerate function from $(\mathcal{N}, \mathcal{A}, \mathcal{M}, \mathcal{T})$ to $\{\perp\}$. We use the superscript $\pm$ to denote bidirectional access to Π. By abuse of notation the oracle corresponding to a function (like E, Π etc.) is denoted by that function itself.

Definition 1. *Consider any AEAD scheme AE_Π defined over $(\mathcal{K}, \mathcal{N}, \mathcal{A}, \mathcal{M}, \mathcal{T})$ with the random permutation Π as it's underlying primitive. The AEAD advantage of an adversary $\mathscr{A}$ against AE_Π is defined as*

$$\mathbf{Adv}^{\mathrm{AEAD}}_{\mathsf{AE}_\Pi}(\mathscr{A}) := \left| \Pr_{\substack{K \leftarrow_\$ \mathcal{K} \\ \Pi^\pm}} \left[\mathscr{A}^{\mathsf{E}_K, \mathsf{D}_K, \Pi^\pm} = 1 \right] - \Pr_{\Gamma, \Pi^\pm} \left[\mathscr{A}^{\Gamma, \perp, \Pi^\pm} = 1 \right] \right|,$$

where $\mathscr{A}$'s response after its interaction with E_K, D_K, and $\Pi^\pm$ is denoted by $\mathscr{A}^{\mathsf{E}_K, \mathsf{D}_K, \Pi^\pm}$. Similarly, $\mathscr{A}^{\Gamma, \perp, \Pi^\pm}$ denotes $\mathscr{A}$'s response after its interaction with Γ, $\perp$, and $\Pi^\pm$.

In this paper, we only consider adversaries which do not make any repetitive or redundant queries. Let q_e and q_d denote the number of queries to E_K and D_K respectively. Let σ_e and σ_d denote the sum of input (associated data and message) lengths across all encryption and decryption queries respectively. Any adversary making q_p primitive calls, q_e encryption queries, q_d decryption queries with a total of at most σ_e and σ_d blocks of encryption and decryption queries is called a $(q_p, q_e, q_d, \sigma_e, \sigma_d)$-adversary or simply (q_p, σ)-adversary, where $\sigma := \sigma_e + \sigma_d$.

2.2 Coefficients H Technique

Consider any deterministic yet computationally bounded adversary $\mathscr{A}$ using a black box type interaction with one of two oracles $\mathcal{O}_0$ and $\mathcal{O}_1$ and trying to differentiate between them. The query-response tuple associated with $\mathscr{A}$'s interaction with its oracle is called its transcript. A transcript ω may also contain any other information that the oracle decides to reveal to the distinguisher at the end of the game's query-response phase. This expanded definition of transcript will be taken into consideration. Suppose Θ_1 (res. Θ_0) denotes the random transcript variable for $\mathscr{A}$'s interaction with $\mathcal{O}_1$ (res. $\mathcal{O}_0$). The *interpolation probability* of ω with regard to $\mathcal{O}$ is the probability of obtaining a given transcript ω in the security game with an oracle $\mathcal{O}$. Since $\mathscr{A}$ is deterministic, this probability only depends on the transcript ω and the oracle $\mathcal{O}$. A transcript ω is said to be *attainable* if $\Pr\left[\Theta_0 = \omega\right] > 0$. In this paper, $\mathcal{O}_1 = (\mathsf{E_K}, \mathsf{D_K}, \mathsf{\Pi^{\pm}})$ and $\mathcal{O}_0 = (\mathsf{\Gamma}, \mathsf{bot}, \mathsf{\Pi^{\pm}})$ and the adversary is trying to distinguish $\mathcal{O}_1$ from $\mathcal{O}_0$ in the AEAD sense. We now state the coefficient H technique(or simply the H-technique), a simple yet powerful tool developed by Patarin [21]. A proof of this theorem can be found in a number of papers including [13,19,22].

Theorem 1 (H-technique [21,22]). *Let Ω be the set of all transcripts. For some $\epsilon_{\mathsf{bad}}, \epsilon_{\mathsf{ratio}} > 0$, suppose there is a set $\Omega_{\mathsf{bad}} \subseteq \Omega$ satisfying the following:*

- $\Pr\left[\Theta_0 \in \Omega_{\mathsf{bad}}\right] \le \epsilon_{\mathsf{bad}}$;
- *For any $\omega \notin \Omega_{\mathsf{bad}}$, ω is attainable and*

$$\frac{\Pr\left[\Theta_1 = \omega\right]}{\Pr\left[\Theta_0 = \omega\right]} \ge 1 - \epsilon_{\mathsf{ratio}}.$$

Then the distinguishing advantage for any adversary $\mathscr{A}$ can be bounded as

$$\mathbf{Adv}_{\mathcal{O}_1}^{\mathrm{dist}}(\mathscr{A}) \le \epsilon_{\mathsf{bad}} + \epsilon_{\mathsf{ratio}}.$$

2.3   Multi-chain Graph

In this section, we revisit the multi-chain graph structure and an important result that bounds the number multi-chains as discussed in Chakraborty et al. [11].

Multi-chain Graph. Let $\Theta = \{(U_1, V_1), \ldots, (U_t, V_t)\}$ be a list of pairs of b-bit elements such that $U_1, \ldots, U_t$ are distinct and $V_1, \ldots, V_t$ are distinct. For any such list of pairs, we write $\mathsf{domain}(\Theta) = \{U_1, \ldots, U_t\}$ and $\mathsf{range}(\Theta) = \{V_1, \ldots, V_t\}$. Let $\mathcal{L}$ be a linear function over b bits with the transformation matrix L. Given such a list Θ and a linear transformation matrix L, we define a labeled directed graph G_{Θ}^L (call it a multi-chain graph) over the set of vertices $\mathsf{range}(\Theta)$. Given $V_i, V_j \in G_{\Theta}^L$ and $X \in \{0,1\}^b$, we draw an X labeled directed edge $V_i \xrightarrow{X} V_j$ in the graph iff   .

$$L \cdot V_i \oplus X = U_j.$$

We can similarly extend this to a label walk $\mathcal{W}$ from a node W_0 to W_k as

$$\mathcal{W} : W_0 \xrightarrow{X_1} W_1 \xrightarrow{X_2} W_2 \cdots \xrightarrow{X_k} W_k$$

and simply denote it as $W_0 \xrightarrow{X} W_k$ where $X = (X_1, \ldots, X_k)$. Here k is the length of the walk.

Multi-chain. Let G_{Θ}^L be any multi-chain graph as defined above. Given any fixed levels $(X_1, \ldots, X_k)$, we say the set of k length walks $\{\mathcal{W}_i : W_0^i \xrightarrow{(X_1, \ldots, X_k)} W_k^i\}$ form an *multi-chain* if and only if $W_k^i = W_k^j$ for all i, j. Note that a multi-chain is a set of walks and if $\mathcal{W}$ is a multi-chain then so is any subset of $\mathcal{W}$. The following lemma bounds the number of multi-chains of any length.

Lemma 1. *Consider the set of all multi-chains in G_Θ^L of length k. Let Γ_k denote the size of the largest of all such multi-chains of length k. If L is invertible, then*

$$\mu_t := \max_{k>0} \mathsf{Ex}\left[\frac{\Gamma_k}{k}\right] \leq 1.$$

The proof of this lemma can be found in [11].

3 Full-Rate-Transform-then-Permute AEAD

3.1 Revisiting Transform-then-Permute Paradigm

Let us first revisit the Transform-then-Permute construction introduced by Chakraborty et al. [11]. We assume that the underlying primitive of the construction is a b bit public permutation and r is the rate of message/associated data injection. Let κ, ν denote the size of the key and the nonce respectively. For simplicity, we assume $\kappa < b, \nu = b - \kappa, r \leq b$.

The construction takes a nonce N, an associated data A and a message M as input. We define a formatting function Fmt that maps any (A, M) to $(B_1, \ldots, B_{a+m}) \in (\{0,1\}^b)^{a+m}$ where $a := \lceil |A|/r \rceil$ and $m := \lceil |M|/r \rceil$, such that given any two tuples $(A, M) \neq (A', M')$ and $\mathsf{Fmt}(A, M) = (B_1, \ldots, B_{a+m})$ and $\mathsf{Fmt}(A', M') = (B'_1, \ldots, B'_{a'+m'})$, we have

1. $(B'_1, \ldots, B'_a) \neq (B_1, \ldots, B_a)$ whenever $A \neq A'$ and $a \leq a'$.
2. $(B'_{a+1}, \ldots, B'_{a+m}) \neq (B_{a+1}, \ldots, B_{a+m})$, whenever $A = A'$ and $m \leq m'$.

We consider the Sponge-type construction which takes state output Y_i and data input B_i and generate next state input X_{i+1} and the data output C_i using a linear feedback function $\mathcal{E} : \{0,1\}^b \times \{0,1\}^b \to \{0,1\}^b \times \{0,1\}^r$. This can alternatively be represented using a transformation matrix E as follows:

$$\begin{bmatrix} X_{i+1} \\ C_i \end{bmatrix} = E \cdot \begin{bmatrix} Y_i \\ B_i \end{bmatrix}.$$

Chakraborty et al. [11] considered a special type of Sponge based construction, dub them as Transform-then-Permute, where the transformation matrix is of the form

$$E = \begin{bmatrix} \star & \star \\ [I_r \ 0_{r\times(b-r)}] & [I_r \ 0_{r\times(b-r)}] \end{bmatrix}.$$

It is easy to see that accordingly the decryption transformation matrix D would also have the same form as E. A pictorial description of the Transform-then-Permute construction is depicted in Fig. 1., most of the Sponge-based AEAD designs such as ASCON, and Beetle belongs to this category.

We say that a Transform-then-Permute AEAD has Full-rate if $r = b$. It is well known that a Full-Rate-Transform-then-Permute construction is not secure:

Proposition 1. *Any full-rate Transform-then-Permute AEAD is insecure.*

Interested readers are referred to the full version [10] for the concrete proof of the Proposition.

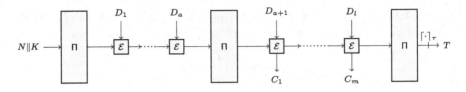

Fig. 1. Schematic of the Transform-then-Permute AEAD mode. $\mathsf{Fmt}(A, M) = (D_1, \ldots, D_l)$. The data outputs during the associated data processing are ignored.

3.2 Full-Rate-Transform-then-Permute AEAD with Extra-State

We now define a Full-Rate-Transform-then-Permute (frTtP in short) AEAD mode which uses an s-bit extra secret state. The necessity of this extra state is evident from Proposition 1.

General Structure of AEAD with Extra-State. As before considering a frTtP encryption protocol with a permutation Π of state size b bits, key size κ, nonce size $b - \kappa$ and tag size τ.

☐ **Initialization:** Given any encryption query of the form (N, A, M), the encryption algorithm first applies a formatting function Fmt that maps any (A, M) to $(B_1, \ldots, B_{a+m}) \in (\{0,1\}^b)^{a+m}$, where the first a (≥ 1) blocks are generated from A. The format function should ensure that given any two tuples $(A, M) \neq (A', M')$ and $\mathsf{Fmt}(A, M) = (B_1, \ldots, B_{a+m})$ and $\mathsf{Fmt}(A', M') = (B'_1, \ldots, B'_{a'+m'})$, we have

(i) $(B_1, \ldots, B_a) \neq (B'_1, \ldots, B'_{a'})$, if and only if $A \neq A'$.
(ii) $(B_1, \ldots, B_{a+m}) \neq (B'_1, \ldots, B'_{a'+m'})$, if and only if $(A, M) \neq (A', M')$.

It is easy to see the following is a simple example of a format function satisfying the restrictions:

$$\mathsf{Fmt}(A, M) := \mathsf{ozs}(A) \parallel \mathsf{oozs}(M) \parallel \langle |A| \rangle_{b/2} \parallel \langle |M| \rangle_{b/2}.$$

Here ozs and oozs means 10^* and optional 10^* padding to make the blocks multiple of b bits.

In addition, we define $X_0 = K \parallel N$; $Y_0 = \Pi(X_0)$. The algorithm uses an extra-state initialization protocol to generate the initial extra-state $S_0 = \rho \circ \Pi(K \parallel N)$, where ρ can be any linear function from $\{0,1\}^b$ to $\{0,1\}^s$ with rank s. A trivial choice of ρ is $\rho(B) = \lfloor B \rfloor_s$.

☐ **Associated Data and Message Processing:** For $i \in [1, a + m]$ and a linear feedback function $\mathcal{E} : \{0,1\}^{2b+s} \to \{0,1\}^{2b+s}$, the algorithm recursively calculates Y_i, S_i, C_i as follows:

$$(X_i, S_i, C_i) = \mathcal{E}(Y_{i-1}, S_{i-1}, B_i); \quad Y_i = \Pi(X_i).$$

Alternatively, we can represent the feedback function via a transformation matrix as given below:

$$\begin{bmatrix} X_i \\ S_i \\ C_i \end{bmatrix} = \begin{bmatrix} E_1 & E_2 & E_3 \\ E_4 & E_5 & E_6 \\ E_7 & E_8 & E_9 \end{bmatrix} \begin{bmatrix} Y_{i-1} \\ S_{i-1} \\ B_i \end{bmatrix} .$$

Accordingly, there should exist a decryption transformation matrix D such that

$$\begin{bmatrix} X_i \\ S_i \\ B_i \end{bmatrix} = \begin{bmatrix} D_1 & D_2 & D_3 \\ D_4 & D_5 & D_6 \\ D_7 & D_8 & D_9 \end{bmatrix} \begin{bmatrix} Y_{i-1} \\ S_{i-1} \\ C_i \end{bmatrix} \Leftrightarrow \begin{bmatrix} X_i \\ S_i \\ C_i \end{bmatrix} = E \cdot \begin{bmatrix} Y_{i-1} \\ S_{i-1} \\ B_i \end{bmatrix} .$$

☐ **Ciphertext and Tag generation:** Finally the protocol outputs $\lfloor C_{a+1} \| \cdots \| C_{a+l} \rfloor_{|M|}$ as the ciphertext and $\lfloor Y_{a+l} \rfloor_\tau$ as the tag.
    We represent the generic structure in Fig. 2.

Understanding the frTtP Class. Now we define a special class of full rate Transform-then-Permute, dub frTtp, where we impose the following four conditions on the feedback encryption and decryption matrices:

(C1) $E_9 = D_9 = I_b$, (C2) $E_6 = D_6 = 0$, (C3) $E_7 = D_7 = I_b$, (C4) $E_2 = E_3 \cdot E_8 \ (\neq 0)$.

Note that the above restrictions ensure the following (via simple linear algebraic calculations):

$$D_1 = E_1 \oplus E_3, \ D_2 = 0, \text{ and } D_i = E_i, \ \forall i = 3, \ldots, 9.$$

This simplified feedback function for frTtP is depicted in Fig. 3, where the initial extra state is calculated as $S_0 = \rho(Y_0)$, for some linear function $\rho : \{0,1\}^b \rightarrow \{0,1\}^s$ of rank s.

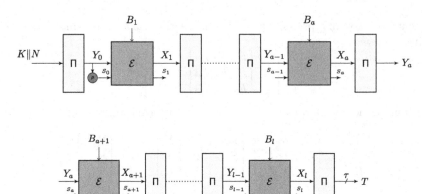

Fig. 2. A frTtP AEAD with extra state. Here $(B_1, \ldots, B_l) = \mathsf{Fmt}(A, M)$. $\rho : \{0,1\}^b \to \{0,1\}^s$ is a linear function of rank s.

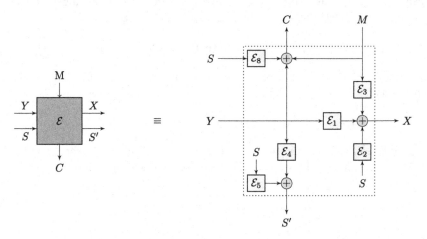

Fig. 3. Simplified Representation of an frTtP feedback function.

3.3   Rationale of the Assumptions on the Feedback Function

In this section, we justify our choices for the encryption and decryption subma-trices.

☐ **Choice on the Feedback Function.** To begin with, let us first state the following proposition that provides a few necessary conditions for the Encryption and Decryption Feedback Functions:

Proposition 2. *If $\mathcal{E}$ and $\mathcal{D}$ are the encryption and decryption feedback functions of a secured frTtP construction, then $\mathcal{E}, \mathcal{D}$ must satisfy the following conditions.*

(i) $\mathsf{rank}(E_9) = \mathsf{rank}(D_9) = b$,
(ii) $\mathsf{rank}(E_8) = \mathsf{rank}(D_8) \neq 0$,
(iii) $\mathsf{rank}([E_7\ E_8])$, $\mathsf{rank}([D_7\ D_8]) = b$.

Proof. Condition (i) follows from the correctness of the construction and the observation that if $\mathsf{rank}(E_9) \neq b$, then there exists $M \neq M'$ such that $E_9 \cdot M = E_9 \cdot M'$, and hence the decryption function will not be deterministic. $\mathsf{rank}(D_9) = b$ follows from a similar argument. Conditions (ii) and (iii) are necessary from a security perspective. Condition (ii) follows from the fact that if $\mathsf{rank}(E_8) = 0$ or $\mathsf{rank}(D_8) = 0$ then the internal Y state values are completely determined and hence the adversary can forge the construction in the same way as the frTtP construction with no extra-state. For condition 3, suppose $\mathsf{rank}([E_7\ E_8]) \neq b$. Then, there exists a non zero vector γ such that $\gamma \cdot ([E_7\ E_8]) = 0$. Hence, $\gamma \cdot C = \gamma \cdot M$ with probability 1.

The above necessary conditions are incorporated to define a new simplified sufficient assumptions on the feedback function $\mathcal{E}$:

(C1) $E_9 = D_9 = I_b$, (C2) $E_6 = D_6 = 0$, (C3) $E_7 = D_7 = I_b$, (C4) $E_2 = E_3 \cdot E_8 \ (\neq 0)$.

Now let us try to justify the above-mentioned assumptions. To justify condition (C1), observe that with since $\mathsf{rank}(E_9) = b$ one can simply define $M' = E_9 \cdot M$, and proceed with that. For (C2), observe that since M is known, it doesn't contribute to the randomness of the extra state and hence taking $E_6 = 0$ doesn't affect the security of the AEAD scheme. $D_6 = 0$ follows from $E_6 = 0$ and assumption (i). Note that (C3) and (C4) takes care of the necessary conditions (iii) and (ii) respectively. However, they are stronger assumptions than necessary condition (iii) and (ii) respectively, which are used in simplifying the overall calculations for the special class of general frTtP feedback functions. Note that condition (C4) essentially ensure that $D_2 = 0$, i.e., during decryption the permutation input does not depend on the extra state. This condition helps in achieving the desired bound. As a consequence, we do not have any matching attacks on frTtP to justify (C3), (C4). Nonetheless, as we will see in Sect. 4, many full-rate feedback functions used in popular constructions such as ORANGE-ZEST, the one used in COFB or Beetle satisfy both these conditions. Moreover, in the feedback functions used in Transform-then-Permute constructions without an extra state, (C3) is a necessary condition.

☐ **Choice on the Initial Extra-state Generation.** Consider a frTtP construction with extra state size s and linear feedback function $\mathcal{E}$ as defined above. Note that during associated data processing no information is leaked. Hence, for an encryption query say (N, A, M), if a many blocks of associated data are processed via format function, then it is not necessary to generate the extra-state values $S_0, \ldots, S_{a-1}$. Infact, even if $S_i = 0$ for all $0 \leq i \leq a - 1$, the adversary cannot compute Y_a. So, a possible choice of defining the extra state is to define it via Y_a or nonce N. The following proposition suggests that simply generating it through (i) a linear function on Y_a or (ii) a linear function on N does not suffice.

Proposition 3. *For any encryption query (N, A, M), let a many blocks be processed due to associated data A via the format function. If (i) S_a is independent or linearly dependent on N, or (ii) $S(a)$ is a linear function of Y_a, then there exists a forging adversary against the frTtP construction.*

Proof. For part (i), assume that S_a^i is independent of N^i. Now, suppose an adversary makes two encryption queries $(N^1, A, M) \neq (N^2, A, M)$, and corresponding responses are $(C^1, T^1), (C^2, T^2)$. Let $\mathsf{Fmt}(A, M) = (B_1, \ldots, B_a, B_{a+1})$. It is easy to see that $S_a^1 = S_a^2$, and hence, $Y_a^1 = Y_a^2 \oplus C^1 \oplus C^2$. This implies $X_{a+1}^1 = B_1 \cdot Y_a^1 \oplus B_2 \cdot S_a^1 \oplus B_3 \cdot C^1 = B_1 Y_a^2 \oplus B_2 \cdot S_a^1 \oplus B_1 (C^1 \oplus C^2) \oplus B_3 \cdot C^1$. Hence, if an adversary choses C^* in such a way that $B_3 \cdot (C^* \oplus C^1) = B_1 \cdot (C^1 \oplus C^2)$ then (N^2, A^2, C^*, T^1) is a valid forgery. A similar analysis goes through if S_a is linearly dependent on N. In that case, we have $S_a^1 \oplus S_a^2 = F \cdot (N^1 \oplus N^2)$ where F is some $s \times \nu$ linear matrix. Here (N^2, A^2, C^*, T^1) would be a valid forgery, where $B_3 \cdot (C^* \oplus C^1) = B_1 \cdot (C^1 \oplus C^2 \oplus E_8 \cdot F \cdot (N^1 \oplus N^2)) \oplus B_2 \cdot F \cdot (N^1 \oplus N^2)$.

For part (ii), let us assume $S_a = \rho(Y_a)$. Then, $(I_b \oplus E_8 \cdot \rho) \cdot Y_a = C_{a+1} \oplus B_{a+1}$. Now, if $\mathsf{rank}(I_b \oplus E_8 \cdot \rho) = b$, then Y_a can be calculated as $(I_b \oplus E_8 \cdot \rho)^{-1} \cdot (C_{a+1} \oplus B_{a+1})$. If $\mathsf{rank}((I_b \oplus E_8 \cdot \rho)) < b$, then there exists vector γ such that $\gamma \cdot (I_b \oplus E_8 \cdot \rho) = 0$ which implies $\gamma \cdot C_{a+1} = \gamma \cdot B_{a+1}$ with probability 1.

Now, assuming the underlying primitive Π is the only nonlinear component, a natural choice for the initial extra-state would be $\rho \circ \Pi \circ \rho'(N, K)$, where $\rho : \{0, 1\}^b \to \{0, 1\}^s$, $\rho' : \{0, 1\}^\kappa \times \{0, 1\}^\nu \to \{0, 1\}^b$ are two linear functions. A straightforward choice for $\rho'(N, K)$ would be $K \| N$, which in fact is used in many popular AEAD protocols such as CoFB [9]. However, this doesn't work if no block is processed in the associated data (e.g., empty-associated data) due to the above Proposition. However, as mentioned in our format function, it always generates one associate data block to ensure that at least one block is processed during the associated data, and take $\rho'(N, K) = K \| N$.

4   Security of frTtP AEAD with Extra State

In this section, we bound the advantage of any AEAD adversary against the frTtP construction defined in Sect. 3. Consider an frTtP construction with the encryption and decryption feedback functions $\mathcal{E}$ and $\mathcal{D}$ respectively. Alongside, we consider a linear function $\rho : \{0, 1\}^b \to \{0, 1\}^s$ of rank s for processing the initial extra-state. Before proceeding to the exact proposition statement, we define a notation for multi-collision as follows: Let $X_1, \ldots, X_\mu \xleftarrow{\text{wor}} \mathcal{D}$ where $|\mathcal{D}| = \beta$ and $\beta \geq 4$. Let $\mathsf{mc}_{\mu,\beta}$ denote the maximum multicollision random variable for the sample i.e., $\mathsf{mc}_{\mu,\beta} = \max_a |\{i : X_i = a\}|$. We define $\mathsf{mcoll}(\mu, \beta) := \mathsf{Ex}\left[\mathsf{mc}_{\mu,\beta}\right]$. Given this definition, we now state our main proposition as follows.

Proposition 4. *The AEAD advantage of all adversaries making q_p many primitive queries, a total of σ_e blocks in encryption queries, and a total of σ_d blocks in decryption queries against an frTtP construction with s bit extra-state as defined above, can be bounded as follows*

$$\mathbf{Adv}_{frTtP}^{AEAD}(q_p, \sigma_e, \sigma_d) \leq \frac{q_p}{2^\kappa} + \frac{9\sigma_e q_p}{2^{r_{12}}} + \frac{2\sigma_e^2}{2^{r_{12}}} + \frac{2\sigma_e^2}{2^{r_{45}+r_{45}'-s}} + \frac{q_p \sigma_d}{2^{r_d}} + \frac{2\sigma_d}{2^\tau} + \frac{\sigma_d \sigma_e}{2^{r_d+r_3-b}}$$

$$+ \frac{q_p \mathsf{mcoll}(\sigma_e, 2^\tau)}{2^{b-\tau}} + \frac{2q_p \mathsf{mcoll}(\sigma_e, 2^{b+s-r_{45}-r_8})}{2^{r_{45}+r_{12}+r_8-s-b}}$$

$$+ \frac{3\sigma_d(\sigma_e + \sigma_d + q_p)}{2^{r_d}} + \frac{\sigma_d q_p \mathsf{mcoll}(\sigma_e, 2^{b+s-r_{45}-r_8})}{2^{r_{12}+r_d+r_8-r_{45}-b-s}},$$

where

$$r_3 := \mathsf{rank}(E_3); \quad r_8 = \mathsf{rank}(E_8); \quad r_{12} := \mathsf{rank}\left(E_1 \oplus E_2 \cdot \rho\right); \quad r_d := \mathsf{rank}(D_1);$$

$$r_{45} := \mathsf{rank}\left((E_4 \cdot E_8 \oplus E_5)^s\right); \quad r_{45}' = \min_{j \leq \ell} \left\{\mathsf{rank}(I_s \oplus (E_4 \cdot E_8 \oplus E_5)^j)\right\}.$$

Here ℓ denotes the maximum allowed message length.

The above proposition gives a generic security bound on the security of frTtP. Now let us consider a special simplified class of frTtP, call it frTtP+, with the following additional restrictions:

$$\mathsf{rank}(E_3) = \mathsf{rank}(r_{12}) = \mathsf{rank}(D_1) = b, \quad rank(E_8) = s, \quad E_4 \cdot E_8 \oplus E_5 = \alpha \cdot I_s,$$

where α is a primitive element in $GF(2^s)$. As we will see, we can construct several efficient authenticated cipher construction following frTtP+ paradigm. Now we state a simplified result on the security of this new class of frTtP constructions.

Theorem 2. *The AEAD advantage of any adversary against an authenticated encryption construction following frTtP+ paradigm making q_p many primitive queries, a total of σ blocks in encryption and decryption queries can be bounded by*

$$\mathbf{Adv}_{frTtP+}^{AEAD}(q_p, \sigma) \leq \frac{q_p}{2^\kappa} + \frac{2\sigma}{2^\tau} + \frac{4\tau q_p}{2^{b-\tau}} + \frac{8\sigma q_p + 3\sigma^2}{2^b} + \frac{2\sigma^2 + 12(b-s)q_p}{2^s},$$

where $\sigma \leq \min\{2^{b-s}, 2^\tau\}$, $\ell < 2^s$.

Proof. First, observe that the restrictions on the encryption matrices, by definition, ensure $r_3 = r_{12} = r_d = b$, $r_8 = s$. In addition, note that $E_4 \cdot E_8 \oplus E_5 = \alpha \cdot I_s$ implies that (i) $r_{45} := \mathsf{rank}\left((E_4 \cdot E_8 \oplus E_5)^s\right) = \mathsf{rank}(\alpha^s \cdot I_s) = s$ and (ii) $r_{45}' = \min_{j \leq \ell} \left\{\mathsf{rank}(I_s \oplus (E_4 \cdot E_8 \oplus E_5)^j)\right\} = \min_{j \leq \ell}\{\mathsf{rank}((\alpha^j \oplus 1) \cdot I_s)\} = s$. This follows from the fact that α is a primitive element in $GF(2^s)$, and $\ell < 2^s$. Next, we simplify all the terms involving mcoll by the following result: $\mathsf{mcoll}(\mu, 2^\beta) \leq 4\beta$, for any μ, β with $\mu \leq 2^\beta$ [11]. We can apply this result as we assume $\sigma \leq \min\{2^{b-s}, 2^\tau\}$. Finally, the Theorem follows from Proposition 4 as we simplify all the terms and use $\sigma = \sigma_e + \sigma_d$.

4.1 Security of Modified ORANGE-ZEST

In this section, we discuss the security of the modified variant of ORANGE-ZEST, as proposed in [12]. Note that the construction uses the format function satisfying the definition and the initial extra secret state is generated by $\rho(\Pi(K\|N))$, where $\rho(X) = \lfloor X \rfloor_s$. Now let us look at the feedback function of the design. Note that the feedback function remains the original one, and it is given as follows.

$$E_{\text{ORANGE-ZEST}} = \begin{bmatrix} \begin{bmatrix} I_{b-s} \oplus A^{-1} & 0_{(b-s)\times s} \\ 0_{s\times(b-s)} & 0_{s\times s} \end{bmatrix} & \begin{bmatrix} 0_{(b-s)\times s} \\ I_s \end{bmatrix} & I_b \\ \begin{bmatrix} 0_{s\times(b-s)} & \alpha\cdot I_s \end{bmatrix} & 0_s & 0_{s\times b} \\ I_b & \begin{bmatrix} 0_{(b-s)\times s} \\ I_s \end{bmatrix} & I_b \end{bmatrix},$$

where $A_{b-s} = \begin{bmatrix} 0_{(b-s-1)\times 1} & I_{b-s} \\ 1 & 0_{1\times(b-s)} \end{bmatrix}$.

It is easy to verify that the above feedback function along with the modified format function satisfies the definition of frTtP construction. Moreover, the feedback function satisfies (i) $\text{rank}(r_{12}) = b$, (ii) $\text{rank}(D_1) = b$, (iii) $E_4 \cdot E_8 \oplus E_5 = \alpha \cdot I_s$, and hence belongs to the frTtP+ family. Hence, applying Theorem 2, we obtain the security of ORANGE-ZEST$_{\text{mod}}$:

$$\mathbf{Adv}^{\text{AEAD}}_{\text{ORANGE-ZEST}_{\text{mod}}}(\sigma, q_p) \leq \frac{q_p}{2^\kappa} + \frac{2\sigma}{2^\tau} + \frac{4\tau q_p}{2^{b-\tau}} + \frac{8\sigma q_p + 3\sigma^2}{2^b} + \frac{2\sigma^2 + 12(b-s)q_p}{2^s}.$$

Remark 1. We would like to point out that Dobraunig et al. [15] mounted a forgery attack on the original construction, i.e., ORANGE-ZEST, exploiting the property that the extra-state doesn't depend on the nonce under certain cases (to be precise, for the case of empty associated data), which is a necessary condition as discussed in Proposition 3. However, the attack is not applicable on the ORANGE-ZEST$_{\text{mod}}$ as it follows the proper formatting, as mentioned. This result shows that the weakness was only due to the initial extra-state generation protocol, not the weakness of the underlying feedback function. Also, we would like to highlight that the security becomes void if $\tau = b$, as evident from the bound, justifying a matching attack as reported by Khairallah et al. [18].

4.2 (In)security of Full Rate Sponge-Duplex and Oribatida

In this subsection, we discuss the full-rate version of conventional Sponge-duplex which uses some extra-state. The corresponding feedback function can be represented by E_{duplex}.

$$E_{\text{duplex}} = \begin{bmatrix} I_b \star I_b \\ \star \; \star \; \star \\ I_b \star I_b \end{bmatrix}, \quad E_{\text{Oribatida}} = \begin{bmatrix} I_b & 0_{b\times b} & I_b \\ \alpha\cdot I_b & 0 & 0 \\ I_b & I_b & I_b \end{bmatrix}.$$

In [7], Bhattacharjee et al. designed Oribatida, a variant of Sponge-duplex with extra-state to achieve integrity security in the RUP setting, where plaintexts may be released before verification. Now let us look at the design of Oribatida when we make it full-rate. The feedback function of a full-rate Oribatida construction can be represented as $E_{\text{Oribatida}}$.

Now let us consider a full rate Transform-then-Permute construction that uses an instantiation of E_{duplex} (and consequently $E_{\text{Oribatida}}$) as the underlying feedback function. By simple linear algebra, one can show that for such a feedback function, we have $D_1 = 0$, which essentially says, during decryption, Y_{i+1} does not depend on X_i. This can be exploited by an adversary $\mathscr{A}$ to mount a forgery attack. Let us assume $\mathscr{A}$ makes an encryption query (N, A, M) such that $\text{Fmt}(A, M) = (B_1, \ldots, B_a, B_{a+1})$, and the corresponding response is (C, T), where $|C| = b$. Now, $\mathscr{A}$ can choose an A' such that $\text{Fmt}(A', M) = (B_1, \ldots, B_{a-1}, B'_a, B_{a+1})$, where $B'_a \neq B_a$ and makes a forging of the form (N, A', C, T). As S_a is generated using $N, K, B_1, \ldots, B_{a-1}$ which are the same in both the encryption and decryption queries. As $D_1 = 0$, for both the encryption and decryption queries, we have $X_{a+1} = D_2 \cdot S_a \oplus D_3 \cdot C$ validating (N, A', C, T) to be a valid forgery. This attack shows the insecurity of a full-rate variant of conventional Sponge-duplex (and consequently Oribatida), even when an additional extra state in incorporated.

4.3 frTtP with Combined and Beetle Feedback

Now let us look at what happened if we use a combined feedback function (as in CoFB) [9], or use a full-rate version of Beetle feedback [8] incorporating extra state. The combined and the full-rate Beetle feedback function (dubbed as Beetle-fb) can be represented as below:

$$E_{\text{combined}} = \begin{bmatrix} G & \begin{bmatrix} 0_s \\ I_s \end{bmatrix} & I_b \\ 0_{s \times b} & \alpha \cdot I_s & 0_{s \times b} \\ I_b & 0_{b \times s} & I_b \end{bmatrix}, \quad E_{\text{Beetle-fb}} = \begin{bmatrix} \rho_1 \star I_b \\ \star \; \star \; \star \\ I_b \star I_b \end{bmatrix}.$$

For combined feedback, G is a square matrix of size b, such that both G and $G \oplus I_b$ are non-singular. On the other hand, Beetle-fb considers a family of feedback functions with ρ_1 is a square matrix of size b such that both ρ_1 and $\rho_1 \oplus I_b$ are non-singular. Hence, we can visualize that combined feedback is essentially one instantiation from the more generalized Beetle-fb family of feedback functions.

Observe that $E_{\text{Beetle-fb}}$ satisfies the conditions $\text{rank}(D_1) = \text{rank}(\rho_1 \oplus I) = b$. Next, we consider a sub-family of Beetle-fb where the sub matrices E_2, E_4, E_5, E_6, E_8 satisfies the condition $E_8 = E_2$, $\text{rank}(E_2) = s$, $E_6 = 0$, and $E_4 \cdot E_8 \oplus E_5 = \alpha \cdot I_s$. Let us consider the family of frTtP constructions which uses a feedback function from this new sub-family of Beetle-fb, and call them fr-Beetle family of constructions. It is easy to see that, by the above definition, any fr-Beetle construction belongs to frTtP+ class. Thus, applying Theorem 2, we obtain

$$\mathbf{Adv}^{\text{AEAD}}_{\text{fr-Beetle}}(\sigma, q_p) \leq \frac{q_p}{2^\kappa} + \frac{2\sigma}{2^\tau} + \frac{4\tau q_p}{2^{b-\tau}} + \frac{8\sigma q_p + 3\sigma^2}{2^b} + \frac{2\sigma^2 + 12(b-s)q_p}{2^s}.$$

Now let us look at some efficient instantiation of Beetle-fb. More precisely, we look for the choices of E_2, E_4, E_5, E_6, E_8 satisfying the above properties. Interestingly, the choices for E_2, E_4, E_5, E_6 in E_{combined} satisfy the last three properties. If we modify the combined feedback function by defining $E_8 = E_2$, we will immediately obtain an efficient instantiation of Beetle-fb. So, let us consider this modified feedback matrix:

$$E_{\text{combined}+} = \begin{bmatrix} G & \begin{bmatrix} 0_{\frac{b}{2}} \\ I_{\frac{b}{2}} \end{bmatrix} & I_b \\ 0_{\frac{b}{2} \times b} & \alpha \cdot I_{\frac{b}{2}} & 0_{\frac{b}{2} \times b} \\ I_b & \begin{bmatrix} 0_{\frac{b}{2}} \\ I_{\frac{b}{2}} \end{bmatrix} & I_b \end{bmatrix}.$$

We dub an frTtP the construction with $E_{\text{combined}+}$ feedback function as fr-COFB. As mentioned already, fr-COFB belongs to the fr-Beetle family of frTtP constructions, and hence obtain the same security bound. The construction fr-COFB is depicted in Fig. 4.

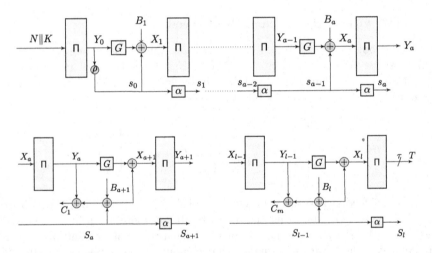

Fig. 4. fr-CoFB. Here $(B_1, \ldots, B_l) = \text{Fmt}(A, M)$ and $\rho : \{0,1\}^b \to \{0,1\}^s$ is a linear function of rank s. In the diagram, $B_i \oplus S_{i-1}$ represents the bit-wise xor of B_i and $0^{|B_i| - |S_{i-1}|} \| S_{i-1}$.

Discussion: Let us look at the original Beetle [8] construction. Assuming message injection rate of r bits and state size of b bits, Chakraborty et al. [11] showed that Beetle achieves a security of $\mathcal{O}\left(\frac{\sigma + q_p}{2^{b-r}} + \frac{\sigma q_p}{2^{2b-2r}} + \frac{\sigma q_p^2}{2^{2b-r}}\right)$. Thus, assuming $q_p \sigma \ll 2^b$, Beetle construction is secure with a data absorption rate

$r \ll b - \log(\sigma + q_p)$. Now with our new proposal fr-Beetle, we obtain the full rate with the same level of security at the cost of an extra state of size $s \gg \log(\sigma^2 + q_p)$ bits.

5 Proof of Theorem 2

5.1 Description of the Ideal World

The ideal world responds to the encryption queries, decryption queries, and primitive queries in the online phase as follows:

(1) ON PRIMITIVE QUERY (W_i, dir_i): The ideal world simulates $\Pi^{\pm}$ query honestly. In particular, if $\mathsf{dir}_i = 1$, it sets $U_i \leftarrow W_i$ and returns $V_i = \Pi(U_i)$. Similarly, when $\mathsf{dir}_i = -1$, it sets $V_i \leftarrow W_i$ and returns $U_i = \Pi^{-1}(V_i)$.

(2) ON ENCRYPTION QUERY $Q_i := (\mathsf{N}_i, \mathsf{A}_i, \mathsf{M}_i)$: It first defines

$$(B_{i,1} \ldots B_{i,a_i}, B_{i,a_i+1}, \ldots, B_{i,l_i}) := \mathsf{Fmt}(\mathsf{A}_i, \mathsf{M}_i)$$

where a_i represents the number of blocks of size b bits generated using associated data A_i. It then, samples $Y_{i,0}, \ldots, Y_{i,l_i} \leftarrow_\$ \{0,1\}^b$. For all $1 \le j \le l_i$, it then calculates

$$S_{i,j} = \begin{cases} \rho \cdot Y_{i,0}, & \text{if } j = 0 \\ E_5^{j-1} \cdot (E_4 \oplus E_5 \cdot \rho) \cdot Y_{i,0} \oplus \bigoplus_{k=1}^{j-1} E_5^{j-1-k} \cdot E_4 \cdot Y_{i,k}, & \text{otherwise.} \end{cases}$$

Next, it computes

$$C_{i,j} = Y_{i,j-1} \oplus E_8 \cdot S_{i,j-1} \oplus B_{i,j}, \ \forall \ a_i + 1 \le j \le l_i.$$

Finally, it returns $(\mathsf{C}_i, \mathsf{T}_i)$, where $\mathsf{C}_i = \lfloor C_{i,a_i+1} \| \cdots \| C_{i,l_i} \rfloor_{|\mathsf{M}_i|}$, $\mathsf{T}_i = \lfloor Y_{i,l_i} \rfloor_\tau$.

(3) ON DECRYPTION QUERY $Q_i := (\mathsf{N}_i^*, \mathsf{A}_i^*, \mathsf{C}_i^*, \mathsf{T}_i^*)$: We only consider non-trivial decryption queries, and the ideal world always aborts (returns the abort symbol $\bot$) for any such query.

OFFLINE PHASE OF IDEAL WORLD. After completion of oracle interaction (the above three types of queries possibly in an interleaved manner), the ideal oracle sets $\mathbb{E}, \mathbb{D}, \mathbb{P}$ to denote the sets of all the query indices corresponding to the encryption, decryption, and primitive queries respectively. Let $|\mathbb{E}| = q_e, |\mathbb{D}| = q_d, |\mathbb{P}| = q_p$.

☐ Extended Transcripts (Encryption Queries). Now we describe the extended transcript for the encryption queries. It samples $K \leftarrow_\$ \{0,1\}^\kappa$. For all $i \in \mathbb{E}$ and $j \subset [0, l_i]$, we define

$$X_{i,j} = \begin{cases} K \| \mathsf{N}_i, & \text{if } j = 0 \\ E_1 \cdot Y_{i,j-1} \oplus E_2 \cdot S_{i,j-1} \oplus E_3 \cdot B_{i,j}, & \text{otherwise.} \end{cases}$$

☐ **Extended Transcripts (Decryption Queries).** Now we describe an extended transcript for the decryption queries. Given any decryption query $(N_i^*, A_i^*, C_i^*, T_i^*)$, $i \in \mathbb{D}$, let $(B_{i,1}^*, \ldots, B_{i,a_i^*}^*)$ are the blocks generated corresponding to A^*, and $(C_{i,1}^* \ldots, C_{l_i^*-a_i^*}^*) \xleftarrow{b} C_i^*$. Now, we define an integer p_i as follows.

- If $\forall i' \in \mathbb{E}$, $N_i^* \neq N_{i'}$, define $p_i = -1$.
- Else, consider $i' \in \mathbb{E}$, such that $N_i^* = N_{i'}$. Since the adversary is nonce respecting there exists a unique i'.
  - If $a_{i'} \leq a_i^*$, define p_i to be the length of the maximum blockwise common prefix of $(B_{i,1}^*, \ldots, B_{i,a_i^*}^*, C_{i,1}^* \ldots, C_{l_i^*-a_i^*}^*)$ and $(B_{i',1}, \ldots, B_{i',a_i^*}, C_{i',l_{i'}-a_i^*}, \ldots, C_{i',l_{i'}})$.

  - Else, $a_i^* < a_{i'}$. Using the extended encryption transcript, for all $j \in [a_i^* + 1, a_{i'}]$ define,

  $$C_{i,j} = Y_{i,j-1} \oplus E_8 \cdot S_{i,j-1} \oplus B_{i,j}.$$

    Finally define, p_i to be the length of the maximum block-wise common prefix of $(B_{i,1}^*, \ldots, B_{i,a_i^*}^*, C_{i,1}^* \ldots, C_{l_i^*-a_i^*}^*)$ and $(B_{i',1}, \ldots, B_{i',a_i^*}, C_{i',a_i^*+1}, \ldots, C_{l_{i'}})$.

Further, for all $i \in \mathbb{D}$ and $0 \leq j \leq p_i$, we define the internal states of the ith decryption query as follows: $X_{i,j}^* = X_{i',j}$, $Y_{i,j}^* = Y_{i',j}$, $S_{i,j}^* = S_{i,j}$. In addition, we compute

$$X_{i,p_i+1}^* = \begin{cases} E_1 \cdot Y_{i,p_i}^* \oplus E_2 \oplus S_{i,p_i}^* \oplus E_3 \cdot B_{i,p_i+1}^*, & \text{if } p_i < a_i^* \\ D_1 \cdot Y_{i,p_i}^* \oplus D_3 \cdot C_{i,l_i-p_i}^*, & \text{otherwise} \end{cases}$$

and $S_{i,p_i+1}^* = S_{i',p_i+1}$. Note that by property of Fmt function, $X_{i,p_i+1}^* \neq X_{i',p_i+1}^*$. However, it might collide with a permutation query, i.e., $(X_{i,p_i+1}^*, \star, \star) \in \omega_p$. To handle this case, we now consider multi-chain graph $G_{\omega_p}^L$, where $L = D_1$. Note that it is possible to apply the multi-chain graph as $D_2 = 0$ (justifying our choice of (C4) for frTtP). Let us assume $x_{i,j} := E_3 \cdot C_{i,j-a_i^*}^*$ for all $i \in \mathbb{D}$, $j \in [a_i^*, l_i^*]$. If $a_i^* \leq p_i$ using $Y_{p_i+1}^*$, we consider all possible labeled walks

$$Y_{p_i+1}^* \xrightarrow{(x_{i,p_i+2}, \ldots, x_{i,j})} Y_{i,k}^*.$$

Let $j_{\max}$ denote the maximum of all such j values. Now, we define a new integer p_i' in the following way:

$$p_i' = \begin{cases} p_i, & \text{if } p_i \leq a_i^* \text{ or } (X_{i,p_i+1}^*, \star, \star) \notin \omega_p \\ j_{\max}, & \text{otherwise.} \end{cases}$$

Finally, we define

$$X_{i,p_i'+1}^* = \begin{cases} E_1 \cdot Y_{i,p_i'}^* \oplus E_2 \oplus S_{i,p_i'}^* \oplus E_3 \cdot B_{i,p_i'+1}^* & \text{if } p_i < a_i^*. \\ D_1 \cdot Y_{i,p_i'}^* \oplus D_3 \cdot C_{i,l_i-p_i'}^* & \text{otherwise.} \end{cases}$$

☐ Extended Adversarial Transcripts. The overall transcript of the adversary consists of $\omega = (\omega_e, \omega_d, \omega_p)$, where the primitive, encryption, and decryption transcripts are given as follows:

$$\omega_p = (U_i, V_i, \pm)_{i\in\mathbb{P}}$$
$$\omega_e = (N_i, A_i, M_i, X_{i,j}, Y_{i,j}, S_{i,j}, C_i, T_i)_{i\in\mathbb{E}, j\in[l_i]}$$
$$\omega_d = (N_i^*, A_i^*, C_i^*, T_i^*, X_{i,j}^*, Y_{i,j}^*, S_{i,j}^*, \bot)_{i\in\mathbb{D}, j\in[p_i'+1]}.$$

5.2 Defining and Bounding Bad Transcripts in Ideal World

We now consider some bad events that may occur due to the primitive, encryption and decryption transcript.

BAD1: $\exists \, (U, \star, \star) \in \omega_p : \; K = \lceil U \rceil_\kappa$.
BAD2: $\exists \, (i,j) \neq (i',j')$ such that $S_{i,j} = S_{i',j'}$, where $i \in \mathbb{E}$, $j \in [l_i]$, $i' \in \mathbb{E}$, $j' \in [l_{i'}]$.
BAD3: $\exists \, i \in \mathbb{E}, j \in [l_i]$ such that $(\star, Y_{i,j}, \star) \in \omega_p$.
BAD4: $\exists \, i \in \mathbb{E}, \; j \in [l_i]$ such that $(X_{i,j}, \star, \star) \in \omega_p$.
BAD5: $\exists \, (i,j) \neq (i',j')$ such that $Y_{i,j} = Y_{i',j'}$, where $i \in \mathbb{E}$, $j \in [l_i]$, $i' \in \mathbb{E}$, $j' \in [l_{i'}]$.
BAD6: $\exists \, (i,j) \neq (i',j')$ such that $X_{i,j} = X_{i',j'}$, where $i \in \mathbb{E}$, $j \in [l_i]$, $i' \in \mathbb{E}$, $j' \in [l_{i'}]$.
BAD7: $\exists \, i \in \mathbb{D}, p_i' = l_i^*$ and $\lfloor Y_{i,l_i}^* \rfloor_\tau = T_i^*$.
BAD8: $\exists \, i \in \mathbb{D}, i' \in \mathbb{E}, \; j' \in [l_{i'}]$ such that $X_{i,p_i'+1}^* = X_{i',j}$, where $p_i' \leq l_i^* - 1$.

We would like to point out that the first six events broadly represent some collisions in the internal states during encryption and primitive queries. Such a collision essentially induces a collision in the permutation input or output and makes the transcript permutation incompatible which can be used to perform privacy attacks. Hence, we call them bad events. The last two bad events are due to the decryption queries and can lead to forgery attacks.

Now we use the following lemma to upper bound the probability of the bad events.

Lemma 2. *Let us define* BAD = BAD1 $\cup \cdots \cup$ BAD8. *We can bound the probability of* BAD *as follows.*

$$\Pr[\mathsf{BAD}] \leq \frac{q_p}{2^\kappa} + \frac{9\sigma_e q_p}{2^{r_{12}}} + \frac{2\sigma_e^2}{2^{r_{12}}} + \frac{\sigma_e^2}{2^{r_{45}+r_{45}'-s}} + \frac{q_p\mathsf{mcoll}(\sigma_e, 2^\tau)}{2^{b-\tau}}$$
$$+ \frac{q_p\mu_{q_p}\sigma_d}{2^{r_d}} + \frac{\sigma_e + q_p}{2^{r_d}} + \frac{\sigma_d\sigma_e}{2^{r_d+r_3-b}} + \frac{q_p\mathsf{mcoll}(\sigma_e, 2^{b+s-r_{45}-r_8})}{2^{r_{12}+r_{45}+r_8-b-s}}$$
$$+ \frac{\sigma_d q_p\mathsf{mcoll}(\sigma_e, 2^{b+s-r_{45}-r_8})}{2^{r_{12}+r_d+r_8+r_{45}-b-s}} + + \frac{q_p\mathsf{mcoll}(\sigma_e, 2^{b+s-r_{45}-r_8})}{2^{r_{45}+r_8-s}}.$$

The proof of the Lemma can be found in the full version [10].

5.3 Good Transcript Analysis and Completion of the Proof

In the online phase, the AE encryption, decryption, and direct primitive queries are faithfully responded to based on $\Pi^{\pm}$. Like the ideal world, after the completion of interaction, the real world returns all X-values Y-values and S-values corresponding to the encryption queries only and all the derived X, Y, S values corresponding to the decryption queries.

Lemma 3. *Let Θ_0 and Θ_1 denote the random transcript variable obtained in the ideal and real worlds, respectively. For any good transcript $\omega = (\omega_p, \omega_e, \omega_d)$, we have*

$$\frac{\Pr\left[\Theta_1 = w\right]}{\Pr\left[\Theta_0 = w\right]} \geq 1 - \left(\frac{2q_d}{2^\tau} + \frac{2\sigma_d(\sigma + q_p)}{2^{r_d}}\right).$$

Here we briefly discuss an informal proof sketch of the lemma. The tuples ω_e is permutation compatible and disjoint from ω_p. So, the union of tuples $\omega_e \cup \omega_p$ also remains permutation compatible. Now, in the real world, a decryption query may return M_i which is not $\perp$. However, since a good transcript always aborts on a decryption query, we need to bound the probability of this event. Suppose for all $0 \leq j \leq p_i'$, $Y_{i,j}^*$, $S_{i,j}^*$ and $X_{i,j+1}^*$ are defined as before. Now, for all $i \in \mathbb{D}$, we have either $p_i' = l_i - 1$ and $(X_{i,m_i}^*, \star \| T_i^*) \in \omega_p \cup \omega_e$ (call it a Type-1 decryption query) or $p_i' < l_i - 1$ but $X_{i,p_i'+1}^* \notin \omega_p \cup \omega_e$ (call it a Type-2 decryption query). Type-1 decryption queries are taken care of in bad events. To be precise, such queries are already rejected due to BAD6. For the Type-2 decryption query, observe that $X_{i,p_i'+1}^*$ is fresh i.e. it has never been queried before by the adversary. So, $\Pi(X_{i,p_i'+1}^*)$ would be random over a large set. This would ensure a high probability that such decryption queries will also be rejected. The formal proof can be found in the full version [10].

Proof of Theorem 2: Finally the proof of the theorem is complete as we apply Lemma 2 and Lemma 3 in Theorem 1.

5.4 Conclusion and Future Direction

In this paper, we introduce a class of full-rate Sponge-type constructions called the frTtP by introducing an extra-state as compensation for increasing the size of the rate part. We further extend the result to show that a sub-class of the constructions, called frTtP+, achieves security up to $D \ll 2^{s/2}, T \ll 2^s$, where s is the size of the extra-state (in bits). Consequently, we have shown that ORANGE-ZEST$_{\mathrm{mod}}$ and a family of constructions following Beetle-like feedback functions belongs to the frTtP+ class, and hence, achieve the desired security. Extending the result for a more general class of constructions (beyond the frTtP class) and designing a more efficient full-rate Transform-then-Permute than ORANGE-ZEST$_{\mathrm{mod}}$ or fr-COFB can be considered as an interesting open problem. In fact, one can investigate whether a hybrid feedback function (as used in HyENA), can be used efficiently to construct a full rate Transform-then-Permute. Finally, using an extra state may lead to the necessity of increased protection against a wide variety of side-channel attacks. A concrete side channel analysis of frTtP schemes is an important open problem and is left for future research.

References

1. AlTawy, R., et al.: SpoC. Submission to NIST LwC Standardization Process (Round 2) (2019)
2. Bertoni, G., Daemen, J., Peeters, M., Van Assche, G.: Sponge functions. In: Proceedings of the ECRYPT Hash Workshop 2007 (2007)
3. Bertoni, G., Daemen, J., Peeters, M., Van Assche, G.: On the indifferentiability of the sponge construction. In: Smart, N. (ed.) EUROCRYPT 2008. LNCS, vol. 4965, pp. 181–197. Springer, Heidelberg (2008). https://doi.org/10.1007/978-3-540-78967-3_11
4. Bertoni, G., Daemen, J., Peeters, M., Van Assche, G.: Sponge-based pseudorandom number generators. In: Mangard, S., Standaert, F.-X. (eds.) CHES 2010. LNCS, vol. 6225, pp. 33–47. Springer, Heidelberg (2010). https://doi.org/10.1007/978-3-642-15031-9_3
5. Bertoni, G., Daemen, J., Peeters, M., Van Assche, G.: Duplexing the sponge: single-pass authenticated encryption and other applications. In: Miri, A., Vaudenay, S. (eds.) SAC 2011. LNCS, vol. 7118, pp. 320–337. Springer, Heidelberg (2012). https://doi.org/10.1007/978-3-642-28496-0_19
6. Bertoni, G., Daemen, J., Peeters, M., Van Assche, G.: On the security of the keyed sponge construction. In: Proceedings of the Symmetric Key Encryption Workshop 2011 (2011)
7. Bhattacharjee, A., List, E., López, C.M., Nandi, M.: The oribatida family of lightweight authenticated encryption schemes. Submission to NIST LwC Standardization Process (Round 2) (2019)
8. Chakraborti, A., Datta, N., Nandi, M., Yasuda, K.: Beetle family of lightweight and secure authenticated encryption ciphers. IACR Trans. Cryptogr. Hardw. Embed. Syst. 2018(2), 218–241 (2018)
9. Chakraborti, A., Iwata, T., Minematsu, K., Nandi, M.: Blockcipher-based authenticated encryption: how small can we go? In: Fischer, W., Homma, N. (eds.) CHES 2017. LNCS, vol. 10529, pp. 277–298. Springer, Cham (2017). https://doi.org/10.1007/978-3-319-66787-4_14
10. Chakraborty, B., Datta, N., Nandi, M.: Designing full-rate sponge based AEAD modes. Cryptology ePrint Archive, Paper 2023/1673 (2023). https://eprint.iacr.org/2023/1673
11. Chakraborty, B., Jha, A., Nandi, M.: On the security of sponge-type authenticated encryption modes. IACR Trans. Symmetric Cryptol. 93–119 (2020)
12. Chakraborty, B., Nandi, M.: ORANGE. Submission to NIST LwC Standardization Process (Round 2) (2019)
13. Chen, S., Steinberger, J.: Tight security bounds for key-alternating ciphers. In: Nguyen, P.Q., Oswald, E. (eds.) EUROCRYPT 2014. LNCS, vol. 8441, pp. 327–350. Springer, Heidelberg (2014). https://doi.org/10.1007/978-3-642-55220-5_19
14. Dobraunig, C., Eichlseder, M., Mendel, F., Schäffer, M.: ASCON v1.2: lightweight authenticated encryption and hashing. J. Cryptol. 34, 1–42 (2021)
15. Dobraunig, C., Mendel, F., Mennink, B.: Round 1 official comments: ORANGE. Submission to NIST LwC Standardization Process (2019)
16. Jovanovic, P., Luykx, A., Mennink, B.: Beyond $2^{c/2}$ security in sponge-based authenticated encryption modes. In: Sarkar, P., Iwata, T. (eds.) ASIACRYPT 2014. LNCS, vol. 8873, pp. 85–104. Springer, Heidelberg (2014). https://doi.org/10.1007/978-3-662-45611-8_5

110 B. Chakraborty et al.

17. Jovanovic, P., Luykx, A., Mennink, B., Sasaki, Y., Yasuda, K.: Beyond conventional security in sponge-based authenticated encryption modes. J. Cryptol. **32**(3), 895–940 (2019)
18. Kairallah, M.M.M., Rohit, R., Sarkar, S.: Round 2 official comments: ORANGE. Submission to NIST LwC Standardization Process (2019)
19. Mennink, B., Neves, S.: Encrypted Davies-Meyer and its dual: towards optimal security using mirror theory. In: Katz, J., Shacham, H. (eds.) CRYPTO 2017. LNCS, vol. 10403, pp. 556–583. Springer, Cham (2017). https://doi.org/10.1007/978-3-319-63697-9_19
20. Mennink, B., Reyhanitabar, R., Vizár, D.: Security of full-state keyed sponge and duplex: applications to authenticated encryption. In: Iwata, T., Cheon, J.H. (eds.) ASIACRYPT 2015. LNCS, vol. 9453, pp. 465–489. Springer, Heidelberg (2015). https://doi.org/10.1007/978-3-662-48800-3_19
21. Patarin, J.: Etude des Générateurs de Permutations Pseudo-aléatoires Basés sur le Schéma du DES. Ph.D. thesis, Université de Paris (1991)
22. Patarin, J.: The "Coefficients H" technique. In: Avanzi, R.M., Keliher, L., Sica, F. (eds.) SAC 2008. LNCS, vol. 5381, pp. 328–345. Springer, Heidelberg (2009). https://doi.org/10.1007/978-3-642-04159-4_21

Towards Minimizing Tweakable Blockcipher-Based Generalized Feistel Networks

Yuqing Zhao[1,2] and Chun Guo[1,2,3(✉)]

[1] School of Cyber Science and Technology, Shandong University,
Qingdao, Shandong, China
yqzhao@mail.sdu.edu.cn, chun.guo@sdu.edu.cn
[2] Key Laboratory of Cryptologic Technology and Information Security of Ministry
of Education, Shandong University, Qingdao 266237, Shandong, China
[3] Shandong Research Institute of Industrial Technology, Jinan 250102,
Shandong, China

Abstract. A generalized Feistel network (GFN) is a classical approach to constructing a blockcipher from pseudorandom functions (PRFs). Recently, Nakaya and Iwata (ToSC, 2022) formalized tweakable blockcipher (TBC)-based counterparts of type-1, type-2, and type-3 GFNs. This paper studies minimizing the number of TBC calls in such GFN variants. Motivated by the so-called extended GFNs of Berger et al. (IEEE TC, 2016) and Zhao et al. (CANS 2023), we consider TBC-based type-2 GFN and replace the blockwise shuffle with a block-oriented linear diffusion layer. We show that when this diffusion layer is moderately strong, 4 TBC-based GFN rounds are sufficient for CCA security, which is independent of the number of lines. This provides a much more efficient approach to TBC-based enciphering schemes.

Keywords: Tweakable blockcipher · Generalized Feistel network · H-coefficient technique · Provable security

1 Introduction

(Generalized) Feistel Networks. The well-known Feistel blockciphers, including the Data Encryption Standard (DES) [12], rely on the Feistel permutation $\mathrm{Feistel}^F(A, B) := (B, A \oplus F(B))$ defined upon a domain-preserving round function $F : \{0,1\}^n \to \{0,1\}^n$. To increase flexibility, this structure has been generalized to various flavors, including: (i) *contracting Feistel networks* that employ contracting round functions $G : \{0,1\}^m \to \{0,1\}^n$, $m > n$ [36]; (ii) *expanding Feistel networks* that employ expanding round functions $F : \{0,1\}^n \to \{0,1\}^m$ [36]; (iii) *alternating Feistel networks* [1,23] that alternate the invocations of contracting and expanding round functions; and (iv) *type-1, type-2, and type-3 Feistel networks* [43] that use n-bit to n-bit round functions to create a dn-bit blockcipher for some $d > 2$ (we call them *multi-line GFNs*). A particular advantage of multi-line GFNs is the well-compatibility with round functions of small domains, which eases the design of round functions. In extreme cases [40,41], designers can simply use a keyed S-box as the round function.

© The Author(s), under exclusive license to Springer Nature Switzerland AG 2024
A. Chattopadhyay et al. (Eds.): INDOCRYPT 2023, LNCS 14459, pp. 111–133, 2024.
https://doi.org/10.1007/978-3-031-56232-7_6

The provable security of Feistel networks and their variants was initiated by Luby and Rackoff [22]. The approach is to model the round functions as pseudorandom functions (PRFs) and to analyze the structural features of the designs. With this model, Luby and Rackoff proved CCA security for 4-round balanced Feistel networks, and subsequent works extended this direction to refine results [16,31,33,35] or to cover the aforementioned generalized Feistel networks [1,3,6,16,23,27,32,43].

Tweakable Blockcipher-Based GFNs. At CRYPTO 2002, Liskov et al. [19] formalized the notion of *tweakable blockciphers (TBCs)*. A TBC is a family of independent blockciphers indexed by a *tweak*. TBCs can be constructed via modes of operation of classical blockciphers: see [38] and the references therein. They can also be designed from the scratch, with famous designs such as Skinny [2] and Deoxys-BC [17].

The additional tweak input raises much more possibilities for the design of TBC-based modes. Importantly to us, Minematsu at FSE 2009 [24] initiated constructing (wide) blockciphers from TBCs. His construction leverages universal hash functions and somewhat resembles the Naor-Reingold construction [32]. Almost concurrently, Coron et al. [10] proposed a TBC-based Feistel network $\text{TFeistel}_K^{\widetilde{E}}(A, B) := \big(\widetilde{E}_K(A, B), A\big)$, which instantiates a $2n$-bit permutation from a TBC $\widetilde{E} : \mathcal{K} \times \{0,1\}^n \times \{0,1\}^n \to \{0,1\}^n$ with n-bit blocks and tweaks. Coron et al. proved (birthday) CCA security up to $O(2^{n/2})$ queries for 2 rounds and (beyond-birthday) CCA security up to $O(2^n)$ queries for 3 rounds.

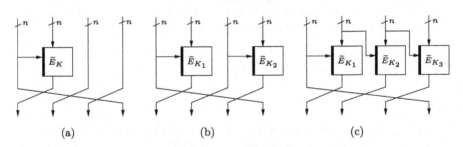

Fig. 1. Round functions of TBC-based GFNs. (a) Type-1 (b) Type-2 (c) Type-3

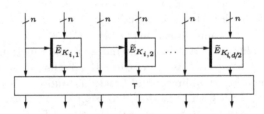

Fig. 2. Extended TBC-based type-2 GFN (ETGFN) with d lines.

Inspired by the various types of GFNs, it is natural to consider TBC-based GFN variants. In this respect, the TBC-based contracting Feistel variant has been proposed by Minematsu [25] and extended by Nakamichi and Iwata [28,29]. While TBC-based multi-line GFN variants have been formalized by Nakaya and Iwata [30], we refer to Fig. 1 for depiction and Table 1 for their proven results.

Note that the multi-line GFN variants use TBCs of n-bit blocks and n-bit tweaks, while the contracting variants require TBCs with long tweaks (or tweak extension methods [18,25]). On the positive side, the contracting variants achieved faster diffusion.

Our Results. Recently, NIST has mentioned its intent to call for a standard for wide blockcipher modes[1]. TBC-based GFNs provide potential solutions. In this respect, it is necessary to seek to minimize the number of TBC calls, which was also (implicitly) mentioned in [30].

A very recent work of Zhao et al. [42] (which gained inspirations from [4,14]) proposed to replace the shuffles in type-2 GFNs with a strong (linear) diffusion layer. By an impossibility result of Nandi [31], any type-2 GFN variant needs at least 4 rounds. Whereas Zhao et al. [42] proved CCA security for their strong type-2 GFN variant with 5 rounds. This clearly demonstrated the power of strong diffusion layers in minimizing the number of block function invocations.

Motivated by this, we consider employing a strong diffusion layer T in the TBC-based type-2 GFN as well. The obtained variant is illustrated in Fig. 2. We call this scheme *extended TBC-based type-2 generalized Feistel networks (ETGFNs)*.

As our main result, we prove CCA security for 4 rounds, regardless of d the number of lines in the construction. We identify conditions on the diffusion layer T that are sufficient for this CCA security proof and list several candidates. Unsurprisingly, the conditions are more relaxed than the analogs in structures using domain-preserving block functions [14,42], and this in turn demonstrates the power of TBCs in constructing modes of operations.

Discussion. By an impossibility result of Bhaumik et al. [5, Sect. 3], any dn-bit SPRPs built from TBCs n-bit blocks and n-bit tweaks and linear diffusion functions must make at least $(3d - 1)/2$ TBC calls. By this, TBC-based type-2 GFNs need at least 3 rounds to achieve CCA security. Our proof method only yields a slightly inferior result of 4 rounds, and we leave characterizing the security of 3 rounds as a challenging open question.

Bhaumik et al. [5, Sect. 4] also proposed a TBC-based SPRP ZCZ that makes about $3d/2$ TBC calls, which is optimal by their impossibility result. Meanwhile, ZCZ admits variable input length. Our current provable result is thus less efficient and flexible than ZCZ (and other TBC-based variable-input-length wide tweakable blockcipher constructions [8,9,11,26,39] as well). On the other hand,

---

[1] https://www.nist.gov/news-events/events/2023/10/third-nist-workshop-block-cipher-modes-operation-2023.

Table 1. Summary of previous and our results on TBC-based multi-line GFN variants. "Model" shows the attack model. "Construction" is a blockcipher with dn-bit blocks, except for TBC-based Feistel which is a $2n$-bit blockcipher. In the table, q denotes the number of queries, and in the results of [37], $t \geq 1$ is a parameter that specifies the number of rounds.

Model	Construction	Security bound	# of rounds	Reference
SPRP	Feistel	$O\left(q^2/2^{2n}\right)$	3	[10]
		$2\left(\frac{2q}{t+1}\left(\frac{30q}{2^{2n}}\right)^t\right)^{1/2}$	$4t+2$	[37]
PRP	Type-1 GFN	$dq^2/2^n$	$2d-2$	[30]
		$d^2q^2/2^{2n}$	$3d-2$	
SPRP	Type-1 GFN	$d^2q^2/2^n$	d^2-2d+2	
		$d^3q^2/2^{2n}$	d^2-d+2	
	Type-2 GFN	$d^2q^2/2^n$	d	
		$d^3q^2/2^{2n}$	$d+2$	
	Type-3 GFN	$d^2q^2/2^n$	d	
		$d^3q^2/2^{2n}$	$d+1$	
SPRP	ETGFN	$dq^2/2^n$	4	Theorem 1

our new TBC-based GFN variant has an iterative structure which is similar to Nakaya and Iwata's constructions [30]. These TBC-based GFNs thus provide a new approach to designing wide blockciphers via "prove-then-prune" [15]. Namely, one may instantiate the TBC-based GFNs with a round-reduced Skinny or Deoxys-BC, and compensate by iterating more rounds. By this, the iterative nature of the TBC-based GFNs is fully utilized. In fact, the Simpira family [13] of permutations was the result of such an idea regarding classical GFNs. We leave further exploring this idea for future work.

Organization. The rest of this paper is organized as follows. Section 2 presents notations and definitions. Then, we formally introduce extended TBC-based type-2 GFN in Sect. 3. The CCA security proof of our extended GFN variant is given in Sect. 4. We finally conclude in Sect. 5.

2   Preliminaries

2.1   Notation

For a finite set $\mathcal{S}$, $s \xleftarrow{\$} \mathcal{S}$ denotes the procedure of selecting an element from $\mathcal{S}$ uniformly at random, and assigning it to s. The set of all the bit strings of n bits is written as $\{0,1\}^n$, and for a bit string x, $|x|$ denotes its length in bits. For two integers a and b with $a \leq b$, we let $[a..b] = \{a, a+1, \ldots, b\}$.

We view n as a cryptographic security parameter and let $\mathbb{F} := GF(2^n)$, which is identified with $\{0,1\}^n$. For any positive integer d, a dn-bit string $x \in \{0,1\}^{dn}$

is also viewed as a *column vector* in $\mathbb{F}^d$. Indeed, strings and column vectors are just two sides of the same coin. Let x be a *column vector* in $\mathbb{F}^d$, then x'^2 is a row vector obtained by transposing x. Throughout the remaining, depending on the context, the same notation, e.g., x, may refer to both a string and a column vector, without additional highlight. In the same vein, the concatenation $x\|y$ is also "semantically equivalent" to the column vector $\binom{x}{y}$.

In this respect, for $x \in \mathbb{F}^d$, we denote the j-th entry of x (for $j \in \{1, ..., d\}$) by $x[j]$ and define $x[a..b] := (x[a], \ldots, x[b])$ for any integers $1 \le a < b \le d$. Let's assume that d is an even number. We define $x[\text{even}] := (x[2], x[4], \ldots, x[d])$ and $x[\text{odd}] := (x[1], x[3], \ldots, x[d-1])$.

For a keyed function $F : \mathcal{K} \times \mathcal{X} \to \mathcal{Y}$, where $\mathcal{K}$ is the key space, $\mathcal{X}$ is the input space, and $\mathcal{Y}$ is the output space, the output $Y \in \mathcal{Y}$ for a key $k \in \mathcal{K}$ and an input $X \in \mathcal{X}$ is written as $Y = F_K(X)$ or $Y = F[K](X)$. For a key $K \in \mathcal{K}$, if $F_K(\cdot)$ is a permutation over $\mathcal{X}$, its inverse permutation is written as $F_K^{-1}(\cdot)$ or $F^{-1}[K](\cdot)$.

Let T be a matrix. We denote by T_{OE} the submatrix composed of odd rows and even columns of matrix T, by T_{OO} the submatrix composed of odd rows and odd columns of matrix T, by T_{EE} the submatrix composed of even rows and even columns of matrix T, and by T_{EO} the submatrix composed of even rows and odd columns of matrix T.

A Useful Operator on the (Linear) Diffusion Layer. We will frequently write $M \in \mathbb{F}^{d \times d}$ in the block form of 4 submatrices in $\mathbb{F}^{d/2 \times d/2}$. For this, we follow the convention using U, B, L, R for *upper, bottom, left,* and *right* resp., i.e.,

$$M = \begin{pmatrix} M_{\text{UL}} & M_{\text{UR}} \\ M_{\text{BL}} & M_{\text{BR}} \end{pmatrix}.$$

We use brackets, i.e., $(M^{-1})_{\text{XX}}$, $\text{XX} \in \{\text{UL, UR, BL, BR}\}$, to distinguish submatrices of M^{-1} (the inverse of M) from M_{XX}^{-1}, the inverse of M_{XX}.

As per our convention, we view $u, v \in \mathbb{F}^d$ as column vectors. During the proof, we will need to derive the "second halves" $u_2 := u[d/2 + 1..d]$ and $v_2 := v[d/2 + 1..d]$ from the "first halves" $u_1 := u[1..d/2], v_1 := v[1..d/2]$, and the equality $v = \mathsf{T} \cdot u$. To this end, we follow [14] and define an operator on T:

$$\widehat{\mathsf{T}} := \begin{pmatrix} \mathsf{T}_{\text{UR}}^{-1} \cdot \mathsf{T}_{\text{UL}} & \mathsf{T}_{\text{UR}}^{-1} \\ \mathsf{T}_{\text{BR}} \cdot \mathsf{T}_{\text{UR}}^{-1} \cdot \mathsf{T}_{\text{UL}} \oplus \mathsf{T}_{\text{BL}} & \mathsf{T}_{\text{BR}} \cdot \mathsf{T}_{\text{UR}}^{-1} \end{pmatrix}, \tag{1}$$

which satisfies

$$v = \mathsf{T} \cdot u \iff \begin{pmatrix} u_2 \\ v_2 \end{pmatrix} = \widehat{\mathsf{T}} \cdot \begin{pmatrix} u_1 \\ v_1 \end{pmatrix}.$$

2.2 Blockciphers and Tweakable Blockciphers

A blockcipher (BC) is a keyed permutation $E : \mathcal{K} \times \{0,1\}^n \to \{0,1\}^n$, where for any key $K \in \mathcal{K}$, $E_K(\cdot)$ is a permutation over $\{0,1\}^n$. Here, $\mathcal{K}$ is the key space

---

[2] In many papers, it is also denoted as x^T.

and n is the block length. If $C \in \{0,1\}^n$ is a ciphertext for a key $K \in \mathcal{K}$ and a plaintext $M \in \{0,1\}^n$, we have $C = E_K(M)$ in encryption and $M = E_K^{-1}(C)$ in decryption. In what follows, we write n-BC for a blockcipher with the block length of n bits.

A tweakable blockcipher (TBC) [20,21] is a keyed permutation $\widetilde{E} : \mathcal{K} \times \{0,1\}^t \times \{0,1\}^n \to \{0,1\}^n$ that has an additional input called a tweak. For any key $K \in \mathcal{K}$ and any tweak $T \in \{0,1\}^t$, $\widetilde{E}_K(T, \cdot)$ is a permutation over $\{0,1\}^n$. Here, $\mathcal{K}$ is the key space, t is the tweak length, and n is the block length. If $C \in \{0,1\}^n$ is a ciphertext for a key $K \in \mathcal{K}$, a tweak $T \in \{0,1\}^t$, and a plaintext $M \in \{0,1\}^n$, then we have $C = \widetilde{E}_K(T, M)$ in encryption and $M = \widetilde{E}_K^{-1}(T, C)$ in decryption. We write (t, n)-TBC for a TBC with the tweak length of t bits and the block length of n bits.

We write $\mathsf{Perm}(n)$ for the set of all the permutations over $\{0,1\}^n$. A random permutation π is a permutation that is chosen uniformly at random from $\mathsf{Perm}(n)$, i.e., $\pi \xleftarrow{\$} \mathsf{Perm}(n)$. We write $\widetilde{\mathsf{Perm}}(t, n)$ for the set of all the functions $\widetilde{P} : \{0,1\}^t \times \{0,1\}^n \to \{0,1\}^n$ such that, for any tweak $T \in \{0,1\}^t$, $\widetilde{P}(T, \cdot) \in \mathsf{Perm}(n)$. A (t, n)-tweakable random permutation $((t, n)$-TRP) $\widetilde{P}$ is a function that is chosen uniformly at random from $\widetilde{\mathsf{Perm}}(t, n)$, i.e., $\widetilde{P}$ is a (t, n)-TRP if $\widetilde{P} \xleftarrow{\$} \widetilde{\mathsf{Perm}}(t, n)$. For a (t, n)-TRP $\widetilde{P}$, for any tweak $T \in \{0,1\}^t$, $\widetilde{P}(T, \cdot)$ is a random permutation over $\{0,1\}^n$, and we write $\widetilde{P}^{-1}(T, \cdot)$ for its inverse permutation. On the other hand, for any input $X \in \{0,1\}^n$, $\widetilde{P}(\cdot, X)$ can be viewed as a random function from $\{0,1\}^t$ to $\{0,1\}^n$.

2.3 Security Definition and H-Coefficient Technique

In this paper, we consider the security of a blockcipher E as a strong PRP (SPRP) [22]. An SPRP-adversary $\mathcal{A}$ is given oracle access to $E_K(\cdot)$ and $E_K^{-1}(\cdot)$ in the real world, and it is given oracle access to $\pi(\cdot)$ and $\pi^{-1}(\cdot)$ in the ideal world. We define SPRP-advantage as follows [22]:

$$\mathbf{Adv}_E^{\mathrm{sprp}}(\mathcal{A}) = \left| \Pr[\mathcal{A}^{E_K(\cdot), E_K^{-1}(\cdot)} = 1] - \Pr[\mathcal{A}^{\pi(\cdot), \pi^{-1}(\cdot)} = 1] \right|.$$

The probabilities are taken over the randomness of $\mathcal{A}$, K, and π. An adversary $\mathcal{A}$ in the SPRP notion is in a chosen-ciphertext-attack (CCA) setting, and we may call it a CCA-adversary or an SPRP-adversary.

In our security proof, we heavily make use of the H-coefficient technique [7, 34]. Let $\mathcal{R}$ (resp. $\mathcal{R}^{-1}$) be the real world oracle defined by a blockcipher E (resp. E^{-1}), and let $\mathcal{I}$ (resp. $\mathcal{I}^{-1}$) be the ideal world oracle defined by a random permutation π (resp. π^{-1}). For an adversary $\mathcal{A}$ that makes at most q queries, a transcript θ records the interaction between $\mathcal{A}$ and the oracles, i.e., it contains all the queries of $\mathcal{A}$ and responses from the oracles. Let $\Theta_{\mathcal{R}}$ be the probability distribution of transcript θ when $\mathcal{A}$ interacts with $\mathcal{R}$ and $\mathcal{R}^{-1}$ in the real world, and $\Theta_{\mathcal{I}}$ be the probability distribution of θ when $\mathcal{A}$ interacts with $\mathcal{I}$ and $\mathcal{I}^{-1}$ in the ideal world. An attainable transcript is a transcript θ that satisfies $\Pr[\Theta_{\mathcal{I}} =$

$\theta] > 0$, i.e., it has a nonzero probability in the ideal world. Let $\mathcal{T}$ be the set of all the attainable transcripts.

With the notation above, the H-coefficient technique is the following lemma.

Lemma 1. *Consider a deterministic adversary $\mathcal{A}$. Let $\mathcal{T} = \mathcal{T}_{\text{good}} \cup \mathcal{T}_{\text{bad}}$ be a partition of the set of attainable transcripts $\mathcal{T}$. Assume that there exists ε_1 such that for any $\theta \in \mathcal{T}_{\text{good}}$, one has*

$$\frac{\Pr[\Theta_{\mathcal{R}} = \theta]}{\Pr[\Theta_{\mathcal{I}} = \theta]} \geq 1 - \varepsilon_1,$$

and that there exists ε_2 such that $\Pr[\Theta_{\mathcal{I}} \in \mathcal{T}_{\text{bad}}] \leq \varepsilon_2$. Then $\mathbf{Adv}_E^{\text{sprp}}(\mathcal{A}) \leq \varepsilon_1 + \varepsilon_2$.

3 Definition of Extended TBC-Based Type-2 GFN

In this section, we formalize extended TBC-based type-2 generalized Feistel networks (ETGFNs). An ETGFN is obtained from classical PRF-based type-2 GFN by using an (n,n)-TBC to define a round function and replacing the blockwise shuffle with a block-oriented linear diffusion layer. By iterating the round function for r times and diffusion layer for $r-1$ times, we obtain a dn-BC, where r is the number of rounds and d is the number of input/output blocks. The encryption round function, the decryption round function, the r-round encryption function, and the r-round decryption function of ETGFN are written as $\varPhi_d$, $\varPhi_d^{-1}$, $\mathcal{E}_{r,d}$ and $\mathcal{E}_{r,d}^{-1}$, respectively, which are defined as follows.

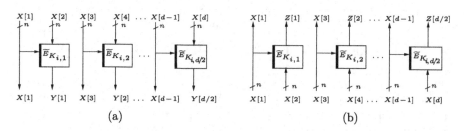

Fig. 3. (a) $\varPhi_d[\widetilde{E}_{K_{i,1}}, \ldots, \widetilde{E}_{K_{i,d/2}}](X[1]\| \ldots \|X[d]) = (X[1]\|Y[1]\| \ldots \|X[d-1]\|Y[d/2])$, where $Y[\ell] = \widetilde{E}_{K_{i,\ell}}(X[2\ell - 1], X[2\ell])$. (b) $\varPhi_d^{-1}[\widetilde{E}_{K_{i,1}}, \ldots, \widetilde{E}_{K_{i,d/2}}](X[1]\| \ldots \|X[d]) = (X[1]\|Z[1]\| \ldots \|X[d-1]\|Z[d/2])$, where $Z[\ell] = \widetilde{E}_{K_{i,\ell}}^{-1}(X[2\ell - 1], X[2\ell])$.

Let $d \geq 4$, $r \geq 1$, where d is even, $\widetilde{E}$ be an (n,n)-TBC, and $K_{1,1}, \ldots, K_{r,d/2}$ be $rd/2$ independent keys of $\widetilde{E}$. We first define the encryption round function $\varPhi_d$ of an extended TBC-based type-2 GFN. It is a permutation over $\{0,1\}^{dn}$ that takes $X \in \{0,1\}^{dn}$ as input. Internally, Φ_d uses $d/2$ TBCs $\widetilde{E}_{K_{i,1}}, \ldots, \widetilde{E}_{K_{i,d/2}}$, where $i \in \{1, 2, \ldots, r\}$, and is defined as

$$\Phi_d[\widetilde{E}_{K_{i,1}}, \ldots, \widetilde{E}_{K_{i,d/2}}](X[1]\| \ldots \|X[d])$$
$$= \Big(X[1]\|\widetilde{E}_{K_{i,1}}(X[1], X[2])\| \ldots \|X[d-1]\|\widetilde{E}_{K_{i,d/2}}(X[d-1], X[d])\Big).$$

See Fig. 3(a) for an illustration. Based on the output of TBCs, the following notation is established:

$$\widetilde{E}_{K_{i,[1..d/2]}}\big(X[\text{odd}], X[\text{even}]\big)$$
$$:= \Big(\widetilde{E}_{K_{i,1}}(X[1], X[2])\|\widetilde{E}_{K_{i,2}}(X[3], X[4])\| \ldots \|\widetilde{E}_{K_{i,d/2}}(X[d-1], X[d])\Big). \quad (2)$$

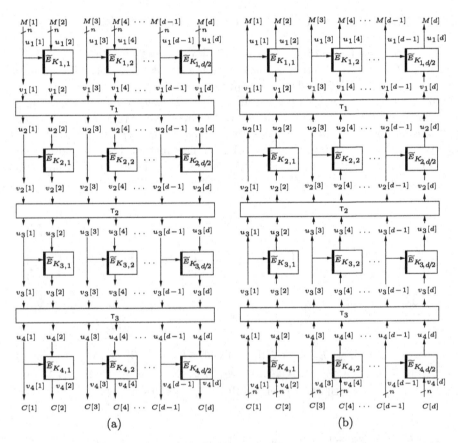

Fig. 4. (a) $\mathcal{E}_{4,d}[\widetilde{E}_{K_{1,1}}, \ldots, \widetilde{E}_{K_{4,d/2}}](M) = C$. (b) $\mathcal{E}_{4,d}^{-1}[\widetilde{E}_{K_{1,1}}, \ldots, \widetilde{E}_{K_{4,d/2}}](C) = M$.

The decryption round function Φ_d^{-1} uses decryption function $\widetilde{E}_{K_{i,1}}^{-1}, \ldots, \widetilde{E}_{K_{i,d/2}}^{-1}$ of $\widetilde{E}_{K_{i,1}}, \ldots, \widetilde{E}_{K_{i,d/2}}$ and is defined as

$$\Phi_d^{-1}[\widetilde{E}_{K_{i,1}}, \ldots, \widetilde{E}_{K_{i,d/2}}](X[1]\| \ldots \|X[d])$$
$$= \left(X[1]\|\widetilde{E}_{K_{i,1}}^{-1}(X[1], X[2])\| \ldots \|X[d-1]\|\widetilde{E}_{K_{i,d/2}}^{-1}(X[d-1], X[d]) \right).$$

See Fig. 3(b). And we introduce the notation:

$$\widetilde{E}_{K_{i,[1..d/2]}}^{-1}(X[\mathrm{odd}], X[\mathrm{even}])$$
$$:= \left(\widetilde{E}_{K_{i,1}}^{-1}(X[1], X[2])\|\widetilde{E}_{K_{i,2}}^{-1}(X[3], X[4])\| \ldots \|\widetilde{E}_{K_{i,d/2}}^{-1}(X[d-1], X[d]) \right). \quad (3)$$

The r-round encryption function $\mathcal{E}_{r,d}$ of an extended TBC-based type-2 GFN is a dn-BC that takes $M \in \{0,1\}^{dn}$ as input, as shown in Fig. 4(a). It uses r encryption round functions $\Phi_d[\widetilde{E}_{K_{i,1}}, \ldots, \widetilde{E}_{K_{i,d/2}}]$ for $i = \{1, 2, \ldots, r\}$ and $r-1$ diffusion layer $\mathsf{T}_1, \mathsf{T}_2, \ldots, \mathsf{T}_{r-1}$. The output of the $\mathcal{E}_{r,d}[\widetilde{E}_{K_{1,1}}, \ldots, \widetilde{E}_{K_{r,d/2}}](M)$ is computed as follows:

- Let $u_1 := M$.
- for $i = 1, 2, \ldots, r-1$ do:
  1. $v_i := \Phi_d[\widetilde{E}_{K_{i,1}}, \ldots, \widetilde{E}_{K_{i,d/2}}](u_i)$.
  2. $u_{i+1} := \mathsf{T}_i \cdot v_i$.
- $v_r := \Phi_d[\widetilde{E}_{K_{r,1}}, \ldots, \widetilde{E}_{K_{r,d/2}}](u_r)$.
- Outputs $C := v_r$.

The r-round decryption function $\mathcal{E}_{r,d}^{-1}$ is defined in an obvious way by using r decryption round functions $\Phi_d^{-1}[\widetilde{E}_{K_{i,1}}, \ldots, \widetilde{E}_{K_{i,d/2}}]$ for $i = r, r-1, \ldots, 1$ and $r-1$ diffusion layer $\mathsf{T}_{r-1}^{-1}, \mathsf{T}_{r-2}^{-1}, \ldots, \mathsf{T}_1^{-1}$, as shown in Fig. 4(b). Given input $C \in \{0,1\}^{dn}$, the output of the $\mathcal{E}_{r,d}^{-1}[\widetilde{E}_{K_{1,1}}, \ldots, \widetilde{E}_{K_{r,d/2}}](C)$ is computed as follows:

- Let $v_r := C$.
- for $i = r, r-1, \ldots, 2$ do:
  1. $u_i := \Phi_d^{-1}[\widetilde{E}_{K_{i,1}}, \ldots, \widetilde{E}_{K_{i,d/2}}](v_i)$.
  2. $v_{i-1} := \mathsf{T}_{i-1}^{-1} \cdot u_i$.
- $u_1 := \Phi_d^{-1}[\widetilde{E}_{K_{1,1}}, \ldots, \widetilde{E}_{K_{1,d/2}}](v_1)$.
- Outputs $M := u_1$.

In Sect. 4, we prove the security of $\mathcal{E}_{4,d}[\widetilde{P}_{1,1}, \ldots, \widetilde{P}_{4,d/2}]$ that is obtained by replacing each (n,n)-TBC $\widetilde{E}_{K_{x,y}}$ with an (n,n)-TRP $\widetilde{P}_{x,y}$, where $\widetilde{P}_{1,1}, \ldots, \widetilde{P}_{4,d/2}$ are $2d$ independent (n,n)-TRPs. We write $\mathcal{E}_{4,d}$ for $\mathcal{E}_{4,d}[\widetilde{P}_{1,1}, \ldots, \widetilde{P}_{4,d/2}]$ and $\mathcal{E}_{4,d}^{-1}$ for $\mathcal{E}_{4,d}^{-1}[\widetilde{P}_{1,1}, \ldots, \widetilde{P}_{4,d/2}]$.

4   Birthday SPRP Security at 4 Rounds

We will prove SPRP security of $\mathcal{E}_{4,d}$, extended TBC-based type-2 generalized Feistel network, where we use $2d$ independent (n, n)-TRPs $\widetilde{P}_{1,1}, \ldots, \widetilde{P}_{4,d/2}$ and *a single diffusion layer* T. Formally,

$$\mathcal{E}_{4,d}(X) := \varPhi_d[\widetilde{P}_{4,1}, \ldots, \widetilde{P}_{4,d/2}](\mathsf{T} \cdot (\varPhi_d[\widetilde{P}_{3,1}, \ldots, \widetilde{P}_{3,d/2}]($$
$$\mathsf{T} \cdot (\varPhi_d[\widetilde{P}_{2,1}, \ldots, \widetilde{P}_{2,d/2}](\mathsf{T} \cdot (\varPhi_d[\widetilde{P}_{1,1}, \ldots, \widetilde{P}_{1,d/2}](X))))))). \quad (4)$$

Using a single diffusion layer simplifies both the construction and the notations. Recall from our convention that $\mathsf{T}_{\mathrm{OE}}, \ldots, (\mathsf{T}^{-1})_{\mathrm{EE}}$ constitute the eight submatrices of T and T^{-1}. In fact, $(\mathsf{T}^{-1})_{\mathrm{OE}}, \ldots, (\mathsf{T}^{-1})_{\mathrm{EE}}$ can be derived from $\mathsf{T}_{\mathrm{OE}}, \ldots, \mathsf{T}_{\mathrm{EE}}$, but the expressions are too complicated to use.

We next characterize the properties on T that are sufficient for security.

Definition 1 (Good Linear Diffusion Layer for 4 Rounds). *An invertible matrix* $\mathsf{T} \in \mathbb{F}^{d \times d}$ *is good if the 4 induced matrices* $\mathsf{T}_{\mathrm{EE}}, \ \mathsf{T}_{\mathrm{OE}}^{-1} \cdot \mathsf{T}_{\mathrm{OO}} \cdot \mathsf{T}_{\mathrm{OE}}, \ \mathsf{T}_{\mathrm{EE}} \cdot \mathsf{T}_{\mathrm{OE}}^{-1} \cdot (\mathsf{T}^{-1})_{\mathrm{OE}}, \ and \ (\mathsf{T}^{-1})_{\mathrm{EE}} \ contain \ no \ zero \ entries.$

To justify the soundness of this definition, we list several candidates in Appendix A. With such a good linear transformation T, we have the following theorem on 4-round extended TBC-based type-2 GFN.

Theorem 1. *Fix* $d \geq 4$, *where* d *is even, and let* $\widetilde{P}_{1,1}, \ldots, \widetilde{P}_{4,d/2}$ *be* $2d$ *independent* (n, n)-*TRPs and* $\mathcal{E}_{4,d} = \mathcal{E}_{4,d}[\widetilde{P}_{1,1}, \ldots, \widetilde{P}_{4,d/2}]$ *be the extended TBC-based type-2 GFN defined in Eq. (4). Assume* $q \leq 2^n/2$, *then for any SPRP-adversary* $\mathcal{A}$ *that makes* q *queries, we have*

$$\mathbf{Adv}_{\mathcal{E}_{4,d}}^{\mathrm{sprp}}(\mathcal{A}) \leq \frac{2dq^2}{2^n} + \frac{0.5q^2}{2^{dn}}. \quad (5)$$

In Theorem 1, Eq. (5) shows birthday-bound security, which is obtained from Lemma 2, Lemma 3 and the H-coefficient technique (Lemma 1).

All the remaining of this section devotes to proving Theorem 1. We first present the definition of the oracles in Sect. 4.1. We then complete the two steps *defining and analyzing bad transcripts* and *analyzing good transcripts* in Sect. 4.2 and 4.3 resp.

We consider a CCA-adversary $\mathcal{A}$ that interacts with the real world oracles $\mathcal{R}$ and $\mathcal{R}^{-1}$, or with the ideal world oracles $\mathcal{I}$ and $\mathcal{I}^{-1}$. Without loss of generality, $\mathcal{A}$ is assumed to be deterministic, makes exactly q queries, does not repeat the same query, and does not make a redundant query, i.e., if $\mathcal{A}$ makes an encryption query M to obtain C, then it does not subsequently make a decryption query C, and vice versa.

4.1   Definition of the Oracles

The oracles $\mathcal{R}$ and $\mathcal{R}^{-1}$ represent the real world and implement $\mathcal{E}_{4,d}$ and $\mathcal{E}_{4,d}^{-1}$, and the oracles $\mathcal{I}$ and $\mathcal{I}^{-1}$ represent the ideal world and implement π and π^{-1}.

In the real world, the definitions of $\mathcal{R}$ and $\mathcal{R}^{-1}$ are in Algorithms 1 and 2. For the i-th query $M^{(i)} \in \{0,1\}^{dn}$ to $\mathcal{R}$, we compute $C^{(i)} \in \{0,1\}^{dn}$ by following the definition of $\mathcal{E}_{4,d}$, and return it to $\mathcal{A}$. We also save internal states $\left(u_2^{(i)} \| v_2^{(i)}[\text{even}] \| u_3^{(i)} \| v_3^{(i)}[\text{even}]\right) \in \{0,1\}^{3dn}$ that are maintained through S in Algorithms 1, which is the $3dn$-bit strings about the internal random tweakable permutations in $\mathcal{R}$, and return S to $\mathcal{A}$ after making all the q queries and before $\mathcal{A}$ returns the decision bit. $\mathcal{R}^{-1}$ is similarly defined, and for a decryption query $C^{(i)}$, $\mathcal{A}$ obtains $M^{(i)}$ during the interaction and $\left(u_2^{(i)} \| v_2^{(i)}[\text{even}] \| u_3^{(i)} \| v_3^{(i)}[\text{even}]\right)$ after making all the queries.

Algorithm 1: Procedure of $\mathcal{R}$ for the i-th query

    Input : $M^{(i)} \in \{0,1\}^{dn}$
    Output: $C^{(i)} \in \{0,1\}^{dn}$

1 $u_1^{(i)} \leftarrow M^{(i)}$;
2 **for** $x = 1,2,3,4$ **do**
3   **for** $y = 1,2,...,d/2$ **do**
4          $v_x^{(i)}[2y-1] \leftarrow u_x^{(i)}[2y-1]$;
5          $v_x^{(i)}[2y] \leftarrow \widetilde{P}_{x,y}(u_x^{(i)}[2y-1], u_x^{(i)}[2y])$;
6      $u_{x+1}^{(i)} \leftarrow \mathsf{T} \cdot v_x^{(i)}$;
7 $C^{(i)} \leftarrow v_4^{(i)}$;
8 **return** $C^{(i)}$;

9 $\mathsf{S} \leftarrow \mathsf{S} \| u_2^{(i)} \| v_2^{(i)}[\text{even}] \| u_3^{(i)} \| v_3^{(i)}[\text{even}]$;

Algorithm 2: Procedure of $\mathcal{R}^{-1}$ for the i-th query

---

    Input : $C^{(i)} \in \{0,1\}^{dn}$
    Output: $M^{(i)} \in \{0,1\}^{dn}$

1 $v_4^{(i)} \leftarrow C^{(i)}$;
2 **for** $x = 4,3,2,1$ **do**
3   **for** $y = 1,2,...,d/2$ **do**
4          $u_x^{(i)}[2y-1] \leftarrow v_x^{(i)}[2y-1]$;
5          $u_x^{(i)}[2y] \leftarrow \widetilde{P}_{x,y}^{-1}(v_x^{(i)}[2y-1], v_x^{(i)}[2y])$;
6      $v_{x-1}^{(i)} \leftarrow \mathsf{T}^{-1} \cdot u_x^{(i)}$;
7 $M^{(i)} \leftarrow u_1^{(i)}$;
8 **return** $M^{(i)}$;

9 $\mathsf{S} \leftarrow \mathsf{S} \| u_2^{(i)} \| v_2^{(i)}[\text{even}] \| u_3^{(i)} \| v_3^{(i)}[\text{even}]$;

In the ideal world, we define $\mathcal{I}$ and $\mathcal{I}^{-1}$ as in Algorithms 3 and 4. For the i-th query, $\mathcal{I}$ and $\mathcal{I}^{-1}$ generate dummy internal states $\left(u_2^{(i)} \| v_2^{(i)}[\text{even}] \| u_3^{(i)} \|\right.$

$v_3^{(i)}$[even]), and record them into S. After $\mathcal{A}$ makes q queries, S is given to $\mathcal{A}$. We generate $v_1^{(i)}$[even] and $u_4^{(i)}$[even] by simulating $\widetilde{P}_{1,1}, \ldots, \widetilde{P}_{1,d/2}$ and $\widetilde{P}_{4,1}^{-1}, \ldots, \widetilde{P}_{4,d/2}^{-1}$. Then, we generate dummy internal states $(u_2^{(i)} \| v_2^{(i)}$[even] $\| u_3^{(i)} \| v_3^{(i)}$[even]) by $v_1^{(i)}$[even] and $u_4^{(i)}$[even]. And we use $\mathcal{S}_{1,x}^{(i)}$ ($\mathcal{S}_{4,x}^{(i)}$, resp.) as the record of the output values that share the same tweak in $\widetilde{P}_{1,x}$ ($\widetilde{P}_{4,x}^{-1}$, resp.) in the first $i-1$ queries, for $x \in [1..d/2]$. The simulation uses the lazy-sampling, as shown in Algorithms 3 and 4, and the probability distribution of $(v_1^{(1)}$[even]$, u_4^{(1)}$[even]$), \ldots, (v_1^{(q)}$[even]$, u_4^{(q)}$[even]$)$ is equal to the one in the real world.

Algorithm 3: Procedure of $\mathcal{I}$ for the i-th query

    Input : $M^{(i)} \in \{0,1\}^{dn}$
    Output: $C^{(i)} \in \{0,1\}^{dn}$

1 $C^{(i)} \leftarrow \pi(M^{(i)})$;

2 $u_1^{(i)} \leftarrow M^{(i)}, v_1^{(i)}$[odd] $\leftarrow u_1^{(i)}$[odd];

3 $v_4^{(i)} \leftarrow C^{(i)}, u_4^{(i)}$[odd] $\leftarrow v_4^{(i)}$[odd];

4 **for** $x = 1, 2, \ldots, d/2$ **do**   // Simulating $\widetilde{P}_{1,x}$ and obtaining $v_1^{(i)}$[even]

5     $\mathcal{S}_{1,x}^{(i)} \leftarrow \{v_1^{(j)}[2x] \mid j < i \wedge u_1^{(i)}[2x-1] = u_1^{(j)}[2x-1]\}$;

6   **if** $\exists j < i, (u_1^{(i)}[2x-1], u_1^{(i)}[2x]) = (u_1^{(j)}[2x-1], u_1^{(j)}[2x])$ **then**

7         $v_1^{(i)}[2x] \leftarrow v_1^{(j)}[2x]$;

8   **else** $v_1^{(i)}[2x] \xleftarrow{\$} \{0,1\}^n \setminus \mathcal{S}_{1,x}^{(i)}$;

9 $u_2^{(i)} \leftarrow \mathsf{T} \cdot v_1^{(i)}$;

10 **for** $y = 1, 2, \ldots, d/2$ **do**   // Simulating $\widetilde{P}_{4,x}^{-1}$ and obtaining $u_4^{(i)}$[even]

11     $\mathcal{S}_{4,y}^{(i)} \leftarrow \{u_4^{(j)}[2y] \mid j < i \wedge v_4^{(i)}[2y-1] = v_4^{(j)}[2y-1]\}$;

12   **if** $\exists j < i, (v_4^{(i)}[2y-1], v_4^{(i)}[2y]) = (v_4^{(j)}[2y-1], v_4^{(j)}[2y])$ **then**

13         $u_4^{(i)}[2y] \leftarrow u_4^{(j)}[2y]$;

14   **else** $u_4^{(i)}[2y] \xleftarrow{\$} \{0,1\}^n \setminus \mathcal{S}_{4,y}^{(i)}$;

15 $v_3^{(i)} \leftarrow \mathsf{T}^{-1} \cdot u_4^{(i)}, u_3^{(i)}$[odd] $\leftarrow v_3^{(i)}$[odd];

16 $v_2^{(i)}$[even] $\leftarrow \mathsf{T}_{\mathrm{OE}}^{-1} \cdot \mathsf{T}_{\mathrm{OO}} \cdot u_2^{(i)}$[odd] $\oplus \mathsf{T}_{\mathrm{OE}}^{-1} \cdot v_3^{(i)}$[odd];

17 $u_3^{(i)}$[even] $\leftarrow (\mathsf{T}_{\mathrm{EE}} \cdot \mathsf{T}_{\mathrm{OE}}^{-1} \cdot \mathsf{T}_{\mathrm{OO}} \oplus \mathsf{T}_{\mathrm{EO}}) \cdot u_2^{(i)}$[odd] $\oplus \mathsf{T}_{\mathrm{EE}} \cdot \mathsf{T}_{\mathrm{OE}}^{-1} \cdot v_3^{(i)}$[odd];

18 **return** $C^{(i)}$;

19 S $\leftarrow$ S$\|u_2^{(i)}\|v_2^{(i)}$[even]$\|u_3^{(i)}\|v_3^{(i)}$[even];

4.2 Bad Transcripts and Bad Probability

The adversary $\mathcal{A}$ is given all the internal states in S after it makes q queries, and the interaction between the oracle and $\mathcal{A}$ can be summarized as the following transcript θ:

Algorithm 4: Procedure of $\mathcal{I}^{-1}$ for the i-th query

Input : $C^{(i)} \in \{0,1\}^{dn}$
Output: $M^{(i)} \in \{0,1\}^{dn}$

1   $M^{(i)} \leftarrow \pi^{-1}(C^{(i)})$;

2   $u_1^{(i)} \leftarrow M^{(i)}, v_1^{(i)}[\text{odd}] \leftarrow u_1^{(i)}[\text{odd}]$;

3   $v_4^{(i)} \leftarrow C^{(i)}, u_4^{(i)}[\text{odd}] \leftarrow v_4^{(i)}[\text{odd}]$;

4   **for** $x = 1, 2, ..., d/2$ **do**   // Simulating $\widetilde{P}_{1,x}$ and obtaining $v_1^{(i)}[\text{even}]$

5      $S_{1,x}^{(i)} \leftarrow \{v_1^{(j)}[2x] \mid j < i \wedge u_1^{(i)}[2x-1] = u_1^{(j)}[2x-1]\}$;

6   **if** $\exists j < i, (u_1^{(i)}[2x-1], u_1^{(i)}[2x]) = (u_1^{(j)}[2x-1], u_1^{(j)}[2x])$ **then**

7         $v_1^{(i)}[2x] \leftarrow v_1^{(j)}[2x]$;

8   **else** $v_1^{(i)}[2x] \xleftarrow{\$} \{0,1\}^n \setminus S_{1,x}^{(i)}$;

9   $u_2^{(i)} \leftarrow \mathsf{T} \cdot v_1^{(i)}$;

10   **for** $y = 1, 2, ..., d/2$ **do**   // Simulating $\widetilde{P}_{4,x}^{-1}$ and obtaining $u_4^{(i)}[\text{even}]$

11      $S_{4,y}^{(i)} \leftarrow \{u_4^{(j)}[2y] \mid j < i \wedge v_4^{(i)}[2y-1] = v_4^{(j)}[2y-1]\}$;

12   **if** $\exists j < i, (v_4^{(i)}[2y-1], v_4^{(i)}[2y]) = (v_4^{(j)}[2y-1], v_4^{(j)}[2y])$ **then**

13         $u_4^{(i)}[2y] \leftarrow u_4^{(j)}[2y]$;

14   **else** $u_4^{(i)}[2y] \xleftarrow{\$} \{0,1\}^n \setminus S_{4,y}^{(i)}$;

15   $v_3^{(i)} \leftarrow \mathsf{T}^{-1} \cdot u_4^{(i)}, u_3^{(i)}[\text{odd}] \leftarrow v_3^{(i)}[\text{odd}]$;

16   $v_2^{(i)}[\text{even}] \leftarrow \mathsf{T}_{\mathrm{OE}}^{-1} \cdot \mathsf{T}_{\mathrm{OO}} \cdot u_2^{(i)}[\text{odd}] \oplus \mathsf{T}_{\mathrm{OE}}^{-1} \cdot v_3^{(i)}[\text{odd}]$;

17   $u_3^{(i)}[\text{even}] \leftarrow (\mathsf{T}_{\mathrm{EE}} \cdot \mathsf{T}_{\mathrm{OE}}^{-1} \cdot \mathsf{T}_{\mathrm{OO}} \oplus \mathsf{T}_{\mathrm{EO}}) \cdot u_2^{(i)}[\text{odd}] \oplus \mathsf{T}_{\mathrm{EE}} \cdot \mathsf{T}_{\mathrm{OE}}^{-1} \cdot v_3^{(i)}[\text{odd}]$;

18   **return** $M^{(i)}$;

19   $\mathsf{S} \leftarrow \mathsf{S} \| u_2^{(i)} \| v_2^{(i)}[\text{even}] \| u_3^{(i)} \| v_3^{(i)}[\text{even}]$;

$$\theta = \Big((M^{(1)}, C^{(1)}, u_2^{(1)} \| v_2^{(1)}[\text{even}] \| u_3^{(1)} \| v_3^{(1)}[\text{even}]), \dots,$$

$$(M^{(q)}, C^{(q)}, u_2^{(q)} \| v_2^{(q)}[\text{even}] \| u_3^{(q)} \| v_3^{(q)}[\text{even}]) \Big). \tag{6}$$

The adversary does not repeat a query, so we have $(M^{(i)}[1], \dots, M^{(i)}[d]) \neq (M^{(j)}[1], \dots, M^{(j)}[d])$ and $(C^{(i)}[1], \dots, C^{(i)}[d]) \neq (C^{(j)}[1], \dots, C^{(j)}[d])$ for any $1 \leq j < i \leq q$.

In the real world, let us focus on $\widetilde{P}_{x,y}$ for $x \in [1..4]$ and $y \in [1..d/2]$, and let $X^{(i)}, T^{(i)}$ and $Y^{(i)}$ be the input, tweak, and output of the i-th query, respectively. We observe that in the i-th and j-th queries, if $X^{(i)} = X^{(j)}$ and $T^{(i)} = T^{(j)}$ hold, then we must have $Y^{(i)} = Y^{(j)}$, and similarly, if we have $Y^{(i)} = Y^{(j)}$ and $T^{(i)} = T^{(j)}$, then $X^{(i)} = X^{(j)}$ holds.

In the ideal world, for $\widetilde{P}_{x,y}$ with $x = 1, 4$ and $y \in [1..d/2]$, i.e., for TPRs that are used in the simulation, it has the same input-tweak-output relation as in the real world. However, for $\widetilde{P}_{x,y}$ with $x = 2, 3$ and $y \in [1..d/2]$ that are not used in the simulation, this may not be the case. That is, the output can be different even though it takes the same input and tweak, or the input can be different when it takes the same output and tweak. In other words, in the ideal world, TRPs $\widetilde{P}_{x,y}$ with $x = 2, 3$ and $y \in [1..d/2]$ are not used in the simulation, and there are conditions on these TRPs that can hold only in the ideal world. Our definition of the set of bad transcripts consists of all such transcripts θ.

Furthermore, our definition of the set of bad transcripts captures collisions on $u_2[2x-1]\|u_2[2x]$, $u_2[2x-1]\|v_2[2x]$, $u_3[2x-1]\|u_3[2x]$ and $u_3[2x-1]\|v_3[2x]$ from two distinct queries for $x \in [1..d/2]$.

We use these conditions to define $\mathcal{T}_{\text{bad}}$, the set of bad transcripts.

Definition 2. *An attainable transcript θ in Eq. (6) is bad, if at least one of the following conditions is fulfilled:*

- *(B-1) Bad at $\widetilde{P}_{2,x}$ for $x \in [1..d/2]$:*

$$1 \leq \exists j < \exists i \leq q,\ u_2^{(i)}[2x-1]\|u_2^{(i)}[2x] = u_2^{(j)}[2x-1]\|u_2^{(j)}[2x];$$

- *(B-2) Bad at $\widetilde{P}_{2,x}$ for $x \in [1..d/2]$:*

$$1 \leq \exists j < \exists i \leq q,\ u_2^{(i)}[2x-1]\|v_2^{(i)}[2x] = u_2^{(j)}[2x-1]\|v_2^{(j)}[2x];$$

- *(B-3) Bad at $\widetilde{P}_{3,x}$ for $x \in [1..d/2]$:*

$$1 \leq \exists j < \exists i \leq q,\ u_3^{(i)}[2x-1]\|u_3^{(i)}[2x] = u_3^{(j)}[2x-1]\|u_3^{(j)}[2x];$$

- *(B-4) Bad at $\widetilde{P}_{3,x}$ for $x \in [1..d/2]$:*

$$1 \leq \exists j < \exists i \leq q,\ u_3^{(i)}[2x-1]\|v_3^{(i)}[2x] = u_3^{(j)}[2x-1]\|v_3^{(j)}[2x];$$

Otherwise we say that θ is good. We denote $\mathcal{T}_{\text{good}}$, resp. $\mathcal{T}_{\text{bad}}$ the set of good, resp. bad transcripts.

To understand the conditions, consider a good transcript θ and let's see some properties (informally). First, since neither the 1st nor the 2nd condition is fulfilled, the input strings[3] to the $\widetilde{P}_{2,x}$ (resp. $\widetilde{P}_{2,x}^{-1}$) induced by the queries are distinct for $x \in [1..d/2]$. Second, since neither the 3rd nor the 4th condition is fulfilled, the input strings to the $\widetilde{P}_{3,x}$ (resp. $\widetilde{P}_{3,x}^{-1}$) induced by the queries are distinct for $x \in [1..d/2]$. These ensure that $\widetilde{P}_{2,x}$ and $\widetilde{P}_{3,x}$ satisfy dq new and distinct equations.

We now present the upper bound on the probability of getting bad transcripts in the ideal world.

Lemma 2. *When $q \leq 2^n/2$, we have*

$$\Pr[\Theta_{\mathcal{I}} \in \mathcal{T}_{\text{bad}}] \leq \frac{2dq^2}{2^n}. \tag{7}$$

Proof. To bound $\Pr[\Theta_{\mathcal{I}} \in \mathcal{T}_{\text{bad}}]$, we consider the probability of each condition in turn, i.e., $\Pr[(B - \ell)]$ for $\ell = 1, 2, 3, 4$.

---

[3] For $\widetilde{P}_{2,x}$, the input strings are inputs and tweaks. For $\widetilde{P}_{2,x}^{-1}$, the input strings are outputs and tweaks.

Condition (B-1). Let's start by considering condition (B-1). We focus on the i-th and j-th queries, with $1 \leq j < i \leq q$, and fix $x \in [1..d/2]$. Then we derive the upper bound on

$$\Pr\left[u_2^{(i)}[2x-1]\|u_2^{(i)}[2x] = u_2^{(j)}[2x-1]\|u_2^{(j)}[2x]\right] \leq \Pr\left[u_2^{(i)}[2x] = u_2^{(j)}[2x]\right].$$

The occurrence of event $u_2^{(i)}[2x-1]\|u_2^{(i)}[2x] = u_2^{(j)}[2x-1]\|u_2^{(j)}[2x]$ implies the simultaneous occurrence of events $u_2^{(i)}[2x-1] = u_2^{(j)}[2x-1]$ and $u_2^{(i)}[2x] = u_2^{(j)}[2x]$. We use the trivial bound of $\Pr\left[u_2^{(i)}[2x-1] = u_2^{(j)}[2x-1] \mid u_2^{(i)}[2x] = u_2^{(j)}[2x]\right] \leq 1$, as our objective is to compute the upper bound. Hence, we consider $\Pr\left[u_2^{(i)}[2x] = u_2^{(j)}[2x]\right]$. From the design of ETGFN, we have

$$u_2^{(i)}[\text{even}] = \mathsf{T}_{\text{EO}} \cdot X^{(i)}[\text{odd}] \oplus \mathsf{T}_{\text{EE}} \cdot \widetilde{P}_{1,[1..d/2]}\left(X^{(i)}[\text{odd}], X^{(i)}[\text{even}]\right),$$

where the definition of $\widetilde{P}_{1,[1..d/2]}\left(X^{(i)}[\text{odd}], X^{(i)}[\text{even}]\right)$ is similar to that of Eq. (2). In detail,

$$u_2^{(i)}[2x] = t_{2x,1} \cdot X^{(i)}[1] \oplus t_{2x,2} \cdot \widetilde{P}_{1,1}(X^{(i)}[1], X^{(i)}[2]) \oplus \ldots$$
$$\oplus t_{2x,d-1} \cdot X^{(i)}[d-1] \oplus t_{2x,d} \cdot \widetilde{P}_{1,d/2}(X^{(i)}[d-1], X^{(i)}[d]),$$

where $t_{2x,y}$ represents the element located at the $2x$-th row and y-th column of matrix T, for $y = 1, 2, \ldots, d$.

As assumed before, $\mathcal{A}$ does not repeat a query. This means $X^{(i)}[j_0] \neq X^{(j)}[j_0]$ for some $j_0 \in \{1, 2, \ldots, d\}$. Without loss of generality, we assume j_0 is the minimum index. If j_0 is an odd number, $\widetilde{P}_{1,(j_0+1)/2}(X^{(i)}[j_0], X^{(i)}[j_0 + 1])$ is a "fresh" computation, the value of which is uniform in 2^n values. Since every entry in the $((j_0 + 1)/2)$-th column of T_{EE} is nonzero (see Definition 1), the probability to have $u_2^{(i)}[2x] = u_2^{(j)}[2x]$ is equal to the probability that $\widetilde{P}_{1,(j_0+1)/2}(X^{(i)}[j_0], X^{(i)}[j_0 + 1])$ equals some fixed value. So we have,

$$\Pr\left[u_2^{(i)}[2x] = u_2^{(j)}[2x]\right] = \frac{1}{2^n}. \tag{8}$$

If j_0 is an even number, $\widetilde{P}_{1,j_0/2}(X^{(i)}[j_0 - 1], X^{(i)}[j_0])$ is a "fresh" computation. So after conditioning on the value of $\widetilde{P}_{1,j_0/2}(X^{(j)}[j_0 - 1], X^{(j)}[j_0])$, the value of $\widetilde{P}_{1,j_0/2}(X^{(i)}[j_0 - 1], X^{(i)}[j_0])$ is uniform in $\geq 2^n - q$ possibilities. Because every entry in the $(j_0/2)$-th column of T_{EE} is nonzero, we have,

$$\Pr\left[u_2^{(i)}[2x] = u_2^{(j)}[2x]\right] \leq \frac{1}{2^n - q}. \tag{9}$$

From Eqs. (8) and (9), we have $\Pr\left[u_2^{(i)}[2x] = u_2^{(j)}[2x]\right] \leq \frac{1}{2^n - q}$.

Summing over $x \in \{1, 2, \ldots, d/2\}$, $i \in \{2, 3, \ldots, q\}$, $j \in \{1, 2, \ldots, i-1\}$, we reach

$$\Pr[(\text{B-1})] \leq \sum_{x \in [1..d/2]} \sum_{i \in [2..q]} \sum_{j \in [1..i-1]} \Pr\left[u_2^{(i)}[2x] = u_2^{(j)}[2x]\right]$$

$$\leq \sum_{i \in [2..q]} \frac{d/2(i-1)}{2^n - q} \leq \frac{0.25dq^2}{2^n - q}. \tag{10}$$

Condition (B-2). Next, consider (B-2). We focus on $x \in [1..d/2]$ and the i-th and j-th queries, with $1 \leq j < i \leq q$, and evaluate

$$\Pr\left[u_2^{(i)}[2x-1] \| v_2^{(i)}[2x] = u_2^{(j)}[2x-1] \| v_2^{(j)}[2x]\right] \leq \Pr\left[v_2^{(i)}[2x] = v_2^{(j)}[2x]\right].$$

We use the trivial bound of $\Pr\left[u_2^{(i)}[2x-1] = u_2^{(j)}[2x-1] \mid v_2^{(i)}[2x] = v_2^{(j)}[2x]\right] \leq 1$, which is the same as the bound specified in condition (B-1). Hence, we consider $\Pr\left[v_2^{(i)}[2x] = v_2^{(j)}[2x]\right]$. Emanating from the design of ETGFN, we obtain the following:

$$\begin{pmatrix} u_3^{(i)}[1] \\ u_3^{(i)}[2] \\ \cdots \\ u_3^{(i)}[d] \end{pmatrix} = \mathsf{T} \cdot \begin{pmatrix} v_2^{(i)}[1] \\ v_2^{(i)}[2] \\ \cdots \\ v_2^{(i)}[d] \end{pmatrix} \Leftrightarrow \begin{pmatrix} u_3^{(i)}[\text{odd}] \\ u_3^{(i)}[\text{even}] \end{pmatrix} = \underbrace{\begin{pmatrix} \mathsf{T_{OO}} & \mathsf{T_{OE}} \\ \mathsf{T_{EO}} & \mathsf{T_{EE}} \end{pmatrix}}_{\mathsf{T_1}} \cdot \begin{pmatrix} v_2^{(i)}[\text{odd}] \\ v_2^{(i)}[\text{even}] \end{pmatrix},$$

and

$$\begin{pmatrix} v_2^{(i)}[\text{even}] \\ u_3^{(i)}[\text{even}] \end{pmatrix} = \widehat{\mathsf{T}_1} \cdot \begin{pmatrix} v_2^{(i)}[\text{odd}] \\ u_3^{(i)}[\text{odd}] \end{pmatrix} = \widehat{\mathsf{T}_1} \cdot \begin{pmatrix} u_2^{(i)}[\text{odd}] \\ v_3^{(i)}[\text{odd}] \end{pmatrix}. \tag{11}$$

By Eq. (1), it can be seen $v_2^{(i)}[\text{even}]$ is written as

$$v_2^{(i)}[\text{even}] = \mathsf{T_{OE}^{-1}} \cdot \mathsf{T_{OO}} \cdot \left(\mathsf{T_{OO}} \cdot X^{(i)}[\text{odd}] \oplus \mathsf{T_{OE}} \cdot \widetilde{P}_{1,[1..d/2]}\left(X^{(i)}[\text{odd}], X^{(i)}[\text{even}]\right)\right)$$

$$\oplus \mathsf{T_{OE}^{-1}} \cdot \left((\mathsf{T}^{-1})_{\text{OO}} \cdot Y^{(i)}[\text{odd}] \oplus (\mathsf{T}^{-1})_{\text{OE}} \cdot \widetilde{P}_{4,[1..d/2]}^{-1}\left(Y^{(i)}[\text{odd}], Y^{(i)}[\text{even}]\right)\right)$$

$$= \mathsf{T_{OE}^{-1}} \cdot \mathsf{T_{OO}} \cdot \mathsf{T_{OE}} \cdot \widetilde{P}_{1,[1..d/2]}(X^{(i)}[\text{odd}], X^{(i)}[\text{even}]) \tag{12}$$

$$\oplus \mathsf{T_{OE}^{-1}} \cdot (\mathsf{T}^{-1})_{\text{OE}} \cdot \widetilde{P}_{4,[1..d/2]}^{-1}(Y^{(i)}[\text{odd}], Y^{(i)}[\text{even}]) \oplus g_1(X^{(i)}[\text{odd}], Y^{(i)}[\text{odd}]),$$

where g_1 is a function of $X^{(i)}[\text{odd}]$ and $Y^{(i)}[\text{odd}]$ and is independent of the first two terms, the definition of $\widetilde{P}_{4,[1..d/2]}^{-1}(Y^{(i)}[\text{odd}], Y^{(i)}[\text{even}])$ is similar to that of Eq. (3).

Because $\mathcal{A}$ does not repeat a query, we have $X^{(i)}[j_0] \neq X^{(j)}[j_0]$ for some $j_0 \in \{1, 2, \ldots, d\}$. We also assume j_0 is the minimum index. Similar to the analysis in the condition (B-1), $\widetilde{P}_{1,(j_0+1)/2}(X^{(i)}[j_0], X^{(i)}[j_0+1])$ or $\widetilde{P}_{1,j_0/2}(X^{(i)}[j_0-1], X^{(i)}[j_0])$ is a "fresh" computation. Due to every entry in $\mathsf{T_{OE}^{-1}} \cdot \mathsf{T_{OO}} \cdot \mathsf{T_{OE}}$ is nonzero, we have $\Pr\left[v_2^{(i)}[2x] = v_2^{(j)}[2x]\right] \leq \frac{1}{2^n - q}$.

Summing over $x \in \{1, 2, \ldots, d/2\}$, $i \in \{2, 3, \ldots, q\}$, $j \in \{1, 2, \ldots, i - 1\}$, we reach

$$\Pr[(\text{B-2})] \leq \sum_{x \in [1..d/2]} \sum_{i \in [2..q]} \sum_{j \in [1..i-1]} \Pr\left[v_2^{(i)}[2x] = v_2^{(j)}[2x]\right]$$

$$\leq \sum_{i \in [2..q]} \frac{d/2(i-1)}{2^n - q} \leq \frac{0.25 dq^2}{2^n - q}. \tag{13}$$

Condition (B-3). We now consider (B-3) and employ an analogous argument as in the case of (B-2). For the case of $1 \leq j < i \leq q$, fix $x \in [1..d/2]$. We derive the upper bound on

$$\Pr\left[u_3^{(i)}[2x-1] \| u_3^{(i)}[2x] = u_3^{(j)}[2x-1] \| u_3^{(j)}[2x]\right] \leq \Pr\left[u_3^{(i)}[2x] = u_3^{(j)}[2x]\right].$$

We use the trivial bound of $\Pr\left[u_3^{(i)}[2x-1] = u_3^{(j)}[2x-1] \mid u_3^{(i)}[2x] = u_3^{(j)}[2x]\right] \leq 1$. Subsequently, our attention is directed towards $\Pr\left[u_3^{(i)}[2x] = u_3^{(j)}[2x]\right]$. We refer to Eq. (11) for the expression of $u_3^{(i)}[\text{even}]$, and by Eq. (1), it can be seen $u_3^{(i)}[\text{even}]$ is written as

$$u_3^{(i)}[\text{even}] = \left(\mathsf{T}_{\text{EE}} \cdot \mathsf{T}_{\text{OE}}^{-1} \cdot \mathsf{T}_{\text{OO}} \oplus \mathsf{T}_{\text{EO}}\right) \cdot \left(\mathsf{T}_{\text{OO}} \cdot X^{(i)}[\text{odd}] \oplus \mathsf{T}_{\text{OE}} \cdot \widetilde{P}_{1,[1..d/2]}(X^{(i)}[\text{odd}], X^{(i)}[\text{even}])\right)$$

$$\oplus \mathsf{T}_{\text{EE}} \cdot \mathsf{T}_{\text{OE}}^{-1} \cdot \left((\mathsf{T}^{-1})_{\text{OO}} \cdot Y^{(i)}[\text{odd}] \oplus (\mathsf{T}^{-1})_{\text{OE}} \cdot \widetilde{P}_{4,[1..d/2]}^{-1}(Y^{(i)}[\text{odd}], Y^{(i)}[\text{even}])\right)$$

$$= \left(\mathsf{T}_{\text{EE}} \cdot \mathsf{T}_{\text{OE}}^{-1} \cdot \mathsf{T}_{\text{OO}} \oplus \mathsf{T}_{\text{EO}}\right) \cdot \mathsf{T}_{\text{OE}} \cdot \widetilde{P}_{1,[1..d/2]}(X^{(i)}[\text{odd}], X^{(i)}[\text{even}]) \tag{14}$$

$$\oplus \mathsf{T}_{\text{EE}} \cdot \mathsf{T}_{\text{OE}}^{-1} \cdot (\mathsf{T}^{-1})_{\text{OE}} \cdot \widetilde{P}_{4,[1..d/2]}^{-1}(Y^{(i)}[\text{odd}], Y^{(i)}[\text{even}]) \oplus g_2(X^{(i)}[\text{odd}], Y^{(i)}[\text{odd}]),$$

where g_2 is a function of $X[\text{odd}]$ and $Y[\text{odd}]$ and is independent of the first two terms.

Because $\mathcal{A}$ does not repeat a query, we have $Y^{(i)}[j_0] \neq Y^{(j)}[j_0]$ for some $j_0 \in \{1, 2, \ldots, d\}$. We also assume j_0 is the minimum index. If j_0 is an odd number, $\widetilde{P}_{4,(j_0+1)/2}^{-1}(Y^{(i)}[j_0], Y^{(i)}[j_0 + 1])$ is a "fresh" computation, the value of which is uniform in 2^n values. Since every entry in the $((j_0 + 1)/2)$-th column of $\mathsf{T}_{\text{EE}} \cdot \mathsf{T}_{\text{OE}}^{-1} \cdot (\mathsf{T}^{-1})_{\text{OE}}$ is nonzero (see Definition 1), we have,

$$\Pr\left[u_3^{(i)}[2x] = u_3^{(j)}[2x]\right] = \frac{1}{2^n}. \tag{15}$$

If j_0 is an even number, $\widetilde{P}_{4,j_0/2}^{-1}(Y^{(i)}[j_0 - 1], Y^{(i)}[j_0])$ is a "fresh" computation. After conditioning on the value of $\widetilde{P}_{4,j_0/2}^{-1}(Y^{(j)}[j_0 - 1], Y^{(j)}[j_0])$, the value of $\widetilde{P}_{4,j_0/2}^{-1}(Y^{(i)}[j_0 - 1], Y^{(i)}[j_0])$ is uniform in $\geq 2^n - q$ possibilities. Because every entry in the $(j_0/2)$-th column of $\mathsf{T}_{\text{EE}} \cdot \mathsf{T}_{\text{OE}}^{-1} \cdot (\mathsf{T}^{-1})_{\text{OE}}$ is nonzero, we have,

$$\Pr\left[u_3^{(i)}[2x] = u_3^{(j)}[2x]\right] \leq \frac{1}{2^n - q}. \tag{16}$$

From Eqs. (15) and (16), we have $\Pr\left[u_3^{(i)}[2x] = u_3^{(j)}[2x]\right] \leq \frac{1}{2^n-q}$.

Summing over $x \in \{1, 2, \ldots, d/2\}$, $i \in \{2, 3, \ldots, q\}$, $j \in \{1, 2, \ldots, i-1\}$, we reach

$$\Pr[(\text{B-3})] \leq \sum_{x \in [1..d/2]} \sum_{i \in [2..q]} \sum_{j \in [1..i-1]} \Pr\left[u_3^{(i)}[2x] = u_3^{(j)}[2x]\right]$$
$$\leq \sum_{i \in [2..q]} \frac{d/2(i-1)}{2^n - q} \leq \frac{0.25dq^2}{2^n - q}. \tag{17}$$

Condition (B-4). Lastly, for the condition (B-4), we focus on $x \in [1..d/2]$ and the i-th and j-th queries, with $1 \leq j < i \leq q$. Then we derive the upper bound on

$$\Pr\left[u_3^{(i)}[2x-1]\|v_3^{(i)}[2x] = u_3^{(j)}[2x-1]\|v_3^{(j)}[2x]\right] \leq \Pr\left[v_3^{(i)}[2x] = v_3^{(j)}[2x]\right].$$

We also use the trivial bound of $\Pr\left[u_3^{(i)}[2x-1] = u_3^{(j)}[2x-1] \mid v_3^{(i)}[2x] = v_3^{(j)}[2x]\right] \leq 1$. Hence, we focus on $\Pr\left[v_3^{(i)}[2x] = v_3^{(j)}[2x]\right]$. Arising from the design of the ETGFN, we have

$$v_3^{(i)}[\text{even}] = (\mathsf{T}^{-1})_{\text{EO}} \cdot Y^{(i)}[\text{odd}] \oplus (\mathsf{T}^{-1})_{\text{EE}} \cdot \widetilde{P}_{4,[1..d/2]}^{-1}\big(Y^{(i)}[\text{odd}], Y^{(i)}[\text{even}]\big).$$

Because $\mathcal{A}$ does not repeat a query, we have $Y^{(i)}[j_0] \neq Y^{(j)}[j_0]$ for some $j_0 \in \{1, 2, \ldots, d\}$. We also assume j_0 is the minimum index. Similar to the analysis in the condition (B-3), $\widetilde{P}_{4,(j_0+1)/2}^{-1}(Y^{(i)}[j_0], Y^{(i)}[j_0 + 1])$ or $\widetilde{P}_{4,j_0/2}^{-1}(Y^{(i)}[j_0 - 1], Y^{(i)}[j_0])$ is a "fresh" computation. Due to every entry in $(\mathsf{T}^{-1})_{\text{EE}}$ is nonzero, we have $\Pr\left[v_3^{(i)}[2x] = v_3^{(j)}[2x]\right] \leq \frac{1}{2^n-q}$.

Summing over $x \in \{1, 2, \ldots, d/2\}$, $i \in \{2, 3, \ldots, q\}$, $j \in \{1, 2, \ldots, i-1\}$, we reach

$$\Pr[(\text{B-4})] \leq \sum_{x \in [1..d/2]} \sum_{i \in [2..q]} \sum_{j \in [1..i-1]} \Pr\left[v_3^{(i)}[2x] = v_3^{(j)}[2x]\right]$$
$$\leq \sum_{i \in [2..q]} \frac{d/2(i-1)}{2^n - q} \leq \frac{0.25dq^2}{2^n - q}. \tag{18}$$

Comprehensive Analysis. When $q \leq 2^n/2$, summing over Eqs. (10), (13), (17) and (18), we reach Eq. (7):

$$\Pr[\Theta_{\mathcal{I}} \in \mathcal{T}_{\text{bad}}] = \sum_{\ell \in [1..4]} \Pr[(B-\ell)] \leq \frac{dq^2}{2^n - q} \leq \frac{2dq^2}{2^n}.$$

$\square$

4.3 Analysis of Good Transcripts

Here, we prove the following lemma regarding a good transcript $\theta \in \mathcal{T}_{\text{good}}$.

Lemma 3. *For any $\theta \in \mathcal{T}_{\text{good}}$, we have*

$$\frac{\Pr[\Theta_{\mathcal{R}} = \theta]}{\Pr[\Theta_{\mathcal{I}} = \theta]} \geq 1 - \frac{0.5q^2}{2^{dn}}.$$

Proof. We consider $\theta \in \mathcal{T}_{\text{good}}$. In the real world, for $x \in [1..4]$ and $y \in [1..d/2]$, we define $Q_{x,y}^{(1)} = \emptyset$ and for $2 \leq i \leq q$,

$$Q_{x,y}^{(i)} = \{j \mid j < i \text{ and } (i\text{-th tweak of } \widetilde{P}_{x,y}) = (j\text{-th tweak of } \widetilde{P}_{x,y})\}.$$

$Q_{x,y}^{(i)}$ is the set of all indices $1 \leq j < i$ such that the i-th tweak of $\widetilde{P}_{x,y}$ is the same as the j-th tweak of $\widetilde{P}_{x,y}$. Then we have

$$\Pr[\Theta_{\mathcal{R}} = \theta] = \prod_{x \in [1..4]} \prod_{y \in [1..d/2]} \prod_{i \in [1..q]} \frac{1}{2^n - |Q_{x,y}^{(i)}|}$$

$$\geq \frac{1}{2^{dnq}} \cdot \left(\prod_{y \in [1..d/2]} \prod_{i \in [1..q]} \frac{1}{2^n - |Q_{1,y}^{(i)}|} \right) \left(\prod_{y \in [1..d/2]} \prod_{i \in [1..q]} \frac{1}{2^n - |Q_{4,y}^{(i)}|} \right).$$

In the ideal world, for $x = 1, 4$ and $y \in [1..d/2]$ and for $1 \leq i \leq q$, we define $Q_{x,y}^{(i)}$ as in the real world. Then we have

$$\Pr[\Theta_{\mathcal{I}} = \theta] = \left(\prod_{i \in [1..q]} \frac{1}{2^{dn} - (i-1)} \right)$$

$$\cdot \left(\prod_{y \in [1..d/2]} \prod_{i \in [1..q]} \frac{1}{2^n - |Q_{1,y}^{(i)}|} \right) \left(\prod_{y \in [1..d/2]} \prod_{i \in [1..q]} \frac{1}{2^n - |Q_{4,y}^{(i)}|} \right).$$

Then, we have

$$\frac{\Pr[\Theta_{\mathcal{R}} = \theta]}{\Pr[\Theta_{\mathcal{I}} = \theta]} \geq \prod_{i \in [1..q]} \frac{2^{dn} - (i-1)}{2^{dn}} = \prod_{i \in [1..q]} \left(1 - \frac{i-1}{2^{dn}}\right) \geq 1 - \frac{0.5q^2}{2^{dn}},$$

and we obtain Lemma 3. □

5 Conclusion

As shown in [30], if we focus on SPRP security, type-2 GFN has the smallest number of TBC calls among type-1/2/3 GFN. Moreover, due to the parallel processing capability of $d/2$ TBCs for both encryption and decryption, type-2 GFN exhibits more efficiency. Therefore, we explore the possibility of minimizing the calls of TBC in type-2 GFN. Motivated by SPN and the so-called extended GFNs, we replace the blockwise shuffle with a block-oriented linear diffusion layer and study the TBC-based counterpart of this construction, which we call extended TBC-based type-2 GFN (ETGFN). With a moderately strong layer, we prove birthday-bound security at 4 rounds. The number of TBC calls for multi-line GFN is summarized in Table 1. We point out that our construction has the smallest number of TBC calls and is independent of the number of lines.

As open questions, we leave the security of 3-round ETGFN and the analysis to obtain stronger security bounds by increasing the number of rounds. Also, as mentioned in Sect. 1 (discussion), we leave further exploring "prove-then-prune" as an interesting future work. This paper focuses on the indistinguishability notion, and the analysis in the indifferentiability framework would be interesting.

Acknowledgments. We thank the anonymous reviewers for their invaluable comments and suggestions, which helped us improve the manuscript. Yuqing Zhao and Chun Guo were partly supported by the Program of Qilu Young Scholars (Grant No. 61580089963177) of Shandong University.

A   Candidate Good Diffusion Layers for Definition 1

Using the primitive polynomial $x^8 + x^4 + x^3 + x^2 + 1$, two candidates for $n = 8$ and $d = 8, 16$ respectively are as follows.

$$
\begin{pmatrix}
0x37 & 0x8E & 0x7F & 0xB9 & 0x8C & 0xC9 & 0x3D & 0x06 \\
0x2A & 0x80 & 0x09 & 0xF3 & 0x31 & 0x91 & 0xFE & 0x0F \\
0x73 & 0xEB & 0x4C & 0x9C & 0x25 & 0x60 & 0xD4 & 0xD8 \\
0x8C & 0x79 & 0x5E & 0x0F & 0xAC & 0x97 & 0x22 & 0xBC \\
0x28 & 0x0F & 0x34 & 0x15 & 0x1F & 0xA5 & 0x2F & 0x92 \\
0x8D & 0x3D & 0x3D & 0x47 & 0xE1 & 0x0D & 0x25 & 0x02 \\
0x51 & 0xFF & 0xBA & 0x59 & 0xF3 & 0xBF & 0x8B & 0x8B \\
0x82 & 0xF7 & 0x25 & 0xA1 & 0xCF & 0xAB & 0x8D & 0x19
\end{pmatrix} ,
$$

$$
\begin{pmatrix}
0xCA & 0xAF & 0xA2 & 0xBB & 0x56 & 0xF7 & 0xFB & 0xA2 & 0xD2 & 0x86 & 0xA5 & 0xAA & 0x05 & 0x10 & 0x29 & 0xE2 \\
0xC2 & 0x83 & 0x97 & 0x11 & 0x57 & 0xD3 & 0x7D & 0xA5 & 0x51 & 0xB9 & 0x37 & 0x74 & 0x02 & 0x40 & 0xAE & 0xDD \\
0xB8 & 0x82 & 0x5B & 0x64 & 0xAE & 0xE6 & 0xFC & 0xDB & 0x36 & 0x49 & 0x64 & 0x46 & 0xE1 & 0xB0 & 0x1E & 0x79 \\
0x84 & 0x2F & 0x9B & 0xE9 & 0x45 & 0xAB & 0x98 & 0x47 & 0x35 & 0x26 & 0x7C & 0x3C & 0x41 & 0xC2 & 0xBB & 0x82 \\
0xA7 & 0xC9 & 0x6D & 0x99 & 0xF1 & 0x1E & 0x83 & 0x40 & 0xB6 & 0xFE & 0x6F & 0x30 & 0xD5 & 0xC8 & 0xB1 & 0x0C \\
0x79 & 0x56 & 0xE3 & 0x4D & 0x55 & 0x5A & 0x9E & 0xF6 & 0xF8 & 0xF1 & 0x60 & 0xEF & 0x71 & 0x53 & 0x8D & 0xCE \\
0x9F & 0xFC & 0xFE & 0xD5 & 0xFE & 0xC9 & 0x5E & 0xD8 & 0xAD & 0x53 & 0x5D & 0x55 & 0xDF & 0xEB & 0x03 & 0x39 \\
0x03 & 0x24 & 0x1B & 0xF8 & 0xDF & 0x0C & 0xA4 & 0x25 & 0x35 & 0xB2 & 0x60 & 0x22 & 0x92 & 0x65 & 0x9A & 0x3F \\
0xF8 & 0xFB & 0x18 & 0x0C & 0xB9 & 0xEE & 0x38 & 0x81 & 0xE1 & 0x4C & 0x86 & 0xBE & 0x06 & 0xCD & 0x0F & 0xA9 \\
0x5D & 0xBE & 0xC2 & 0x94 & 0x67 & 0x5D & 0x27 & 0xC1 & 0x77 & 0x05 & 0x92 & 0x4C & 0xCB & 0xC6 & 0x05 & 0xCC \\
0x4C & 0x69 & 0xCD & 0x13 & 0xD0 & 0x90 & 0xCD & 0x61 & 0x8F & 0x18 & 0x14 & 0x59 & 0x8C & 0x2C & 0x97 & 0xFB \\
0xE7 & 0x32 & 0xFF & 0x8E & 0x09 & 0x7E & 0xE1 & 0x6A & 0x89 & 0x52 & 0x3F & 0x52 & 0x1E & 0xBB & 0x24 & 0x6E \\
0xC3 & 0xA7 & 0x2F & 0xFB & 0xEC & 0xF1 & 0x07 & 0xB2 & 0x40 & 0x34 & 0x70 & 0x81 & 0xBE & 0xF5 & 0xE0 & 0x37 \\
0xE5 & 0xBB & 0x26 & 0xDA & 0x28 & 0x09 & 0x5A & 0xFE & 0x27 & 0xA0 & 0x65 & 0x8D & 0xD5 & 0x43 & 0x14 & 0xCB \\
0xBE & 0xED & 0x5B & 0xE8 & 0x27 & 0x57 & 0x15 & 0xA6 & 0x9E & 0x10 & 0x69 & 0x58 & 0xBA & 0x46 & 0xD0 & 0xB1 \\
0x24 & 0x5D & 0x2A & 0xB1 & 0x29 & 0x58 & 0xF8 & 0xD0 & 0x93 & 0x37 & 0xA1 & 0x52 & 0xFC & 0x53 & 0xED & 0xAF
\end{pmatrix} .
$$

Using the primitive polynomial $x^{11} + x^2 + 1$ a candidate for $n = 11$ and $d = 8$ is as follows:

$$
\begin{pmatrix}
0x22A & 0x308 & 0x7B4 & 0x406 & 0x1D3 & 0x66A & 0x02F & 0x507 \\
0x153 & 0x61A & 0x6A0 & 0x4A1 & 0x618 & 0x689 & 0x17A & 0x4A2 \\
0x663 & 0x167 & 0x773 & 0x7D2 & 0x64C & 0x751 & 0x441 & 0x2F8 \\
0x144 & 0x39D & 0x6F5 & 0x563 & 0x0B3 & 0x365 & 0x133 & 0x3AB \\
0x434 & 0x2EF & 0x44F & 0x7DE & 0x1B0 & 0x7E9 & 0x422 & 0x730 \\
0x47F & 0x3DB & 0x07F & 0x161 & 0x060 & 0x7C2 & 0x65F & 0x746 \\
0x704 & 0x18F & 0x410 & 0x1C3 & 0x0ED & 0x551 & 0x7F4 & 0x111 \\
0x2E9 & 0x53E & 0x36E & 0x76D & 0x464 & 0x1D2 & 0x661 & 0x002
\end{pmatrix} .
$$

We have also found plenty of candidates for other parameters, which are however omitted for the sake of space.

References

1. Anderson, R., Biham, E.: Two practical and provably secure block ciphers: BEAR and LION. In: Gollmann, D. (ed.) FSE 1996. LNCS, vol. 1039, pp. 113–120. Springer, Heidelberg (1996). https://doi.org/10.1007/3-540-60865-6_48
2. Beierle, C., et al.: The SKINNY family of block ciphers and its low-latency variant MANTIS. In: Robshaw, M., Katz, J. (eds.) CRYPTO 2016, Part II. LNCS, vol. 9815, pp. 123–153. Springer, Heidelberg (2016). https://doi.org/10.1007/978-3-662-53008-5_5
3. Bellare, M., Ristenpart, T., Rogaway, P., Stegers, T.: Format-preserving encryption. In: Jacobson, M.J., Rijmen, V., Safavi-Naini, R. (eds.) SAC 2009. LNCS, vol. 5867, pp. 295–312. Springer, Heidelberg (2009). https://doi.org/10.1007/978-3-642-05445-7_19
4. Berger, T.P., Francq, J., Minier, M., Thomas, G.: Extended generalized feistel networks using matrix representation to propose a new lightweight block cipher: lilliput. IEEE Trans. Comput. 65(7), 2074–2089 (2016). https://doi.org/10.1109/TC.2015.2468218
5. Bhaumik, R., List, E., Nandi, M.: ZCZ – achieving n-bit SPRP security with a minimal number of tweakable-block-cipher calls. In: Peyrin, T., Galbraith, S. (eds.) ASIACRYPT 2018, Part I. LNCS, vol. 11272, pp. 336–366. Springer, Cham (2018). https://doi.org/10.1007/978-3-030-03326-2_12
6. Black, J., Rogaway, P.: Ciphers with arbitrary finite domains. In: Preneel, B. (ed.) CT-RSA 2002. LNCS, vol. 2271, pp. 114–130. Springer, Heidelberg (2002). https://doi.org/10.1007/3-540-45760-7_9
7. Chen, S., Steinberger, J.: Tight security bounds for key-alternating ciphers. In: Nguyen, P.Q., Oswald, E. (eds.) EUROCRYPT 2014. LNCS, vol. 8441, pp. 327–350. Springer, Heidelberg (2014). https://doi.org/10.1007/978-3-642-55220-5_19
8. Chen, Y.L., Luykx, A., Mennink, B., Preneel, B.: Efficient length doubling from tweakable block ciphers. IACR Trans. Symmetric Cryptol. 2017(3), 253–270 (2017). https://doi.org/10.13154/tosc.v2017.i3.253-270
9. Chen, Y.L., Mennink, B., Nandi, M.: Short variable length domain extenders with beyond birthday bound security. In: Peyrin, T., Galbraith, S. (eds.) ASIACRYPT 2018, Part I. LNCS, vol. 11272, pp. 244–274. Springer, Cham (2018). https://doi.org/10.1007/978-3-030-03326-2_9
10. Coron, J.-S., Dodis, Y., Mandal, A., Seurin, Y.: A domain extender for the ideal cipher. In: Micciancio, D. (ed.) TCC 2010. LNCS, vol. 5978, pp. 273–289. Springer, Heidelberg (2010). https://doi.org/10.1007/978-3-642-11799-2_17
11. Dutta, A., Nandi, M.: Tweakable HCTR: a BBB secure tweakable enciphering scheme. In: Chakraborty, D., Iwata, T. (eds.) INDOCRYPT 2018. LNCS, vol. 11356, pp. 47–69. Springer, Cham (2018). https://doi.org/10.1007/978-3-030-05378-9_3
12. Feistel, H., Notz, W.A., Smith, J.L.: Some cryptographic techniques for machine-to-machine data communications. Proc. IEEE 63(11), 1545–1554 (1975)
13. Gueron, S., Mouha, N.: Simpira v2: a family of efficient permutations using the AES round function. In: Cheon, J.H., Takagi, T. (eds.) ASIACRYPT 2016, Part I. LNCS, vol. 10031, pp. 95–125. Springer, Heidelberg (2016). https://doi.org/10.1007/978-3-662-53887-6_4
14. Guo, C., Standaert, F., Wang, W., Wang, X., Yu, Y.: Provable security of SP networks with partial non-linear layers. IACR Trans. Symmetric Cryptol. 2021(2), 353–388 (2021). https://doi.org/10.46586/tosc.v2021.i2.353-388

15. Hoang, V.T., Krovetz, T., Rogaway, P.: Robust authenticated-encryption AEZ and the problem that it solves. In: Oswald, E., Fischlin, M. (eds.) EUROCRYPT 2015, Part I. LNCS, vol. 9056, pp. 15–44. Springer, Heidelberg (2015). https://doi.org/10.1007/978-3-662-46800-5_2

16. Hoang, V.T., Rogaway, P.: On generalized feistel networks. In: Rabin, T. (ed.) CRYPTO 2010. LNCS, vol. 6223, pp. 613–630. Springer, Heidelberg (2010). https://doi.org/10.1007/978-3-642-14623-7_33

17. Jean, J., Nikolic, I., Peyrin, T., Seurin, Y.: The deoxys AEAD family. J. Cryptol. **34**(3), 31 (2021). https://doi.org/10.1007/s00145-021-09397-w

18. Jha, A., List, E., Minematsu, K., Mishra, S., Nandi, M.: XHX – a framework for optimally secure tweakable block ciphers from classical block ciphers and universal hashing. In: Lange, T., Dunkelman, O. (eds.) LATINCRYPT 2017. LNCS, vol. 11368, pp. 207–227. Springer, Cham (2019). https://doi.org/10.1007/978-3-030-25283-0_12

19. Liskov, M., Rivest, R.L., Wagner, D.: Tweakable block ciphers. In: Yung, M. (ed.) CRYPTO 2002. LNCS, vol. 2442, pp. 31–46. Springer, Heidelberg (2002). https://doi.org/10.1007/3-540-45708-9_3

20. Liskov, M., Rivest, R.L., Wagner, D.: Tweakable block ciphers. J. Cryptol. **24**(3), 588–613 (2011). https://doi.org/10.1007/s00145-010-9073-y

21. Liskov, M.D., Rivest, R.L., Wagner, D.A.: Tweakable block ciphers. In: Yung, M. (eds.) CRYPTO 2002. LNCS, vol. 2442, pp. 31–46. Springer, Heidelberg (2002). https://doi.org/10.1007/3-540-45708-9_3. https://api.semanticscholar.org/CorpusID:126254492

22. Luby, M., Rackoff, C.: How to construct pseudorandom permutations from pseudorandom functions. SIAM J. Comput. **17**(2), 373–386 (1988)

23. Lucks, S.: Faster Luby-Rackoff ciphers. In: Gollmann, D. (ed.) FSE 1996. LNCS, vol. 1039, pp. 189–203. Springer, Heidelberg (1996). https://doi.org/10.1007/3-540-60865-6_53

24. Minematsu, K.: Beyond-birthday-bound security based on tweakable block cipher. In: Dunkelman, O. (ed.) FSE 2009. LNCS, vol. 5665, pp. 308–326. Springer, Heidelberg (2009). https://doi.org/10.1007/978-3-642-03317-9_19

25. Minematsu, K.: Building blockcipher from small-block tweakable blockcipher. Des. Codes Cryptogr. **74**(3), 645–663 (2015). https://doi.org/10.1007/s10623-013-9882-8

26. Minematsu, K., Iwata, T.: Building blockcipher from tweakable blockcipher: extending FSE 2009 proposal. In: Chen, L. (ed.) IMACC 2011. LNCS, vol. 7089, pp. 391–412. Springer, Heidelberg (2011). https://doi.org/10.1007/978-3-642-25516-8_24

27. Morris, B., Rogaway, P., Stegers, T.: How to encipher messages on a small domain. In: Halevi, S. (ed.) CRYPTO 2009. LNCS, vol. 5677, pp. 286–302. Springer, Heidelberg (2009). https://doi.org/10.1007/978-3-642-03356-8_17

28. Nakamichi, R., Iwata, T.: Iterative block ciphers from tweakable block ciphers with long tweaks. IACR Trans. Symm. Cryptol. **2019**(4), 54–80 (2019). https://doi.org/10.13154/tosc.v2019.i4.54-80

29. Nakamichi, R., Iwata, T.: Beyond-birthday-bound secure cryptographic permutations from ideal ciphers with long keys. IACR Trans. Symm. Cryptol. **2020**(2), 68–92 (2020). https://doi.org/10.13154/tosc.v2020.i2.68-92

30. Nakaya, K., Iwata, T.: Generalized feistel structures based on tweakable block ciphers. IACR Trans. Symmetric Cryptol. **2022**(4), 24–91 (2022). https://doi.org/10.46586/tosc.v2022.i4.24-91

31. Nandi, M.: On the optimality of non-linear computations of length-preserving encryption schemes. In: Iwata, T., Cheon, J.H. (eds.) ASIACRYPT 2015, Part II. LNCS, vol. 9453, pp. 113–133. Springer, Heidelberg (2015). https://doi.org/10.1007/978-3-662-48800-3_5

32. Naor, M., Reingold, O.: On the construction of pseudorandom permutations: Luby-Rackoff revisited. J. Cryptol. **12**(1), 29–66 (1999). https://doi.org/10.1007/PL00003817

33. Patarin, J.: Security of random feistel schemes with 5 or more rounds. In: Franklin, M. (ed.) CRYPTO 2004. LNCS, vol. 3152, pp. 106–122. Springer, Heidelberg (2004). https://doi.org/10.1007/978-3-540-28628-8_7

34. Patarin, J.: The "coefficients H" technique. In: Avanzi, R.M., Keliher, L., Sica, F. (eds.) SAC 2008. LNCS, vol. 5381, pp. 328–345. Springer, Heidelberg (2009). https://doi.org/10.1007/978-3-642-04159-4_21

35. Sadeghiyan, B., Pieprzyk, J.: A construction for super pseudorandom permutations from a single pseudorandom function. In: Rueppel, R.A. (ed.) EUROCRYPT 1992. LNCS, vol. 658, pp. 267–284. Springer, Heidelberg (1993). https://doi.org/10.1007/3-540-47555-9_23

36. Schneier, B., Kelsey, J.: Unbalanced feistel networks and block cipher design. In: Gollmann, D. (ed.) FSE 1996. LNCS, vol. 1039, pp. 121–144. Springer, Heidelberg (1996). https://doi.org/10.1007/3-540-60865-6_49

37. Shen, Y., Guo, C., Wang, L.: Improved security bounds for generalized Feistel networks. IACR Trans. Symm. Cryptol. **2020**(1), 425–457 (2020). https://doi.org/10.13154/tosc.v2020.i1.425-457

38. Shen, Y., Standaert, F.: Optimally secure tweakable block ciphers with a large tweak from n-bit block ciphers. IACR Trans. Symmetric Cryptol. **2023**(2), 47–68 (2023). https://doi.org/10.46586/tosc.v2023.i2.47-68

39. Shrimpton, T., Terashima, R.S.: A modular framework for building variable-input-length tweakable ciphers. In: Sako, K., Sarkar, P. (eds.) ASIACRYPT 2013, Part I. LNCS, vol. 8269, pp. 405–423. Springer, Heidelberg (2013). https://doi.org/10.1007/978-3-642-42033-7_21

40. Suzaki, T., Minematsu, K., Morioka, S., Kobayashi, E.: TWINE: a lightweight block cipher for multiple platforms. In: Knudsen, L.R., Wu, H. (eds.) SAC 2012. LNCS, vol. 7707, pp. 339–354. Springer, Heidelberg (2013). https://doi.org/10.1007/978-3-642-35999-6_22

41. Wu, W., Zhang, L.: LBlock: a lightweight block cipher. In: Lopez, J., Tsudik, G. (eds.) ACNS 2011. LNCS, vol. 6715, pp. 327–344. Springer, Heidelberg (2011). https://doi.org/10.1007/978-3-642-21554-4_19

42. Zhao, Y., Guo, C., Wang, W.: Towards Minimizing Non-linearity in Type-II Generalized Feistel Networks. Cryptology ePrint Archive, Report 2023/1295 (2023). https://eprint.iacr.org/2023/1295. To appear at CANS 2023

43. Zheng, Y., Matsumoto, T., Imai, H.: On the construction of block ciphers provably secure and not relying on any unproved hypotheses. In: Brassard, G. (ed.) CRYPTO 1989. LNCS, vol. 435, pp. 461–480. Springer, New York (1990). https://doi.org/10.1007/0-387-34805-0_42

The Patching Landscape of Elisabeth-4 and the Mixed Filter Permutator Paradigm

Clément Hoffmann[2]($\boxtimes$)(ID), Pierrick Méaux[1]($\boxtimes$)(ID), and François-Xavier Standaert[2]($\boxtimes$)(ID)

[1] University of Luxembourg, Esch-sur-Alzette, Luxembourg
[2] UCLouvain, ICTEAM/ELEN/Crypto Group, Ottignies-Louvain-la-Neuve, Belgium
`clement.hoffmann@uclouvain.be`

Abstract. Filter permutators are a family of stream cipher designs that are aimed for hybrid homomorphic encryption. While originally operating on bits, they have been generalized to groups at Asiacrypt 2022, and instantiated for evaluation with the TFHE scheme which favors a filter based on (negacyclic) Look Up Tables (LUTs). A recent work of Gilbert et al., to appear at Asiacrypt 2023, exhibited (algebraic) weaknesses in the Elisabeth-4 instance, exploiting the combination of the 4-bit negacyclic LUTs it uses as filter. In this article, we explore the landscape of patches that can be used to restore the security of such designs while maintaining their good properties for hybrid homomorphic encryption. Starting with minimum changes, we observe that just updating the filter function (still with small negacyclic LUTs) is conceptually feasible, and propose the resulting Elisabeth-b4 design with three levels of NLUTs. We then show that a group permutator combining two different functions in the filter can simplify the analysis and improve performances. We specify the Gabriel instance to illustrate this claim. We finally propose to modify the group filter permutator paradigm into a mixed filter permutator, which considers the permutation of the key with elements in a group and a filter outputting elements in a different group. We specify the Margrethe instance as a first example of mixed filter permutator, with key elements in $\mathbb{F}_2$ and output in $\mathbb{Z}_{16}$, that we believe well-suited for recent fully homomorphic encryption schemes that can efficiently evaluate larger (not negacyclic) LUTs.

1 Introduction

In recent years, symmetric designs purposed for hybrid homomorphic encryption have been a topic of considerable research attention. They enable a client requiring homomorphic computations to exclusively rely on symmetric cryptography, delegating the computationally intensive homomorphic tasks to the server. Examples include LowMC [ARS+15], Kreyvium [CCF+16], FLIP [MJSC16], FiLIP [MCJS19], Rasta [DEG+18], Hera [CHK+21], Rubato [HKL+22], Chagri [AMT22] or Pasta [DGH+23]. In this paper we are concerned with the Filter Permutator (FP) paradigm, which is a type of stream ciphers leveraging the possibility to perform a part of its computations in a non homomorphic manner.

The FP paradigm has been extended to groups at Asiacrypt 2022 – leading to the Group Filter Permutator (GFP) paradigm – and the Elisabeth family of such ciphers was introduced as a first example by Cosseron et al. in [CHMS22].

© The Author(s), under exclusive license to Springer Nature Switzerland AG 2024
A. Chattopadhyay et al. (Eds.): INDOCRYPT 2023, LNCS 14459, pp. 134–156, 2024.
https://doi.org/10.1007/978-3-031-56232-7_7

In a work to appear at Asiarcypt 2023, Gilbert et al. demonstrated that Elisabeth-4, an instance of the Elisabeth family, has less than the 128 bits of security claimed. They show that the algebraic system derived from Elisabeth-4 has an insufficient number of monomials over $\mathbb{F}_2$ that leads to a linearization attack which, after various optimizations, is able to retrieve the secret key in 2^{88} operations. There are two main factors allowing this vulnerability. The first one is the size of Elisabeth-4's inner function, which only takes 5 elements of $\mathbb{Z}_{16}$ as input. The second one is the combination of Negacyclic Look-Up Tables (NLUTs) and the modular addition over $\mathbb{Z}_{16}$ which results in the least significant bit of the output having a weak algebraic structure.

In this article, we present three methods for modifying the Elisabeth-4 instance, necessitating increasingly substantial alterations to the original scheme. The first method, Elisabeth-b, essentially involves expanding the parameters' size. It entails an increase in both the number of inputs to the inner function and the number of inner functions that are summed. In the second method, Gabriel, we combine different inner functions in the sum in order to harness the benefits of each, thus enabling us to mitigate performance overheads. Finally, we introduce the Mixed Filter Permutator (MFP) paradigm, which combines different groups. We then present Margrethe, a family of stream ciphers that displays how the MFP avoids the weaknesses highlighted in [GHBJR23].

2 Preliminaries

Notations. For two integers a and b such that $a \leq b$ we denote by $[a, b]$ the set $\{a, a + 1, \ldots, b - 1, b\}$ and $[b]$ for $[1, b]$.

2.1 Boolean Functions and Cryptography

We introduce some key concepts of Boolean functions and relevant cryptographic criteria, and their generalizations for functions from $\mathbb{G}^n$ to $\mathbb{G}$ where $\mathbb{G}$ is a group (with operation denoted "+"). We refer to [Car21] for the notions on Boolean functions, and [CHMS22] for the generalizations we consider.

Definition 1 ((Vectorial) Boolean function). *An (n, m) vectorial Boolean function F is a function from $\mathbb{F}_2^n$ to $\mathbb{F}_2^m$. When $m = 1$ the $(n, 1)$ vectorial Boolean function is referred to as a Boolean function and the space of n-variable Boolean function is denoted by $\mathcal{B}_n$. The coordinate functions of F are the m Boolean functions f_i which associate for each $x \in \mathbb{F}_2^n$ the i-th binary output of $F(x)$. The component functions of F are the $2^m - 1$ non-trivial linear combinations of the coordinate functions of F.*

Definition 2 (Algebraic Normal Form (ANF) and degree). *The Algebraic Normal Form of a Boolean function f is its n-variable polynomial representation over $\mathbb{F}_2$ (belonging to $\mathbb{F}_2[x_1, \ldots, x_n]/(x_1^2 + x_1, \ldots, x_n^2 + x_n)$):*

$$f(x) = \sum_{I \subseteq [n]} a_I \left(\prod_{i \in I} x_i \right) = \sum_{I \subseteq [n]} a_I x^I,$$

where $a_I \in \mathbb{F}_2$. The algebraic degree of f is defined as: $\deg(f) = \max_{\{I \mid a_I = 1\}} |I|$ (with the convention that $\deg(0) = 0$).

Definition 3 (Algebraic Immunity). *The algebraic immunity of a Boolean function* $f \in \mathcal{B}_n$, *denoted by* $\mathsf{AI}(f)$, *is defined as:*

$$\mathsf{AI}(f) = \min_{g \neq 0}\{\deg(g) \mid fg = 0 \text{ or } (f+1)g = 0\}.$$

The Boolean function g is called an annihilator of f (or f + 1).

Definition 4 (Balancedness and Resiliency). *A Boolean function* $f \in \mathcal{B}_n$ *is balanced if* $|\{x \mid f(x) = 0\}| = |\{x \mid f(x) = 1\}|$. *The function f is m-resilient if any of its restrictions obtained by fixing at most m of its variables is balanced. We denote by* $\mathsf{res}(f)$ *the maximum resiliency (also called resiliency order) m of f and set* $\mathsf{res}(f) = -1$ *if f is unbalanced.*

Definition 5 (Nonlinearity). *For* $f \in \mathcal{B}_n$ *the nonlinearity* NL *of f is the minimum Hamming distance between f and all the affine functions in* $\mathcal{B}_n$:

$$\mathsf{NL}(f) = \min_{g,\,\deg(g) \leq 1}\{d_H(f,g)\},$$

where $d_H(f,g) = \#\{x \in \mathbb{F}_2^n \mid f(x) \neq g(x)\}$ *is the Hamming distance between f and g; and with* $g(x) = a \cdot x + \varepsilon$, $a \in \mathbb{F}_2^n, \varepsilon \in \mathbb{F}_2$ *where · denotes the inner product.*

We denote by NL^d *the order-d nonlinearity of f, the minimum Hamming distance between f, and all functions of degree at most d.*

We recall the notion of direct sum and some properties of the functions obtained with this construction.

Definition 6 (Direct Sum). *Let* $f \in \mathcal{B}_n$ *and* $g \in \mathcal{B}_m$, *f and g operating on different variables, the direct sum h of f and g is defined by:*

$$h(x,y) = \mathsf{DS}(f,g) = f(x) + g(y), \quad \text{where } x \in \mathbb{F}_2^n \text{ and } y \in \mathbb{F}_2^m.$$

Lemma 1 (Boolean direct sum properties (e.g. [MJSC16] Lemma 3)). *Let* $h = \mathsf{DS}(f,g)$ *be the direct sum of f and g n and m-variable Boolean functions respectively. Then* $\mathsf{DS}(f,g)$ *has the following cryptographic properties:*

1. *Degree:* $\deg(h) = \max(\deg(f), \deg(g))$.
2. *Algebraic immunity:* $\max(\mathsf{AI}(f), \mathsf{AI}(g)) \leq \mathsf{AI}(h) \leq \mathsf{AI}(f) + \mathsf{AI}(g)$.
3. *Resiliency:* $\mathsf{res}(h) = \mathsf{res}(f) + \mathsf{res}(g) + 1$.
4. *Nonlinearity:* $\mathsf{NL}(h) = 2^m\mathsf{NL}(f) + 2^n\mathsf{NL}(g) - 2\mathsf{NL}(f)\mathsf{NL}(g)$.

Lemma 2 ([Méa22] Lemma 5). *Let* $n, m \in \mathbb{N}^*$, *f and g be Boolean functions in n and m. If* $\mathsf{AI}(f) < \deg(g)$ *then* $\mathsf{AI}(\mathsf{DS}(f,g)) > \mathsf{AI}(f)$.

Lemma 3 ([Méa22] Lemma 6). *Let* $t \in \mathbb{N}^*$, *and* $f_1, \ldots, f_t$ *be t Boolean functions, if for* $r \in [t]$ *there exists r different indexes* $i_1, \cdots, i_r$ *of* $[t]$ *such that* $\forall j \in [r], \deg(f_{i_j}) \geq j$ *then* $\mathsf{AI}(\mathsf{DS}(f_1, \ldots, f_t)) \geq r$.

The notions defined above on Boolean functions can easily be extended to functions from $\mathbb{G}^n$ to $\mathbb{G}$, we denote these extended notions by the subscript $\mathbb{G}$.

Definition 7 (Cryptographic criteria over $\mathbb{G}$). *For a function f from $\mathbb{G}^n$ to $\mathbb{G}$ we denote:*

- $\deg_{\mathbb{G}}(f)$ *the degree over $\mathbb{G}$. It corresponds to the minimum degree over the polynomial representations of f in the polynomial ring $(\mathbb{G}, \cdot)[x_1, \ldots, x_n]$ when such representations exist.*
- $\mathrm{res}_{\mathbb{G}}(f)$ *the resiliency order over $\mathbb{G}$. f is balanced if and only if:*

$$\forall a \in \mathbb{G} : |\{x \mid f(x) = a\}| = |\mathbb{G}|^{n-1}.$$

  f is m-resilient if all the sub-functions obtained by fixing up to m variables are balanced.
- $\mathrm{NL}_{\mathbb{G}}(f)$ *the nonlinearity over $\mathbb{G}$. The nonlinearity is taken as the minimum Hamming distance between f and the affine functions: $a_0 + \sum_{i=1}^{n} a_i x_i$ where the a_i describe $\mathbb{G}^{n+1}$. $\mathrm{NL}_{\mathbb{G}}^d(f)$ denotes the order-d nonlinearity over $\mathbb{G}$.*

We also denote $\mathrm{DS}_{\mathbb{G}}(f, g)$ the direct sum $f(x) + g(y)$ where $x \in \mathbb{G}^n$ and $y \in \mathbb{G}^m$ and f and g are two n-variable and m-variable functions defined on distinct variables, the "+" here designs the operation of $\mathbb{G}$.

Functions from $\mathbb{G}^n$ to $\mathbb{G}$ (when $\mathbb{G}$ is only a group and not a field) are referred to as polyfunctions when they admit a polynomial representation. This formalism has been used lately in cryptography in the context of bootstrapping for fully homomorphic encryption [GIKV23]. In the following, we recall a result on the number of polyfunctions:

Property 1 (Number of polyfunctions, [SHW23], Proposition 26). The number of d-variable polynomial with coefficients in $\mathbb{Z}_{p^m}$ is:

$$\psi_d(p^m) = \prod_{\substack{\boldsymbol{k} \in \mathbb{N}^d \\ e_p(\boldsymbol{k}) < m}} p^{m - e_p(\boldsymbol{k})},$$

where $e_p(\boldsymbol{k}) := \max\{x \in \mathbb{N} : p^x | \boldsymbol{k}!\}$ and $\boldsymbol{k}! := \prod_{i=1}^{d} k_i!$.

2.2 Group Filter Permutator Paradigm

Paradigm. The Group Filter Permutator, illustrated in Fig. 1, is defined by a group $\mathbb{G}$ with operation noted $+$, a forward secure PRNG, a key size N, a subset size n, and a filtering function f from $\mathbb{G}^n$ to $\mathbb{G}$. To encrypt m elements of $\mathbb{G}$ under a secret key $K \in \mathbb{G}^N$, the public parameters of the PRNG are chosen and then the following process is executed for each key stream s_i (for $i \in [m]$):

- The PRNG is updated, its output determines a subset, a permutation, and a length-n vector of $\mathbb{G}$.
- the n-element subset S_i is chosen as a subset of the N key elements,
- the n-to-n permutation P_i is chosen,

- a $\mathbb{G}^n$ vector is uniformly sampled. This vector is named whitening and denoted by w_i,
- the key stream element s_i is computed as $s_i = f(P_i(S_i(K))+w_i)$, where $+$ denotes the element-wise addition of $\mathbb{G}$.

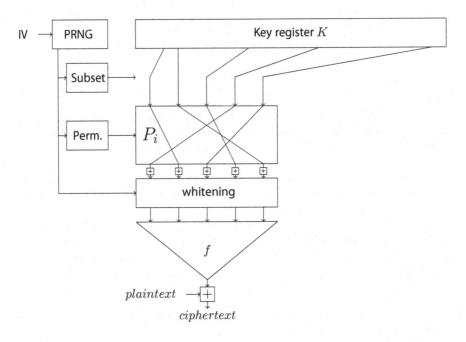

Fig. 1. The Group Filter Permutator design

Precision on the Forward Secure PRNG. We instantiate the forward secure PRNG following the Bellare Yee construction [BY03]. The high-level idea of the construction is to use an IV as an AES key. With this key, two AES encryptions are performed, one producing pseudo-random bits, the other providing a fresh AES key to iterate the process. More precisely, the IV provides two constants C_0, C_1, and for each K_i, $\mathrm{AES}_{K_i}(C_0) = K_{i+1}$ and $\mathrm{AES}_{K_i}(C_1)$ provides 128 bits of randomness.

The subset and permutation are computed at the same time using an aborted Knuth shuffling. To create a size-n permuted subset from the N-element key, we shuffle n elements of the key using Knuth shuffling [FY53, Knu97], and stop the algorithm after the first n iterations.[1]. Finally, n last random nibbles are used to generate the whitening. If the number of requested ciphertexts $m < 2^\lambda$ requires more pseudo-random bits than the PRNG can provide in a secure manner, a new instance is used with new constants.

Security Model. In the paradigms we consider in this article, the subsets, permutations and whitening are chosen from a forward secure PRNG to prevent malleability, and as

---

[1] See the Elisabeth implementation for an example.

before for the GFP paradigm we assume that no weaknesses come from the randomness generation and the analysis can focus on the properties of the filter. As a consequence, the IV of the PRNG is considered known by the adversary, but not chosen. In the context of HHE, where only an honest but curious model is considered for the server, it is classical to assume that the client sets the public values. The honest but curious model is justified by the fact that FHE in not IND-CCA (INDistinguishability against Chosen Ciphertext Attack), therefore we consider that the same server that does not temper with the homomorphic protocol will not temper with the symmetric encryption part. In this context, we do not consider chosen-IV attacks.

We consider the security in the known plaintext/ciphertext pairs setting. In order to assess the security of the schemes presented in this article, we verify that attacks that apply against the paradigms used have a cost of more than $2^\lambda = 2^{128}$ in either memory or operations and a data cost of more than $2^{\frac{\lambda}{2}} = 2^{64}$.

Elisabeth-4. Elisabeth has been introduced in [CHMS22]. It is a stream cipher family following the GFP paradigm that has been designed to take advantage of the most native functions of the TFHE [CGGI16] fully homomorphic encryption scheme. Accordingly, it relies on modular additions and negacyclic LUTs, where for a LUT of size n even negacyclic means that for $x \in [0, n/2 - 1]$ $f(i) = -f(i+n/2) \mod n$. For efficiency reasons, the modulo chosen for Elisabeth is 16 according to the state of the art on TFHE, which gives the instance Elisabeth-4.

Elisabeth-4 is the instance of the GFP with the following particularities: $\mathbb{G} = \mathbb{Z}_{16}$, $N = 256$, $n = 60$, the filtering function f is the direct sum of 12 times the 5-to-1 function g described in Fig. 2. The 8 particular negacyclic LUTs of Elisabeth-4 have been chosen at random, they are explicitly given in [CHMS22] Appendix B and the generation protocol is available at https://github.com/princess-elisabeth/sboxes_generation.

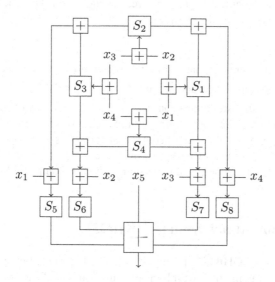

Fig. 2. Elisabeth-4's 5-to-1 inner function.

2.3 GFP and Security Analysis

We summarize the attacks considered on the GFP and in particular on Elisabeth-b4 in [CHMS22], for more details we refer to Section 4 of the same reference.

The GFP generalizes the Improved Filter Permutator (IFP) paradigm from [MCJS19] to any group, therefore the main attacks considered are the ones applying to the IFP and adaptations to the particular group used. These attacks are regrouped in [MCJS19] as algebraic-like attacks and correlation-like attacks that can be combined with guess and determine strategies. They are built on top of former analyses on similar paradigms, filter permutator [MJSC16] further analyzed in [DLR16,CMR17] and Goldreich's pseudorandom generator [Gol00]. The complexity of the different attacks on IFP is derived from parameters of the filtering function f, seen over $\mathbb{G}$ or over another group $\mathbb{G}'$ if this representation is favorable to the attacker. In particular for Elisabeth-4, $\mathbb{G} = \mathbb{Z}_{16}$ and various attacks take advantage of the representation over $\mathbb{F}_2$.

We now give a list of the considered attacks and their complexity. We do not give a detailed description of these attacks nor the methods used to compute their complexities and refer to [CHMS22] for these explanations.

- **Algebraic attacks** [CM03] aims at solving a system of equations in which variables are the key elements. Its complexity is $\mathcal{O}(D)^\omega$ where $D = \sum_{i=1}^{\text{AI}(f)} \binom{N}{i}$, where N is the number of variables of the system, f is the filter function and ω is the exponent appearing in the complexity for solving a linear system. We use $\omega = \log(7)$.
- **Fast algebraic attack** [Cou03] was introduced as an attack against filtered Linear-feedback shift register. While this attack does not apply to the GFP, we make the conservative choice of considering it to provide a lower bound on the algebraic attack complexity. This covers the fact that algebraic attacks may be more efficient than the complexity given above, depending on the structure of the system (i.e. [Fau99,Fau02]). Its complexity is $\mathcal{O}(D\log^2(D) + N \cdot D\log(D))$ where $D = \sum_{i=1}^{\text{AI}(f)+1} \binom{N}{i}$.
- **Correlation attack** approximates the filter function by a linear one and solves the resulting linear system. Their complexity is $\mathcal{O}\left(((|\mathbb{G}|^n)/\text{NL}_{\mathbb{G}}^d) \cdot N^\omega\right)$.
- **Guess and determine strategies** consist of guessing some key elements and applying the previous attack on the resulting subfunctions. This strategy is only worth it if the cost of guessing key elements is compensated by the simplicity of the remaining system. Performing l guesses brings an overhead of $|\mathbb{G}|^l \binom{N}{l}$, and we consider the complexity of the simpler remaining system (which is computed by exhausting all subfunctions).

3   Linearization Attack from [GHBJR23]

The key observation of [GHBJR23] is that for Elisabeth-4, at the binary level, the coordinate function of f giving the lowest bit is the direct sum of 12 functions that depends only on 5 chunks of 4 bits[2], with 4 of them interacting in a nonlinear way. There-fore, calling this Boolean function g, it holds that monomials in its ANF cannot have monomials with variables coming from more than 4 different chunks. The number of monomials appearing in the ANF of g is therefore at most:

$$E = \binom{N}{c} 2^{c \cdot t},$$

where N is the number of ($\mathbb{Z}_{16}$) key elements, c the number of chunks, and t the number of bits per chunk. Moreover, since the term $2^{c \cdot t}$ corresponds to the number of mono-mials in $c \cdot t$ variables, it can be reduced if the function has a lower degree or an extra structure. Accordingly, a linearization attack in complexity E^ω is doable instead of the usual bound D^ω where $D = \sum_{i=0}^{\deg(g)} \binom{Nt}{i}$ is the number of all monomials of degree up to $\deg(g)$ in Nt binary variables. As shown in [GHBJR23], filtering the equations allows us to improve the time complexity of the attack. Taking only the equations such that $N' < N$ key variables appear leads to a smaller value of E, at the cost of a higher data complexity.

Since the number of monomials is consequently smaller than the numbers of mono-mials in Nt variables of degree up to $\deg(g)$, the complexity of this attack by lin-earization is lower than in the general case, and attacks using the sparsity of the system improve upon this bound. For Elisabeth-4, it gives the following bound on E:

$$E \leq \binom{256}{4} 2^{4 \cdot 4} \approx 2^{43.38},$$

which leads to a linearization attack in complexity $E^\omega \approx 2^{122}$ considering $\omega = \log 7$, already giving a successful attack in the security model of [CHMS22].

The smaller-than-expected number of monomials comes from the small number of chunks mixed in a nonlinear fashion, the use of NLUTs, and the linearity of $\mathbb{Z}_{16}$ addition over $\mathbb{F}_2$. More precisely, we will show that a bigger value of c, the number of chunks, quickly increases the number of monomials and makes the attack impracticable. Then, the ANF of the coordinate function of the lowest bit produced by an NLUT does not contain monomials with the highest input bit (Proposition 2 [GHBJR23]). Finally, note that due to the addition modulo 2^t, the next coordinate Boolean function already has a more complex equation. The linear addition of the m outputs is combined with the m choose 2 quadratic terms from these outputs for a direct sum of m components (12 in the case of Elisabeth-4) brought by the carry value. From these three remarks, we can deduce that the attack based on the low number of monomials comes from the specific choice of parameters for Elisabeth-4, and does not apply to the GFP model in general.

Exact Number of Monomials. Looking at the Boolean function of the lowest bit in the particular 5-to-1 function used in [CHMS22], we can count experimentally the number of monomials over $\mathbb{F}_2$ that can be generated from 4 different chunks (looking at the 4! possible permutations). In practice, we obtain that 32535 such monomials can be

---

[2] each chunk corresponding to a $\mathbb{Z}_{16}$ element of the key.

obtained from the chosen function, compared to 64839 if we consider the number of monomials of degree at most 12 in 16 variables. This number is a lower bound on the number of monomials appearing in the algebraic system obtained by an adversary since the addition of the whitening can create more monomials, but not of a higher degree nor involving variables from different chunks. Accordingly, we consider $E' = E/2$ to estimate a more accurate number of monomials, and display in Table 3 the minimal size of N necessary to have enough monomials to avoid the linearization attack, based on the number of chunks c combined in the nonlinear (over $\mathbb{Z}_{16}$) part of the filter and ω.

Table 3 appears to indicate that small modifications of Elisabeth-4's parameter would be sufficient to avoid the attack (while keeping the same Gaussian elimination exponent ω and the same number of chunks $c = 4$, doubling the key-size would give a high enough complexity). However, the filtering function of Elisabeth relies on a direct sum (over $\mathbb{Z}_2^t$) which gives a sparse system of equations. Therefore lower complexity algorithms may apply and considering a smaller value of ω is recommended. For instance, as exhibited by the analysis of [GHBJR23] on Elisabeth-4, using block Wiedemann algorithm [Cop94] to take advantage of the sparsity of the system allows to decrease the attack complexity from 2^{122} to 2^{88}. Accordingly, we consider the conservative choice of $\omega = 2$ for the linearization attack, which leads to considering functions mixing 6 chunks instead of 4 for a number of variables lower than 500.

Table 1. Minimal value of N such that $E'^\omega > 2^{128}$.

c	$\omega = 2$	$\omega = \log(7)$
4	10783	446
5	1336	106
6	344	44

Accordingly, we present alternative instances of Elisabeth-4 in the following. First, starting with minimal changes we modify the inner function of the filter to mix more chunks nonlinearly, which gives the cipher Elisabeth-b. Then, we consider a filter combining two different functions, in order to improve performance and use each function to counter particular attacks. We specify the Gabriel instance to illustrate this claim. Finally, we propose to extend the group filter permutator paradigm into a mixed filter permutator, which considers the key elements in one group, and the output of the filter in another group. We propose the Margrethe instance as a first example of mixed filter permutator, where the key elements are bits, and the outputs are elements of $\mathbb{Z}_{16}$. We analyze the security of these schemes first following the strategy of [CHMS22] summarized in Sect. 2.3. Then, we study their security relative to the attack of [GHBJR23], providing a bound on the number of monomials and considering the conservative bound of $\omega = 2$ for the algebraic attack and the linearization attack from [GHBJR23].

4   Elisabeth-b

Design Motivation and Description. The design approach behind Elisabeth-b consists of using a filter based on the direct sum of the same base function (as for Elisabeth-4) but with a base function mixing more chunks. Mixing enough chunks in a nonlinear part allows to prevent a too-low number of monomials in one of the 15 Boolean functions obtained when the keystream is written as equations over $\mathbb{F}_2$. Based on Table 3 a base function mixing 6 chunks (in a nonlinear manner) is sufficient to reach $\lambda = 128$ bits of security.

Contrarily to the 5-to-1 function of Elisabeth-4, 2 levels of NLUTs are not sufficient to mix 6 chunks of 4 bits since a 4-variable Boolean function has degree at most 4 in its inputs. Then, at least 3 levels of $NLUT$ are necessary. Since we are not aware of designs combining 6 entries, or 3 layers of Sboxes in the literature, we propose the 7-to-1 function over $\mathbb{Z}_{16}$ for Elisabeth-b, that mixes relations modulo 2 and modulo 3 to mix the 6 chunks over the three levels. Similarly to Elisabeth, another variable is simply added, to ensure the balancedness of the output, independently of the choice of the NLUTs.

It gives the following instance, Elisabeth-b4 is the GFP paradigm instantiated with:

- $\mathbb{G} = \mathbb{Z}_{16}$,
- $N = 512$,
- the filter function $f(x_1, \cdots, x_{98})$ the direct sum of 14 times the 7-to-1 function g,
- the 7-to-1 function g described in Algorithm 1,

NLUTs Generation. We generate the NLUTs for all the schemes with the same guidelines. In Python, the protocol starts by hashing a character string (for instance, in the case of Elisabeth-b4, the string "Welcome to Gabriel" is used) with SHA256 (in the scripts, using the hashlib module). This hash is then used as the seed of a PRNG (randint from Python, seeded with random.seed), which is then used to generate hex numbers (using random.randrange(16)). Each number encodes an image of an NLUT, so in order to generate n NLUTS we need $8n$ hex numbers (only the first half of the NLUT is random, the other half can be determined from the previous values).

The script used for generating the NLUTs can be found in the git repository.

Security Analysis. Using the security analysis from [CHMS22] and the analysis on the number of monomials to avoid the attack from [GHBJR23], we verify that the filter satisfies different properties over $\mathbb{Z}_{16}$ and $\mathbb{F}_2$.

- g is not a polyfunction,
- the nonlinearity of g is sufficiently high, even after guessing $\mathbb{Z}_{16}$ variables,
- g is balanced,
- the 15 component Boolean functions of g have a sufficient algebraic immunity, even after guessing $\mathbb{F}_2$ variables,
- the 15 component Boolean functions of g have a sufficient nonlinearity, even after guessing $\mathbb{F}_2$ variables,

---

Algorithm 1: Elisabeth-b4 7-to-1 function.

] **input** : $(x_1, x_2, x_3, x_4, x_5, x_6, x_7) \in \mathbb{Z}_{16}^7$
output: $z \in \mathbb{Z}_{16}$
begin
  for i in range(3) **do**
    $x_{2i+2} = x_{2i+2} + x_{2i+1}$
  for i in range(6) **do**
    // First round of NLUTs
    $y_{i+1} = S_{i+1}(x_{i+1})$
  for i in range(3) **do**
    $z_{2i+1} = y_{2i \bmod 6} + y_{2i+1}$
    $z_{2i+2} = y_{2i+5 \bmod 6} + y_{2i+2}$
  for i in range(6) **do**
    $z_{i+1} = z_{i+1} + x_{i+3 \bmod 6}$
    // Second round of NLUTs
    $z_{i+1} = S_{i+7}(z_{i+1})$
  for i in range(2) **do**
    $t_{3i+1} = z_{3i+1} + z_{3i+2} + z_{3i+3}$
    $t_{3i+2} = z_{3i+2} + z_{3i+4 \bmod 6}$
    $t_{3i+3} = z_{3i+3} + z_{3i+4 \bmod 6} + y_{3i+1}$
  $t_1 = t_1 + x_6$
  $t_2 = t_2 + x_5$
  $t_3 = t_3 + x_4$
  $t_4 = t_4 + x_2$
  $t_5 = t_5 + x_1$
  $t_6 = t_6 + x_3$
  $z = x_7$
  for i in range(6) **do**
    // Third round of NLUTs
    $u_{i+1} = S_{i+13}(t_{i+1})$
    $z = z + u_{i+1}$
  return z

- the 15 component Boolean functions of g are balanced,
- the 15 component Boolean functions of g have enough monomials in their ANF, even after guessing $\mathbb{F}_2$ variables.

Since Elisabeth-b4 uses a 7-to-1 inner function it leads to different challenges than Elisabeth-4 where the inner function is 5-to-1. Both for the analysis over $\mathbb{Z}_{16}$ and $\mathbb{F}_2$ the computations needed to determine the cryptographic parameters are possible in reasonable time for Elisabeth-4 (looking at the properties of function from $\mathbb{G}^4$ to $\mathbb{G}$ or Boolean functions in 16 variables), whereas standard algorithms require a consequent amount of time or data to perform the computations on functions from $\mathbb{G}^6$ to $\mathbb{G}$ or 24-variable Boolean functions.

Tools and Methods for Computing the Boolean Function Properties. In order to compute the Boolean properties of Elisabeth-b4's filter function, we used the Sagemath Boolean function module for the nonlinearity, and the DahuHunting Python DahuHunting Python (from [DMR23]) to compute the degree[3].

While those libraries are the fastest we found to perform those computations, the timings were not good enough to compute the degree for each of the $4 \times \binom{24}{4}$ 22-variable subfunctions after 2 variable guesses. To speed up the computation, we use a trick to not compute the complete algebraic normal form of every subfunction.

The main idea of this technique is to first sort the ANF of the n-variable functions in $n+1$ lists, each one containing the indexes of all the monomials with a given degree. We can focus solely on the list that corresponds to the monomials with the highest degree, and then use the guess and determine approach exclusively for those highest-degree monomials. Finding a high-degree monomial is sufficient to determine the degree of the subfunctions.

All the scripts used for generating Elisabeth-b4's negacyclic look-up tables, computing Elisabeth-b4's security parameters as well as an implementation of the stream-ciphers filter inner function can be found in the following repository: https://anonymous.4open.science/r/elisabeth_patch_scriptsB27F.

Degree and Nonlinearity of the Component Boolean Functions. We perform the analysis over $\mathbb{F}_2$ using a similar method as in [CHMS22] to analyze Elisabeth-4 through the analysis of Beth-b4, the scheme obtained by replacing some of the modulo 16 addition with bitwise XOR. The addition of the independent variable in the 7-to-1 function (5-to-1 in Elisabeth-4) is replaced, and the 13 additions (11 in Elisabeth-4) inside the direct sum are replaced. This simplification allows us to use the results on direct sums to bound the properties of the Boolean functions on the filter in 98×4 variables from 24-variable functions. Since the addition over Z_{16} seen as a vectorial Boolean function corresponds to the (length-4) bitwise XOR plus functions of higher degree (corresponding to the carry values), it is natural to think that the cryptographic properties of the simplified version are weaker than the original one. Accordingly, we introduce Beth-b4, the variation of Elisabeth-b4 in which some $\mathbb{Z}_{16}$ additions are replaced by a bitwise XOR, we conjecture Elisabeth-b4, is at least as secure as Beth-b4.

Formally, the filter function of Beth-b4 can be written as the function ϕ from $\mathbb{G}^{98}$ to $\mathbb{F}_2^4$ defined as:

$$\phi(x_1, \ldots, x_{98}) = \bigoplus_{i=1}^{14} g(x_{7i-6}, x_{7i-5}, x_{7i-4}, x_{7i-3}, x_{7i-2}, x_{7i-1}, 0) \bigoplus_{i=1}^{14} x_{7i},$$

where $\bigoplus$ denotes the bitwise XOR on length-4 vectors.

First, we determine the cryptographic properties of the 15 component Boolean functions of the nonlinear part of g, and the properties of their sub-functions up to guessing two variables, in Table 2. Additionally, we compute the number of monomials of the 15 component functions. Then, we derive lower bounds on the properties of ϕ using

---

[3] The degree is not directly implemented in DahuHunting, but can be computed from the Algebraic Normal Form.

the direct sum properties (using Lemma 1 various times for the resiliency and non-linearity, and Lemma 2 and Lemma 3 for the algebraic immunity from the degree). Finally, we derive the complexity bounds on the different attacks, following the analysis of Sect. 2.3. The results on ϕ on the resiliency and nonlinearity are displayed in Table 3. For the algebraic immunity, applying Lemma 3 we get that with no guesses the AI is at least 14, and the same bound applies when none of the 14 parts receive more than two guesses. When there are more than 3 guessed variables in one of the 14 functions we cannot conclude with the same arguments since Table 2 stops at two guesses (for computational reasons since computing the degree of all the sub-functions or the AI of the 24-variable functions is time and resources consuming) so this function could be null and therefore not used in the chain of Lemma 3. In this case, we can still obtain AI $\geq 14 - \lfloor \ell/3 \rfloor$ by applying directly the lemma, but we expect this bound to be too conservative. Instead, we rely on the assumption that the 24-variable functions have AI at least 8, that is $n/3$ (note that the AI of a random function is close to $n/2$: for all $a < 1$ when n tends to infinity $\mathrm{AI}(f)$ is almost surely larger than $n/2 - \sqrt{(n/2)\ln(n/2a\ln 2)}$ [Did06,CM13]). In this case, if one function did not receive guesses and 6 others received less than 3 guesses, applying Lemma 2 already guarantees an AI of at least 14 (the function that did not receive a guess gives an AI of at least 8, and then the degree of the 6 others allow to increase one by one). We note also that both the complexity of the algebraic attack and fast algebraic attack as described in Sect. 2.3 (considering the worst possible case for the fast algebraic attack, taking $\deg(g) = 1$ and $\deg(h) = \mathrm{AI}(f) + 1$) is already larger than 128-bits of security for an AI value of 12 and up to 28 variables guessed. The scripts used to compute the complexities given in Table 3 can be found in the git repository.

Table 2. Minimum cryptographic parameters of the nonlinear part of the inner function g seen as a (24,4)-vectorial Boolean function after fixing up to ℓ binary inputs.

ℓ	0	1	2
deg	23	21	20
res	0	-1	-1
NL	8355114	4172710	2083058

Table 3. Beth-b4 minimal parameter bounds up to ℓ fixed binary inputs and complexity estimations (in bits). The complexity estimations come from Sect. 2.3, when ℓ lies in an interval the complexity given is the minimal one for this interval of guesses.

Number of guesses ℓ	0	$[1,13]$	$[15,27]$
Resiliency	13	$13 - \ell$	-1
NL/$2^{4n-\ell 7}$	$\geq 0.5 - 10^{-28}$	$\geq 0.5 - 10^{-25}$	$\geq 0.5 - 10^{-23}$
Correlation attack complexity (bits)	$\gg 128$	$\gg 128$	$\gg 128$

Bound on the Number of Monomials of the Component Boolean Functions. To avoid the attack from [GHBJR23], we counted the number of monomials in the ANF of the component of the inner function. The minimum number of monomials among all of the component Boolean functions is at least $22^{22.95}$. Note that a random polynomial would have 22^{23} monomials, the number of monomials then appears close to optimal.

Applying the analysis from Subsect. 3, for the given instance Elisabeth-b4, the number of monomials is bounded by $E \geq \binom{512}{6} 2^{22.95}$. With the assumption that the exponent from the Gaussian elimination is only $\omega = 2$ (the assumption is conservative, the complexity of used algorithms being usually closer to $\log(7)$), the linearization attack complexity is already $E^{\omega} \geq 2^{134}$.

Distance to Polyfunctions. To begin with, we prove that g is not a polyfunction. We then use a code-based argument to prove that the probability that g is close to a polyfunction (in a sense that is formalized below) is negligible. Having g close to a low degree polynomial would expose the scheme to correlation attacks. Since the negacyclic look-up tables were chosen using a hash function, the argument also asserts that the authors could not have put a backdoor by choosing a function g close to a low-degree polynomial.

Proposition 1. *Let f a function from $\mathbb{G}^6$ to $\mathbb{G}$. If $f(8,8,8,8,8,8) - f(0,0,0,0,0,0) \notin \{0,8\}$ then f is not a polynomial of $(\mathbb{G}, \cdot)[x_1, \ldots, x_6]$.*

Proof. All monomials of $(\mathbb{G}, \cdot)[x_1, \ldots, x_6]$ have the form $x_1^{i_1} x_2^{i_2} x_3^{i_3} x_4^{i_4}, x_5^{i_5}, x_6^{i_6}$ where i_1 to i_6 are positive integers. In $(0,0,0,0,0,0)$ all non-trivial monomials give 0. In $(8,8,8,8,8,8)$ if $\sum_{k=1}^{6} i_k > 1$ the monomial gives 0, else the remaining non trivial monomials give 8. Let p be a polynomial of $(\mathbb{G}, \cdot)[x_1, \ldots, x_6]$, its value in $(0,0,0,0,0,0)$ will be the constant coefficient a_0. In $(8,8,8,8,8,8)$, since $2 \times 8 = 0$, every non-trivial monomial multiplied by its constant also gives 0 or 8, and so does their sum. Therefore $p(8,8,8,8,8,8)$ is either equal to a_0 or $a_0 + 8$, allowing to conclude. $\square$

For the instance of Elisabeth-b4 we have $g(8,8,8,8,8,8) - g(0,0,0,0,0,0) = 13 \neq 0, 8$, therefore by Proposition 1 g is not a polyfunction therefore no classical algebraic attack on $\mathbb{G}$ can be applied.

With the parameters of the scheme, using Proposition 1, the number of 6-variable polyfunctions in $\mathbb{G}$ is 2^{4992} (the script used for computing this number can also be found in the public git while the total number of functions is $16^{16^6} = 2^{262144}$). It allows us to derive the following result to thwart the attacks based on low degree approximation of the filter:

Proposition 2. *Let $\mathcal{G}$ be the space of functions from $\mathbb{G}^6$ to $\mathbb{G}$, the probability that a function f taken uniformly at random in $\mathcal{G}$ agrees with a polynomial of $(\mathbb{G}, \cdot)[x_1, x_2, x_3, x_4, x_5, x_6]$ on at least half of the inputs is lower than 2^{-128}.*

Proof. For this proof, we use the formalism of error-correcting codes. Each function in $\mathcal{G}$ can be uniquely represented by its vector of outputs in $\mathbb{G}^{16^6}$, and the polynomials of $(\mathbb{G}, \cdot)[x_1, x_2, x_3, x_4]$ form a linear code of the vector-space $(\mathbb{G}, \cdot)^{16^6}$ with parameter $[n, k, d]$. The code dimension k is given by Proposition 1 while the length n is 2^{24}. We then determine an upper bound on the number of elements of $\mathbb{G}^{16^4}$ at Hamming distance lower than $n/2 + 1$, and compare this bound with the number of functions in $\mathcal{G}$ to conclude.

We use the standard packing/covering bound to determine the maximum number of elements covered by the union of Hamming balls of radius $n/2$ centered around the code elements (here the polynomials). Each ball contains $B = \sum_{i=0}^{n/2}(|\mathbb{G}| - 1)^i \binom{n}{i}$ elements, giving a total number of $2^{4992}B$ elements: the upper bound reached if no ball intersect.

Finally, since there are $(16)^{16^6} = 2^{2^{26}}$ elements in $\mathcal{G}$ we can compute the proportion p of functions coinciding on at least half of their inputs with a polynomial:

$$p \le \frac{2^{4992}B}{2^{2^{26}}} = \frac{\sum_{i=0}^{2^{23}} 15^i \binom{2^{24}}{i}}{2^{2^{26}-4992}} \le \frac{15^{2^{23}} \cdot 2^{2^{24}}}{2^{2^{26}-4992}}$$

$$\le \frac{16^{2^{23}} \cdot 2^{2^{24}}}{2^{2^{26}-4992}} \le 2^{2^{25}+2^{24}-2^{26}+4992} = 2^{-2^{24}+4992} < 2^{-128}.$$

$\square$

This property asserts that the amount of functions from $\mathbb{G}^6$ to $\mathbb{G}$ that can be well approximated by a polynomial is negligible, which rules out algebraic attacks that would try to solve a noisy polynomial system.

To sum up, Elisabeth-b4 provides a patch while simply increasing the filter inner function of Elisabeth-4. Since this scheme is purposed to be ran homomorphically on a server, it is relevant to consider a parallel implementation and the time cost of the keystream will be dominated by the evaluation time of the NLUTs. In the end, the performance overhead for this secure instance is limited to one extra layer of (negacyclic) look-up tables.

5   Gabriel

Design Motivation and Description. The design approach for Gabriel consists of having two different branches for the filter function, which we call the left and right parts. The left part uses functions mixing a relatively low number of chunks, like Elisabeth-b4, and is designed to tackle the different attacks described in [CHMS22]. The right part is aimed to mix more chunks to provide sufficiently enough monomials in the algebraic system, to tackle the attack presented in [GHBJR23] or variants. Therefore the two parts consist of direct sums of a function g_L and a function g_R (for Left and Right) and are combined in a direct sum to give the filter f.

Differently to Elisabeth-b's design where all base functions mix 7 chunks, this approach gives a trade-off, allowing some base functions cheaper to compute. For security, the number of inner functions mixing more chunks needs to be sufficient to provide enough monomials in all equations. It can be guaranteed if this number is high enough even after taking into consideration that the adversary can guess a bounded number of variables (fixing the values of the monomials containing it), or build a system where some key variables never appear, and therefore no monomials in these variables either (by doing preprocessing, since the IV is given by the user and the selected subsets are publicly derived from the IV).

We could propose an instance of this design strategy only using the 5-to-1 function of Elisabeth-b4 and its compositions. We detail why this natural idea is not the best option, in our opinion. In this case, g_L is only the 5-to-1 function, in direct sum 12 times in order to use the results on the properties over $\mathbb{Z}_{16}$ and $\mathbb{F}_2$ already proven or computed in [CHMS22]. We consider g_R as (direct sums of) the composition of the 5-to-1 function g, that is $g(y_1, y_2, y_3, y_4, x_5)$ where y_1, y_2, y_3, y_4 are the output of g applied on different key entries. Consequently, it gives a 21-to-1 function over $\mathbb{Z}_{16}$ admitting g as a subfunction. This alternative has the advantage of limiting the number of different NLUTs used in the full scheme and allows an easier implementation from Elisabeth-b4. On the downside, very few results are known relatively to the cryptographic parameters of the composition of functions (even for Boolean functions), making it difficult to obtain meaningful bounds on the properties of g_R. Using experimental approaches to determine the parameters of such function is also beyond the computation limits, the nonlinear part gives Boolean functions in 80 variables making it challenging to store the truth tables of size 2^{80} and impossible to compute the parameters (these algorithms are polynomial in the truth table size, therefore exponential in the number of variables).

Accordingly, for simplicity of analysis, we consider instantiating the left part with the inner function of Elisabeth and the right part with the inner function of Elisabeth-b. Based on the analysis of [CHMS22], the attacks based on the nonlinearity, algebraic immunity and degree over $\mathbb{F}_2$, and the attacks based on the polynomial representation and distance to low degree polyfunctions over $\mathbb{Z}_{16}$ take a complexity higher than the threshold of 2^{128}. Then, based on the analysis of Sect. 4, the number of monomials produced by the 7-to-1 function of Gabriel-4 is sufficient to prevent the linearization attack. It gives the following proposed instance, Gabriel-4 is the GFP paradigm instantiated with:

- $\mathbb{G} = \mathbb{Z}_{16}$,
- $N = 512$,
- the filter function $f(x_1, \cdots, x_{110})$ the direct sum of 8 times the 5-to-1 function g_L and 10 times the 7-to-1 function g_R,
- the 5-to-1 function is the inner function of Elisabeth-4, described in Fig. 2, with the same NLUTs,
- the 7-to-1 function is the function g described in Algorithm 1, with the same NLUTs as Elisabeth-b4.

Security Analysis. Since Gabriel-4 uses two branches to tackle both the attacks studied in [CHMS22] and [GHBJR23], the analysis is straightforward form the one of Elisabeth-4 and Elisabeth-b4. The filter function f is a direct sum containing 8 times the inner function of Elisabeth-4, accordingly the properties of f are at least the ones of Elisabeth-b4's inner function, studied in [CHMS22]. Using the properties on the direct sum of these 8 functions and 6 form, the part taken from Elisabeth-b4 gives better cryptographic parameters than the ones of the filter of Elisabeth-4. Thereafter, no attack described in Sect. 2.3 from [CHMS22] leads to a complexity lower than 2^{128}. Regarding the attack from [GHBJR23], the right part mixes non-linearly 6 chunks, which is sufficient to thwart the attack from the analysis of Sect. 3. The 10 7-to-1 functions in the direct sum prevent further guess and determine techniques to guess $\mathbb{Z}_{16}$ variables to get only equations where only 4 chunks are mixed.

Mixing two different kinds of branches leads to better performances than Elisabeth-b4. While the number of NLUT layers remains 3 because of the Elisabeth-b4 branches, the total number of NLUT to evaluate decreases.

6   Margrethe and Mixed Filter Permutators

The small number of monomials in the ANF of the Boolean functions of Elisabeth-4's filter is strongly linked to the constraints of using only 4-bits NLUTs and the addition modulo 16. The inner function mixes a fixed number of $\mathbb{G}$-variables (c), therefore, in the representation over $\mathbb{F}_2$, no monomials with variables from more than c chunks can appear in the ANF (due to the linearity of the Least Significant Bit (LSB) of the addition modulo 16 when considered over $\mathbb{F}_2$). On the contrary, if the filter acts at the bit level it breaks this structure and the bound of binary variables from at most c different chunks. Moreover, as proven in [GHBJR23] the negacyclic property implies that the Boolean function corresponding to the LSB cannot have monomials depending on the Most Significant Bit of the input[4].

Accordingly, we consider a different setting with two differences, the LUTs do not need to be negacyclic and the function can operate at the bit level, that is the (binary) key variables are not in the same chunks for each application of the filter. On the HHE side, removing these restrictions is sound since recent techniques allow the evaluation of larger LUTs, as witnessed by the line of works on full domain functional bootstrapping [KS23,CZB+22,MHWW23]. Furthermore, diverse techniques using changes of representation are getting more common (e.g. [CDPP22]).

Since the group filter permutator is defined only on one group, we introduce the more general paradigm of Mixed Filter Permutator (MFP) in Sect. 6.1, and a stream cipher design named Margrethe in Sect. 6.2. We propose an instance, Margrethe-$18 - 4$, and analyze its security in Sect. 6.3.

6.1   Mixed Filter Permutator Paradigm

The Mixed Filter Permutator (MFP) paradigm differs from the GFP by the use of two (potentially) different groups. Therefore, the nature of the inputs and the outputs can be different, and the evaluation of the cipher can be performed over different structures. Mixing operations from two different structures to get a secure cryptographic primitive is a principle that has been used implicitly for example in AES [DR20], and explicitly more recently [BIP+18,DMMS21,DGH+21,HMM+23]. We give the paradigm description in the following:

---

[4] Denoting f the Boolean function corresponding to the LSB of an n-variable NLUT, the negacyclicity implies $f(x) = f(x + (1, 0_{n-1}))$, hence the sensitivity of f in the MSB is null so no monomial containing the MSB appear in the ANF of f.

Paradigm. The MFP is defined by two groups $\mathbb{G}_1$ and $\mathbb{G}_2$ with operation noted $+_1$ and $+_2$, a forward secure PRNG, a key size N, a subset size n, and a filtering function f from $\mathbb{G}_1^n$ to $\mathbb{G}_2$. To encrypt m elements of $\mathbb{G}_2$ under a secret key $K \in \mathbb{G}_1^N$, the public parameters of the PRNG are chosen and then the following process is executed for each key stream s_i (for $i \in [m]$):

- The PRNG is updated, its output determines a subset, a permutation, and a length-n vector of $\mathbb{G}_1$.
- the n-element subset S_i is chosen over N-element key,
- the n to n permutation P_i is chosen,
- the vector, called whitening and denoted w_i, from $\mathbb{G}_1^n$ is chosen,
- the key stream element s_i is computed as $s_i = f(P_i(S_i(K)) +_1 w_i)$, where $+_1$ denotes the element-wise addition of $\mathbb{G}_1$.

Security. Due to the strong similarity with the GFP paradigm, we consider the same attacks detailed in Sect. 2.3, namely algebraic attacks, [GHBJR23]'s attack, correlation attacks, and their variants using guess and determine strategies. Since in the MFP the operations are performed over two groups, the attacks have to be considered relative to these two representations. We remark that it is very similar to the two perspectives considered for the GFP, where attacks are considered on $\mathbb{G}$ and on G', such as $\mathbb{Z}_{16}$ and $\mathbb{F}_2$ for the instance Elisabeth-4. Thereafter, we study the same kind of attacks for the security of stream ciphers following the MFP paradigm, focusing on particular algebraic and correlation attacks depending on the particular choice of $\mathbb{G}_1$ and $\mathbb{G}_2$.

6.2 Margrethe

We define the family of stream cipher Margrethe and give an instance, Margrethe-18-4, analyzed in Sect. 6.3. Margrethe's design is characterized by the use of $\mathbb{F}_2$ for $\mathbb{G}_1$, and the group $\mathbb{Z}_{2^\ell}$ with $\ell \in \mathbb{N}^*$ for $\mathbb{G}_2$. The filter function is obtained by the direct sum over $\mathbb{G}_2$ of t times a function obtained by a look-up table following the general principle of her cousin Elisabeth. Using look-up tables from a bits to b bits, we denote by Margrethe-a-b the instance associated with this LUT size.

We propose the following instance, Margrethe-18-4 is the MFP paradigm instantiated with:

- $\mathbb{G}_1 = \mathbb{F}_2$,
- $\mathbb{G}_2 = \mathbb{Z}_{16}$,
- $N = 2048$,
- the filter function is the function f from $\mathbb{F}_2^n$ to $\mathbb{Z}_{16}$, that uses the inner function g, a function from $\mathbb{F}_2^{18}$ to $\mathbb{Z}_{16}$.
  $g : \mathbb{F}_2^{18} \mapsto \mathbb{Z}_{16}$ is given by a LUT.
  $f : \mathbb{F}_2^{308} \mapsto \mathbb{Z}_{16}$ is defined by:

$$f(x_1, \ldots, x_{308}) = \sum_{i=0}^{13} g(x_{22i+1}, \ldots, x_{22i+18}) + \mathbb{Z}_{16}\left(\sum_{k=0}^{3} 2^k x_{22i+19+k}\right),$$

where the symbols Σ and $+$ denote the addition modulo 16 and $\mathbb{Z}_{16}(x_1, x_2, x_3, x_4)$ denotes the element of $\mathbb{Z}_{16}$ with binary representation (x_1, x_2, x_3, x_4).

Note that the key of Margrethe-18-4 has the same size as the one of Elisabeth-b4. Indeed, key elements of Margrethe-18-4 are elements of $\mathbb{F}_2$ while Elisabeth-b4 key elements are in $\mathbb{Z}_{16}$.

The LUT generation is similar to the NLUTs generation for Elisabeth-b (Sect. 4). In Python, the protocol starts by hashing the character string "Welcome to Magrethe" with SHA256 (in the scripts, using the hashlib module). This hash is then used as the seed of a PRNG, which is then used to generate hex numbers. Each number encodes an image of a LUT, so in order to generate the 18-to-4 LUT we need 2^{18} hex numbers. The script used for generating the LUT can be found in the git repository.

6.3 Security Analysis

For the security of Margrethe-18-4 we consider the attacks only coming from the representation over $\mathbb{F}_2$ for two reasons. First, the function g acts at the bit level therefore the result of f cannot be written in terms of key elements of $\mathbb{Z}_{16}$ as for Elisabeth-4. Then, as proven in Proposition 2, the probability for a function taken at random to be close (in Hamming distance) to a polyfunction over $\mathbb{Z}_{16}$ is negligible, making it difficult for an adversary to take advantage of the representation over $\mathbb{Z}_{16}$.

Consequently, we analyze the security of Margrethe-18-4 by considering the different attacks based on its representation as a vectorial Boolean function. Similarly to the analysis of Sect. 4, we study the properties of a simplification of Margrethe-18-4 that we call Mag-18-4 where the additions modulo 16 are replaced by a length-4 bitwise XOR, allowing to study the property of the filter function from the properties of direct sums. As previously, we conjecture that Margrethe-18-4, is at least as secure as Mag-18-4. More formally, the filter function of Mag-4 is F from $\mathbb{F}_2^{308}$ to $\mathbb{F}_2^4$ is defined by:

$$F(x_1, \ldots, x_{308}) = \bigoplus_{i=1}^{13} G(x_{22i+1}, \ldots, x_{22i+18}) \bigoplus_{i=1}^{13} (x_{22i+19}, x_{22i+20}, x_{22i+21}, x_{22i+22}),$$

where $\bigoplus$ denotes the (length-4) bitwise XOR and G the interpretation of g of a vectorial Boolean function rather than a function from $\mathbb{F}_2^{18}$ to $\mathbb{Z}_{16}$.

First, we determine the properties of the 15 component functions of G relative to the Boolean cryptographic criteria of resilience, degree and nonlinearity. We also study the same properties for their sub-functions up to guessing two variables. Additionally, we compute the number of monomials of the 15 component functions. Then, we give lower bounds on the properties of F using the direct sum properties (see Lemma 1, Lemma 2 and Lemma 3). Finally, we derive the complexity bounds on the different attacks, following the analysis of Sect. 2.3. We give the parameters of G in Table 4 and the ones of F and the security bounds in Table 5. For the complexity of the algebraic and fast algebraic attacks, we follow the same strategy as in Sect. 4. Without any guess, the AI of the components of F is at least 14 using Lemma 3, the same bound is valid when no more than 2 guesses are in the same function. If at least 3 variables are guessed in one function, this one cannot be considered in the chain of Lemma 3, accordingly the bound gives AI $\geq 14 - \lfloor \ell/3 \rfloor$. If we assume the components of G have AI at least $18/3 = 6$, if one function did not receive guesses and 8 others received less than 3 guesses applying Lemma 2 already guarantees an AI of at least 14. In the end, for an AI value of 12 and up to 28 variables guessed, following the formulas of Sect. 2.3, the bound on the algebraic attacks complexity is 2^{288} and 2^{128} for the one on the fast algebraic attack complexity.

The minimal number of monomials observed in the ANF for a component of G is $2^{\log 16.9983}$, approximately half of the monomials as for a random 18-variable Boolean function. Accordingly, the number of monomials to consider for a linearization attack is the standard one from Sect. 2.3 rather than Sect. 3.

Table 4. Minimum cryptographic parameters for the function G seen as a vectorial Boolean function after fixing up to ℓ binary inputs

ℓ	0	1	2
deg	17	16	15
res	-1	-1	-1
NL	129757	64516	32013

Table 5. Mag-18-4 minimal parameter bounds up to ℓ fixed binary inputs and complexity estimations (in bits). The complexity estimations come from Sect. 2.3, when ℓ lies in an interval the complexity given is the minimal one for this interval of guesses.

Number of guesses ℓ	0	$[1, 13]$	$[14, 27]$
Resiliency	13	$13 - \ell$	-1
NL/$2^{4n-\ell}$	$\geq 0.5 - 10^{-28}$	$\geq 0.5 - 10^{-25}$	$\geq 0.5 - 10^{-23}$
Correlation attack complexity (bits)	$\gg 128$	$\gg 128$	$\gg 128$

7 Conclusion and Open Question

To sum up, we studied different patches to the Elisabeth-4 cryptosystem. Elisabeth-b4 brings minimal changes by simply modifying the filter inner function by combining more elements. Gabriel-4 reduces the performance overhead by integrating two different branches in the direct sum of the filter. Leaving the GFP paradigm, Margrethe-18-4 fixes the root of the issue raised by [GHBJR23] by mixing the key at the bit level.

On the one hand Elisabeth-b4 and Gabriel-4 use the same operations as Elisabeth-4, with the same size of negacyclic LUT. Their homomorphic implementations would therefore be similar to the one of Elisabeth-4, with slightly different homomorphic parameters. On the other hand, the homomorphic evaluation of instances from Margrethe is left as an open problem. We believe that the structure of the mixed filter permutator can exploit recent advancements in FHE research using full domain, bigger, LUTs [KS23, CZB+22, MHWW23] or by mixing different representations [CDPP22].

References

[AMT22]   Ashur, T., Mahzoun, M., Toprakhisar, D.: Chaghri - a fhe-friendly block cipher. In: Yin, H., Stavrou, A., Cremers, C., Shi, E. (eds.), Proceedings of the 2022 ACM SIGSAC Conference on Computer and Communications Security, CCS 2022, Los Angeles, CA, USA, 7–11 November 2022, pp. 139–150. ACM (2022)

[ARS+15] Albrecht, M.R., Rechberger, C., Schneider, T., Tiessen, T., Zohner, M.: Ciphers for MPC and FHE. In: Oswald, E., Fischlin, M. (eds.) Advances in Cryptology – EUROCRYPT 2015. EUROCRYPT 2015. LNCS, vol. 9056, pp. 430–454. Springer, Berlin, Heidelberg (2015). https://doi.org/10.1007/978-3-662-46800-5_17

[BIP+18] Boneh, D., Ishai, Y., Passelegue, A., Sahai, A., Wu, D.J.: Exploring crypto dark matter: - new simple PRF candidates and their applications. In: Beimel, A., Dziembowski, S. (eds.) Theory of Cryptography. TCC 2018. LNCS, vol. 11240, pp. 699–729. Springer, Cham (2018). https://doi.org/10.1007/978-3-030-03810-6_25

[BY03]   Bellare, M., Yee, B.: Forward-security in private-key cryptography. In: Joye, M. (ed.) Topics in Cryptology – CT-RSA 2003. CT-RSA 2003. LNCS, vol. 2612, pp. 1–18. Springer, Berlin, Heidelberg (2003). https://doi.org/10.1007/3-540-36563-X_1

[Car21]   Carlet, C.: Boolean Functions for Cryptography and Coding Theory. Cambridge University Press, Cambridge (2021)

[CCF+16] Canteaut, A., et al.: Stream ciphers: a practical solution for efficient homomorphic-ciphertext compression. In: Peyrin, T. (eds.) Fast Software Encryption. FSE 2016. LNCS, vol. 9783, pp. 313–333. Springer, Berlin, Heidelberg (2016). https://doi.org/10.1007/978-3-662-52993-5_16

[CDPP22] Cong, K., Das, D., Park, J., Pereira, H.V.: Sortinghat: efficient private decision tree evaluation via homomorphic encryption and transciphering. In: Yin, H., Stavrou, A., Cremers, C., Shi, E. (eds.), Proceedings of the 2022 ACM SIGSAC Conference on Computer and Communications Security, CCS 2022, Los Angeles, CA, USA, 7–11 November 2022, pp. 563–577. ACM (2022)

[CGGI16] Chillotti, I., Gama, N., Georgieva, M., Izabachene, M.: Faster fully homomorphic encryption: bootstrapping in less than 0.1 seconds. In: Cheon, J., Takagi, T. (eds.) Advances in Cryptology – ASIACRYPT 2016. ASIACRYPT 2016. LNCS, vol. 10031, pp. 3–33. Springer, Berlin, Heidelberg (2016). https://doi.org/10.1007/978-3-662-53887-6_1

[CHK+21] Cho, J., et al.: Transciphering framework for approximate homomorphic encryption. In: Tibouchi, M., Wang, H. (eds.) Advances in Cryptology – ASIACRYPT 2021. ASIACRYPT 2021. LNCS, vol. 13092, pp. 640–669. Springer, Cham (2021). https://doi.org/10.1007/978-3-030-92078-4_22

[CHMS22] Cosseron, O., Hoffmann, C., Meaux, P., Standaert, F.X.: Towards case-optimized hybrid homomorphic encryption - featuring the elisabeth stream cipher. In: Agrawal, S., Lin, D. (eds.) Advances in Cryptology – ASIACRYPT 2022. ASIACRYPT 2022. LNCS, vol. 13793, pp. 32–67. Springer, Cham (2022). https://doi.org/10.1007/978-3-031-22969-5_2

[CM03]   Courtois, N.T., Meier, W.: Algebraic attacks on stream ciphers with linear feedback. In: Biham, E. (ed.) Advances in Cryptology – EUROCRYPT 2003. EUROCRYPT 2003. LNCS, vol. 2656, pp. 345–359. Springer, Berlin, Heidelberg (2003). https://doi.org/10.1007/3-540-39200-9_21

[CM13]   Carlet, C., Merabet, B.: Asymptotic lower bound on the algebraic immunity of random balanced multi-output Boolean functions. Adv. Math. Commun. **7**, 197–217 (2013)

[CMR17] Carlet, C., Méaux, P., Rotella, Y.: Boolean functions with restricted input and their robustness; application to the FLIP cipher. IACR Trans. Symmetric Cryptol. **3**, 2017 (2017)

[Cop94] Coppersmith, D.: Solving homogeneous linear equations over gf(2) via block wiedemann algorithm. Math. Comput. **62**(205), 333–350 (1994)

[Cou03] Courtois, N.T.: Fast Algebraic Attacks on Stream Ciphers with Linear Feedback. In: Boneh, D. (ed.) Advances in Cryptology – CRYPTO 2003. CRYPTO 2003. LNCS, vol. 2729, pp. 176–194. Springer, Berlin, Heidelberg (2003). https://doi.org/10.1007/978-3-540-45146-4_11

[CZB+22] Clet, P.E., Zuber, M., Boudguiga, A., Sirdey, R., Gouy-Pailler, C.: Putting up the swiss army knife of homomorphic calculations by means of TFHE functional bootstrapping. IACR Cryptol. ePrint Arch., p. 149 (2022)

[DEG+18] Dobraunig, C., et al.: Rasta: a cipher with low anddepth and few ANDs per bit. In: Shacham, H., Boldyreva, A. (ed.) Advances in Cryptology – CRYPTO 2018. CRYPTO 2018. LNCS, vol. 10991, pp. 662–692 . Springer, Cham (2018). https://doi.org/10.1007/978-3-319-96884-1_22

[DGH+21] Dinur, I., et al.: MPC-friendly symmetric cryptography from alternating moduli: candidates, protocols, and applications. In: Malkin, T., Peikert, C. (eds.) Advances in Cryptology – CRYPTO 2021. CRYPTO 2021. LNCS, vol. 12828, pp. 517–547. Springer, Cham (2021). https://doi.org/10.1007/978-3-030-84259-8_18

[DGH+23] Dobraunig, C., Grassi, L., Helminger, L., Rechberger, C., Schofnegger, M., Walch, R.: Pasta: a case for hybrid homomorphic encryption. IACR Trans. Cryptogr. Hardw. Embed. Syst. **2023**(3), 30–73 (2023)

[Did06] Didier, F.: A new upper bound on the block error probability after decoding over the erasure channel. IEEE Trans. Inf. Theory **52**(10), 4496–4503 (2006)

[DLR16] Duval, S., Lallemand, V., Rotella, Y.: Cryptanalysis of the FLIP family of stream ciphers. In: Robshaw, M., Katz, J. (eds.) Advances in Cryptology – CRYPTO 2016. CRYPTO 2016. LNCS, vol. 9814, pp. 457–475. Springer, Berlin, Heidelberg (2016). https://doi.org/10.1007/978-3-662-53018-4_17

[DMMS21] Duval, S., Méaux, P., Momin, C., Standaert, F.-X.: Exploring crypto-physical dark matter and learning with physical rounding towards secure and efficient fresh rekeying. IACR Trans. Cryptogr. Hardw. Embed. Syst. **2021**(1), 373–401 (2021)

[DMR23] Dupin, A., Méaux, P., Rossi, M.: On the algebraic immunity - resiliency trade-off, implications for goldreich's pseudorandom generator. Des. Codes Cryptogr. **91**(9), 3035–3079 (2023)

[DR20] Daemen, J., Rijmen, V.: The Design of Rijndael - The Advanced Encryption Standard (AES), 2nd edn. Springer, Information Security and Cryptography. Springer, Berlin, Heidelberg (2020). https://doi.org/10.1007/978-3-662-04722-4

[Fau99] Faugère, J.-C.: A new efficient algorithm for computing groebner bases. J. Pure Appl. Algebra **139**, 61–88 (1999)

[Fau02] Faugère, J.-C.: A new efficient algorithm for computing Grobner bases without reduction to zero. In: Workshop on Application of Groebner Bases 2002, Catania, Spain (2002)

[FY53] Fisher, R.A., Yates, F.: Statistical Tables for Biological, Agricultural and Medical Research. Hafner Publishing Company, London (1953)

[GHBJR23] Gilbert, H., Heim Boissier, R., Jean, J., Reinhard, J.R.: Cryptanalysis of Elisabeth-4. In: Guo, J., Steinfeld, R. (eds.) Advances in Cryptology – ASIACRYPT 2023. ASIACRYPT 2023. LNCS, vol. 14440, pp. 256–284. Springer, Singapore (2023). https://doi.org/10.1007/978-981-99-8727-6_9

[GIKV23] Geelen, R., Iliashenko, I., Kang, J., Vercauteren, F.: On polynomial functions modulo p^e and faster bootstrapping for homomorphic encryption. In: Hazay, C., Stam, M. (eds.) Advances in Cryptology – EUROCRYPT 2023. EUROCRYPT 2023. LNCS, vol. 14006, pp. 257–286. Springer, Cham (2023). https://doi.org/10.1007/978-3-031-30620-4_9

[Gol00] Goldreich, O.: Candidate one-way functions based on expander graphs. In: Goldreich, O. (ed.) Studies in Complexity and Cryptography. Miscellanea on the Interplay between Randomness and Computation. LNCS, vol. 6650, pp. 76–87. Springer, Berlin, Heidelberg (2000). https://doi.org/10.1007/978-3-642-22670-0_10

[HKL+22] Ha, J., Kim, S., Lee, B., Lee, J., Son, M.: Rubato: noisy ciphers for approximate homomorphic encryption. In: Dunkelman, O., Dziembowski, S. (eds.) Advances in Cryptology – EUROCRYPT 2022. EUROCRYPT 2022. LNCS, vol. 13275, pp. 581–610. Springer, Cham (2022). https://doi.org/10.1007/978-3-031-06944-4_20

[HMM+23] Hoffmann, C., Meaux, P., Momin, C., Rotella, Y., Standaert, F.X., Udvarhelyi, B.: Learning with physical rounding for linear and quadratic leakage functions. In: Handschuh, H., Lysyanskaya, A. (eds.) Advances in Cryptology – CRYPTO 2023. CRYPTO 2023. LNCS, vol. 14083, pp. 410–439. Springer, Cham (2023). https://doi.org/10.1007/978-3-031-38548-3_14

[Knu97] Knuth, D.E.: Seminumerical Algorithms, volume 2 of The Art of Computer Programming, third edition. Addison-Wesley Professional, Boston, November 1997

[KS23] Kluczniak, K., Schild, L.: FDFB: full domain functional bootstrapping towards practical fully homomorphic encryption. IACR Trans. Cryptogr. Hardw. Embed. Syst. **2023**(1), 501–537 (2023)

[MCJS19] Meaux, P., Carlet, C., Journault, A., Standaert, F.X.: Improved filter permutators for efficient FHE: better instances and implementations. In: Hao, F., Ruj, S., Sen Gupta, S. (eds.) Progress in Cryptology – INDOCRYPT 2019. INDOCRYPT 2019. LNCS, vol. 11898, pp. 68–91. Springer, Cham (2019). https://doi.org/10.1007/978-3-030-35423-7_4

[Méa22] Méaux, P.: On the algebraic immunity of direct sum constructions. Discret. Appl. Math. **320**, 223–234 (2022)

[MHWW23] Ma, S., Huang, T., Wang, A., Wang, X.: Fast and accurate: efficient full-domain functional bootstrap and digit decomposition for homomorphic computation. IACR Cryptol. ePrint Arch., p. 645 (2023)

[MJSC16] Meaux, P., Journault, A., Standaert, FX., Carlet, C.: Towards stream ciphers for efficient FHE with low-noise ciphertexts. In: Fischlin, M., Coron, J.S. (eds.) Advances in Cryptology – EUROCRYPT 2016. EUROCRYPT 2016. LNCS, vol. 9665, pp. 311–343. Springer, Berlin, Heidelberg (2016). https://doi.org/10.1007/978-3-662-49890-3_13

[SHW23] Specker, E., Hungerbühler, N., Wasem, M.: The ring of polyfunctions over z/nz. Commun. Algebra **51**(1), 116–134 (2023)

Elliptic Curves, Zero-Knowledge Proof, Signatures

Generating Supersingular Elliptic Curves over $\mathbb{F}_p$ with Unknown Endomorphism Ring

Youcef Mokrani[(✉)] and David Jao[ORCID]

Department of Combinatorics and Optimization, University of Waterloo,
Waterloo, ON N2L 3G1, Canada
{ymokrani,djao}@uwaterloo.ca

Abstract. A number of supersingular isogeny based cryptographic protocols require the endomorphism ring of the initial elliptic curve to be either unknown or random in order to be secure. To instantiate these protocols, Basso et al. recently proposed a secure multiparty protocol that generates supersingular elliptic curves defined over $\mathbb{F}_{p^2}$ of unknown endomorphism ring as long as at least one party acts honestly. However, there are many protocols that specifically require curves defined over $\mathbb{F}_p$, for which the Basso et al. protocol cannot be used. Also, the simple solution of using a signature scheme such as CSI-FiSh or SeaSign for proof of knowledge either requires extensive precomputation of large ideal class groups or is too slow for everyday applications.

In this paper, we present CSIDH-SCG, a new multiparty protocol that generates curves of unknown endomorphism ring defined over $\mathbb{F}_p$. This protocol relies on CSIDH-ROIP, a new CSIDH based proof of knowledge. We also present CSIDH-CR, a multiparty algorithm that be used in conjunction with CSIDH-SCG to generate a random curve over $\mathbb{F}_p$ while still keeping the endomorphism ring unknown.

Keywords: Elliptic curves · Supersingular curves · CSIDH · Multiparty computation

1 Introduction

Recent attacks on SIDH by Castryck, Decru, Maino, Martindale and Robert [7,15,19] have shown that torsion point information can be enough to find an isogeny between two supersingular elliptic curves. It follows that, for an isogeny based scheme to be secure, it must avoid giving too much information about its elliptic curves and isogenies.

One such piece of information is the endomorphism ring of the starting elliptic curve. In fact, the first break by Wouter Castryck and Thomas Decru [7] exploits this knowledge. We also note that Petit's torsion point attacks on SIDH [18] also need a known endomorphism ring. Since many attacks on isogeny based schemes make use of the endomorphism ring, it stands to reason that, unless necessary, a cryptographic protocol should avoid working on elliptic curves with

© The Author(s), under exclusive license to Springer Nature Switzerland AG 2024
A. Chattopadhyay et al. (Eds.): INDOCRYPT 2023, LNCS 14459, pp. 159–174, 2024.
https://doi.org/10.1007/978-3-031-56232-7_8

a known endomorphism ring. In addition, a number of existing schemes require a supersingular curve of unknown endomorphishm ring. To solve this issue, Basso et al. [3] proposed a multiparty protocol that generates a supersingular elliptic curve defined over $\mathbb{F}_{p^2}$ as long as at least one participant acts honestly.

Although the Basso et al. protocol solves the problem in general, there remain a number of schemes that explicitly require a supersingular curve of unknown endomorphism ring defined over $\mathbb{F}_p$. As mentioned in [3], some examples of such protocols include CSIDH-based Verifiable Delay Functions [11], as well as Delay Encryption algorithms [6] that need to start with a random curve over $\mathbb{F}_p$. Such curves are also required for some Oblivious Transfer protocols [14] and dual mode PKE [1]. For these curves, the protocol found in [3] cannot be directly applied, as random walks in the supersingular isogeny graph have a negligible probability of ending on a curve defined over $\mathbb{F}_p$. Basso et al. [3] mention possible solutions, but they all come with important issues as they either leak too much information, require specific parameter sets or are too inefficient for everyday use. One of these possible solution would be to use the CSI-RAShi Distributed Key Generation protocol [4], but this would require parameter sets for which the ideal class group is known, and there is currently no known method to find such parameters in polynomial time. Another possible solution was proposed by Moriya, Takashima and Tagaki [16]. However, its security proof only deals with honest but curious participants (prover and verifier), and does not take into account the case where malicious adversaries send malformed data. Finally, a recent paper by Atapoor et al. [2] presents a distributed key generation protocol for CSIDH, but this situation differs from our scenario in that we are not trying to retain collective knowledge of any associated secret key.

In this paper, we present CSIDH-SCG, a new multiparty protocol that generates supersingular elliptic curves defined over $\mathbb{F}_p$ of unknown endomorphism ring as long as at least one of the participating parties is honest. CSIDH-SCG does not require the knowledge of any ideal class group, is efficient even for large groups, and resists active adversaries. We also present CSIDH-CR, a multiparty protocol that uses the secure curve returned by CSIDH-SCG to generate a random supersingular elliptic curve of unknown endomorphism ring.

Section 2 presents the basic definitions and assumptions used in this paper. Past results related to this problem can be found in Sect. 3. We present two new CSIDH based zero-knowledge proofs in Sect. 4. Section 5 presents CSIDH-SCG, a new multiparty protocol generating curves of unknown endomorphism ring over $\mathbb{F}_p$, as well as CSIDH-ASCG, a variation avoiding the use of a random oracle function at the cost of an additional round of interaction. In Sect. 6, we present CSIDH-CR, which is used in addition to CSIDH-SCG or CSIDH-ASCG for cases where the desired curve needs to be random. Finally, Sect. 7 contains a brief summary of the results and possible avenues of further work.

2   Definitions and Assumptions

In this section, we present the various definitions and assumptions that are used at different points in this paper, as well as heuristics to justify said assumptions.

Since the protocols in this paper work with elliptic curves defined over $\mathbb{F}_p$ when the associated ideal class group is unknown, we start by presenting the necessary definitions for the sampling method used in CSIDH [8].

Notation 2.1. *Let p be a prime number and let E be a known supersingular elliptic curve defined over $\mathbb{F}_p$. We denote by $\mathcal{C}$ the ideal class group of the endomorphism ring of E.*

Notation 2.2. *In cases where $\mathcal{C}$ is unknown, let $\mathcal{S} = \{\mathfrak{l}_1, \ldots, \mathfrak{l}_t\} \subseteq \mathcal{C}$ denote a generating set of $\mathcal{C}$ consisting of prime ideals of small (relatively prime) norm.*

Definition 2.3. *Let B be a positive integer. Define* CSIDHSAMPLE($\mathcal{S}, B$) *to be the procedure which outputs a random element of $\mathcal{C}$ using the following algorithm:*

CSIDHSAMPLE($\mathcal{S}, B$)

$(e_1, \ldots, e_t) \leftarrow_\$ [-B, B]^t$

$\mathfrak{a} \leftarrow \prod_{i=1}^{t} \mathfrak{l}_i^{e_i}$

return $\mathfrak{a}$

We also need a basic notation for a set of nonces.

Notation 2.4. *Let N denote a known large set of nonces.*

Similarly to Basso et al. [3], we use a chain of secret isogenies to obtain an elliptic curve of unknown endomorphism ring. This idea is based on Wesolowski's theorem.

Theorem 2.5 ([20, §7], [21, **Cor. 5**]). *Let $\mathbb{F}$ be either $\mathbb{F}_p$ or $\mathbb{F}_{p^2}$ for some prime p. Let IsogenyPath be the problem where, given two supersingular elliptic curves E and F defined over $\mathbb{F}$, one must compute an isogeny $\phi : E \to F$. Let EndRing be the problem of computing the endomorphism ring of a supersingular elliptic curve E over $\mathbb{F}$. Then, assuming that the generalized Riemann hypothesis holds, IsogenyPath and EndRing can be polynomially reduced to each other.*

Since the goal of this paper is to generate elliptic curves of unknown endomorphism ring, we have to assume that computing such a ring is hard. With the above theorem, we will often use the problems of computing an isogeny and the problem of computing an endomorphism ring interchangeably and therefore also assume that computing an isogeny between two curves is hard. To be more precise, the isogeny problem on which we base our protocol requires a stronger assumption.

Assumption 2.6 (CSIDH [8]**).** *Let E be a supersingular elliptic curve defined over $\mathbb{F}_p$ of unknown endomorphism ring. Given $\mathfrak{a} \star E$ and $\mathfrak{b} \star E$ with unknown $\mathfrak{a}, \mathfrak{b} \in \mathcal{C}$, the CCISDH problem is to compute $\mathfrak{a} \star \mathfrak{b} \star E$. We assume that this problem is hard.*

The above assumption is required for (and equivalent to) the one-way security of CSIDH, and is therefore already widely accepted for many protocols that work with isogenies over $\mathbb{F}_p$. We also note that Assumption 2.6 implies that computing isogenies between two curves over $\mathbb{F}_p$ is hard.

Our second assumption is an adaptation of the Knowledge-of-Exponent Assumption (KEA) that was first described by Damgård [9]. This assumption was then used to create a classical Diffie-Hellman protocol by Wu and Stinson [22]. We denote this new assumption as the CSIDH Knowledge of Exponent assumption, or CSIDHKoE for short.

Assumption 2.7 (CSIDHKoE). *Let E be a supersingular elliptic curve defined over $\mathbb{F}_p$. For any probabilistic polynomial-time (PPT) $\boldsymbol{A}$ that takes as input E and $\mathfrak{b} \star E$, where $\mathfrak{b}$ is sampled using* CSIDHSAMPLE$(\mathcal{S}, B)$, *and which produces as output a pair of supersingular elliptic curves over $\mathbb{F}_p$ (F, F'), there exists probabilistic polynomial-time extractor $\boldsymbol{E}$ which takes the same input and outputs the same pair (F, F'), along with $\mathfrak{a}$, such the probability that $F' = \mathfrak{b} \star F$ and $\mathfrak{a} \star E \neq F$ is negligible.*

In other words, we assume that there is no pair of supersingular elliptic curves over $\mathbb{F}_p$ such that computing an isogeny between them is hard but computing a random CSIDH exchange involving them is easy. The original KEA is used in the context of classical Diffie-Hellman, while this new version is applied to CSIDH. While this assumption is new in the context of post-quantum cryptography, there is currently no known way to attack it.

Our third assumption simply states that we have access to a function H with some strong security properties. This same assumption is used by Basso et al. [3] for generating a multiparty protocol for secure curves over $\mathbb{F}_{p^2}$.

Assumption 2.8 (Secure Commitment Scheme Assumption). *As in [3], we assume the existence of a function H which is a statistically hiding and computationally binding commitment scheme on the set of binary strings. Denote by $\mathcal{H}$ the codomain of H.*

The above three assumptions are enough to obtain an efficient protocol generating supersingular elliptic curves of unknown endomorphism rings. However, we can obtain a more efficient protocol using a random oracle function.

Assumption 2.9 (Random Oracle Assumption). *Assume that W is a random oracle function. Let $\mathcal{W}$ be the codomain of W. We also assume that $\mathcal{W}$ is large enough for collisions to be unfeasible to find and that the mapping of any input by W can be efficiently computed.*

To be more precise, for this paper, we require that, for any function f with domain X, the problem of distinguishing between $(f(x), W(x))$ for a random $x \in X$ and $(f(x), r)$, where r is a uniformly random element of $\mathcal{H}$ is equivalent to the problem of computing $x \in X$ when given $f(x)$.

In cases where we use H or W on arbitrary data, we implicitly assume that this data is encoded in the form of a binary string using a suitable encoding scheme.

The above assumptions are required for most of our protocols. These assumptions are enough for all our protocols except CSIDH-CR. Hence, if we only need to generate supersingular elliptic curves with unknown endomorphism rings over $\mathbb{F}_p$, we do not need to invoke Assumption 2.10 below. On the other hand, for our results on random curve generation (Sect. 6), we require the following assumption, which was considered in the SeaSign paper [10, p.766].

Assumption 2.10. *The output distribution of* CSIDHSAMPLE$(\mathcal{S}, B)$ *is indistinguishable from the uniform distribution on the ideal class group* $\mathcal{C}$ *of* End(E).

3   Existing Solutions

In this section, we present a brief overview of possible ways to generate supersingular elliptic curves of unknown endomorphism ring that were either presented or mentioned in previous papers.

3.1   Signature Schemes

In the model used by Basso et al. [3], we have n parties $\mathcal{P}_1, \ldots, \mathcal{P}_n$ whose goal is to generate a supersingular elliptic curve of unknown endomorphism ring. A starting elliptic curve E_0 is provided. Their idea is to have each party $\mathcal{P}_i$ in turn compute a random isogeny $\phi_i : E_{i-1} \to E_i$ and publish E_i. If all parties act honestly and do not share their secret isogenies, then the endomorphism ring is unknown to them all by Theorem 2.5.

Of course, the above is not sufficient against dishonest adversaries, as nothing stops $\mathcal{P}_n$ from choosing the curve of their choice as E_n and lying about their isogeny. To solve this issue, [3] proposed having each party prove their knowledge of their claimed isogeny by publishing a Fiat-Shamir signature of a zero-knowledge proof.

Using a signature as a proof of knowledge has the advantage of keeping the number of required interactions to a minimum. The one proposed in [3], in particular, is fast enough for the desired application over $\mathbb{F}_{p^2}$. However, the zero-knowledge proof proposed in [3] reveals the degree of the secret isogeny. While it does not create issues when working over $\mathbb{F}_{p^2}$, this leaks too much information when dealing with isogenies defined over $\mathbb{F}_p$. This is because isogenies defined over $\mathbb{F}_p$ are usually sampled using CSIDHSAMPLE. Therefore, the isogeny degree can be used to efficiently compute $|e_i|$ and this knowledge massively reduces the possible key space. Because of this, a different signature scheme would need to be used to prove the knowledge of the claimed isogenies.

Basso et al. mention the possible use of either SeaSign [10] or CSI-FiSh [5] as possible replacement signatures. While both work in theory, they each come with issues limiting their practical applications. While SeaSign is a zero-knowledge signature, its current computation times are way too long to be used for everyday applications. However, it is worth noting that, in cases where a single secure curve needs to be generated by parties that can afford to wait multiple hours, for example when generating secure parameters for a scheme, using SeaSign is a possible solution.

On the other hand, CSI-FiSh can potentially be both zero-knowledge and efficient. It is worth noting that such a construction would lead to a protocol similar to CSI-RAShi [4]. However, it requires full knowledge of the ideal class group associated with the chosen parameter set. Currently, the parameter sets for which the ideal class group is known are pretty limited and, as discussed by Panny [17], computing new ones would require an extensive amount of computation even with access to a quantum computer. While recent results presented in the SCALLOP paper [12] have expanded the number of parameter sets that can be used today, the complexity of finding new ones is still super-polynomial.

3.2   Multiparty Key Generation

Two other possible ideas to generate secure curves defined over $\mathbb{F}_p$ were proposed by Moriya, Takashima and Tagaki [16]. However, the adversarial model in that paper is honest but curious, and this creates issues when trying to adapt their techniques when dealing with active adversaries.

The core idea of the protocol, given n parties $\mathcal{P}_1, \ldots, \mathcal{P}_n$, is to have n chains of isogenies that all loop over the same set of ideal class group elements so that, for each party, there is a chain where they are the last to apply their group action. This then implies that each party can trust the security of one chain and that, since the commutativity of the ideal class group implies that all chains end at the same curve, they can all trust the security of the final curve.

While the security of the above scheme is not proven against active adversaries in [16], such a proof might be possible. However, another issue is that the number of required interactions grows quadratically in proportion to the number of parties, making the scheme inefficient when working with a large number of parties.

4   A New Zero-Knowledge Proof

As mentioned in the previous session, there is currently no known fast CSIDH based signature scheme that works with any parameter set without heavy precomputation or the use of a quantum computer. To get around this issue, we propose replacing the signature part of the multiparty protocol from Basso et al. [3] with an interactive zero-knowledge proof. The following proposal, CSIDH Random Oracle Interactive Proof (CSIDH-ROIP), requires Assumptions 2.6, 2.9 and 2.7 for its security proof. Note that, in contrast to Moriya et al. [16], we allow for malicious parties.

Definition 4.1 (CSIDH-ROIP). *Let a prover* P *know a secret* $\mathfrak{a} \in \mathcal{C}$ *with associated public data* $(E, E_{\mathcal{P}} := \mathfrak{a} \star E)$. *The goal of the following protocol is for* $\mathcal{P}$ *to prove their knowledge of* $\mathfrak{a}$ *to a verifier* $\mathcal{V}$ *without leaking any extra information. CSIDH-ROIP consists of the followings steps:*

1. *Challenge:* $\mathcal{V}$ *sends the challenge curve* $E_\mathcal{V}$.
2. *Response:* $\mathcal{P}$ *computes* $\mathfrak{a} \star E_\mathcal{V}$ *and publishes a masked version of it using a random oracle function.*
3. *Verification:* $\mathcal{V}$ *verifies* $\mathcal{P}$*'s answer.*

Challenge($\mathcal{P}$, $\mathcal{V}$, E, $E_\mathcal{P}$)	*Response($\mathcal{P}$, $\mathcal{V}$, E, $E_\mathcal{P}$, $\mathfrak{a}$, $E_\mathcal{V}$)*
$\mathfrak{b} \leftarrow \text{CSIDHSAMPLE}(\mathcal{S}, B)$	$E_{\mathcal{P},\mathcal{V}} \leftarrow \mathfrak{a} \star E_\mathcal{V}$
$E_\mathcal{V} \leftarrow \mathfrak{b} \star E$	$M_{\mathcal{P},\mathcal{V}} \leftarrow W(\mathcal{P}, E_{\mathcal{P},\mathcal{V}})$
return $E_\mathcal{V}$	**return** $E_{\mathcal{P},\mathcal{V}}$

Verification($\mathcal{P}$,$\mathcal{V}$,E,$E_\mathcal{P}$,$E_\mathcal{V}$,$\mathfrak{b}$,$M_{\mathcal{P},\mathcal{V}}$)
$M'_{\mathcal{P},\mathcal{V}} \leftarrow W(\mathcal{P}, \mathfrak{b} \star E_\mathcal{P})$
if $M'_{\mathcal{P},\mathcal{V}} = M_{\mathcal{P},\mathcal{V}}$: **return true**
else : **return false**

The correctness of CSIDH-ROIP comes from the fact that $\mathfrak{a} \star (\mathfrak{b} \star E) = \mathfrak{b} \star (\mathfrak{a} \star E)$. Its soundness and zero-knowledge properties come from the following theorems.

Theorem 4.2. *Given Assumption 2.9, CSIDH-ROIP is zero-knowledge.*

Proof. Recall that W is a random oracle function defined in Assumption 2.9 whose input is arbitrary data. We can separate the adversarial strategies as the prover in two cases. Either they send a challenge $E_\mathcal{V}$ for which they know the associated $\mathfrak{b}$ or they choose a curve for whose associated group element is unknown and not feasibly computable.

In the former case, for any chosen $\mathfrak{b}$, the associated simulator is simple as an honest verifier knows the correct answer without needing to interact with the prover. In the latter case, we use the fact that W is a random oracle function to simulate transcripts indistinguishable from honest ones. This time, the adversary can use any method they desire to sample $E_\mathcal{V}$, but they do not gain any usable information from the answer as it is masked by W.

As we have two cases, we present two simulators. SIMULATOR1 represents the first case while SIMULATOR2 represents the second. The simulator returns a challenge-response pair with the same distribution as that of an honest exchange.

Simulator1($\mathcal{P}$, E, $E_\mathcal{P}$, $\mathfrak{b}$)	Simulator2($\mathcal{P}$, E, $E_\mathcal{P}$, $E_\mathcal{V}$)
$E_\mathcal{V} \leftarrow \mathfrak{b} \star E$	Challenge $\leftarrow E_\mathcal{V}$
Challenge $\leftarrow E_\mathcal{V}$	$M_{\mathcal{P},\mathcal{V}} \leftarrow_{\$} W$
$M_{\mathcal{P},\mathcal{V}} \leftarrow W(\mathcal{P}, \mathfrak{b} \star E_\mathcal{P})$	Response $\leftarrow E_{\mathcal{P},\mathcal{V}}$
Response $\leftarrow E_{\mathcal{P},\mathcal{V}}$	**return** (Challenge, Response)
return (Challenge, Response)	

Theorem 4.3. *CSIDH-ROIP is computationally sound under Assumptions 2.6, 2.9 and 2.7.*

Proof. Let $\mathcal{A}$ be a probabilistic polynomial-time (PPT) algorithm able to generate valid responses $M_{\mathcal{P},\mathcal{V}}$ for random challenges $E_{\mathcal{V}}$. Since W is a random oracle function, computing a valid $M_{\mathcal{P},\mathcal{V}}$ is equivalent to computing the correct $E_{\mathcal{P},\mathcal{V}}$ associated with the given challenge. This implies the existence of a PPT algorithm $\mathcal{A}'$ capable of generating $E_{\mathcal{P},\mathcal{V}}$ when given $E_{\mathcal{V}}$. By Assumption 2.7, this then implies the existence a PPT extractor capable of computing a valid witness $\mathfrak{a}$.

4.1 Avoiding the Random Oracle Model

As mentioned in Sect. 2, it is possible to generate supersingular elliptic curves over $\mathbb{F}_p$ with unknown endomorphism ring without having to rely on a random oracle function. The core idea of this trick is to remark that, without the use of W, CSIDH-ROIP is still honest verifier zero-knowledge. Therefore, we only need a way to deal with dishonest verifiers.

CSIDH-ROIP does so by publishing information unusable by dishonest verifiers, but this is not the only way to proceed. Another way is to have the verifiers prove that they were honest at the end of the proof. This can be done by requiring verifiers to publish a masked commitment for their challenge using a statistically hiding and computationally binding protocol H.

Of course, an adversary might still lie about their commitment and not care about being found cheating, as they still obtain information from the prover's answer. However, when it comes to using our proof scheme to generate secure curves, once the honest parties detect that someone cheated, the guilty party can then simply be removed from the protocol and the honest parties can then restart the protocol using new random values. As long as the new values are independent of the previous ones, the adversary gains no information by cheating.

Definition 4.4 (CSIDH-AIP). *Let a prover $\mathcal{P}$ know a secret $\mathfrak{a} \in \mathcal{C}$ with associated public data $(E, E_{\mathcal{P}} := \mathfrak{a} \star E)$. The goal of the following protocol is for $\mathcal{P}$ to prove their knowledge of $\mathfrak{a}$ to a verifier $\mathcal{V}$ without leaking any extra information. CSIDH-AIP consists of the followings steps:*

1. *Challenge: $\mathcal{V}$ sends the challenge curve $E_{\mathcal{V}}$ and commits the associated $\mathfrak{b}$ using H.*
2. *Response: $\mathcal{P}$ computes $\mathfrak{a} \star E_{\mathcal{V}}$ and publishes it.*
3. *Verification1: $\mathcal{V}$ verifies $\mathcal{P}$'s answer and, if it is valid, publishes $\mathfrak{b}$.*
4. *Verification2: $\mathcal{P}$ verifies that $\mathcal{V}$'s commitment coincides with the challenge.*

The details of the protocol are as follows:

Challenge$(\mathcal{V}, E)$	Response$(\mathcal{P}, \mathfrak{a}, E_\mathcal{V})$
$\mathfrak{b} \leftarrow \text{CSIDHSAMPLE}(\mathcal{S}, B)$	$E_{\mathcal{P},\mathcal{V}} \leftarrow \mathfrak{a} \star E_\mathcal{V}$
$E_\mathcal{V} \leftarrow \mathfrak{b} \star E$	**return** $E_{\mathcal{P},\mathcal{V}}$
$r \leftarrow_\$ N$	
$C \leftarrow H(\mathfrak{b}, r)$	
return $(E_\mathcal{V}, C)$	

Verification1$(\mathcal{V}, E_\mathcal{P}, \mathfrak{b}, r, E_{\mathcal{P},\mathcal{V}})$	Verification2$(\mathcal{P}, E, E_\mathcal{V}, \mathfrak{b}, r, C)$
$E'_{\mathcal{P},\mathcal{V}} \leftarrow \mathfrak{b} \star E_\mathcal{P}$	**if** $C \neq H(\mathfrak{b}, r)$: **return false**
if $E'_{\mathcal{P},\mathcal{V}} = E_{\mathcal{P},\mathcal{V}}$: **return** $(\mathfrak{b}, r)$	$E'_\mathcal{V} \leftarrow \mathfrak{b} \star E$
else : **return false**	**if** $E'_\mathcal{V} \neq E_\mathcal{V}$: **return false**
	return true

The proof fails if either Verification1 or Verification2 returns **false** and succeeds otherwise.

The correctness of CSIDH-AIP holds for the same reason as CSIDH-ROIP's. It is also sound and zero-knowledge, as the following theorems show.

Theorem 4.5. *Given Assumption 2.8, if it does not abort, CSIDH-AIP is zero-knowledge.*

Proof. Since H is computationally binding, $\mathcal{V}$ must compute $E_\mathcal{V}$ by first choosing a valid $\mathfrak{b}$. By the same reasoning, they must also choose their nonce r in advance. The following simulator returns a challenge-response-verification1 triple for any challenge constructed using $(\mathfrak{b}, r)$.

Simulator$(E, E_\mathcal{P}, \mathfrak{b}, r)$
$E_\mathcal{V} \leftarrow \mathfrak{b} \star E$
$C \leftarrow H(\mathfrak{b}, r)$
Challenge $\leftarrow (E_\mathcal{V}, C)$
$E_{\mathcal{P},\mathcal{V}} \leftarrow \mathfrak{b} \star E_\mathcal{P}$
Response $\leftarrow E_{\mathcal{P},\mathcal{V}}$
Verification1 $\leftarrow (\mathfrak{b}, r)$
return (Challenge, Response, verification1)

Theorem 4.6. *CSIDH-AIP is computationally sound under Assumptions 2.6, 2.7 and 2.8.*

Proof. Since H is statistically hiding, from $\mathcal{P}$'s point of view C can be replaced with a random value and, therefore, gives no advantage when it comes to beating the soundness property. Without being able to make use of this extra information, successfully generating a valid CSIDH-AIP proof becomes equivalent to generating $\mathfrak{a} \star \mathfrak{b} \star E$ when given $\mathfrak{b} \star E$ and $\mathfrak{a} \star E$.

By Assumption 2.7, if there was a PPT algorithm capable of doing so, then there would be another PPT algorithm capable of solving the CCISDH problem, which we assume is hard.

5   Secure Curve Generation

With the help of CSIDH-ROIP, we can now present our multiparty protocol for generating supersingular elliptic curves defined over $\mathbb{F}_p$ with unknown endomorphism ring. As its security mostly relies on CSIDH-ROIP, this new protocol, which we call CSIDH Secure Curve Generator (CSIDH-SCG), requires Assumptions 2.6 2.7 and 2.9, but not Assumption 2.10.

Definition 5.1 (CSIDH-SCG). *Let $\mathcal{P}_1, \ldots, \mathcal{P}_n$ be n parties that want to generate a supersingular elliptic curve over $\mathbb{F}_p$ with unknown endomorphism ring. Let E_0 be a known supersingular curve over the same field. CSIDH-SCG consists of the following steps.*

- **CurveGen:** *For i from 1 to n, party $\mathcal{P}_i$ computes an ideal class group element $\mathfrak{a}_i$, saves it, and publishes $E_i := \mathfrak{a}_i \star E_{i-1}$.*
- **Challenge:** *For $j \in [n] \setminus \{i\}$, $\mathcal{P}_i$ sends a CSIDH-ROIP challenge $E_{i,j}$ to $\mathcal{P}_j$ and saves the associated $\mathfrak{b}_{i,j}$.*
- **Response:** *After received a challenge from every $\mathcal{P}_j$ such that $j \in [n] \setminus \{i\}$, $\mathcal{P}_i$ publishes a CSIDH-ROIP response $M'_{i,j}$ for each of them.*
- **Verification:** *After receiving responses to all their challenges, $\mathcal{P}_i$ checks them for validity. If all the responses are correct, $\mathcal{P}_i$ published **true** and then accepts E_n as the final curve if every other party also publishes **true**. Otherwise, $\mathcal{P}_i$ publishes **false** and aborts the entire protocol.*
- **Abort:** *If at any point, a party published **false**, the protocol is aborted and every party must publish all their computed values. Any dishonest parties are thereby revealed.*

The algorithms for each step are as follows:

CurveGen($\mathcal{P}_i$, E_{i-1})	Challenge($\mathcal{P}_i$, $\mathcal{P}_j$, E_{j-1})
$\mathfrak{a}_i \leftarrow \text{CSIDHSAMPLE}(\mathcal{S}, B)$	$\mathfrak{b}_{i,j} \leftarrow \text{CSIDHSAMPLE}(\mathcal{S}, B)$
$E_i \leftarrow \mathfrak{a}_i \star E_{i-1}$	$E_{i,j} \leftarrow \mathfrak{b}_{i,j} \star E_{j-1}$
return E_i	**return** $E_{i,j}$

Response($\mathcal{P}_i$, $\mathcal{P}_j$, $\mathfrak{a}_i$, $E_{j,i}$)	Verification($\mathcal{P}_i$, $\mathcal{P}_j$, $M'_{i,j}$, E_j)
$M'_{j,i} \leftarrow W(\mathcal{P}_i, \mathfrak{a}_i \star E_{j,i})$	$M''_{i,j} \leftarrow W(\mathcal{P}_j, \mathfrak{b}_{i,j} \star E_j)$
return $M'_{j,i}$	**if** $M''_{i,j} = M'_{i,j}$: **return true**
	else : **return false**

Notation 5.2 (Secure Curve Generation Adversary). *For a secure curve generation multiparty protocol with n parties, the adversary is denoted $\mathcal{A}_{SCG}$. The goal of $\mathcal{A}_{SCG}$ is to compute the endomorphism ring of the final curve E_n. $\mathcal{A}_{SCG}$ is able to take control of all parties but one, say $\mathbb{P}_i$. They can try to be dishonest during the multiparty protocol. However, they fail if the protocol is aborted.*

Theorem 5.3. *Given Assumptions 2.6, 2.7, and 2.9, CSIDH-SCG is secure against $\mathcal{A}_{SCG}$ adversaries.*

Proof. During CSIDH-SCG, every party must prove knowledge of their $\mathfrak{a}_i$ using $n-1$ parallel CSIDH-ROIP proofs. By Theorem 4.2, $\mathcal{A}_{SCG}$ gains no information about $\mathfrak{a}_i$. By Theorem 4.3, $\mathcal{A}_{SCG}$ cannot lie about any of their $\mathfrak{a}_j$. Since $\mathcal{A}_{SCG}$ knows every $\mathfrak{a}_j$ expect for $\mathfrak{a}_i$, computing the endomorphism ring of E_n is equivalent to computing the endomorphism ring of E_i. However, by Theorem 2.5, doing so is equivalent to computing $\mathfrak{a}_i$, which is hard given Assumption 2.6.

In addition to being secure, CSIDH-SCG is also efficient, even when considering large groups. This is true for both the computation time and the number of interactions. In practice, the CSIDH parameters are chosen so that both sampling and group actions are computed efficiently. W is also assumed to be efficiently computable. In CSIDH-SCG, the number of times each party must compute a group action grows linearly in terms of the number of participants. The same is true for the number of times each party must call the functions W and CSIDHSAMPLE.

When it comes to the number of operations, CSIDH-SCG is constructed in a way that every step of the proof can be done in parallel with every other party. Because of this, each party is only required to publish once for each of the four steps of CSIDH-SCG, making the total number of interactions linear in terms of the number of participants.

5.1   Generating Secure Curves Without a Random Oracle

In cases where we want to avoid needing a random oracle, we can use CSIDH-AIP instead of CSIDH-ROIP.

Definition 5.4 (CSIDH-ASCG). *Let $\mathcal{P}_1, \ldots, \mathcal{P}_n$ be n parties that want to generate a supersingular elliptic curve over $\mathbb{F}_p$ with unknown endomorphism ring. Let E_0 be a known supersingular curve over the same field. CSIDH-ASCG consists of the following steps.*

- *CurveGen: For i from 1 to n, party $\mathcal{P}_i$ computes an ideal class group element $\mathfrak{a}_i$, saves it, and publishes $E_i := \mathfrak{a}_i \star E_{i-1}$.*
- *Challenge: For $j \in [n] \setminus \{i\}$, $\mathcal{P}_i$ sends a CSIDH-AIP challenge $(E_{i,j}, C_{i,j})$ to $\mathcal{P}_j$ and saves the associated $(\mathfrak{b}_{i,j}, r_{i,j})$.*
- *Response: After received a challenge from every $\mathcal{P}_j$ such that $j \in [n] \setminus \{i\}$, $\mathcal{P}_i$ publishes a CSIDH-AIP response $E'_{i,j}$ for each of them.*
- *Verification1: After receiving responses to all their challenges, $\mathcal{P}_i$ checks them for validity. If all the responses are correct, $\mathcal{P}_i$ publishes all their pairs $(\mathfrak{b}_{i,j}, r_{i,j})$. Otherwise, $\mathcal{P}_i$ publishes **false** and aborts the entire protocol.*
- *Verification2: After receiving every pair $(\mathfrak{b}_{j,i}, r_{j,i})$, $\mathcal{P}_i$ verifies that each pair agrees with their challenges. If that is the case for every pair, $\mathcal{P}_i$ publishes **true**. Otherwise, $\mathcal{P}_i$ publishes **false** and aborts the entire protocol. If every party publishes **true**, E_n is accepted as the curve of unknown endomorphism ring.*

- **Abort:** *If at any point, a party published* **false**, *the protocol is aborted and every party must publish all their computed values. Any dishonest parties are thereby revealed.*

The algorithms for each step are as follows:

$\underline{CurveGen(\mathcal{P}_i,\ E_{i-1})}$ $\underline{Challenge(\mathcal{P}_i,\ \mathcal{P}_j,\ E_{j-1})}$

$\mathfrak{a}_i \leftarrow \text{CSIDHSAMPLE}(\mathcal{S}, B)$ $\mathfrak{b}_{i,j} \leftarrow \text{CSIDHSAMPLE}(\mathcal{S}, B)$

$E_i \leftarrow \mathfrak{a}_i \star E_{i-1}$ $E_{i,j} \leftarrow \mathfrak{b}_{i,j} \star E_{j-1}$

return E_i $r_{i,j} \leftarrow_{\$} N$

                           $C_{i,j} \leftarrow H(\mathfrak{b}_{i,j}, r_{i,j})$

                           return $(E_{i,j}, C_{i,j})$

$\underline{Response(\mathcal{P}_i,\ \mathcal{P}_j,\ \mathfrak{a}_i,\ E_{j,i})}$ $\underline{Verification1(\mathcal{P}_i,\ \mathcal{P}_j,\ E'_{i,j},\ E_j,\ \mathfrak{b}_{i,j},\ r_{i,j})}$

$E'_{j,i} \leftarrow \mathfrak{a}_i \star E_{j,i}$ $E''_{i,j} \leftarrow \mathfrak{b}_{i,j} \star E_j$

return $E'_{j,i}$ **if** $E''_{i,j} = E'_{i,j}$: **return** $(\mathfrak{b}_{i,j}, r_{i,j})$

                        else : **return false**

$\underline{Verification2(\mathcal{P}_i,\ \mathcal{P}_j,\ C_{j,i},\ \mathfrak{b}_{j,i},\ r_{j,i},\ E_{i-1},\ E_{j,i})}$

if $C_{j,i} \neq H(\mathfrak{b}_{j,i}, r_{j,i})$: **return false**

$E'''_{j,i} \leftarrow \mathfrak{b}_{j,i} \star E_{i-1}$

if $E'''_{j,i} \neq E_{j,i}$: **return false**

return true

Theorem 5.5. *Given Assumptions 2.6, 2.8, and 2.7, CSIDH-ASCG is secure against $\mathcal{A}_{SCG}$ adversaries.*

Proof. During CSIDH-ASCG, every party must prove knowledge of their $\mathfrak{a}_i$ using $n - 1$ parallel CSIDH-AIP proofs. By Assumption 2.8, since H is statistically hiding, $\mathcal{A}_{SCG}$ gains no information from $\mathcal{P}_i$ publishing $C_{i,j}$. This implies that $\mathcal{A}_{SCG}$ must prove knowledge of its $\mathfrak{a}_j$ using CSIDH-AIP without any extra information. By Theorem 4.5, since the adversary looses if they cause an abort, $\mathcal{A}_{SCG}$ gains no information about $\mathfrak{a}_i$. By Theorem 4.6, $\mathcal{A}_{SCG}$ cannot lie about any of their $\mathfrak{a}_j$.

Since $\mathcal{A}_{SCG}$ knows every $\mathfrak{a}_j$ expect for $\mathfrak{a}_i$, computing the endomorphism ring of E_n is equivalent to computing the endomorphism ring of E_i. However, by Theorem 2.5, doing so is equivalent to computing $\mathfrak{a}_i$, which is hard given Assumption 2.6.

While efficient and secure, CSIDH-ASCG comes with two disadvantages compared CSIDH-SCG. First, since there are two verification steps in CSIDH-AIP, CSIDH-ASCG requires one more round of interaction than CSIDH-SCG. The other issue is that CSIDH-AIP is only zero-knowledge if there are no aborts. Because of this, every time a dishonest party is found, the entire process needs to be restarted with new random values. In practice, needing a random oracle function assumption is usually worth it for the efficiency gains of CSIDH-SCG.

6   Curve Randomizer

CSIDH-SCG allows us to generate a supersingular elliptic curve of unknown endomorphism ring. However, E_n is not uniformly random, as $\mathcal{P}_n$ has some control over what the final curve is. This limitation also appears in the protocol proposed by Basso et al. [3]. However, some protocols (for example Delay Encryption [6]) explicitly ask for a uniformly random (or at least nearly uniform) curve over $\mathbb{F}_p$. In order to generate curves that are compatible with the security proofs of such schemes, we require a way to convert a secure elliptic curve into a random and secure elliptic curve, which justifies the need of the following protocol.

The core idea of CSIDH Curve Randomizer (CSIDH-CR) is to use a multiparty commitment scheme to generate some random data and then convert that data into a random isogeny whose codomain is chosen as the random curve. CSIDH-CR is the only protocol in this paper that requires Assumption 2.10. On the other hand, it does not require Assumption 2.6.

Definition 6.1 (CSIDH-CR). *Let $\mathcal{P}_1, \ldots, \mathcal{P}_n$ be n parties that want to generate a random supersingular elliptic curve over $\mathbb{F}_p$. Let E_n be supersingular elliptic curve of unknown endomorphism ring defined over the same field. CSIDH-SCG consists of the following steps.*

- **RandomSample:** *Each $\mathcal{P}_i$ samples a random $\mathfrak{a}'_i$ and nonce r'_i and commits $H(\mathfrak{a}'_i, r'_i)$.*
- **CurveComp:** *Once every commitment has been published, each party publishes their pair $(\mathfrak{a}'_i, r'_i)$. If every pair corresponds to their commitment, the final curve is chosen to be $(\prod_{i=1}^n \mathfrak{a}'_i) \star E_n$.*
- **Abort:** *If any of the pairs do not correspond to their commitments, the protocol aborts and any dishonest parties are thereby revealed.*

The algorithms for each step are as follows:

RandomSample($\mathcal{P}_i$)
***
$\mathfrak{a}'_i \leftarrow \text{CSIDHSAMPLE}(\mathcal{S}, B)$
$r'_i \leftarrow_\$ N$
$C'_i \leftarrow H(\mathfrak{a}'_i, r'_i)$
return C'_i

CurveComp($E_n, (C'_1, \ldots, C'_n), ((\mathfrak{a}'_1, r'_1), \ldots, (\mathfrak{a}'_1, r'_1)))$
***
$F \leftarrow E_n$
for $j \in [n]$:
$\quad$ **if** $C'_j \neq H(\mathfrak{a}'_j, r'_j)$: **return false**
$\quad F \leftarrow \mathfrak{a}'_j \star F$
return F

Notation 6.2 (Curve Randomizer Adversary). *Given a curve randomizer multiparty protocol with n parties and whose initial curve E_n has an unknown endomorphism ring, the adversary is denoted $\mathcal{A}_{CR}$. The goal of $\mathcal{A}_{CR}$ is either to guess the final curve F before starting the scheme or to compute its endomorphism ring. $\mathcal{A}_{CR}$ is able to take control of all parties but one, say $\mathbb{P}_i$. They can try to be dishonest during the multiparty protocol. However, they fail if the protocol is aborted.*

Theorem 6.3. *Given Assumptions 2.8 and 2.10, CSIDH-CR is secure against $\mathcal{A}_{CR}$ adversaries.*

Proof. By Assumption 2.8, H is statistically hiding and $\mathcal{A}_{CR}$ gains no information from C_i'. By the same assumption, since H is a binding commitment scheme, $\mathcal{A}_{CR}$ must choose their $\mathfrak{a}_j'$ before knowing anything about $\mathfrak{a}_i'$. Once the $\mathfrak{a}_j'$ have been chosen, set $F' := \left(\prod_{j \in [n] \setminus \{j\}} \mathfrak{a}_j' \right) \star E_n$.

By Assumption 2.10, $\mathfrak{a}_i'$ is indistinguishable from a uniformly random ideal class group element. Since $\mathcal{C}$ is an abelian group, we have that $F = \mathfrak{a}_i' \star F'$. Therefore, $\mathfrak{a}_i'$ is indistinguishable from random, and so is F. Also, since every $\mathfrak{a}_i'$ is revealed at the end of the protocol, computing the endomorphism ring of F is equivalent to computing the endomorphism ring of E_n, which is hard. ∎

7   Conclusion

CSIDH-SCG enables efficient generation of supersingular elliptic curves defined over $\mathbb{F}_p$ with unknown endomorphism ring. In analogy to the work of Basso et al. for curves over $\mathbb{F}_{p^2}$ [3], the total number of interactions required for CSIDH-SCG grows linearly in terms of the number of participants. Currently, the greatest limitation of CSIDH-SCG is the need of a Knowledge of Exponent Assumption. While such an assumption is sometimes used in classical schemes, this new CSIDH variant requires further study on its security.

The other limitation of CSIDH-SCG is its need of a random oracle function. However, this can be dealt with by using CSIDH-ASCG. While this alternative is strictly less efficient, the increase in computation time is only a single additional round of interactions. In cases where a new secure curve must be generated often or some parties are slow to interact, one might prefer to use CSIDH-SCG. The decision on which of the two schemes to use depends on details of the application and one's comfort in using a random oracle assumption. Given an additional assumption on the randomness of CSIDH samples, CSIDH-CR makes it possible for the generated curve to be random.

It is worth mentioning that the curve randomizer structure can also be adapted to generate random supersingular elliptic curves defined over $\mathbb{F}_{p^2}$. Using the Ramanujan property of supersingular isogeny graphs defined over $\mathbb{F}_{p^2}$, Jao, Miller and Venkatesan [13] showed that the codomain of a random isogeny of large enough degree is indistinguishable from random. It is therefore possible to use a multiparty protocol to generate random data, which can then be converted into a random isogeny in order to obtain a random supersingular elliptic curve. We leave the implementation of such a protocol for future work.

By itself, CSIDH-ROIP is a secure and efficient zero-knowledge proof that can be used with any CSIDH parameter sets. While its structure makes it so that it cannot be used in a signature scheme, its large challenge space makes it so that a single run of CSIDH-ROIP is enough to achieve levels of security comparable to CSIDH given the CSIDHKoE assumption.

Acknowledgments. We would like to thank Andrea Basso for helpful comments on a draft version of this paper. This research was supported by NSERC Alliance Consortia Quantum Grant ALLRP 578463–2022.

References

1. Alamati, N., De Feo, L., Montgomery, H., Patranabis, S.: Cryptographic group actions and applications. In: Moriai, S., Wang, H. (eds.) ASIACRYPT 2020. LNCS, vol. 12492, pp. 411–439. Springer, Cham (2020). https://doi.org/10.1007/978-3-030-64834-3_14
2. Atapoor, S., Baghery, K., Cozzo, D., Pedersen, R.: Practical robust DKG protocols for CSIDH. In: Tibouchi, M., Wang, X. (eds.) ACNS 2023. LNCS, vol. 13906, pp. 219–247. Springer, Cham (2023). https://doi.org/10.1007/978-3-031-33491-7_9
3. Basso, A., et al.: Supersingular curves you can trust. In: Hazay, C., Stam, M. (eds.) EUROCRYPT 2023. LNCS, vol. 14005, pp. 405–437. Springer, Cham (2023). https://doi.org/10.1007/978-3-031-30617-4_14
4. Beullens, W., Disson, L., Pedersen, R., Vercauteren, F.: CSI-RAShi: distributed key generation for CSIDH. In: Cheon, J.H., Tillich, J.-P. (eds.) PQCrypto 2021. LNCS, vol. 12841, pp. 257–276. Springer, Cham (2021). https://doi.org/10.1007/978-3-030-81293-5_14
5. Beullens, W., Kleinjung, T., Vercauteren, F.: CSI-FiSh: efficient isogeny based signatures through class group computations. In: Galbraith, S.D., Moriai, S. (eds.) ASIACRYPT 2019. LNCS, vol. 11921, pp. 227–247. Springer, Cham (2019). https://doi.org/10.1007/978-3-030-34578-5_9
6. Burdges, J., De Feo, L.: Delay encryption. In: Canteaut, A., Standaert, F.-X. (eds.) EUROCRYPT 2021. LNCS, vol. 12696, pp. 302–326. Springer, Cham (2021). https://doi.org/10.1007/978-3-030-77870-5_11
7. Castryck, W., Decru, T.: An efficient key recovery attack on SIDH. In: Hazay, C., Stam, M. (eds.) EUROCRYPT 2023. LNCS, vol. 14008, pp. 423–447. Springer, Cham (2023). https://doi.org/10.1007/978-3-031-30589-4_15
8. Castryck, W., Lange, T., Martindale, C., Panny, L., Renes, J.: CSIDH: an efficient post-quantum commutative group action. In: Peyrin, T., Galbraith, S. (eds.) ASIACRYPT 2018, Part III. LNCS, vol. 11274, pp. 395–427. Springer, Cham (2018). https://doi.org/10.1007/978-3-030-03332-3_15
9. Damgård, I.: Towards practical public key systems secure against chosen ciphertext attacks. In: Feigenbaum, J. (ed.) CRYPTO 1991. LNCS, vol. 576, pp. 445–456. Springer, Heidelberg (1992). https://doi.org/10.1007/3-540-46766-1_36
10. De Feo, L., Galbraith, S.D.: SeaSign: compact isogeny signatures from class group actions. In: Ishai, Y., Rijmen, V. (eds.) EUROCRYPT 2019. LNCS, vol. 11478, pp. 759–789. Springer, Cham (2019). https://doi.org/10.1007/978-3-030-17659-4_26

11. De Feo, L., Masson, S., Petit, C., Sanso, A.: Verifiable delay functions from supersingular isogenies and pairings. In: Galbraith, S.D., Moriai, S. (eds.) ASIACRYPT 2019. LNCS, vol. 11921, pp. 248–277. Springer, Cham (2019). https://doi.org/10.1007/978-3-030-34578-5_10

12. Feo, L.D., et al.: SCALLOP: scaling the CSI-FiSh. In: Boldyreva, A., Kolesnikov, V. (eds.) PKC 2023. LNCS, vol. 13940, pp. 345–375. Springer, Cham (2023). https://doi.org/10.1007/978-3-031-31368-4_13

13. Jao, D., Miller, S.D., Venkatesan, R.: Expander graphs based on GRH with an application to elliptic curve cryptography. J. Number Theory **129**(6), 1491–1504 (2009). https://doi.org/10.1016/j.jnt.2008.11.006

14. Lai, Y.-F., Galbraith, S.D., Delpech de Saint Guilhem, C.: Compact, efficient and UC-secure isogeny-based oblivious transfer. In: Canteaut, A., Standaert, F.-X. (eds.) EUROCRYPT 2021. LNCS, vol. 12696, pp. 213–241. Springer, Cham (2021). https://doi.org/10.1007/978-3-030-77870-5_8

15. Maino, L., Martindale, C.: An attack on SIDH with arbitrary starting curve. Cryptology ePrint Archive, Paper 2022/1026 (2022)

16. Moriya, T., Takashima, K., Takagi, T.: Group key exchange from CSIDH and its application to trusted setup in supersingular isogeny cryptosystems. In: Liu, Z., Yung, M. (eds.) Inscrypt 2019. LNCS, vol. 12020, pp. 86–98. Springer, Cham (2020). https://doi.org/10.1007/978-3-030-42921-8_5

17. Panny, L.: CSI-FiSh really isn't polynomial-time. https://yx7.cc/blah/2023-04-14.html

18. Petit, C.: Faster algorithms for isogeny problems using torsion point images. In: Takagi, T., Peyrin, T. (eds.) ASIACRYPT 2017. LNCS, vol. 10625, pp. 330–353. Springer, Cham (2017). https://doi.org/10.1007/978-3-319-70697-9_12

19. Robert, D.: Breaking SIDH in polynomial time. In: Hazay, C., Stam, M. (eds.) EUROCRYPT 2023. LNCS, vol. 14008, pp. 472–503. Springer, Cham (2023). https://doi.org/10.1007/978-3-031-30589-4_17

20. Wesolowski, B.: The supersingular isogeny path and endomorphism ring problems are equivalent. In: 2021 IEEE 62nd Annual Symposium on Foundations of Computer Science (FOCS), pp. 1100–1111 (2021)

21. Wesolowski, B.: Orientations and the supersingular endomorphism ring problem. In: Dunkelman, O., Dziembowski, S. (eds.) EUROCRYPT 2022. LNCS, vol. 13277, pp. 345–371. Springer, Cham (2022). https://doi.org/10.1007/978-3-031-07082-2_13

22. Wu, J., Stinson, D.R.: An efficient identification protocol and the knowledge-of-exponent assumption. Cryptology ePrint Archive, Paper 2007/479 (2007). https://eprint.iacr.org/2007/479

Kummer and Hessian Meet in the Field of Characteristic 2

Sabyasachi Karati[1($\boxtimes$)] and Gourab Chandra Saha[2]

[1] Cryptology and Security Research Unit, Indian Statistical Institute,
Kolkata, India
skarati@isical.ac.in
[2] Applied Statistics Unit, Indian Statistical Institute, Kolkata, India

Abstract. One can compute scalar multiplication on an ordinary short Weierstrass curve defined over a binary field. Also, one can move to the associated binary Kummer line $\mathsf{BKL}_{(1:c)}$, or isomorphic generalized Hessian curve $\mathsf{H}_{(\gamma,\delta)}$ from short Weierstrass, and then compute the scalar multiplication. A generalized Hessian curve provides the best performance of scalar multiplication in $\mathfrak{R}\mathfrak{T}$ coordinates where $\mathfrak{R} = R^3 + S^3$ and $\mathfrak{T} = T^3$ for a point $P = (R : S : T)$ on $\mathsf{H}_{(\gamma,\delta)}$. Montgomery scalar multiplication gives us the nP and $(n+1)P$, again in $\mathfrak{R}\mathfrak{T}$. We propose a method to uniquely obtain the R and S coordinates of nP given $P = (R : S : T)$ and $\mathfrak{R}\mathfrak{T}$ coordinates of nP and $(n+1)P$. Next, we show that $\mathsf{BKL}_{(1:c)}$ can be linked to an isomorphic $\mathsf{H}_{(\gamma,\delta)}$. But small c does not guarantee small γ or δ. First, we introduce two isogenies and their duals: one 2-isogeny between two short Weierstrass curves and one 3-isogeny between two generalized Hessian curves to solve the issue. Using the introduced isogenies, we show that there always exists a generalized Hessian curve $\mathsf{H}_{(\gamma,1)}$ with $\sqrt{\gamma^3(\gamma+1)} = c$ associated with a $\mathsf{BKL}_{(1:c)}$. The obtained $\mathsf{H}_{(\gamma,1)}$ needs $5[\mathsf{M}] + 4[\mathsf{S}] + 1[\mathsf{C}_s]$ field operations for each ladder step of Montgomery scalar multiplication, and the operation count is the smallest one compared to any other curves over a binary field.

Keywords: Binary Kummer line · Generalized Hessian curve · Scalar multiplication · Montgomery ladder · Isogeny

1 Introduction

Elliptic curve cryptography was introduced by Miller [22] and Koblitz [20]. Elliptic curves are widely used to implement real-world public-key cryptographic primitives like Diffie-Hellman Key Exchange and Digital signatures [5] because elliptic curve provides small key sizes and signature sizes. Scalar multiplication is the basic building block of these protocols. We can categorize the scalar multiplications into two types: (i) Fixed-base scalar multiplication, where the point is known and fixed during the setup phase, and (ii) Variable-base scalar multiplication where the point is unknown beforehand. We need one fixed-base and one variable-base scalar multiplication for the Diffie-Hellman key exchange.

© The Author(s), under exclusive license to Springer Nature Switzerland AG 2024
A. Chattopadhyay et al. (Eds.): INDOCRYPT 2023, LNCS 14459, pp. 175–196, 2024.
https://doi.org/10.1007/978-3-031-56232-7_9

On the other hand, key generation and signing of Elliptic Curve Digital Signature Algorithm (ECDSA) [12] each needs one fixed-base scalar multiplication, and verification needs one fixed- and one variable-base scalar multiplication or a multi-scalar multiplication. Improvement in the performance of the scalar multiplication directly affects the computational time of these protocols, and secure and efficient scalar multiplication became a prerequisite for practical deployment.

To improve the performance of scalar multiplication, we must consider several aspects of the target application. First, assume that the underlying system does not support parallelization and possibly resource constraints. In this case, one should choose the form of the curve that takes the minimum number of field operations to perform the curve arithmetic. Second, consider that the underlying system has a parallelization feature. Now, we should choose the curve that provides the best-parallelized performance for the curve arithmetic. Third, the curve should have small curve parameters and a small base point so that the implementation can take advantage of optimized fixed-base scalar multiplication.

Over fields of large prime characteristic, it has been shown that Kummer line [19,23] provides the best x-coordinate-based scalar multiplication if the Single-Instruction-Multiple-Data (SIMD) feature is available. For Diffie-Hellman key exchange, x-coordinate-based arithmetic is sufficient, and one can choose the Kummer Line. But if SIMD is unavailable, Montgomery curve Curve25519 [1] should be chosen because it needs fewer field operations. On the other hand, one can take advantage of the twisted Edwards curve to have the best performance of digital signatures [13]. [14,18] study the connections among these different forms of elliptic curves to facilitate more flexibility in the choice of the curves.

1.1   Our Contribution

In this work, we focus on finite fields of characteristics two. Kummer line $\mathsf{BKL}_{(1:c)}$ over the binary field was proposed by Gaudry and Lubicz [11]. The idea of scalar multiplication using the $\mathsf{BKL}_{(1:c)}$ was studied and [17] develops all the relevant details. $\mathsf{BKL}_{(1:c)}$ is defined by a projection map from a particular form of Weierstrass curve $(\mathsf{BEw}_{(c^4)} : y^2 + xy = x^3 + c^4)$ to a projective line [11,17]. The $\mathsf{BKL}_{(1:c)}$ provides the fastest x-coordinate-based scalar multiplication where parallelization is not available [17]. Hessian curve $\mathsf{H}_{(\delta)}$ was introduced by Smart [26], and [26] further shows that $\mathsf{H}_{(\delta)}$ is suitable for parallelization. Farashahi and Joye proposed the concept of a Generalized Hessian curve $\mathsf{H}_{(\gamma,\delta)}$ [7], and the $\mathsf{H}_{(\delta)}$ becomes a special case with $\gamma = 1$. The twisted Hessian curve $\mathsf{H}^t_{(\rho,\delta)}$ was proposed by Bernstein, Chuengsatiansup, Kohel, and Lange [3], and it is isomorphic to a generalized Hessian curve $\mathsf{H}_{(\gamma,\delta)}$. The generalized Hessian curve provides *unified* formulas of curve arithmetic and has been shown that the formulas are more efficient when the underlying field is of characteristic two.

Field square is significantly faster than field multiplication over a finite field of characteristic 2. Thus, $\mathsf{BKL}_{(1:c)}$ ($5[\mathsf{M}] + 5[\mathsf{S}] + 1[\mathsf{C}_s]$ per bit of scalar[1]) provides

---

[1] $\mathsf{M}, \mathsf{S}, \mathsf{C}$, and C_s denotes a field multiplication, a square, a multiplication by a constant, and a multiplication by a small constant.

faster scalar multiplication than $H_{(\gamma,\delta)}$ ($6[M] + 4[S] + 1[C_s]$ per bit of scalar) if the underlying platform does not support parallelization. On the other hand, [7,26] show that $H_{(\gamma,\delta)}$ is efficiently parallelizable, but effect of parallelization on $BKL_{(1:c)}$ is not known. In this work, we show the connection between $BKL_{(1:c)}$ and $H_{(\gamma,\delta)}$ via isomorphism that supports interoperability between different platforms so that platforms with SIMD feature can take advantage of $H_{(\gamma,\delta)}$ while platforms without SIMD feature can take advantage of $BKL_{(1:c)}$. We also show that we can connect $BKL_{(1:c)}$ to a $H_{(\gamma,1)}$, via isogeny, that supports faster scalar multiplication ($5[M] + 4[S] + 1[C_s]$ per bit of scalar) than the associated $BKL_{(1:c)}$ on platforms without SIMD feature. Therefore, one can take advantage of scalar multiplication of the Kummer line by moving from $BEw_{(c^4)}$ to the Kummer Line $BKL_{(1:c)}$ or generalized Hessian curve by moving to $H_{(\gamma,\delta)}$. A brief description of our contributions is given below.

1. **Retrieve R and S-coordinates of generalized Hessian curve in $\mathfrak{R}\mathfrak{T}$ coordinate system.** To achieve the best possible operation count per bit of the scalar for x-coordinate-based scalar multiplication, a binary Kummer line uses **xz** coordinates and a generalized Hessian uses $\mathfrak{R}\mathfrak{T}$-coordinates where $\mathfrak{R} = R^3 + S^3$ and $\mathfrak{T} = T^3$ for a point $P = (R : S : T)$ on a generalized Hessian curve. Using $\mathfrak{R}\mathfrak{T}$ coordinates with Montgomery scalar multiplication, one computes two points $nP = (R_n : S_n : T_n)$ and $(n+1)P = (R_{n+1} : S_{n+1} : T_{n+1})$, but they are again in $\mathfrak{R}\mathfrak{T}$ coordinate. Recovering R_n and S_n is an important problem but no solution is available in the literature. We propose a method for the first time that allows us to compute unique R_n and S_n from the known details $R, S, T, R_n^3 + S_n^3, T_n^3, R_{n+1}^3 + S_{n+1}^3$, and T_{n+1}^3.

2. **Connecting Binary Kummer/Weierstrass to Binary Hessian using Isomorphism.** One can choose a generalized Hessian curve for scalar multiplication. [7] shows that there exists an isomorphic generalized Hessian curve if the Weierstrass curve has a point of order three and provides the corresponding mapping. But one particular map is not isomorphic. We give the explicit map for binary fields following the map given in [30], and we connect binary Kummer line $BKL_{(1:\sqrt[4]{c})}$ by connecting short Weierstrass curve $BEw_{(c)}$ to a generalized Hessian curve. The obtained isomorphic generalized Hessian curve needs $5[M] + 4[S] + 1[C] + 1[C_s]$ field operations and for large constant it becomes $6[M] + 4[S] + 1[C_s]$.

3. **Connecting Binary Kummer/Weierstrass to Binary Hessian using Isogeny.** First, we propose two isogenies and their duals: (i) ϕ is a 2-isogeny between the short Weierstrass curves $BEw_{(b)}$ and $BEw_{(\sqrt{b})}$, and (ii) ξ is a 3-isogeny between generalized Hessian curves $H_{(a^3,1)}$ and $H_{(\gamma,1)}$ with $\gamma = a + a^2 + a^3$. Next, with the help of ϕ and ξ, we connect binary Kummer line $BKL_{(1:c)}$ by showing that a short Weierstrass curve $BEw_{(c^4)}$ is isogenous to the generalized Hessian curve $H_{(x_3,1)}$ where (x_3, y_3) is a point of order 3 on $BEw_{(c^2)}$. The $H_{(x_3,1)}$ requires $5[M] + 4[S] + 1[C_s]$ field operations for the arithmetic of each ladder step of Montgomery scalar multiplication which is the smallest one for variable-base scalar multiplication compared to the present state-of-the-art over binary fields as given in Table 1.

Table 1. Comparison of operation counts per ladder step for variable base scalar multiplication over different curves

Curve	Coordinates	Operation Count per Ladder Step
Short Weirstrass Curve [27]	XZ	$5[\mathsf{M}] + 6[\mathsf{S}] + 2[\mathsf{C}_s]$
Binary Kummer line [17]	$\mathbf{xz}$	$5[\mathsf{M}] + 5[\mathsf{S}] + 1[\mathsf{C}_s]$
Binary Edwards Curve [17]	$\mathfrak{U}W$	$5[\mathsf{M}] + 4[\mathsf{S}] + 2[\mathsf{C}_s]$
Generalized Hessian Curve [7]	$\mathfrak{RT}$	$5[\mathsf{M}] + 4[\mathsf{S}] + 1[\mathsf{C}] + 1[\mathsf{C}_s]$
Generalized Hessian Curve (this work)	$\mathfrak{RT}$	$\mathbf{5[M] + 4[S] + 1[C_s]}$

4. We go beyond the task of just providing the mappings and optimized arithmetic on the generalized Hessian Curve and propose concrete Kummer lines with small curve parameters and associated binary generalized Hessian curves, and binary Hessian curves with small curve parameters and the associated binary Kummer lines.

2   Background

This section includes brief descriptions of the relevant curves: short Weierstrass curve, binary Kummer Line, and generalized Hessian curve over $\mathbb{F}_{2^m}$ where $\mathbb{F}_{2^m}$ denotes a finite field of characteristic two with 2^m elements for some $m \in \mathbb{N}$.

2.1   Weierstrass Curve

A short Weierstrass affine form of a non-supersingular elliptic curve over $\mathbb{F}_{2^m}$ is given by the Eq. (1) [12]:

$$\mathsf{BEw}_{(a,b)} : \quad y^2 + xy = x^3 + ax^2 + b, \tag{1}$$

where $b \neq 0$. All the points (x, y) of the affine plane $\mathbb{A}^2 = \mathbb{F}_{2^m}^2$ that satisfy the equation of the curve $\mathsf{BEw}_{(a,b)}$ along with a special point $\mathcal{O}$, called point at infinity, form an additive group. The projective form of a short Weierstrass curve is given by the Eq. (2):

$$\mathsf{BEw}_{(a,b)} : \quad Y^2 + XYZ = X^3 + aX^2Z + bZ^3. \tag{2}$$

The affine point (x, y) can be identified with the projective points $(X : Y : Z) \in \mathbb{P}^2 = \mathbb{F}_{2^m}^3 \setminus \{0 : 0 : 0\}$ where $x = X/Z$ and $y = Y/Z$. We say two projective points $(X_1 : Y_1 : Z_1)$ and $(X_2 : Y_2 : Z_2)$ are called equivalent if there exists a non-zero $\lambda \in \mathbb{F}_{2^m}$ such that $X_1 = \lambda X_2$, $Y_1 = \lambda Y_2$ and $Z_1 = \lambda Z_2$. The point at infinity $\mathcal{O}$ is the equivalent class of points with $Z = 0$. In the article, we will use the projective versions mainly because of their practical relevance. But sometimes we use the affine version in proofs for simplicity.

2.2   Binary Kummer Line

In this work, we are specifically interested in short Weierstrass curve $\mathsf{BEw}_{(a,b)}$ with $a = 0$ and we denote it by $\mathsf{BEw}_{(b)}$ whose projective form is given by Eq. (3):

$$\mathsf{BEw}_{(b)} : \quad Y^2 Z + XYZ = X^3 + bZ^3. \tag{3}$$

Let $c \in \mathbb{F}_{2^m}$ and $c \neq 0$. A binary Kummer line $\mathsf{BKL}_{(1:c)}$ is defined over projective space $\mathbb{P}^1$ by mapping $\pi : \mathsf{BEw}_{(c^4)}/\{\pm 1\} \to \mathsf{BKL}_{(1:c)}$ [11,17] as:

$$\pi(P = (X : \cdot : Z)) = \begin{cases} (cZ : X), & \text{if } X \neq 0 \\ (0 : 1), & \text{if } X = 0. \end{cases} \tag{4}$$

For each $c \in \mathbb{F}_{2^m}$, $\mathsf{BEw}_{(c)}$ has a unique $\mathsf{BKL}_{(1:\sqrt[4]{c})}$ because each c has a unique square-root in a binary field.

[11] derives the arithmetic of binary Kummer line over projective space $\mathbb{P}^1$ using theta functions. Each point of $\mathsf{BKL}_{(1:c)}$ in $\mathbb{P}^1$ is defined as $(\mathbf{x}, \mathbf{z}) \in \mathbb{F}_{2^m}^2$. Two points $\mathbf{P} = (\mathbf{x}_1, \mathbf{z}_1)$ and $\mathbf{Q} = (\mathbf{x}_2, \mathbf{z}_2)$ are said to be equivalent if there exists a non-zero $\lambda \in k$ such that $\mathbf{x}_1 = \lambda \mathbf{x}_2$ and $\mathbf{z}_1 = \lambda \mathbf{z}_2$. Let $\mathbf{P} = (\mathbf{x}_1, \mathbf{z}_1)$ and $\mathbf{Q} = (\mathbf{x}_2, \mathbf{z}_2)$ be two points on $\mathsf{BKL}_{(1:c)}$ with known $\mathbf{P} - \mathbf{Q} = (\mathbf{x}, \mathbf{z})$. By Doubling Algorithm $\mathtt{dbl}_k$ and Differential Addition Algorithm $\mathtt{diffAdd}_k$ of Table 2, we compute $2\mathbf{P} = (\mathbf{x}_3, \mathbf{z}_3)$ and $\mathbf{P} + \mathbf{Q} = (\mathbf{x}_4, \mathbf{z}_4)$, respectively. In $\mathsf{BKL}_{(1:c)}$, $\mathbf{I} = (1, 0)$ is an identity and the point $(0, 1)$ is a point of order two [17]. With $\mathbf{z} = 1$, we need $5[\mathsf{M}] + 5[\mathsf{S}] + 1[\mathsf{C}_s]$ to compute $\mathtt{dbl}_k$ and $\mathtt{diffAdd}_k$ in total.

Table 2. Doubling and Differential Addition on Binary Kummer line

$(\mathbf{x}_3, \mathbf{z}_3) = \mathtt{dbl}_k(\mathbf{x}_1, \mathbf{z}_1) :$	$(\mathbf{x}_4, \mathbf{z}_4) = \mathtt{diffAdd}_k(\mathbf{x}_1, \mathbf{z}_1, \mathbf{x}_2, \mathbf{z}_2, \mathbf{x}, \mathbf{z}) :$
$\mathbf{x}_3 = c(\mathbf{x}_1^2 + \mathbf{z}_1^2)^2;$	$\mathbf{x}_4 = \mathbf{z}(\mathbf{x}_1\mathbf{x}_2 + \mathbf{z}_1\mathbf{z}_2)^2;$
$\mathbf{z}_3 = (\mathbf{x}_1\mathbf{z}_1)^2;$	$\mathbf{z}_4 = \mathbf{x}(\mathbf{x}_1\mathbf{z}_2 + \mathbf{x}_2\mathbf{z}_1)^2;$

The inverse mapping $\pi^{-1} : \mathsf{BKL}_{(1:c)} \to \mathsf{BEw}_{(c^4)}/\{\pm 1\}$ maps a point $\mathbf{P}$ of Kummer line $\mathsf{BKL}_{(1:c)}$ to elliptic curve $\mathsf{BEw}_{(c^4)}$ as given by Eq. (5) [11]:

$$\pi^{-1}(\mathbf{P} = (\mathbf{x} : \mathbf{z})) = \begin{cases} (c\mathbf{z} : \cdot : \mathbf{x}), & \text{if } \mathbf{x} \neq 0 \\ \mathcal{O}, & \text{if } \mathbf{x} = 0. \end{cases} \tag{5}$$

Notice that the points on $\mathsf{BEw}_{(c^4)}$ are in $\mathbb{P}^2$. Putting $x = \frac{c\mathbf{z}}{\mathbf{x}}$ in Eq. (3), we can compute the Y-coordinate up to elliptic involution.

But the mapping π alone does not conserve the consistency of the scalar multiplications between Kummer line $\mathsf{BKL}_{(1:c)}$ and the elliptic curve $\mathsf{BEw}_{(c^4)}$. [17] extends the mapping π to $\hat{\pi}$ with the help of the point of order two $T_2 = (0, c^2)$ on $\mathsf{BEw}_{(c^4)}$. Definitions of $\hat{\pi}$ and $\hat{\pi}^{-1}$ are given in Eq. (6).

$$\hat{\pi}(P) = \pi(P + T_2); \text{ and } \hat{\pi}^{-1}(\mathbf{P}) = \pi^{-1}(\mathbf{P}) + T_2. \tag{6}$$

2.3 Binary Generalized Hessian Curve

Hessian curve was introduced by Smart [26], and Joye and Quisquater [16] in 2001. Hessian curve was proposed to achieve side-channel resistant elliptic curve cryptography. Smart [26] also showed that it achieves significant speedups using parallelization. Farashahi and Joye developed the concept of Generalized Hessian Curves in 2010 [7].

Let $\mathbb{F}_{2^m}$ be a finite field, and let $\gamma, \delta \in \mathbb{F}_{2^m}$ such that $\gamma, \delta \neq 0$ and $\delta^3 \neq \gamma$. The projective form of Generalized Hessian Curve [7] is defined by

$$H_{(\gamma,\delta)} : R^3 + S^3 + \gamma T^3 = \delta RST \tag{7}$$

The identity element of $H_{(\gamma,\delta)}$ is $(1 : 1 : 0)$ and it is denoted by $\mathcal{O}_H$. If $P = (R : S : T)$ be a point on $H_{(\gamma,\delta)}$, then $-P = (S : R : T)$.

Let $P_1 = (R_1 : S_1 : T_1)$ and $P_2 = (R_2 : S_2 : T_2)$ be two points on $H_{(\gamma,\delta)}$. In $\mathfrak{RT}$ coordinate system, the points are $P_1 = (\mathfrak{R}_1 : \mathfrak{T}_1) = (R_1^3 + S_1^3 : T_1^3)$ and $P_2 = (\mathfrak{R}_2 : \mathfrak{T}_2) = (R_2^3 + S_2^3 : T_2^3)$. Let $P = P_2 - P_1 = (\mathfrak{R} : 1)$ is given to us. Note that, in $\mathfrak{RT}$ coordinates, $P_1 - P_2 = P_2 - P_1 = (R^3 + S^3 : 1) = (\mathfrak{R} : 1)$. Now we compute $P_3 = 2P_1 = (\mathfrak{R}_3 : \mathfrak{T}_3)$ and $P_4 = P_1 + P_2 = (\mathfrak{R}_4 : \mathfrak{T}_4)$ by $(P_3, P_4) = \mathsf{m\_dbl\_diffAdd}(P_1, P_2, P)$ given in Table 3 and needs $5[M] + 4[S] + 3[C]$.

Table 3. Mixed Doubling and Differential Addition (m_dbl_diffAdd) on $H_{(\gamma,\delta)}$

$A = \mathfrak{R}_1 * \mathfrak{T}_2, B = \mathfrak{R}_2 * \mathfrak{T}_1, C = A * B, D = \mathfrak{R}_1^2, E = \mathfrak{T}_1^2, F = D + \sqrt{\gamma^3(\delta^3 + \gamma)}E,$ $G = \delta^3 C, H = \delta^3 E, \mathfrak{R}_3 = F^2, \mathfrak{T}_3 = D * H, \mathfrak{T}_4 = (A+B)^2, \mathfrak{R}_4 = G + \mathfrak{R} * \mathfrak{T}_4.$

Hessian Curve of [16,26], denoted by H_δ, is a special form of the Generalized Hessian Curve $H_{(\gamma,\delta)}$ with $\gamma = 1$ and is defined as $H_{(\delta)} : R^3 + S^3 + T^3 = \delta RST$. Let $\mathbb{F}_{2^m}\left(\sqrt[3]{\gamma}\right)$ be the smallest extension field containing $\mathbb{F}_{2^m}$ and $\sqrt[3]{\gamma}$. Now we define the map

$$\varphi : H_{(\gamma,\delta)} \longrightarrow H_{(\delta/\sqrt[3]{\gamma})} \tag{8}$$
$$(R : S : T) \mapsto (R : S : \sqrt[3]{\gamma}T)$$

The mapping φ is an isomorphism from $H_{(\gamma,\delta)}$ to $H_{(\delta/\sqrt[3]{\gamma})}$ over $\mathbb{F}_{2^m}\left(\sqrt[3]{\gamma}\right)$. For $H_{(\delta)}$, the total cost for a mixed differential addition and doubling is $5[M] + 4[S] + 2[C]$ field operations. Now, by choosing small δ, one can make δ^3 small, but $\frac{1}{\delta^3}$ may not be a small constant, and the total cost becomes $5[M] + 4[S] + 1[C] + 1[C_s]$ field operations (Table 4).

3 Retrieving the R and S-Coordinates of nP

One of the objectives of this work is to have efficient scalar multiplication either via binary Kummer Line or via generlized Hessian curve. To have resistance against side-channel attacks like Timing attacks, Power attacks, Montgomery

Table 4. Mixed Doubling and Differential Addition (m_dbl_diffAdd) on $H_{(\delta)}$

$$A = \mathfrak{R}_1 * \mathfrak{T}_2, B = \mathfrak{R}_2 * \mathfrak{T}_1, C = A * B, D = \mathfrak{R}_1^2, E = \mathfrak{T}_1^2, F = D + E, G = \delta^3 C, H = \frac{1}{\delta^3}F,$$
$$\mathfrak{R}_3 = (E + H)^2, \mathfrak{T}_3 = D * E, \mathfrak{T}_4 = (A + B)^2, \mathfrak{R}_4 = G + \mathfrak{R} * \mathfrak{T}_4.$$

scalar multiplication is the most popular choice. Montgomery scalar multiplication is an x-coordinate-only scalar multiplication that enables us to achieve constant time scalar multiplication. Montgomery scalar multiplication is also used when resources are limited. In this section, we provide a brief description of Montgomery scalar multiplication. Montgomery ladder-based scalar multiplication achieves the best result in the **xz**-coordinate for binary Kummer line [17], XZ-coordinate for short Weierstrass [27] and $\mathfrak{R}\mathfrak{T}$ for generalized Hessian [7].

Let $P = (R : S : T)$ be a point on the generalized Hessian curve $H_{(\gamma,\delta)}$. Then $P = (\mathfrak{R}, \mathfrak{T})$ with $\mathfrak{R} = R^3 + S^3$ and $\mathfrak{T} = T^3$ in $\mathfrak{R}\mathfrak{T}$ coordinates. We can compute $2P = (\mathfrak{R}_3, \mathfrak{T}_3) = \mathrm{dbl}_H(P)$ as given in [7, Equation 18]. Let $n = \{1, n_{l-2}, \ldots, n_1, n_0\}_2$ be an l-bit long scalar. We can compute $nP = (\mathfrak{R}_n, \mathfrak{T}_n)$ by Montgomery scalar multiplication as given in Table 5. We need one m_dbl_diffAdd for each bit of the scalar. After the i-th iteration, we have $\mathfrak{R}$- and $\mathfrak{T}$-coordinates of $n_i P$ and $(n_i + 1)P$ where $n_i = \{1, n_{l-2}, \ldots, n_i\}_2$. Montgomery ladder scalar multiplication always preserves the invariance $P = (n_i + 1)P - n_i P$.

Table 5. Montgomery Ladder Scalar Multiplication

nP=scalarMult(P, n) :	ladderStep(P_1, P_2, n_i) :
1. Let $n = \{1, n_{l-2}, \ldots, n_0\}$;	1. If $n_i = 0$ then
2. $P_1 = P, R = \mathrm{dbl}_H(P)$;	2. $(P_1, P_2) = $ m_dbl_diffAdd(P_1, P_2, P);
3. For $i = l - 2$ to 0 do	3. Else If $n_i = 1$ then
4. laddarStep(P_1, P_2, n_i)	4. $(P_2, P_1) = $ m_dbl_diffAdd(P_2, P_1, P);
5. End For;	5. End If
6. Return P_1;	

At the end of the ladder step, we have $nP = (\mathfrak{R}_n : \mathfrak{T}_n)$ and $(n + 1)P = (\mathfrak{R}_{n+1} : \mathfrak{T}_{n+1})$ from the P and n in $\mathfrak{R}\mathfrak{T}$ coordinate system. Let $nP = (R_n : S_n : T_n)$ and $(n + 1)P = (R_{n+1} : S_{n+1} : T_{n+1})$, then $\mathfrak{R}_n = R_n^3 + S_n^3$, $\mathfrak{T}_n = T_n^3$, $\mathfrak{R}_{n+1} = R_{n+1}^3 + S_{n+1}^3$ and $\mathfrak{T}_{n+1} = T_{n+1}^3$. Our objective is to retrieve the R_n and S_n explicitly from the known quantities $R, S, T, \mathfrak{R}_n, \mathfrak{T}_n, \mathfrak{R}_{n+1}$ and $\mathfrak{T}_{n+1}$. We use the affine coordinates for simplicity, and our problem reduces to computation of (r_n, s_n) explicitly given $r = R/T$, $s = S/T$, $\mathfrak{r}_n = \mathfrak{R}_n/\mathfrak{T}_n$, and $\mathfrak{r}_{n+1} = \mathfrak{R}_{n+1}/\mathfrak{T}_{n+1}$.

Let the underlying field be $\mathbb{F}_{2^m}$ for some $m \in \mathbb{N}$. During retrieval of (r_n, s_n), we show that there will be two points (if m is odd) or six points (if m is even) if $\mathfrak{r}_n = \mathfrak{r} = r^3 + s^3$ on the curve which satisfy that $(n + 1)P = nP + P$ and $\mathfrak{r}_n = r_n^3 + s_n^3$. Therefore, we need $\mathfrak{r}_n$ *must not be the same as* $\mathfrak{r}$. Theorem 2 shows how we should choose the P and the scalar n to have $\mathfrak{r}_n \neq \mathfrak{r}$. The proof of Theorem 2 uses Lemmas 1 and 2, and Theorem 1. This also shows that we can always explicitly compute (r_n, s_n) for the generalized Hessian curves used in cryptography.

Lemma 1. Let $\mathsf{H}_{(\gamma,\delta)} : r^3 + s^3 + \gamma = \delta rs$ be a generalized Hessian curve in affine coordinate over $\mathbb{F}_{2^m}$. Then, there exists a point of (r, s) on $\mathsf{H}_{(\gamma,\delta)}$ satisfying $r^3 + s^3 = \mathfrak{r}$ if and only if

1. $\mathrm{Tr}\left(\frac{(\gamma+\mathfrak{r})^3}{\mathfrak{r}^2\delta^3}\right) = 0$, and
2. If m is even, then β has cube roots in $\mathbb{F}_{2^m}$, where

$$\beta = \mathfrak{r} \sum_{i=0}^{m-2} \left(\left(\frac{(\gamma+\mathfrak{r})^3}{\mathfrak{r}^2\delta^3}\right)^{2^i} \sum_{j=i+1}^{m-1} \mu^{2^j}\right) \text{ and } \mu \in \mathbb{F}_{2^m} \text{ with } \mathrm{Tr}(\mu) = 1.$$

Proof. Let $\mathrm{Tr}\left(\frac{(\gamma+\mathfrak{r})^3}{\mathfrak{r}^2\delta^3}\right) = 0$. It is equivalent to saying that there exists a non-zero $r' \in \mathbb{F}_{2^m}$ such that $r'^2 + \mathfrak{r}r' + \frac{(\gamma+\mathfrak{r})^3}{\delta^3} = 0$. Now we can rewrite the equation as $r' + \mathfrak{r} + \frac{(\gamma+\mathfrak{r})^3}{r'\delta^3} = 0 \iff r' + \frac{(\gamma+\mathfrak{r})^3}{r'\delta^3} = \mathfrak{r} \iff r' + s' = \mathfrak{r}$, where $s' = \frac{(\gamma+\mathfrak{r})^3}{r'\delta^3}$. If r' is a cube in $\mathbb{F}_{2^m}$ and let $r^3 = r'$, then we have $s' = \left(\frac{(\gamma+\mathfrak{r})^3}{r^3\delta^3}\right) = s^3$ with $s = \frac{\gamma+\mathfrak{r}}{r\delta}$. As a consequence, we get $\gamma + \delta rs = \mathfrak{r} = r' + s' = r^3 + s^3$, that is $r^3 + s^3 + \gamma = \delta rs$.

If m is odd, then roots of the equation $r'^2 + \mathfrak{r}r' + \left(\frac{(\gamma+\mathfrak{r})^3}{\delta^3}\right) = 0$ are $\mathfrak{r} \times$ $\mathsf{HTr}\left(\frac{(\gamma+\mathfrak{r})^3}{\mathfrak{r}^2\delta^3}\right)$ and $\left(\mathfrak{r} \times \mathsf{HTr}\left(\frac{(\gamma+\mathfrak{r})^3}{\mathfrak{r}^2\delta^3}\right) + \mathfrak{r}\right)$ [4]. Also, odd m ensures that every element of $\mathbb{F}_{2^m}$ has a unique cube root, and so do r' and s'. Therefore, (r, s) is a point on the curve $\mathsf{H}_{(\gamma,\delta)}$ such that $r^3 + s^3 = \mathfrak{r}$.

If m is even, then the roots of the equation $r'^2 + \mathfrak{r}r' + \left(\frac{(\gamma+\mathfrak{r})^3}{\delta^3}\right) = 0$ are β and $\beta + \mathfrak{r}$, where $\beta = \mathfrak{r}\sum_{i=0}^{m-2}\left(\left(\frac{(\gamma+\mathfrak{r})^3}{\mathfrak{r}^2\delta^3}\right)^{2^i}\sum_{j=i+1}^{m-1}\mu^{2^j}\right)$ and $\mu \in \mathbb{F}_{2^m}$ with $\mathrm{Tr}(\mu) = 1$ [4]. Let $r' = \beta$, and if β is a cube, then there exists a $r \in \mathbb{F}_{2^m}$ such that $r^3 = r' = \beta$. Therefore $s = \frac{\gamma+\mathfrak{r}}{r\delta} \in \mathbb{F}_{2^m}$. Now we have $s^3 = r^3 + \mathfrak{r} = \beta + \mathfrak{r}$. Therefore, $\beta + \mathfrak{r}$ also be a cube, and there is a point (r, s) on the curve $\mathsf{H}_{(\gamma,\delta)}$ with $r^3 + s^3 = \mathfrak{r}$.

The other direction is trivially true. $\square$

Lemma 2. Let $\mathsf{H}_{(\gamma,\delta)} : r^3 + s^3 + \gamma = \delta rs$ be a generalized Hessian curve over $\mathbb{F}_{2^m}$. Let there be points (r, s) on $\mathsf{H}_{(\gamma,\delta)}$ such that $r^3 + s^3 = \mathfrak{r}$. If m is odd, then there are two such points, else there are six such points.

Proof. 1. Let m be odd. If $r^3 + s^3 = \mathfrak{r}$, the equation $r'^2 + \mathfrak{r}r' + \left(\frac{(\gamma+\mathfrak{r})^3}{\delta^3}\right) = 0$ has roots in $\mathbb{F}_{2^m}$ by Lemma 1 as $\mathrm{Tr}\left(\frac{(\gamma+\mathfrak{r})^3}{\mathfrak{r}^2\delta^3}\right) = 0$. The roots are $\mathfrak{r}\lambda_1$ and $\mathfrak{r}\lambda_1 + \mathfrak{r}$ where $\lambda_1 = \mathsf{HTr}\left(\frac{(\gamma+\mathfrak{r})^3}{\mathfrak{r}^2\delta^3}\right)$. Again, each element of $\mathbb{F}_{2^m}$ has a unique root in $\mathbb{F}_{2^m}$ if m is odd. Therefore, there exists $r_1, r_2 \in \mathbb{F}_{2^m}$ such that $r_1^3 = \mathfrak{r}\lambda_1$ and $r_2^3 = \mathfrak{r}\lambda_1 + \mathfrak{r}$. For each $r_i, i = 1, 2$, there is a unique $s_i = \frac{\gamma+\mathfrak{r}}{r_i\delta}$. Then there are two points $(r_i, s_i), i = 1, 2$ on $\mathsf{H}_{(\gamma,\delta)}$ such that $r_i^3 + s_i^3 = \mathfrak{r}$.

2. If m is even, then the roots of the equation $r'^2 + \mathfrak{r}r' + \left(\frac{(\gamma+\mathfrak{r})^3}{\delta^3}\right) = 0$ are $\mathfrak{r}\lambda_2$

and $\mathfrak{r}\lambda_2 + \mathfrak{r}$ where $\lambda_2 = \sum_{i=0}^{m-2}\left(\left(\frac{(\gamma+\mathfrak{r})^3}{\mathfrak{r}^2\delta^3}\right)^{2^i}\sum_{j=i+1}^{m-1}\mu^{2^j}\right)$ as $\text{Tr}\left(\frac{(\gamma+\mathfrak{r})^3}{\mathfrak{r}^2\delta^3}\right) = 0$

from Lemma 1. Let the roots be $r_1^3 = \mathfrak{r}\lambda_2$ and $r_2^3 = \mathfrak{r}\lambda_2 + \mathfrak{r}$. If $r_1^3 = \mathfrak{r}\lambda_2$ has
a solution in $\mathbb{F}_{2^m}$, then it has three roots as $r_1, r_1\omega$ and $r_1\omega^2$ where ω is a
primitive cube root of unity in $\mathbb{F}_{2^m}$. For each such cube root, we have a point
$(r_1\omega^i, s_{1,i})$ with $s_{1,i} = \frac{\gamma+\mathfrak{r}}{r_1\omega^i\delta}$ for $i = 0, 1, 2$ on $H_{(\gamma,\delta)}$ such that $r_1^3\omega^{3i} + s_{1,i}^3 = \mathfrak{r}$.
In Lemma 1, we have shown that if there are solutions of $r_1^3 = \mathfrak{r}\lambda_2$, there
are solutions for $r_2^3 = \mathfrak{r}\lambda_2 + \mathfrak{r}$. Similarly, we get another three such points
$(r_2\omega^i, s_{2,i})$ where $r_2^3\omega^{3i} + s_{2,i}^3 = \mathfrak{r}$. Therefore, we have six points (r, s) on
$H_{(\gamma,\delta)}$ such that $r^3 + s^3 = \mathfrak{r}$. $\square$

Theorem 1. *Let* $H_{(\gamma,\delta)} : r^3 + s^3 + \gamma = \delta rs$ *be a generalized Hessian curve over*
$\mathbb{F}_{2^m}$. *Let* P *and* Q *be two points on the curve with* $r^3 + s^3 = \mathfrak{r}$. *Then* $Q - P$ *is a*
point of order 3, or the point $2Q$, *or* $2Q + P_3$, *where* P_3 *is a point of order three.*

Proof. We prove the theorem in two parts. First, we proof for $\mathbb{F}_{2^m}$ for some odd
m, and then for even m.

- **Let m be odd.** From Lemma 1, we have two points (r_i, s_i) with $r_i^3 = \mathfrak{r} \times$
  $\text{HTr}\left(\frac{(\gamma+\mathfrak{r})^3}{\mathfrak{r}^2\delta^3}\right) + \mathfrak{r} \times (i-1)$ and $s_i = \frac{\gamma+1}{r_i\delta}$ for $i = 1, 2$ such that $r_i^3 + s_i^3 = \mathfrak{r}$. Let
  $P = (r_1, s_1)$ and $Q = (r_2, s_2)$, then $r_1^3 r_2^3 = \left(\frac{\gamma+\mathfrak{r}}{\delta}\right)^3$, that is $r_1 r_2 = \frac{\gamma+\mathfrak{r}}{\delta}$. Using
  Equation (5) of [7], we get

$$P + Q = (r_1 : s_1 : 1) + (r_2 : s_2 : 1) = \left(r_1 : \frac{\gamma+\mathfrak{r}}{r_1\delta} : 1\right) + \left(r_2 : \frac{\gamma+\mathfrak{r}}{r_2\delta} : 1\right)$$

$$= ((\gamma+\mathfrak{r}) : r_2 r_1 \delta : 0) = ((\gamma+\mathfrak{r}) : (\gamma+\mathfrak{r}) : 0), \text{ where } r_1 r_2 = \frac{\gamma+\mathfrak{r}}{\delta}$$

$$= (1 : 1 : 0).$$

This implies that $P = -Q$. Therefore, $Q - P = 2Q$.

- **Let m be even.** Here $r_1^3 r_2^3 = \left(\frac{\gamma+\mathfrak{r}}{\delta}\right)^3$ and choose r_1, r_2 such that $r_1 r_2 = \frac{\gamma+\mathfrak{r}}{\delta}$.
  From Lemma 2, we have six points $(r_{i,j}, s_{i,j})$ with $r_{i,j} = r_i\omega^j$, $r_i^3 = \mathfrak{r} \times$

$$\sum_{k=0}^{m-2}\left(\left(\frac{(\gamma+\mathfrak{r})^3}{\mathfrak{r}^2\delta^3}\right)^{2^k}\sum_{j=k+1}^{m-1}\mu^{2^j}\right) + \mathfrak{r} \times (i-1) \text{ and } s_{i,j} = \frac{\gamma+\mathfrak{r}}{r_{i,j}\delta} \text{ for all } i = 1, 2 \text{ and}$$

$j = 0, 1, 2$ such that $r_{i,j}^3 + s_{i,j}^3 = \mathfrak{r}$. Let $P = (r_{i,j}, s_{i,j})$ and $Q = (r_{i',j'}, s_{i',j'})$
and $(i, j) \neq (i', j')$. Now we compute $Q - P$ using the projective addition
formulas given in [7] in three different situations.
  1. Let $i = i'$, $j \neq j'$, and $j' = j + j''$ with $|j''| \in \{1, 2\}$. Using Equation (10)
     of [7] and $1/\omega^2 = \omega^3/\omega^2 = \omega$, we get

$$Q - P = (r_{i,j'} : s_{i,j'} : 1) - (r_{i,j} : s_{i,j} : 1) = \left(r_i\omega^{j'} : \frac{\gamma+\mathfrak{r}}{r_i\omega^{j'}\delta} : 1\right) - \left(r_i\omega^j : \frac{\gamma+\mathfrak{r}}{r_i\omega^j\delta} : 1\right)$$

$$= \left(\gamma r_i\omega^j + \frac{\omega^j(\gamma+\mathfrak{r})^3}{r_i^2\delta^3} : \gamma r_i\omega^{j+j''} + \frac{\omega^{j+j''}(\gamma+\mathfrak{r})^3}{r_i^2\delta^3} : 0\right) = (1 : \omega^{j''} : 0),$$

which is a point of order 3.

2. Let $i \neq i'$ and $j + j' \equiv 0 \mod 3$. Now we add the points using the projective addition formula given in Equation (5) of [7], we get

$$P + Q = (r_{i,j} : s_{i,j} : 1) + (r_{i',j'} : s_{i',j'} : 1) = \left(r_{i,j} : \frac{\gamma + \mathfrak{r}}{r_{i,j}\delta} : 1\right) + \left(r_{i',j'} : \frac{\gamma + \mathfrak{r}}{r_{i',j'}\delta} : 1\right)$$

$$= ((\gamma + \mathfrak{r}) : (\gamma + \mathfrak{r}) : 0), \text{ applying } r_{i,j}r_{i',j'} = r_i r_{i'}\omega^{j+j'} = \frac{\gamma + \mathfrak{r}}{\delta}$$

$$= (1 : 1 : 0),$$

Hence, $P = -Q$ that implies $Q - P = 2Q$.

3. Let $i \neq i'$ and $j + j' \not\equiv 0 \mod 3$. Then

$$Q - P = (r_{i',j'}, s_{i',j'}) - (r_{i,j}, s_{i,j}) = (r_{i',j'}, s_{i',j'}) - ((r_{i,-j'} \mod 3, s_{i,-j'} \mod 3) - P_3),$$

where P_3 be a point of order 3 by case 1

$$= 2(r_{i',j'}, s_{i',j'}) + P_3 = 2Q + P_3, \text{ by case 2.}$$

□

Theorem 2. Let $P = (r, s)$ be a point of prime order $p \geq 5$ on a generalized Hessian Curve $H_{(\gamma,\delta)} : r^3 + s^3 + \gamma = \delta rs$ defined over $\mathbb{F}_{2^m}$, and n be a scalar multiple, with $2 \leq n \leq p-2$. Let $nP = (r_n, s_n)$ with $\mathfrak{r}_n = r_n^3 + s_n^3$ and $\mathfrak{r} = r^3 + s^3$. Then $\mathfrak{r}_n \neq \mathfrak{r}$.

Proof. Let $\langle P \rangle$ be the group generated by P. As the order of P is a prime p, every non-identity element of $\langle P \rangle$ has order p. Let $\mathfrak{r} = \mathfrak{r}_n$.

1. P is a point of prime order $p \geq 5$, then $nP - P = (n-1)P$ can not be a point of order 3. Therefore, case 1 of Theorem 1 can not occur.
2. Without loss of generality, assume that $2 \leq n \leq p-2$. Then $3 \leq (n+1) \leq p-1$. Now let $nP - P = 2nP$. Then $(n+1)P = \mathcal{O}_H$, where $3 \leq (n+1) \leq p-1$, which contradicts that the order of the point P is p. Hence, case 2 of Theorem 1 can not happen.
3. Let $nP - P = 2nP + P_3$ where P_3 be a point of order three. Then we have $(n+1)P = -P_3$. Therefore, $(n+1)P$ is also of order p. Therefore, p must be 3, which contradicts the hypothesis. □

3.1   Retrieve R and S Coordinates

Let $P = (r, s)$ be a point on a generalized Hessian Curve $H_{(\gamma,\delta)} : r^3 + s^3 + \gamma = \delta rs$ over a finite field $\mathbb{F}_{2^m}$ where $m \in \mathbb{N}$, and $n \geq 2$ be a scalar. Let $\text{ord}(P) = p \geq 5$ and $\mathfrak{r} = r^3 + s^3 \neq 0$. Using Montgomery scalar multiplication, we compute $\mathfrak{r}_n = r_n^3 + s_n^3$ and $\mathfrak{r}_{n+1} = r_{n+1}^3 + s_{n+1}^3$ in the $\mathfrak{R}\mathfrak{I}$ coordinate where $\mathfrak{r}_n = r_n^3 + s_n^3$ and $\mathfrak{r}_{n+1} = r_{n+1}^3 + s_{n+1}^3$. Notice that $(n+1)P = nP + P$. Applying the addition formula (Equation (3)) of [7] of the generalized Hessian curve, we have $r_{n+1} = \frac{(s^2 r_n + s_n^2 r)}{(rs + r_n s_n)}$ and $s_{n+1} = \frac{(r^2 s_n + r_n^2 s)}{(rs + r_n s_n)}$. Therefore, we can write $\mathfrak{r}_{n+1}^3 =$

$r_{n+1}^3 + s_{n+1}^3 = \frac{(s^2 r_n + s_n^2 r)^3}{(rs + r_n s_n)^3} + \frac{(r^2 s_n + r_n^2 s)^3}{(rs + r_n s_n)^3}$. After simplification, we apply $rs = \frac{\gamma + \tau}{\delta}$ and $r_n s_n = \frac{\gamma + \tau_n}{\delta}$, and we get

$$r_n^6 + \tau r_n^3 = \left(\tau_n^2 r^3 + r^6 \tau_n + \frac{1}{\delta^3}(\tau_n{}^2 + \gamma^2)(\tau + \gamma)\tau + \frac{1}{\delta^3}(\tau_n + \gamma)(\tau^2 + \gamma^2)\tau_n + \frac{\tau_{n+1}(\tau + \tau_n)^3}{\delta^3} \right) / \tau,$$

a quadratic equation in r_n^3. Let the roots be $r_{n,1}^3$ and $r_{n,2}^3$ with $r_{n,1}^3 + r_{n,2}^3 = \tau$.

1. **Let m be odd.** Then, using the relation $r_{n,i}^3 + s_{n,i}^3 = \tau_n$, we have two points $(r_{n,1}, s_{n,1}) = (r_{n,1}, (r_{n,1}^3 + \tau_n)^{1/3})$ and $(r_{n,2}, s_{n,2}) = ((r_{n,1}^3 + \tau)^{1/3}, (r_{n,1}^3 + \tau + \tau_n)^{1/3})$. Let both the points $(r_{n,i}, s_{n,i})$ be on the curve $\mathsf{H}_{(\gamma,\delta)}$. Then, we have

$$r_{n,1} s_{n,1} = r_{n,2} s_{n,2} = \frac{\gamma + \tau_n}{\delta}, \quad \text{from the equation of the curve } \mathsf{H}_{(\gamma,\delta)}$$

$$\implies r_{n,1}^3 s_{n,1}^3 = r_{n,2}^3 s_{n,2}^3 \implies r_{n,1}^3 (r_{n,1}^3 + \tau_n) = (r_{n,1}^3 + \tau)(r_{n,1}^3 + \tau + \tau_n) \implies \tau(\tau + \tau_n) = 0.$$

As $\tau \neq 0$, $\tau = \tau_n$. By Lemma 2, this contradicts the fact $\operatorname{ord}(P) \geq 5$, and it is a contradiction to the assumption that both the points $(r_{n,i}, s_{n,i})$ be on the curve $\mathsf{H}_{(\gamma,\delta)}$. Hence we can compute nP uniquely.

2. **Let m be even.** Using the relation $r_{n,i}^3 + s_{n,i}^3 = \tau_n$, we have six points $(r_{n,i,j}, s_{n,i,j})$ with $r_{n,i,j} = r_{n,i}\omega^j$ for all $i = 1, 2$ and $j = 0, 1, 2$ such that $r_{n,i,j}^3 + s_{n,i,j}^3 = \tau_n$. Let $(r_{n,1,j}, s_{n,1,j})$ and $(r_{n,2,j'}, s_{n,2,j'})$ be two points on $\mathsf{H}_{(\gamma,\delta)}$. Then, we get

$$r_{n,1,j} s_{n,1,j} = r_{n,2,j'} s_{n,2,j'} = \frac{\gamma + \tau_n}{\delta}, \quad \text{from the equation of the curve } \mathsf{H}_{(\gamma,\delta)}$$

$$\implies r_{n,1,j}^3 s_{n,1,j}^3 = r_{n,2,j'}^3 s_{n,2,j'}^3 \implies r_{n,1}^3 s_{n,1}^3 = r_{n,2}^3 s_{n,2}^3$$

$$\implies r_{n,1}^3 (r_{n,1}^3 + \tau_n) = (r_{n,1}^3 + \tau)(r_{n,1}^3 + \tau + \tau_n) \implies \tau(\tau + \tau_n) = 0.$$

As $\tau \neq 0$, $\tau = \tau_n$. By Lemma 2, this contradicts the fact $\operatorname{ord}(P) \geq 5$. Therefore, either $(r_{n,1,j}, s_{n,1,j})$ or $(r_{n,2,j'}, s_{n,2,j'})$ is a point on the curve $\mathsf{H}_{(\gamma,\delta)}$, but not both. Without loss of generality, let us assume that for $(r_{n,1,1}, s_{n,1,1})$ be the point on the curve, and we get three points $(r_{n,1,j}, s_{n,1,j})$, $j = 0, 1, 2$ on the curve $\mathsf{H}_{(\gamma,\delta)}$. By Proposition 1 of [7], these three points are nP, $nP + P_3$ and $nP + 2P_3$, where P_3 is a point of order 3. Among these three points, nP has order p, and the other two have order $3p$. If p is known to us, we can uniquely determine the nP by computing the order of any two.

In order to avoid the computation of the order of a point, we can choose the scalar of the form $3n$. Now we can compute nP, and take any point from the above-mentioned three points. Then we compute $3nP$ with the fixed addition chain $(1,1)$ at the cost of one standard addition and one standard doubling on a generalized Hessian curve. This way we can uniquely compute $3nP$ that is side-channel attack resistant.

For cryptographic purposes, we use $\mathbb{F}_{2^m}$ with m as prime [2,9,15,21]. Except for 2, all the prime numbers are odd, and we can compute the nP uniquely from P, $\tau(nP)$, and $\tau((n+1)P)$ following the algorithm as given in Table 6.

Table 6. Compute nP from P, $\mathfrak{r}(nP)$ and $\mathfrak{r}((n+1)P)$ on $\mathsf{H}_{(\gamma,\delta)}$ over $\mathbf{F}_{2^m}$ with odd m

Input: The point $P = (r,s)$, $\mathfrak{r}(nP)$ and $\mathfrak{r}((n+1)P)$
Output: nP
1.   Set $\mathfrak{r} = r^3 + s^3$.
2.   If $\mathfrak{r} = 0$, Return $\bot$, else Continue;
3.   Parse the input as $\mathfrak{r}_1 = \mathfrak{r}(P), \mathfrak{r}_2 = \mathfrak{r}(nP)$, $\mathfrak{r}_3 = \mathfrak{r}((n+1)P)$
4.   For $i = 1, 2$, Compute $\mathfrak{a}_i = \frac{\mathfrak{r}_i + \gamma}{\delta}$.
5.   Compute $\mathfrak{c} = \left(\mathfrak{r}_2^2 r^3 + \mathfrak{r}_2 r^6 + \mathfrak{a}_1^2 \mathfrak{a}_2 \mathfrak{r}_2 + \mathfrak{a}_1 \mathfrak{a}_2^2 \mathfrak{r}_1 + \frac{\mathfrak{r}_3 (\mathfrak{r}_1 + \mathfrak{r}_2)^3}{\delta^3} \right) / \mathfrak{r}_1^3$
6.   Compute $\mathfrak{h} = \mathfrak{r}_1 \times \text{HalfTrace}(\mathfrak{c})$.
7.   If $\mathfrak{h}(\mathfrak{h} + \mathfrak{r}_2) = \left(\frac{\gamma + \mathfrak{r}_2}{\delta} \right)^3$ (As the cube roots here are unique)
8.   Return $(r_n, s_n) = (\mathfrak{h}^{1/3}, (\mathfrak{h} + \mathfrak{r}_2)^{1/3})$
9.   Else
10.   Return $(r_n, s_n) = ((\mathfrak{h} + \mathfrak{r}_1)^{1/3}, (\mathfrak{h} + \mathfrak{r}_1 + \mathfrak{r}_2)^{1/3})$

4   Moving Between Weierstrass Curve and Generalized Hessian Curve

A short Weierstrass curve $\mathsf{BEw}_{(a,b)}$ has an isomorphic Hessian curve if there exists a point of order three [7,26]. We first move $\mathsf{BEw}_{(a,b)}$ to the triangular form of the Weierstrass curve, and then from the triangular form, the generalized Hessian has been derived. The existence of a triangular form relies on the existence of a point of order three on $\mathsf{BEw}_{(a,b)}$. In this work, our objective is to connect a binary Kummer line to a generalized Hessian curve, thus we will be using the short Weierstrass curve of the form $\mathsf{BEw}_{(b)}$. First, we discuss the existence of a point of order three on short Weierstrass curve $\mathsf{BEw}_{(b)}$. Then we discuss the isomorphic connection of the short Weierstrass curve to the generalized Hessian curve via Triangular form. As the second method, we show that we can connect $\mathsf{BEw}_{(b)}$ to a generalized Hessian curve using two isogenies. This guarantees an isogenous generalized Hessian curve just by choosing a binary Kummer line with a small coefficient c where the associated Weierstrass curve $\mathsf{BEw}_{(b)}$ has a point of order three and the Hessian curve uses the same c in the arithmetic of Montgomery ladder step.

4.1   Moving Between Weierstrass Curve and Triangular Form

Let us start with the definition of the triangular form of the Weierstrass curve.

Definition 1.   *[7] Let $a_3 \in \mathbb{F}_{2^m}$ and $a_3^3(a_3 + 1) \neq 0$. The projective form of the triangular Weierstrass curve over $\mathbb{F}_{2^m}$ is defined by*

$$\mathsf{BEwT}_{(a_3)} : \mathsf{Y}^2 \mathsf{Z} + \mathsf{XYZ} + a_3 \mathsf{YZ}^2 = \mathsf{X}^3. \tag{9}$$

Lemma 3.   *Let $\mathsf{BEw}_{(b)}$ be defined over $\mathbb{F}_{2^m}$ and has a point of order 3 if and only if $x^4 + x^3 + b = 0$ has a root in $\mathbb{F}_{2^m}$, namely x_3, with $\text{Tr}(x_3) = 0$.*

Proof. Let x_3 be a root of $x^4 + x^3 + b = 0$ with $\text{Tr}(x_3) = 0$. There exists a point $P = (x_3, y_3)$ on $\text{BEw}_{(b)}$ over $\mathbb{F}_{2^m}$ if y_3 is a root of the equation $y^2 + x_3 y + (x_3^3 + b) = 0$. y_3 exists if and only if $\text{Tr}\left((x_3^3 + b)/x_3^2\right) = 0$. Using $x_3^4 + x_3^3 + b = 0$ and the characteristic of the underlying field is two, we have $\text{Tr}\left((x_3^3 + b)/x_3^2\right) = \text{Tr}\left(x_3^4/x_3^2\right) = \text{Tr}\left(x_3^2\right) = \text{Tr}(x_3) = 0$. Therefore, $P = (x_3, y_3)$ is a point on $\text{BEw}_{(b)}$ over $\mathbb{F}_{2^m}$. Now, we compute $2P = (x_2, y_2) = \left(x_3^2 + b/x_3^2, x_3^2 + (\eta + 1)x_2\right)$, where $\eta = x_3 + y_3/x_3$. Simplifying x_2 and y_2, we get $x_2 = x_3^2 + b/x_3^2 = (x_3^4 + b)/x_3^2 = x_3^3/x_3^2 = x_3$, and $y_2 = x_3^2 + (\eta + 1)x_2 = x_3^2 + (x_3 + y_3/x_3 + 1)x_3 = x_3 + y_3$. Therefore, we have $2P = (x_3, x_3 + y_3) = -P$, that is $3P = \mathcal{O}$.

Conversely, let $P = (x_3, y_3)$ be a point of order 3 on $\text{BEw}_{(b)}$ over $\mathbb{F}_{2^m}$. Therefore, we have $3P = \mathcal{O}$, that is $2P = -P$. Equating the x-coordinates, we get $x_3^2 + b/x_3^2 = x_3$, that is, x_3 is a root of the equation $x^4 + x^3 + b = 0$. Also, y_3 is a root of the quadratic equation $y^2 + x_3 y + (x_3^3 + b) = 0$, that implies $\text{Tr}(x_3) = 0$. $\square$

Remark 1. One can also find the cardinality of the elliptic curve $\text{BEw}_{(a,b)}$, and if it is divisible by 3, then there is a point of order three by Cauchy's theorem. But this does not give the point explicitly, our approach does.

Lemma 4. *Let* $\text{BEw}_{(b)} : Y^2 Z + XYZ = X^3 + bZ^3$ *be a short Weierstrass elliptic curve defined over* $\mathbb{F}_{2^m}$ *with* $b \neq 0$ *containing a point of order 3, say* $P = (x_3, y_3)$. *Then* $\text{BEw}_{(b)}$ *is isomorphic to the triangular Weierstrass elliptic curve* $\text{BEwT}_{(x_3)} : Y^2 Z + XYZ + x_3 YZ^2 = X^3$ *such that* $x_3(1 + x_3) \neq 0$ *and* P *maps to the point* $(0 : 0 : 1)$.

Proof. Consider the following map $\varrho : \text{BEw}_{(b)} \rightarrow \text{BEwT}_{(x_3)}$ [26]:

$$\varrho(X : Y : Z) \mapsto (\mathsf{X} : \mathsf{Y} : \mathsf{Z}) = (X + x_3 Z : Y + \eta X + x_3^2 Z : Z), \text{ where } \eta = \frac{x_3^2 + y_3}{x_3}. \tag{10}$$

The map ϱ is of the form $(u^2 X + rZ : u^3 Y + u^2 sX + tZ : Z)$ with $u = 1, r = x_3, s = \eta$ and $t = x_3^2$. Therefore, this map forms an isomorphism [25, P. 44]. $\square$

4.2 Moving Between $\text{BEwT}_{(a_3)}$ to $\text{H}_{(\gamma, \delta)}$

[7, Theorem 1] provides similar maps given in Theorem 3. But the map of [7, Theorem 1] that corresponds χ_2 of Theorem 3 does not preserve the identity and thus is not an isomorphism. The next theorem follows the map given in [30] and provides the corrected version that fixes the identity and forms an isomorphism.

Theorem 3. *Let* $\text{BEwT}_{(a_3)} : Y^2 Z + XYZ + a_3 YZ^2 = X^3$ *be a Weierstrass elliptic curve of triangular form with* $a_3 \in \mathbb{F}_{2^m}$ *and* $a_3^3(a_3 + 1) \neq 0$. *Then there exists a Generalized Hessian Curve* $\text{H}_{(\gamma, \delta)} : R^3 + S^3 + \gamma T^3 = \delta RST$ *that is isomorphic to* $\text{BEwT}_{(a_3)}$.

Proof. We prove it in two parts based on whether m is odd or even.

1. **Let m be odd.** Every element of $\mathbb{F}_{2^m}$ has a unique cube root, and thus the $1 + a_3$. Let ν be the cube root of $1 + a_3$. As char $= 2 \neq 3$, $\chi_1 : \mathsf{BEwT}_{(a_3)} \rightarrow \mathsf{H}_{(\delta)} (= \mathsf{H}_{(1,\delta)})$ forms an isomorphism where $\delta = 1/(1 + \nu)$ by the Remark 1 of [7, Theorem 1]. χ_1 is as given below.

$$\left. \begin{array}{l} \chi_1(\mathsf{X} : \mathsf{Y} : \mathsf{Z}) = (R : S : T) \\ R = \nu\mathsf{X} + \mathsf{Y} + a_3\mathsf{Z} \\ S = (1 + \nu)\mathsf{X} + \mathsf{Y} \\ T = (1 + \nu)\mathsf{X} + a_3\mathsf{Z} \end{array} \right\} \tag{11}$$

2. **Let m be even.** Now only one-third of the elements of $\mathbb{F}_{2^m}$ have a cube root, and $\mathsf{BEwT}_{(a_3)}$ is isomorphic to $\mathsf{H}_{(\delta)}$ by χ_1 whenever $a_3 + 1$ has a cube root. The map $\chi_2 : \mathsf{BEwT}_{(a_3)} \rightarrow \mathsf{H}_{(1+a_3,1)}$ forms the isomorphism *when $a_3 + 1$ is not a cube.*

$$\left. \begin{array}{l} \chi_2(\mathsf{X} : \mathsf{Y} : \mathsf{Z}) = (R : S : T) \\ R = \omega^2\mathsf{X} + \mathsf{Y} + a_3\omega\mathsf{Z} \\ S = \omega\mathsf{X} + \mathsf{Y} + a_3\omega^2\mathsf{Z} \\ T = \mathsf{X}, \end{array} \right\} \tag{12}$$

where ω is a primitive cube root of 1. Even m ensures the existence of ω [6]. To resist GHS [10] and JV [15] attacks, finite fields with prime extensions are used for cryptographic purposes. We are interested in the map χ_1 because all primes except 2 are odd.   □

4.3   Moving Between $\mathsf{BEw}_{(b)}$ and $\mathsf{H}_{(\gamma,\delta)}$ via Isomorphism

In this section, we now combine all the individual mappings short Weierstrass to and from triangular Weierstrass, and triangular Weierstrass to and from generalized Hessian. As a result, we achieve the necessary relations among short Weierstrass Curve $\mathsf{BEw}_{(b)}$ and generalized Hessian curve $\mathsf{H}_{(\gamma,\delta)}$.

Theorem 4. $\mathsf{BEw}_{(b)} : Y^2Z + XYZ = X^3 + bZ^3$ *be a short Weierstrass curve such that $x^4 + x^3 + b = 0$ has a root in $\mathbb{F}_{2^m}$, namely x_3, with $\mathrm{Tr}(x_3) = 0$. Let (x_3, y_3) be a point of order 3 of* $\mathsf{BEw}_{(b)}$.

1. *If $1 + x_3$ is a cube in $\mathbb{F}_{2^m}$ with the cube root ν, then $\mathsf{BEw}_{(b)}$ is isomorphic to a Hessian curve $\mathsf{H}_{(\delta=1/(1+\nu))} : R^3 + S^3 + T^3 = \delta RST$ by the mapping*

$$\begin{array}{l} \vartheta_1(X : Y : Z) = (R : S : T) \\ R = (\nu x_3 + x_3^2 + y_3)X + x_3 Y + (\nu x_3^2 + \nu^3 x_3^2)Z \\ S = (x_3 + y_3 + \nu x_3 + x_3^2)X + x_3 Y + (\nu x_3^2 + \nu^3 x_3^2)Z \\ T = x_3(1 + \nu)X + \nu x_3^2 Z. \end{array}$$

2. *If $1 + x_3$ does not have a cube root in $\mathbb{F}_{2^m}$, then $\mathsf{BEw}_{(b)}$ is isomorphic to a generalized Hessian curve $\mathsf{H}_{(\gamma=1+x_3,1)} : R^3 + S^3 + \gamma T^3 = RST$ by the mapping*

$$\vartheta_2(X : Y : Z) = (R : S : T)$$
$$R = (\omega^2 x_3 + x_3^2 + y_3)X + x_3 Y + (x_3^3 + x_3^2)Z$$
$$S = (\omega x_3 + x_3^2 + y_3)X + x_3 Y + (x_3^3 + x_3^2)Z$$
$$T = x_3 X + x_3^2 Z,$$

where ω is a primitive cube root of 1 in $\mathbb{F}_{2^m}$.

Proof. Let $\mathsf{BEw}_{(b)} : Y^2 Z + XYZ = X^3 + bZ^3$ be a short Weierstrass curve. As $x_3 \in \mathbb{F}_{2^m}$ is a root of $x^4 + x^3 + b = 0$ and $\mathrm{Tr}(x_3) = 0$, $\mathsf{BEw}_{(b)}$ has a point of order 3 of the form $(x_3, \cdot)$ by Theorem(3). Let $P = (x_3, y_3)$ be a point order 3 that satisfies $\mathsf{BEw}_{(b)}$. Therefore, $\mathsf{BEw}_{(b)}$ is isomorphic to the triangular form of the Weierstrass elliptic curve $\mathsf{BEwT}_{(x_3)} : \mathsf{Y}^2\mathsf{Z} + \mathsf{XYZ} + x_3\mathsf{YZ}^2 = \mathsf{X}^3$ by mapping ϱ (Eq. (10)) of Lemma 4.

1. If $1 + x_3$ is a cube in $\mathbb{F}_{2^m}$, then $\mathsf{BEwT}_{(x_3)}$ is isomorphic to a Hessian curve $\mathsf{H}_{(\delta=1/(1+\nu))}$ by the mapping χ_1 from Theorem 3, where ν is a cube root of $1 + x_3$. Composition of ϱ and χ_1 gives us $\vartheta_1 = \chi_1 \circ \varrho$, and makes the $\mathsf{BEw}_{(b)}$ isomorphic to a Hessian curve $\mathsf{H}_{(\delta=1/(1+\nu))}$.
2. $1 + x_3 \in \mathbb{F}_{2^m}$ and $1 + x_3$ is not a cube only if the extension degree m of $\mathbb{F}_{2^m}$ is even. Then $\mathsf{BEwT}_{(x_3)}$ is isomorphic to a generalized Hessian curve $\mathsf{H}_{(\gamma=1+x_3,1)}$ by the mapping χ_2 from Theorem 3. Similarly composing ϱ and χ_2, we get $\vartheta_2 = \chi_2 \circ \varrho$, and ϑ_2 makes $\mathsf{BEw}_{(b)}$ isomorphic to a generalized Hessian curve $\mathsf{H}_{(\gamma=1+x_3,1)}$. $\qquad\square$

4.4   Moving Between $\mathsf{BEw}_{(b)}$ and $\mathsf{H}_{(\gamma,\delta)}$ via Isogeny

To move between $\mathsf{BEw}_{(b)}$ and $\mathsf{H}_{(\gamma,\delta)}$, we define two isogenies with their duals: (i) ϕ is a 2-isogeny between $\mathsf{BEw}_{(b)}$ and $\mathsf{BEw}_{(\sqrt{b})}$, and (ii) ξ is a 3-isogeny between $\mathsf{H}_{(a^3,1)}$ and $\mathsf{H}_{(A,1)}$ with $A = a + a^2 + a^3$.

Theorem 5. *Let $\mathsf{BEw}_{(b)} : Y^2 Z + XYZ = X^3 + bZ^3$ be a short Weierstrass curve defined on $\mathbb{F}_{2^m}$ where $b \in \mathbb{F}_{2^m}$ with $b \neq 0$. Then there exists an 2-isogeny $\phi : \mathsf{BEw}_{(b)} \mapsto \mathsf{BEw}_{(\sqrt{b})}$ defined by*

$$\phi(X : Y : Z) = (X^3 + \sqrt{b}XZ^2 : X^2Y + \sqrt{b}X^2Z + \sqrt{b}XZ^2 + \sqrt{b}YZ^2 + bZ^3 : X^2Z),$$

and the dual is

$$\hat{\phi}(\hat{X} : \hat{Y} : \hat{Z}) = (\hat{X}^2 : \hat{Y}^2 : \hat{Z}^2)$$

Proof. For simplicity, we use the affine form of $\mathsf{BEw}_{(b)} : y^2 + xy = x^3 + b$. As every element of $\mathbb{F}_{2^m}$ has a unique square root, $(0, \sqrt{b})$ is the point of order 2 on $\mathsf{BEw}_{(b)}$. Let G be the subgroup of order two generated by $(0, \sqrt{b})$. Using Velu's Formula [29] and Theorem 25.1.6 of [8], We construct an isogeny $\phi_1 : \mathsf{BEw}_{(b)} \to E'$, where $E' : y'^2 + x'y' = x'^3 + \sqrt{b}x' + \sqrt{b} + b$ as

$$\phi_1(x, y) = \left(\frac{x^3 + \sqrt{b}x}{x^2}, \frac{x^2 y + \sqrt{b}(x^2 + x + y)}{x^2} \right).$$

Secondly, we define the isomorphism $\phi_2 : E' \rightarrow \mathsf{BEw}_{(\sqrt{b})}$ as $(x', y') \mapsto (\hat{x} = x', \hat{y} = y' + \sqrt{b})$. The composition of ϕ_1 and ϕ_2 gives us the isogeny ϕ from $\mathsf{BEw}_{(b)}$ to $\mathsf{BEw}_{(\sqrt{b})}$ as

$$\phi(x, y) = (\phi_2 \circ \phi_1)(x, y) = \left(\frac{x^3 + \sqrt{b}x}{x^2}, \frac{x^2 y + \sqrt{b}(x^2 + x + y + \sqrt{b})}{x^2} \right). \quad (13)$$

The dual isogeny $\hat{\phi} : \mathsf{BEw}_{(\sqrt{b})} \rightarrow \mathsf{BEw}_{(b)}$ is defined by $\hat{\phi}(\hat{x}, \hat{y}) = (\hat{x}^2, \hat{y}^2)$, where

$$(\hat{\phi} \circ \phi)(P = (x, y)) = \left(x^2 + \frac{b}{x^2}, y^2 + b + \frac{b}{x^2} + \frac{by^2}{x^4} + \frac{b^2}{x^4} \right)$$

$$= \left(x^2 + \frac{b}{x^2}, x^3 + xy + \frac{b}{x^2} + \frac{yb}{x^3} + \frac{b}{x} \right)$$

$$= \left(x^2 + \frac{b}{x^2}, x^2 + \left(x + \frac{y}{x} \right) \left(x^2 + \frac{b}{x^2} \right) + \left(x^2 + \frac{b}{x^2} \right) \right) = 2P$$

Finally, the projective version of ϕ and $\hat{\phi}$ can be written as $\phi(X : Y : Z) = (X^3 + \sqrt{b}XZ^2 : X^2 Y + \sqrt{b}X^2 Z + \sqrt{b}XZ^2 + \sqrt{b}YZ^2 + bZ^3 : X^2 Z)$ and the dual $\hat{\phi}(\hat{X} : \hat{Y} : \hat{Z}) = (\hat{X}^2 : \hat{Y}^2 : \hat{Z}^2)$. □

The form of the isogeny given by ξ in Theorem 6 appears in [30] without any proof, and also the dual is absent. Without the dual, we can not move in both directions. Theorem 6 completes the isogeny by giving the dual and the required proofs of isogeny and the dual.

Theorem 6. *Let* $\mathsf{H}_{(a^3, 1)} : R^3 + S^3 + a^3 T^3 = RST$ *be a generalized Hessian curve defined over* $\mathbb{F}_{2^m}$ *with* $a \neq 0$. *Then there exists a 3-isogeny* $\xi : \mathsf{H}_{(a^3, 1)} \rightarrow \mathsf{H}_{(A, 1)}$ *where* $A = a + a^2 + a^3$ *defined by*

$$\xi(R : S : T) = (a^2 T^2 S + aR^2 T + S^2 R : a^2 T^2 R + aS^2 T + R^2 S : RST),$$

and the dual is

$$\hat{\xi}(\hat{R} : \hat{S} : \hat{T}) = (a\hat{R}\hat{S}\hat{T} + \hat{S}^3 : (1 + a)\hat{R}\hat{S}\hat{T} + \hat{S}^3 + A\hat{T}^3 : \hat{R}\hat{S}\hat{T} + (1 + a + a^2)\hat{T}^3).$$

Proof. From the rational map $\xi : \mathsf{H}_{(a^3, 1)} \rightarrow \mathsf{H}_{(A, 1)}$, we have

$$\hat{R}^3 + \hat{S}^3 + A\hat{T}^3 + \hat{R}\hat{S}\hat{T} = (aR^2 S^2 T^2 + R^3 S^3 + a^3 R^3 T^3 + a^3 S^3 T^3)(R^3 + S^3 + a^3 T^3 + RST).$$

Therefore, the range of ξ is indeed $\mathsf{H}_{(A, 1)}$. Let $P = (R : S : T)$ and $\xi(P) = (1 : 1 : 0)$. Then we have $RST = 0$, and

1. If $R = 0$, then $a^2 T^2 S = aS^2 T$, i.e. $aT = S$, and $P = (0, a, 1)$.
2. If $S = 0$, then $aR^2 T = a^2 RT^2$, i.e. $R = aT$, and $P = (a, 0, 1)$.
3. If $T = 0$, then $RS^2 = R^2 S$, i.e. $R = S$, and $P = (1, 1, 0)$.

Therefore, the kernel of ξ is $\ker(\xi) = \{(1:1:0),(a:0:1),(0:a:1)\}$, and makes ξ a 3-isogeny.

Now consider the rational map $\hat{\xi} : H_{(A,1)} \to H_{(a^3,1)}$ is defined by

$$\hat{\xi}(\hat{R}:\hat{S}:\hat{T}) = (a\hat{R}\hat{S}\hat{T} + \hat{S}^3 : (1+a)\hat{R}\hat{S}\hat{T} + \hat{S}^3 + A\hat{T}^3 : \hat{R}\hat{S}\hat{T} + (1+a+a^2)\hat{T}^3)$$

The $\hat{\xi}$ fixes the identity, and so $\hat{\xi}$ is an isogeny. Let $P = (R:S:T)$ be a point on $H_{(a^3,1)}$. Let $(\hat{\xi} \circ \xi)(R:S:T) = (R_4:S_4:T_4)$ and $\Delta = R^3 + S^3 + a^3 T^3 + RST$.

1. If $T \neq 0$, then we compute doubling and addition by Equations (6) and (5) of [7]. Let $3P = (R_3:S_3:T_3)$. Now

$$R_3 = T(R_4)$$
$$S_3 = T(S_4 + \Delta(a^3 R^3 T^3 + a^3 S^3 T^3 + R^3 S^3))$$
$$T_3 = T(T_4 + \Delta((1+a)R^2 S^2 T^2 + a^3 RST^4 + R^4 ST + RS^4 T))$$

Therefore, $(\hat{\xi} \circ \xi)(R:S:T) = (R_4:S_4:T_4) = (R_3:S_3:T_3) = 3(R:S:T)$.

2. If $T = 0$, then addition by Equation (5) of [7] fails and produces $(0:0:0)$. In this case, we do addition by Equation (9) of [7]. Notice that, if $T = 0$ both R and S must be non-zero. Let $3P = (R_3':S_3':T_3')$ and we have

$$R_3' = S(R_4)$$
$$S_3' = S(S_4 + \Delta(a^3 R^3 T^3 + a^3 S^3 T^3 + R^3 S^3))$$
$$T_3' = S(T_4 + \Delta((1+a)R^2 S^2 T^2 + a^3 RST^4 + R^4 ST + RS^4 T))$$

Therefore, $(\hat{\xi} \circ \xi)(R:S:T) = (R_4:S_4:T_4) = (R_3':S_3':T_3') = 3(R:S:T)$.

This shows that $\hat{\xi}$ is the dual of ξ. $\qquad\square$

Theorem 7. *Let* $\mathsf{BEw}_{(c^4)} : Y^2 Z + XYZ = X^3 + c^4 Z^3$ *be a short Weierstrass curve on* $\mathbb{F}_{2^m}$ *where* m *is odd and* $c(\neq 0) \in \mathbb{F}_{2^m}$, *and* $\mathsf{BEw}_{(c^4)}$ *has a point of order 3. Then there exists a generalized Hessian curve* $H_{(\gamma,1)}$ *that is isogenous to* $\mathsf{BEw}_{(c^4)}$ *such that* $\sqrt{\gamma^3(\gamma + 1)} = c$.

Proof. By Theorem 5, there exists an isogeny $\phi : \mathsf{BEw}_{(c^4)} \to \mathsf{BEw}_{(c^2)}$. $\mathsf{BEw}_{(c^4)}$ and $\mathsf{BEw}_{(c^2)}$ have the same cardinality by Theorem 1 of [28]. Therefore, $\mathsf{BEw}_{(c^2)}$ has a point of order 3, and let the point be (x_3, y_3). Then by Lemma 3, $x_3^3 + x_3^3 = c^2$. $\mathsf{BEw}_{(c^2)}$ is isomorphic to triangular form $\mathsf{BEwT}_{(x_3)}$ by the mapping ϱ of Lemma 4. The map χ_1 of Theorem 3 makes $\mathsf{BEwT}_{(x_3)}$ isomorphic to the generalized Hessian curve $H_{(1,\delta)}$, where $\delta = \frac{1}{1+\nu}$ and $\nu^3 = 1 + x_3$, for odd m. Again, $H_{(1,\delta)}$ is isomorphic to $H_{(\frac{1}{\delta^3},1)} = H_{((1+\nu)^3,1)}$ through the map φ (Eq. (8)). By Theorem 6, there exists an isogeny $\xi : H_{((1+\nu)^3,1)} \to H_{(\gamma,1)}$, where $\gamma = (1+\nu)+(1+\nu)^2+(1+\nu)^3 = 1+\nu^3 = x_3$. Therefore, the composition $(\xi \circ \varphi \circ \chi_1 \circ \varrho \circ \phi)$ as given in Fig. 1, we get that $H_{(x_3,1)}$ is isogenous to $\mathsf{BEw}_{(c^4)}$ with $\sqrt{\gamma^3(\gamma + 1)} = \sqrt{x_3^4 + x_3^3} = \sqrt{c^2} = c$. $\qquad\square$

Figure 1 shows the diagrammatic view of the compositions of the maps and the connectivity of different curves.

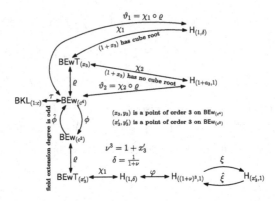

Fig. 1. All the paths of connection between Kummer and Hessian over binary field

4.5   Optimized Arithmetic on $H_{(\gamma,1)}$

The $H_{(\gamma,1)}$ obtained by Theorem 7 requires the smallest number of field operations. Let the $\mathfrak{R}\mathfrak{T}$-coordinates of $P, Q,$ and $P-Q$ on $H_{(\gamma,1)}$ be $(\mathfrak{R}_1, \mathfrak{T}_1), (\mathfrak{R}_2, \mathfrak{T}_2),$ and $(\mathfrak{R}, 1),$ respectively. $\mathfrak{R}\mathfrak{T}$-coordinate of $2P$ and $P+Q$ are $(\mathfrak{R}_3, \mathfrak{T}_3)$ and $(\mathfrak{R}_4, \mathfrak{T}_4)$ respectively. Applying $\delta = 1$ and $\sqrt{\gamma(\gamma^3+1)} = c$ in the arithmetic given in Table 3, we obtain new optimized arithmetic given in Table 7 that needs only $5[M] + 4[S] + 1[C_s]$ operations in total.

5   Concrete Proposal of Curves

In the previous sections, we have established the relations among the Binary Kummer Lines and Hessian Curves. In this section, we apply all the results derived till now to propose the appropriate pair of secure and practical curves which target 128-bit security over three binary fields: $\mathbb{F}_{2^{251}} = \mathbb{F}_2[t]/(t^{251} + t^7 + t^4 + t^2 + 1)$, $\mathbb{F}_{2^{257}} = \mathbb{F}_2[t]/(t^{257} + t^{12} + 1)$, and $\mathbb{F}_{2^{263}} = \mathbb{F}_2[t]/(t^{263} + t^9 + t^6 + t^5 + t^4 + t^3 + t^2 + t + 1)$. We choose these three particular fields as their extension degrees are all primes. Therefore the proposed curves are resistant to GHS attack [9] and JV attack [15].

Table 7. Mixed Doubling and Differential Addition on Binary $H_{(\gamma,1)}$

$A = \mathfrak{R}_1 * \mathfrak{T}_2, B = \mathfrak{R}_2 * \mathfrak{T}_1, C = A * B, D = \mathfrak{R}_1^2, E = \mathfrak{T}_1^2, F = D + \sqrt{\gamma^3(1+\gamma)}E,$
$\mathfrak{R}_3 = F^2, \mathfrak{T}_3 = D * E, \mathfrak{T}_4 = (A+B)^2, \mathfrak{R}_4 = C + \mathfrak{R} * \mathfrak{T}_4$

Table 8. Pairs of Binary Kummer Lines and Binary Generalized Hessian Curves

Field	$\mathsf{BKL}_{(1:c)}$	$\mathsf{H}_{(1,\delta)}$	$(\log_2(p_1), \log_2(p_2))$	(h, h_T)	Bit Security
$\mathbb{F}_{2^{251}}$	$\mathsf{BKL}(\mathbf{251, c_{1,1}})$	$\mathsf{H}(251, 1, \delta_1)$	$(247.4, 248.4)$		123.7
	$\mathsf{BKL}(251, c_{1,2})$	$\mathsf{H}(\mathbf{251, 1, \delta_2})$			
$\mathbb{F}_{2^{257}}$	$\mathsf{BKL}(\mathbf{257, c_{1,3}})$	$\mathsf{H}(257, 1, \delta_3)$	$(253.4, 254.4)$	$(12, 6)$	126.7
	$\mathsf{BKL}(257, c_{1,4})$	$\mathsf{H}(\mathbf{257, 1, \delta_4})$			
$\mathbb{F}_{2^{263}}$	$\mathsf{BKL}(\mathbf{263, c_{1,5}})$	$\mathsf{H}(263, 1, \delta_5)$	$(259.4, 260.4)$		129.7
	$\mathsf{BKL}(263, c_{1,6})$	$\mathsf{H}(\mathbf{263, 1, \delta_6})$			

$\mathbf{c_{1,1}} = \text{0x1bd9}$

$\mathbf{c_{1,2}} = \text{0x760a53a3277bf1f5fe922cc529eb7e95922999fec51f300d6885642c40f067a}$

$\mathbf{c_{1,3}} = \text{0x3c81}$

$\mathbf{c_{1,4}} = \text{0x1073185a9d794b3ab04aade0f1fab315b290030dcce495e1a20afd1998b8557a}$

$\mathbf{c_{1,5}} = \text{0x19001}$

$\mathbf{c_{1,6}} = \text{0x1e491ff78af94d6ee907243a5dbd50922284849715d3ded4c70ff4e97a1d88e16f}$

$\delta_1 = \text{0x4824661fb5bbc013eb1922d62f25db7e83f8553b40bbbfa29d2b333386e621}$

$\delta_2 = \text{0x326}$

$\delta_3 = \text{0x1cdb20e7787e32aa7b48e2287da944fcf4e313ec1bee64a440d047893b95fa2e4}$

$\delta_4 = \text{0x111d2}$

$\delta_5 = \text{0x4db5114bea200f30ccab250fa17cefc93c0eb4991bcf90c0b25d3683be619cf579}$

$\delta_6 = \text{0x9bbd}$

We list six pairs of binary Kummer lines and binary generalized Hessian curves in Table 8. Here, $\mathsf{BKL}(m, c)$ denotes the $\mathsf{BKL}_{(1:c)}$ defined over $\mathbb{F}_{2^m}$ such that the corresponding $\mathsf{BEw}_{(c^4)}$ has a point of order 3, and $\mathsf{H}(m, \gamma, \delta)$ denotes the $\mathsf{H}_{(\gamma,\delta)}$ on the field $\mathbb{F}_{2^m}$ where $\gamma, \delta \neq 0$ and $\delta^3 \neq \gamma$. Observe that all the curves have a cofactor of $h = 12$, and the cofactor of the corresponding twist is $h_T = 6$. Let $\mathsf{BEw}_{(c_{1,1}^4)}$ be the associated short Weierstrass curve to $\mathsf{BKL}(251, c_{1,1})$. From [24] and the existence of a point of order 3, we have that $\mathsf{BEw}_{(c_{1,1}^4)}$ must have a cofactor divisible by 12. Let the cardinality of $\mathsf{BEw}_{(c_{1,1}^4)}$ be $12p_2$, where p_2 is a prime. Then the cardinality of the twist of $\mathsf{BEw}_{(c_{1,1}^4)}$ is $2^{m+1} + 2 - 12p_2 = 2(2^m + 1) - 12p$, and is divisible by 2. Now the chosen m are odd, and thus $2^m \equiv 2 \mod 3$, and $(2^m + 1)$ is divisible by 3. Therefore, the obtained cofactors are optimal. We first choose the smallest curve parameters (c or δ) for the boldfaced curve that achieves appropriate security. Next, we compute the rest of the values in that row. For example, consider the $\mathsf{BKL}(251, c_{1,1})$ and $\mathsf{H}(251, 1, \delta_1)$ pair. Here, we found that $c_{1,1} = \text{0x1bd9}$ is the smallest parameter for which we have optimal cofactor. Then the associated short Weierstrass curve is $\mathsf{BEw}_{(c_{1,1}^4)}$, say (x_3, y_3). By Theorem 4, we have an isomorphic generalized Hessian curve $\mathsf{H}_{(1,\delta_1)}$ where $\delta_1 = 1/(1 + \nu)$ and $\nu^3 = 1 + x_3$. On the other hand, consider the $\mathsf{BKL}(251, c_{1,2})$ and $\mathsf{H}(251, 1, \delta_2)$ pair. The inverse of the isomorphisms reveals that $H_{(\gamma,\delta)}$ is isomorphic to $\mathsf{BEwT}_{(1/\delta_2+1/\delta_2^2+1/\delta_2^3)}$. Furthermore, it can be shown that $\mathsf{BEwT}_{(1/\delta_2+1/\delta_2^2+1/\delta_2^3)}$ has an isomorphic form $\mathsf{BEw}_{(b)}$ with $b = \left(1/\delta_2 + 1/\delta_2^2 + 1/\delta_2^3\right)^3 \left(1 + 1/\delta_2\right)^3$ if $\mathrm{Tr}(\frac{1}{\delta_2^3}) = 0$. Therefore, we choose the δ_2 as the smallest one such that $\mathrm{Tr}(\frac{1}{\delta_2^3}) = 0$ and we have optimal cofactor, and then $c_{1,2}^4 = b$. This way of choosing a generalized Hessian curve is suggested by [7]. But for the curves $H(\cdot, 1, \delta_i)$, $i = 2, 4, 6$, although δ_i^3 is small, $1/\sqrt{\delta_i^3}$ is not. Table 9 includes the binary Kummer line and generalized Hessian curve pairs where we choose the curves $H(\cdot, \gamma_i, 1)$, $i = 1, 2, 3$, following the Theorem 6. The generalized Hessian curves of Table 9 provide the best operation counts mentioned in this article.

Table 9. Binary Generalized Hessian Curves

Field	$H_{(\gamma,1)}$	$c = \sqrt{\gamma^3(\gamma+1)}$	$\mathrm{BKL}_{(1:c)}$	$(\log_2(p_1), \log_2(p_2))$	(h, h_T)	Bit Security
$\mathbb{F}_{2^{251}}$	$H(\mathbf{251}, \gamma_1, \mathbf{1})$	0x1bd9	$\mathrm{BKL}(251, c_{2,1})$	$(247.4, 248.4)$		123.7
$\mathbb{F}_{2^{257}}$	$H(\mathbf{257}, \gamma_2, \mathbf{1})$	0x3c81	$\mathrm{BKL}(257, c_{2,2})$	$(253.4, 254.4)$	$(12, 6)$	126.7
$\mathbb{F}_{2^{263}}$	$H(\mathbf{263}, \gamma_3, \mathbf{1})$	0x19001	$\mathrm{BKL}(263, c_{2,3})$	$(259.4, 260.4)$		129.7

$\gamma_1 =$ 0xee2ff990cc63e819c808dd80c72e8480cce06d2b345f5bed064dc11eac95f7

$\gamma_2 =$ 0x1a60e52cdb75f34dd81a3d06572ad37d97244b45c383fd5af352f3d96bb681c40

$\gamma_3 =$ 0x524570b33dddaa3eb326586232db22913ade54698e379dc9cb03258bfcf6be5571

6   Conclusion

We provide a method to retrieve the R and S coordinates of the point nP obtained by the Montgomery ladder scalar multiplication in $\mathfrak{RT}$ coordinates of a generalized Hessian curve. We connect binary Kummer line to generalized Hessian curve via isomorphism and via isogeny. We also include new concrete proposals of binary Kummer lines and corresponding generalized Hessian curves that target 128-bit security levels and have small constants required for the arithmetic.

References

1. Bernstein, D.J.: Curve25519: New Diffie-Hellman speed records. In: Yung, M., Dodis, Y., Kiayias, A., Malkin, T. (eds.) PKC 2006. LNCS, vol. 3958, pp. 207–228. Springer, Heidelberg (2006). https://doi.org/10.1007/11745853_14
2. Bernstein, D.J.: Batch binary Edwards. In: Halevi, S. (ed.) CRYPTO 2009. LNCS, vol. 5677, pp. 317–336. Springer, Heidelberg (2009). https://doi.org/10.1007/978-3-642-03356-8_19
3. Bernstein, D.J., Chuengsatiansup, C., Kohel, D., Lange, T.: Twisted Hessian curves. In: Lauter, K., Rodríguez-Henríquez, F. (eds.) LATINCRYPT 2015. LNCS, vol. 9230, pp. 269–294. Springer, Cham (2015). https://doi.org/10.1007/978-3-319-22174-8_15

4. Blake, I., Seroussi, G., Smart, N.: Elliptic Curves in Cryptography, vol. 265. Cambridge University Press, Cambridge (1999)
5. Diffie, W., Hellman, M.: New directions in cryptography. IEEE Trans. Inf. Theory **22**(6), 644–654 (1976)
6. Rojalia, F., Mohamad, M.S.B.: An algorithm for finding the cube roots in finite fields. In: 5th International Conference on Computer Science and Computational Intelligence, Procedia Computer Science, vol. 109, pp. 838–844. Elsevier (2021)
7. Farashahi, R.R., Joye, M.: Efficient arithmetic on Hessian curves. In: Nguyen, P.Q., Pointcheval, D. (eds.) PKC 2010. LNCS, vol. 6056, pp. 243–260. Springer, Heidelberg (2010). https://doi.org/10.1007/978-3-642-13013-7_15
8. Galbraith, S.D.: Mathematics of Public Key Cryptography. Cambridge University Press, Cambridge (2012)
9. Galbraith, S.D., Hess, F., Smart, N.P.: Extending the GHS weil descent attack. In: Knudsen, L.R. (ed.) EUROCRYPT 2002. LNCS, vol. 2332, pp. 29–44. Springer, Heidelberg (2002). https://doi.org/10.1007/3-540-46035-7_3
10. Gaudry, P., Hess, F., Smart, N.: Constructive and destructive facets of Weil descent on elliptic curves. J. Cryptol. **15**, 19–46 (2002)
11. Gaudry, P., Lubicz, D.: The arithmetic of characteristic 2 Kummer surfaces and of elliptic Kummer lines. Finite Fields Appl. **15**(2), 246–260 (2009)
12. Hankerson, D., Menezes, A.J., Vanstone, S.: Guide to Elliptic Curve Cryptography, 1st edn. Springer, New York (2010)
13. Hisil, H., Wong, K.K.-H., Carter, G., Dawson, E.: Twisted Edwards curves revisited. In: Pieprzyk, J. (ed.) ASIACRYPT 2008. LNCS, vol. 5350, pp. 326–343. Springer, Heidelberg (2008). https://doi.org/10.1007/978-3-540-89255-7_20
14. Huseyin, H., Joost, R.: On Kummer lines with full rational 2-torsion and their usage in cryptography. ACM Trans. Math. Softw. (TOMS) **45**(4), 1–17 (2019)
15. Joux, A., Vitse, V.: Cover and decomposition index calculus on elliptic curves made practical. In: Pointcheval, D., Johansson, T. (eds.) EUROCRYPT 2012. LNCS, vol. 7237, pp. 9–26. Springer, Heidelberg (2012). https://doi.org/10.1007/978-3-642-29011-4_3
16. Joye, M., Quisquater, J.-J.: Hessian elliptic curves and side-channel attacks. In: Koç, Ç.K., Naccache, D., Paar, C. (eds.) CHES 2001. LNCS, vol. 2162, pp. 402–410. Springer, Heidelberg (2001). https://doi.org/10.1007/3-540-44709-1_33
17. Karati, S.: Binary Kummer line. In: Tibouchi, M., Wang, X. (eds.) ACNS 2023 Part I. LNCS, vol. 13905, pp. 363–393. Springer, Cham (2023). https://doi.org/10.1007/978-3-031-33488-7_14
18. Karati, S., Sarkar, P.: Connecting legendre with Kummer and Edwards. Adv. Math. Commun. **13**(1), 41–66 (2019)
19. Karati, S., Sarkar, P.: Kummer for genus one over prime-order fields. J. Cryptol. **33**, 1–38 (2019). https://doi.org/10.1007/s00145-019-09320-4
20. Koblitz, N.: Elliptic curve cryptosystems. Math. Comp. **48**(177), 203–209 (1987)
21. Menezes, A., Qu, M.: Analysis of the weil descent attack of Gaudry, hess and smart. In: Naccache, D. (ed.) CT-RSA 2001. LNCS, vol. 2020, pp. 308–318. Springer, Heidelberg (1999). https://doi.org/10.1007/3-540-45353-9_23
22. Miller, V.S.: Use of elliptic curves in cryptography. In: Williams, H.C. (ed.) CRYPTO 1985. LNCS, vol. 218, pp. 417–426. Springer, Heidelberg (1986). https://doi.org/10.1007/3-540-39799-X_31
23. Nath, K., Sarkar, P.: Kummer versus montgomery face-off over prime order fields. ACM Trans. Math. Softw. **48**(2), 13:1–13:28 (2022)
24. Pornin, T.: Efficient and complete formulas for binary curves. Cryptology ePrint Archive, Paper 2022/1325 (2022). https://eprint.iacr.org/2022/1325

25. Silverman, J.H.: The Arithmetic of Elliptic Curves, vol. 106. Springer, New York (2009). https://doi.org/10.1007/978-0-387-09494-6

26. Smart, N.P.: The Hessian form of an elliptic curve. In: Koç, Ç.K., Naccache, D., Paar, C. (eds.) CHES 2001. LNCS, vol. 2162, pp. 118–125. Springer, Heidelberg (2001). https://doi.org/10.1007/3-540-44709-1_11

27. Stam, M.: On montgomery-like representations for elliptic curves over $GF(2^k)$. In: Desmedt, Y.G. (ed.) PKC 2003. LNCS, vol. 2567, pp. 240–254. Springer, Heidelberg (2003). https://doi.org/10.1007/3-540-36288-6_18

28. Tate, J.: Endomorphisms of abelian varieties over finite fields. Invent. Math. **2**(2), 134–144 (1966)

29. Vélu, J.: Isogénies entre courbes elliptiques. Comptes-Rendus de l'Académie des Sciences, Série I, 273:238–241, juillet (1971)

30. Wroński, M., Kijko, T.: Arithmetic on generalized hessian curves using compression function and its applications to the isogeny-based cryptography. Publ. Math. Debrecen, 655–682 (2022). https://doi.org/10.5486/PMD.2022.Suppl.7

Synchronized Aggregate Signature Under Standard Assumption in the Random Oracle Model

R. Kabaleeshwaran[1]($\boxtimes$) and Panuganti Venkata Shanmukh Sai[2]

[1] Department of Computer Science and Engineering, Indian Institute of Information Technology Design and Manufacturing Kurnool, Kurnool, India
kabaleesh@iiitk.ac.in
[2] Capgemini Technology Services India Limited, Hyderabad, India

Abstract. An aggregate signature enables to aggregation of multiple signatures generated by different signers on different messages. A synchronized aggregate signature is a special type of aggregate signature in which all the signatures generated at a particular time epoch will be aggregated, where each signer uses the same synchronized clock. So far in the literature, Ahn et al.'s (ACM CCS 2010) synchronized aggregate signature is the only scheme whose security is proved under the standard computational Diffie-Hellman assumption. However, their construction supports only restricted message space. All the other synchronized aggregate signature constructions either use non-standard assumptions or rely on symmetric pairings.

Recently, Tezuka and Tanaka presented a Pointcheval-Sanders signature-based synchronized aggregate signature construction which provides more efficient aggregate signature verification than other existing schemes in the pairing setting. However, their security is proved under the interactive non-standard assumption that turned less trusted. Hence, we construct a synchronized aggregate signature scheme by including a suitable signature component and using Gerbush et al.'s dual-form signature technique we prove its security under standard SXDH assumptions.

Keywords: Synchronized Aggregate signature · Pointcheval-Sanders randomizable signatures · Dual-Form Signature Technique · Standard Assumption

1 Introduction

In digital communication, authentication and data integrity are the two prominent security properties achieved using a digital signature scheme. The digital signature [16,17] is a cryptographic primitive and is analog to the ordinary signing process. An aggregate signature scheme is a digital signature scheme in which anyone can aggregate n signatures generated by n users on n messages into a

---

P. V. S. Sai—Work done at IIITDM Kurnool during his final year B.Tech. project.

A. Chattopadhyay et al. (Eds.): INDOCRYPT 2023, LNCS 14459, pp. 197–220, 2024.
https://doi.org/10.1007/978-3-031-56232-7_10

single short (aggregated) signature. The aggregate signatures are used in many applications such as secure routing protocol [6,28], sensors network [3], secure logging [31], software updates [3] and so on. There are various types of aggregate signature schemes have been proposed, such as sequential aggregate signature [29], identity-based aggregate signature [13], synchronized aggregate signature scheme [3], fault-tolerant aggregate signature [19] and aggregate signature with detecting functionality [22,34].

In the synchronized aggregate signature (SynAS) scheme [13], the time period w is associated with the signature generated by the signers. In SynAS scheme, anyone can aggregate the signatures on different messages into a short aggregate signature, where each signature is generated in the same period. In 2010, Ahn, Green, and Hohenberger [3] revisited the Gentry-Ramzan model and constructed two synchronized aggregate signature schemes based on Hohenberger-Waters' [20] short signatures in the symmetric pairing setting. Their first construction is defined in the standard model, whereas the second one is in the random oracle model. The security of both schemes is proved under the standard computational Diffie-Hellman (CDH) assumption. They have also mentioned how to extend their construction in an efficient asymmetric pairing setting. Although Ahn et al. [3] proposed the SynAS scheme under the standard CDH assumption, their construction has the drawback of limited message length.

Lee et al. [25] proposed an efficient synchronized aggregate signature based on the Camenisch-Lysyanskaya (CL) signature scheme [10]. They have used their 'public key sharing' technique along with Lu et al.'s [28] 'randomness reuse' technique to achieve efficient construction. The *randomness re-use* technique for SynAS scheme ensures that all the signers utilize the same randomness. Whereas the *public key sharing* technique ensures that the aggregation system generates a dedicated public key component and later each signer uses it to aggregate the signatures. The security of Lee et al.'s SynAS scheme is proved under the (interactive) one-time Lysyanskaya Rivest-Sahai-Wolf (OT-LRSW) assumption [25] in the random oracle model. In 2020, Tezuka and Tanaka [38] have given an alternative security proof for Lee et al.'s SynAS scheme under a static assumption in the random oracle model.

In 2022, Tezuka and Tanaka [39], presented a synchronized aggregate signature scheme based on PS signatures. They have used Lu et al.'s [28] randomness re-use technique to maintain the same randomness by all the signers at a time epoch. The linear structure of Pointcheval-Sanders randomizable signature helps them to aggregate the signatures that are generated at a time epoch. However, security of their scheme is proved under Generalized PS assumption [24], which is an interactive assumption.

The interactive assumptions are prevalent in pairing-based cryptography. For example, LRSW assumption [30], PS assumption [33], Generalized PS assumption [24], One-more discrete logarithm assumption [4], Interactive Diffie-Hellman assumptions [1], One-Time LRSW assumption [25]. The efficiency of most of the protocol is achieved using tailor-made interactive assumptions. In particular, most of the efficient pairing-based privacy-preserving cryptosystems are constructed using Pointcheval-Sanders (PS) [33] signature scheme. For example, anonymous credentials [36], direct anonymous attestation [12], group signatures [9,33], blind signatures [15], e-cash [8], sequential aggregate signatures [11], are built based on PS signatures. However, the hardness of the interactive assumptions is proved in the generic group model [32,35] and the generic bilinear group model [7]. Proving the hardness crucially relies on the restrictions posed on the queries performed by the adversary. Most of the time, this is a tedious task and error-prone [5,21,37]. This motivates us to remove the dependency on the interactive assumption and argue the security under the standard assumption.

1.1   Our Contribution

In this work, we present a Pointcheval-Sanders (PS) [33] signature-based synchronized aggregate signature under the standard symmetric external Diffie-Hellman (SXDH) assumption and it is denoted as SynAS. Our construction is inspired by that of Tezuka and Tanaka [39] (which is denoted as PS-SynAS).

1.2   Overview of Techniques

Now we describe the high-level idea of our SynAS construction. Recall the Synchronized Aggregate signature construction of [39] which is based on Pointcheval-Sander's [33] randomizable signature. For the signer holding secret key $(x_i, y_i) \in \mathbb{Z}_p^2$ and his/her public key $(X_i = h^{x_i}, Y_i = h^{y_i}) \in H^2$, the synchronized signature at time epoch ω is a pair $(A_i = H_0(\omega)^{x_i+m_iy_i}, B_i = H_0(\omega))$. Then using Lu et al.'s [28] 'randomness re-use' technique, the synchronized aggregate signature is generated as a pair $(A = \prod_{i=1}^{n} H_0(\omega)^{x_i+m_iy_i}, B = H_0(\omega))$, where all the n signers agree with the same randomness which is guaranteed using the hash function H_0 on the time epoch ω. Tezuka-Tanaka has proved the security of their synchronized aggregate signature scheme under the Generalized Pointcheval-Sanders assumption, which is an interactive assumption.

Now to construct a standard assumption-based construction, we use Gerbush et al.'s [14] dual-form signature technique. In particular, following [11,26,27], we will introduce quasi-adaptive non-interactive zero-knowledge [23] (QA-NIZK) proof component along with the signature component of Tezuka-Tanaka's synchronized aggregate signature construction. In the process, apart from the signature components $A = H_0(w)^{u_1+mv_1}$ and $B = H_0(w)$, we will introduce quasi-adaptive non-interactive zero-knowledge (QA-NIZK) proof component $C = H_0(w)^{u_2+mv_2}$, where the secret components (u_1, u_2) and (v_1, v_2) embeds a suitable linear subspace relation. While proving the unforgeability, the simulator exploits the above-mentioned linear subspace relation to partition the forgery space into two and uses the dual-form signature technique to complete the reduction under the SXDH assumption.

2 Preliminaries

For a prime p, $\mathbb{Z}_p^*$ denotes the set of all non-zero elements from $\mathbb{Z}_p = \{0, 1, \ldots, p-1\}$. We denote $a \xleftarrow{\$} A$ to be an element chosen uniformly at random from the non-empty set A.

2.1 Bilinear Pairing

Definition 1. *A bilinear group generator* $\mathcal{P}$ *is a probabilistic polynomial time (PPT) algorithm which takes the security parameter* λ *as input and outputs* $\Theta = (p, G, H, G_T, e, g, h)$, *where* p *is a prime,* G, H *and* G_T *are the prime order groups and* g *(resp. h) is an arbitrary generator of G (resp. H) and* $e: G \times H \rightarrow G_T$ *is a bilinear map that satisfies,*

(i) **Bilinearity:** *For all* $g_1 \in G, h_1 \in H$, $a, b \in \mathbb{Z}_p^*$, *one has* $e(g_1^a, h_1^b) = e(g_1, h_1)^{ab}$.

(ii) **Non degeneracy:** *If a fixed* $g_1 \in G$ *satisfies* $e(g_1, h_1) = 1$ *for all* $h_1 \in H$, *then* $g_1 = 1$ *and similarly for elements of* H

(iii) **Computability:** *The map e is efficiently computable.*

2.2 Computational Assumptions

We recall the decisional Diffie-Hellman (DDH) assumption in G (denoted as DDH_G) as follows.

Definition 2. *Given* $\Theta = (p, G, H, G_T, e, g, h)$ *along with the elements* g^a, g^b *and* $T = g^{ab+\theta}$ *from G, it is hard to decide whether $\theta = 0$ or not, for $a, b \xleftarrow{\$} \mathbb{Z}_p$.*

In the same way, we can define the DDH assumption in H (denoted as DDH_H).

Definition 3. *If DDH assumption in both G and H holds, then we say that the symmetric external Diffie-Hellman (SXDH) assumption holds in G and H.*

We recall from [18] the double pairing (DBP) assumption in H (denoted as DBP_H) as follows.

Definition 4. *Given* $\Theta = (p, G, H, G_T, e, g, h)$ *along with the elements* h_r *and* h_s *from H, it is hard to compute* $(R, S) \neq (1, 1)$ *from* G^2 *such that* $e(R, h_r) \cdot e(S, h_s) = 1$.

In the same way, we can define the DBP assumption in G (denoted as DBP_G). Abe et al. [2], proved that DBP_G is reducible to DDH_G and DBP_H is reducible to DDH_H.

2.3 Synchronized Aggregate Signature Definition

Synchronized aggregate signature scheme is a special type of aggregate signature scheme. In this scheme, all of the signers have a synchronized time period w and each signer can sign a message at most once for each period w. A set of signatures that are all generated for the same period w can be aggregated into a short signature. The size of an aggregate signature is the same size as an individual signature. Now, we recall the definition of synchronized aggregate signature scheme from [38].

Definition 5 *(Synchronized Aggregate Signature Scheme): A synchronized aggregate signature scheme SAS for a bounded number of periods is a tuple of algorithms (**SAS.Setup, SAS.KeyGen, SAS.Sign, SAS.Verify, SAS.Aggregate, SAS.AggVerify**).*

- **SAS.Setup**$(1^\lambda, 1^W)$: *Given a security parameter λ and the time period bound W, return the public parameter pp. We assume that pp defines the message space $\mathcal{M}_{pp}$.*
- **SAS.KeyGen**(pp): *Given a public parameter pp, return a public key PK and a signing key SK.*
- **SAS.Sign**(pp, SK, m, w): *Given a public parameter pp, a secret key SK, a time period $w \le W$ and a message $m \in \mathcal{M}_{pp}$, return the signature σ.*
- **SAS.Verify**(pp, PK, m, σ): *Given a public parameter pp, a public key PK, a message $m \in \mathcal{M}_{pp}$ and a signature σ, return either 1 (Accept) or 0 (Reject).*
- **SAS.Aggregate**$(pp, (PK_1, \ldots, PK_n), (m_1, \ldots, m_n), (\sigma_1, \ldots, \sigma_n))$: *Given a public parameter pp, a list of public keys $(PK_1, \ldots, PK_n)$, a list of messages $(m_1, \ldots, m_n)$ and a list of signatures $(\sigma_1, \ldots, \sigma_n)$, return either the aggregate signature Σ or $\bot$.*
- **SAS.AggVerify**$(pp, (PK_1, \ldots, PK_n), (m_1, \ldots, m_n), \Sigma)$: *Given a public parameter pp, a list of public keys $(PK_1, \ldots, PK_n)$, a list of messages $(m_1, \ldots, m_n)$ and an aggregate signature Σ, return either 1 (Accept) or 0 (Reject).*

Correctness is satisfied if for all $\lambda, W \in \mathbb{N}$, $pp \leftarrow$ **SAS.Setup**$(1^\lambda, 1^W)$, for any finite sequence of key pairs $(PK_1, SK_1), \ldots, (PK_n, SK_n)$ $\leftarrow$ **SAS.KeyGen**(pp) where PK_i are all distinct, for any time period $w \le W$, for any sequence of messages $(m_1, \ldots, m_n) \in \mathcal{M}_{pp}^n$, $\sigma_i \leftarrow$ **SAS.Sign** (pp, SK_i, m_i, w) for $i \in [n]$, $\Sigma \leftarrow$ **SAS.Aggregate**$(pp, (PK_1, \ldots, PK_n), (m_1, \ldots, m_n), (\sigma_1, \ldots, \sigma_n))$, we have

$$\textbf{SAS.Verify}(pp, PK_i, m_i, \sigma_i) = 1 \text{ for all } i \in [n] \land$$
$$\textbf{SAS.AggVerify}(pp, (PK_1, \ldots, PK_n), (m_1, \ldots, m_n), \Sigma) = 1.$$

The security notion for synchronized aggregate signature is defined using the EUF-CMA security in the *certified-key model*.

Definition 6 *(EUF-CMA Security in the Certified-Key Model). The EUF-CMA security of a synchronized aggregate signature (SAS) scheme in the* *certified-key model is defined by the following game between a challenger C and a PPT adversary A.*

- C *runs* $pp \leftarrow$ **SAS.Setup**$(1^\lambda, 1^W)$, $(PK^*, SK^*) \leftarrow$ **SAS.KeyGen**(pp), *sets* $Q \leftarrow \emptyset$, $L \leftarrow \emptyset$, $w_{ctr} \leftarrow 1$, *and gives* (pp, PK^*) *to* A.
- **Certification Query:** A *is given access (throughout the entire game) to a certification oracle* $\mathcal{O}^{Cert}(\cdot, \cdot)$. *Given an input (PK, SK),* $\mathcal{O}^{Cert}$ *performs the following procedure.*
  - *If the key pair (PK, SK) is valid,* $L \leftarrow L \cup \{PK\}$ *and return "**accept**".*
  - *otherwise return "**reject**".*
    (A must submit key pair (PK, SK) to $\mathcal{O}^{Cert}$ *and get "**accept**" before using PK.)*
- **Signing Query:** A *is given access (throughout the entire game) to a signing oracle* $\mathcal{O}^{Sign}(\cdot, \cdot)$. *Given an input ("**inst**", m),* $\mathcal{O}^{Sign}$ *performs the following procedure. ("**inst**"* $\in$ *{"**skip**", "**sign**"} represent the instruction for* $\mathcal{O}^{Sign}$ *where "**skip**" implies that* A *skips the concurrent period* w_{ctr} *and "**sign**" implies that* A *requires the signature on message m.)*
  - *If* $w_{ctr} \notin W$, *return* $\perp$.
  - *If "**inst**" = "**skip**",* $w_{ctr} \leftarrow w_{ctr} + 1$.
  - *If "**inst**" = "**sign**",* $Q \leftarrow Q \cup \{m\}$, $\sigma \leftarrow$ **SAS.Sign**(pp, SK, m, w), $w_{ctr} \leftarrow w_{ctr} + 1$, *return* σ.
- A *outputs a forgery* $\left((PK_1^*, \ldots, PK_n^*), (m_1^*, \ldots, m_n^*), \Sigma^*\right)$.

A sequential aggregate signature scheme SAS satisfies the EUF-CMA security in the certified-key model if for all PPT adversaries A (denoted as $Adv_{SAS,A}^{UF}$), the following advantage

$$Pr\left[A^{\mathcal{O}^{Cert}(\cdot, \cdot), \mathcal{O}^{Sign}(\cdot, \cdot)}(pp, PK^*) \rightarrow \left((PK_1^*, \ldots, PK_n^*), (m_1^*, \ldots, m_n^*), \Sigma^*\right) : \right.$$

$$\textbf{SAS.AggVerify}\left(pp^*, (PK_1^*, \ldots, PK_n^*), (m_1^*, \ldots, m_n^*), \Sigma^*\right) = 1$$

$$\wedge \text{ For all } j \in [n] \text{ such that } PK_j^* \neq PK^*, \ PK_j^* \in L$$

$$\left. \wedge \text{ For some } j^* \in [n] \text{ such that } PK_{j^*}^* = PK^*, m_{j^*}^* \notin Q \right]$$

is negligible in λ.

3   Synchronized Aggregation Under Standard Assumption

In this section, we describe the synchronized aggregate signature scheme, whose security is proved under the standard SXDH assumption. We denote the resulting construction as SynAS scheme.

3.1 SynAS Construction

The proposed scheme has the following six PPT algorithms:

- **SAS.Setup$(1^\lambda, 1^W)$:** Given the security parameter λ and a time period bound W as input, it returns the public parameters $pp \leftarrow (p, G, H, G_T, e, g, h_z, h_0, H_0, H_1)$, where G, H and G_T are bilinear groups of prime order p with an efficiently computable bilinear pairing map $e : G \times H \rightarrow G_T$, the element g (resp. h_z, h_0) is chosen uniformly at random from G (resp. H) and $H_0 : \mathcal{W} \rightarrow G$, $H_1 : \{0,1\}^* \times \mathcal{W} \rightarrow \mathbb{Z}_p^*$ are the collision resistant hash functions, where $\mathcal{W}$ is time period space.
- **SAS.KeyGen(pp):** Given the public parameter pp as input, it randomly selects u_1, u_2, v_1, v_2 from $\mathbb{Z}_p$, computes $h_1 = h_z^{u_2} h_0^{-u_1}$ and $h_2 = h_z^{v_2} h_0^{-v_1}$. It returns $PK = \{h_1, h_2\}$, $SK = \{u_1, u_2, v_1, v_2\}$.
- **SAS.Sign(pp, SK, m, w):** Given the public parameter pp, a secret key $SK = \{u_1, u_2, v_1, v_2\}$, a message m and a time-period $w \in \mathcal{W}$ with $w \leq W$ as input, it computes $A = H_0(w)^{(u_1 + v_1 H_1(m\|w))}$, $B = H_0(w)$ and $C = H_0(w)^{(u_2 + v_2 H_1(m\|w))}$. Finally, it returns the pair $(m, \sigma = (A, B, C, w))$.
- **SAS.Verify(pp, PK, m, σ):** Given the public parameter pp, a public key $PK = \{h_1, h_2\}$, message m and signature $\sigma = (A, B, C, w)$ as input, it checks whether the message m belongs to $\{0,1\}^*$. If this condition fails to hold, then abort. Otherwise, it computes $H_1(m\|w)$ and returns 1 if the below Equation satisfies,

$$e(A, h_0)e(B, h_1 h_2^{H_1(m\|w)}) \stackrel{?}{=} e(C, h_z). \tag{1}$$

Otherwise, returns 0.
- **SAS.AggSign$(pp, (PK_1, \ldots, PK_n), (m_1, \ldots, m_n), (\sigma_1, \ldots, \sigma_n))$:** Given the signatures $(\sigma_1, \ldots, \sigma_n)$ on messages $(m_1, \ldots, m_n)$ under public keys $(PK_1, \ldots, PK_n)$ as input, it parses the signature σ_i as (A_i, B_i, C_i, w_j), where $A_i = H_0(w_j)^{u_{i1} + v_{i1} H_1(m_i\|w_j)}$, $B_i = H_0(w_j)$ and $C_i = H_0(w_j)^{u_{i2} + v_{i2} H_1(m_i\|w_j)}$ and public key PK_i as (h_{i1}, h_{i2}). It checks that w_1 is equal to w_j for $j = 2$ to n and validates whether each individual signature is valid or not by checking Eq. (1). If the above checks fail to hold, then abort. Otherwise, it sets $w = w_1$ and calculates $B_\Sigma = H_0(w)$ and

$$A_\Sigma = \prod_{i=1}^n A_i = H_0(w)^{\sum_{i=1}^n u_{i1} + \sum_{i=1}^n v_{i1} H_1(m_i\|w)}$$

$$C_\Sigma = \prod_{i=1}^n C_i = H_0(w)^{\sum_{i=1}^n u_{i2} + \sum_{i=1}^n v_{i2} H_1(m_i\|w)}.$$

Finally, it outputs an aggregate signature $\sigma_\Sigma = (A_\Sigma, B_\Sigma, C_\Sigma, w)$.
- **SAS.AggVerify$(pp, (PK_1, \ldots, PK_n), (m_1, \ldots, m_n), \sigma_\Sigma)$:** Given the aggregate signature σ_Σ on messages $(m_1, \ldots, m_n)$ under public keys $(PK_1, \ldots, PK_n)$ as input, it parses the aggregate signature σ_Σ as $(A_\Sigma, B_\Sigma, C_\Sigma, w)$ and public keys as $(PK_1, \ldots, PK_n)$, where $PK_i = (h_{i1}, h_{i2})$. It checks that any

public key PK_i does not appear twice, if so, it outputs 0. Otherwise, it checks the below condition

$$e(A_\Sigma, h_0)e(B_\Sigma, \prod_{i=1}^{n} h_{i1} \prod_{i=1}^{n} h_{i2}^{H_1(m_i\|w)}) \overset{?}{=} e(C_\Sigma, h_z). \tag{2}$$

If the above condition is satisfied, it returns 1. Otherwise, it returns 0.

Correctness can be ensured by checking the Eq. (1) holds for honestly generated signatures and the Eq. (2) holds for honestly generated aggregate signature. To validate the Eq. (1), consider the honestly generated signature $\sigma_i = (A_i, B_i, C_i, w)$ on the message m_i under the $PK_i = (h_{i1} = h_z^{u_{i2}} h_0^{-u_{i1}}, h_{i2} = h_z^{v_{i2}} h_0^{-v_{i1}})$. Then, we have

$$e(A_i, h_0)e(B_i, h_{i1}h_{i2}^{H_1(m_i\|w)}) = e(H_0(w)^{(u_{i1}+v_{i1}H_1(m_i\|w))}, h_0)$$
$$\cdot e(H_0(w), (h_z^{u_{i2}} h_0^{-u_{i1}})(h_z^{v_{i2}} h_0^{-v_{i1}})^{H_1(m_i\|w)})$$
$$= e(H_0(w), h_0^{(u_{i1}+v_{i1}H_1(m_i\|w))})$$
$$\cdot e(H_0(w), h_z^{(u_{i2}+v_{i2}H_1(m_i\|w))} h_0^{(-u_{i1}-v_{i1}H_1(m_i\|w))})$$
$$= e(H_0(w), h_z^{(u_{i2}+v_{i2}H_1(m_i\|w))})$$
$$= e(H_0(w)^{(u_{i2}+v_{i2}H_1(m_i\|w))}, h_z) = e(C_i, h_z).$$

Now, to validate the Eq. (2), consider the honestly generated aggregate signature $\sigma_\Sigma = (A_\Sigma, B_\Sigma, C_\Sigma, w)$ on messages $(m_1, \ldots, m_n)$ under public keys $(PK_1, \ldots, PK_n)$. Then, we have

$$e(A_\Sigma, h_0)e(B_\Sigma, \prod_{i=1}^{n} h_{i1} \prod_{i=1}^{n} h_{i2}^{H_1(m_i\|w)})$$
$$= e(H_0(w)^{\sum_{i=1}^{n} u_{i1}+\sum_{i=1}^{n} v_{i1}H_1(m_i\|w)}, h_0)$$
$$\cdot e(H_0(w), \prod_{i=1}^{n}(h_z^{u_{i2}} h_0^{-u_{i1}}) \prod_{i=1}^{n}(h_z^{v_{i2}} h_0^{-v_{i1}})^{H_1(m_i\|w)})$$
$$= e(H_0(w), h_0^{\sum_{i=1}^{n} u_{i1}+\sum_{i=1}^{n} v_{i1}H_1(m_i\|w)})$$
$$\cdot e(H_0(w), h_z^{\sum_{i=1}^{n} u_{i2}+\sum_{i=1}^{n} v_{i2}H_1(m_i\|w)} \cdot h_0^{-\sum_{i=1}^{n} u_{i1}-\sum_{i=1}^{n} v_{i1}H_1(m_i\|w)})$$
$$= e(H_0(w), h_z^{\sum_{i=1}^{n} u_{i2}+\sum_{i=1}^{n} v_{i2}H_1(m_i\|w)})$$
$$= e(H_0(w)^{\sum_{i=1}^{n} u_{i2}+\sum_{i=1}^{n} v_{i2}H_1(m_i\|w)}, h_z) = e(C_\Sigma, h_z).$$

3.2   Security of SynAS Scheme

In this section, we use the Gerbush et al.'s [14] dual-form signature technique to prove the EUF-CMA security of our SynAS scheme in the certified-key model, under the SXDH assumption.

Partition of Forgery Space: Following [11,26,27], we partition the forgery space into two and define two signing processes. Let $\mathcal{V}$ be the set of all messages

$\overrightarrow{\mathbf{m}}^* = (m_1^*, \ldots, m_n^*)$, public keys $\overrightarrow{\mathbf{PK}}^* = (PK_1^*, \ldots, PK_n^*)$ and synchronized aggregate signature $\sigma_\Sigma^* = (A_\Sigma^*, B_\Sigma^*, C_\Sigma^*, w^*)$ pairs such that they verify under the public keys. We partition the forgery class $\mathcal{V}$ into two disjoint sets $\mathcal{V}_I$ and $\mathcal{V}_{II}$ which are defined as follows:

Type-I: $\mathcal{V}_I = \{(\overrightarrow{\mathbf{PK}}^*, \overrightarrow{m}^*, \sigma_\Sigma^*) \in \mathcal{V} : S_1^* = 1 \text{ and } S_2^* = 1\}$,

Type-II: $\mathcal{V}_{II} = \{(\overrightarrow{\mathbf{PK}}^*, \overrightarrow{m}^*, \sigma_\Sigma^*) \in \mathcal{V} : S_1^* \neq 1 \text{ and } S_2^* \neq 1\}$, where,

$$S_1^* = A_\Sigma^* (B_\Sigma^*)^{\left(-\sum_{i=1}^n u_{i1} - \sum_{i=1}^n v_{i1} H_1(m_i^* \| w^*)\right)} \text{ and}$$

$$S_2^* = (C_\Sigma^*)^{-1}(B_\Sigma^*)^{\left(\sum_{i=1}^n u_{i2} + \sum_{i=1}^n v_{i2} H_1(m_i^* \| w^*)\right)}. \tag{3}$$

From verification Eq. (2), one can simplify derive the relation between S_1^* and S_2^* as follows

$$1 = e(A_\Sigma^*, h_0) e(B_\Sigma^*, \prod_{i=1}^n h_{i1} \prod_{i=1}^n h_{i2}^{H_1(m_i^* \| w^*)}) e(C_\Sigma^*, h_z)^{-1}$$

$$= e(A_\Sigma^*, h_0) e(B_\Sigma^*, \prod_{i=1}^n (h_z^{u_{i2}} h_0^{-u_{i1}}) \prod_{i=1}^n (h_z^{v_{i2}} h_0^{-v_{i1}})^{H_1(m_i^* \| w^*)}) e(C_\Sigma^*, h_z)^{-1}$$

$$= e(A_\Sigma^*, h_0) e\left(B_\Sigma^*, (h_z^{\sum_{i=1}^n u_{i2}} h_0^{-\sum_{i=1}^n u_{i1}})(h_z^{\sum_{i=1}^n v_{i2} H_1(m_i^* \| w^*)} h_0^{-\sum_{i=1}^n v_{i1} H_1(m_i^* \| w^*)})\right)$$

$$\cdot e(C_\Sigma^*, h_z)^{-1}$$

$$= e(A_\Sigma^*, h_0) e(B_\Sigma^*, h_z^{\sum_{i=1}^n u_{i2} + \sum_{i=1}^n v_{i2} H_1(m_i^* \| w^*)} h_0^{(-\sum_{i=1}^n u_{i1} - \sum_{i=1}^n v_{i1} H_1(m_i^* \| w^*))})$$

$$\cdot e(C_\Sigma^*, h_z)^{-1}$$

$$= e(A_\Sigma^* (B_\Sigma^*)^{(-\sum_{i=1}^n u_{i1} - \sum_{i=1}^n v_{i1} H_1(m_i^* \| w^*))}, h_0)$$

$$\cdot e((C_\Sigma^*)^{-1}(B_\Sigma^*)^{(\sum_{i=1}^n u_{i2} + \sum_{i=1}^n v_{i2} H_1(m_i^* \| w^*))}, h_z)$$

$$= e(S_1^*, h_0) e(S_2^*, h_z). \tag{4}$$

Now we argue that the Type-II forgery class is the same as the complement of Type-I forgery class with respect to the forgery space $\mathcal{V}$, i.e., $\mathcal{V}_{II} = \mathcal{V} - \mathcal{V}_I$. In other words, the forgery from $\mathcal{V}_{II}$ ensures that $S_1^* \neq 1$ and $S_2^* \neq 1$ hold. Suppose if $S_1^* = 1$, then from Eq. (4) and from the non-degeneracy of the pairing, S_2^* must be 1. In the same way, suppose if $S_2^* = 1$, then S_1^* must be 1. Hence there is no valid forgery such that (i) $S_1^* = 1$ and $S_2^* \neq 1$ hold or (ii) $S_1^* \neq 1$ and $S_2^* = 1$ hold.

Structure of Forged Signature: Consider the Type-II forgery $(\overrightarrow{\mathbf{PK}}^*, \overrightarrow{m}^*, \sigma_\Sigma^* = (A_\Sigma^*, B_\Sigma^*, C_\Sigma^*, w^*))$ satisfying Eq. (4). From the above discussion, for Type-II forgery $S_1^* \neq 1$ and $S_2^* \neq 1$ hold. If $H_0(w^*) \neq B_\Sigma^*$, then abort the process. Otherwise, re-write the explicit form for A_Σ^* and C_Σ^* as follows. As S_1^* belongs to $G \setminus \{1\}$ and $B_\Sigma^* = H_0(w^*) \in G$, then S_1^* can be written as $H_0(w^*)^{s_1}$, for some s_1 from $\mathbb{Z}_p^*$. Then substituting B_Σ^* and S_1^* in Eq. (3), we get A_Σ^* as follows

$$S_1^* = A_\Sigma^* (B_\Sigma^*)^{(-\sum_{i=1}^n u_{i1} - \sum_{i=1}^n v_{i1} H_1(m_i^* \| w^*))}$$

$$H_0(w^*)^{s_1} = A_\Sigma^* H_0(w^*)^{(-\sum_{i=1}^n u_{i1} - \sum_{i=1}^n v_{i1} H_1(m_i^* \| w^*))}$$

$$A_\Sigma^* = H_0(w^*)^{(\sum_{i=1}^n u_{i1} + \sum_{i=1}^n v_{i1} H_1(m_i^* \| w^*)) + s_1}.$$

Similarly, as $S_2^* \in G \setminus \{1\}$, S_2^* can be written as $S_2^* = H_0(w^*)^{-s}$ for some s from $\mathbb{Z}_p^*$. By substituting B_Σ^* and S_2^* in Eq. (3), we get C_Σ^* as follows

$$S_2^* = (C_\Sigma^*)^{-1} B_\Sigma^{* (\sum_{i=1}^n u_{i2} + \sum_{i=1}^n v_{i2} H_1(m_i^* \| w^*))}$$

$$H_0(w^*)^{-s} = (C_\Sigma^*)^{-1} H_0(w^*)^{(\sum_{i=1}^n u_{i2} + \sum_{i=1}^n v_{i2} H_1(m_i^* \| w^*))}$$

$$C_\Sigma^* = H_0(w^*)^{(\sum_{i=1}^n u_{i2} + \sum_{i=1}^n v_{i2} H_1(m_i^* \| w^*)) + s}.$$

For the forgery signature $\sigma_\Sigma^* = (A_\Sigma^*, B_\Sigma^*, C_\Sigma^*, w^*)$ to be valid, it must satisfy the Eq. (2), then we will get $s_1 = s/\delta_0$ modulo p. Therefore, the Type-II forgery $\sigma_\Sigma^* = (A_\Sigma^*, B_\Sigma^*, C_\Sigma^*, w^*)$ can be written as

$$A_\Sigma^* = H_0(w^*)^{(\sum_{i=1}^n u_{i1} + \sum_{i=1}^n v_{i1} H_1(m_i^* \| w^*)) + (s/\delta_0)}, \quad B_\Sigma^* = H_0(w^*),$$

$$C_\Sigma^* = H_0(w^*)^{(\sum_{i=1}^n u_{i2} + \sum_{i=1}^n v_{i2} H_1(m_i^* \| w^*)) + s},$$

for some s from $\mathbb{Z}_p^*$.

Suppose the forgery is Type-I, then from the definition, both conditions $S_1^* = 1$ and $S_2^* = 1$ hold. From the above explanation for the Type-II forgery case and the above conditions, it is clear that s has to be zero modulo p for Type-I forgery. Hence by substituting $s = 0$ in the above equation, we obtain the desired form of Type-I forgery as defined in our construction.

Two Signing and Aggregate Signing Algorithms: Now we will be defining two algorithms for each of the signing and aggregate signing processes. Let $Sign_A$ be same as the **SAS.Sign** algorithm defined in our construction. Next, we define the following $Sign_B$ algorithm, which the simulator uses only in the unforgeability proof.

$Sign_B (pp, SK \cup \{\delta_0\}, m, w)$: Choose s randomly from $\mathbb{Z}_p$, compute

$$A = H_0(w)^{\left(u_1 + v_1 H_1(m \| w) + s/\delta_0\right)}, \quad B = H_0(w), \quad C = H_0(w)^{\left(u_2 + v_2 H_1(m \| w) + s\right)}$$

and return $(m, \sigma = (A, B, C, w))$.

Let $AggSign_A$ be the same as the **SAS.AggSign** algorithm defined in our construction. Next, we describe the following $AggSign_B$ algorithm, which the simulator uses only in the unforgeability proof.

$AggSign_B (pp, (PK_1, \ldots, PK_n), (m_1, \ldots, m_n), (\sigma_1, \ldots, \sigma_n))$: Parse $\sigma_i = (A_i, B_i, C_i, w_i)$. If $w_1 = w_j$ for $j \in [2, n]$, set $w = w_1$. Otherwise, abort it. Then compute $B_\Sigma = H_0(w)$,

$$A_\Sigma = \prod_{i=1}^{n} A_i = H_0(w)^{\left(\sum_{i=1}^{n} u_{i1} + \sum_{i=1}^{n} v_{i1} H_1(m_i \| w) + (\sum_{i=1}^{n} s_i)/\delta_0\right)},$$

$$C_\Sigma = \prod_{i=1}^{n} C_i = H_0(w)^{\left(\sum_{i=1}^{n} u_{i2} + \sum_{i=1}^{n} v_{i2} H_1(m_i \| w) + \sum_{i=1}^{n} s_i\right)}$$

and return $\sigma_\Sigma = (A_\Sigma, B_\Sigma, C_\Sigma, w)$.

From the verification Eq. (2), the additional element $H_0(w)^{s/\delta_0}$ in A_Σ paired with $h_0 = h_z^{\delta_0}$ is same as an additional element $H_0(w)^s$ in C_Σ paired with h_z. Hence, the signature returned by $Sign_B$ also can be verified under $(PK_1, \ldots, PK_n)$.

Dual-Form Signature Techniques: Now we describe how dual-form signature (DFS) techniques work. This technique uses two signing algorithms, namely $Sign_A$ and $Sign_B$, which return two forms of signatures that are verifiable under the same public key. The security reduction partitions the forgery space into two types, called Type-I and Type-II such that they respectively capture the signature returned by $Sign_A$ and $Sign_B$. While applying the DFS technique on a concrete protocol, the construction uses one of the signing algorithms say $Sign_A$, whereas $Sign_B$ is used only in the security reduction. Security of the scheme is proved using a hybrid argument. In the process, the reduction starts with the original unforgeability game. Firstly, one can show that given $Sign_A$ oracle access, the adversary cannot produce a Type-II forgery. Secondly, the signing oracle queries are changed one by one from $Sign_A$ to $Sign_B$ oracle. These changes are indistinguishable for the adversary. Finally, one can show that given only $Sign_B$ oracle access, the adversary can produce a Type-I forgery with some negligible probability.

Proof Intuition: We prove the unforgeability of our SynAS scheme using a hybrid argument in Theorem 1. Let $Game_R$ be the real EUF-CMA security game in the certified key model, i.e., given the public key, adversary $\mathcal{A}$ makes q_C many certification oracle queries and q many signing oracle queries which are answered using $Sign_A$ and returns an aggregate forgery (from $\mathcal{V}$) on a set of new messages at some time period w^*. Next, we define a new game $Game_0$ which is similar to $Game_R$ except that $\mathcal{A}$ returns a Type-I forgery. The only difference between $Game_R$ and $Game_0$ is that of $\mathcal{A}$ producing a Type-II forgery. Then we prove that (in Lemma 1) under the DBP_H assumption, $\mathcal{A}$ cannot return a Type-II forgery, which ensures that $Game_R$ and $Game_0$ are indistinguishable. In this reduction, simulator $\mathcal{B}$ embeds the DBP instance to generate the public key terms h_z and $h_z^{\delta_0}$. Then by choosing all the other secret exponents, $\mathcal{B}$ can answer for $Sign_A$ queries. Finally, from the Type-II forgery returned by $\mathcal{A}$, $\mathcal{B}$ computes the solution for the DBP_H instance.

Next, we define another game $Game_c$ which is similar to $Game_0$, except that the first c signing queries are answered using $Sign_B$ algorithm. Then we prove that $Game_{c-1}$ and $Game_c$ are indistinguishable under the DDH_G assumption (in Lemma 2). In this reduction, simulator $\mathcal{B}$ embeds one of the terms (say g^b) from the DDH instance to define the secret terms u_{i1} and u_{i2}. Whereas all the other secret terms are chosen in such a way that $\mathcal{B}$ can answer for both $Sign_A$ and $Sign_B$ queries. While answering for c-th signature query, $\mathcal{B}$ embeds the other terms (say g^a and $g^{ab+\theta}$) from the DDH instance. For a given DDH tuple with $\theta = 0$ we are simulating $Game_{c-1}$, whereas for $\theta \neq 0$ we are simulating $Game_c$.

Finally in Lemma 3, we argue that the advantage of $Game_q$ is negligible under the DBP_H assumption. In this reduction, simulator $\mathcal{B}$ embeds the DBP instance to simulate the public key components h_z and $h_z^{\delta_0}$. Then $\mathcal{B}$ defines the exponents u_{i1} and u_{i2} in such a way that $\mathcal{B}$ can only answer for $Sign_B$ oracle queries. In particular, $\mathcal{B}$ defines $u_{i1} = \tilde{u}_{i1} - t/\delta_0$ and $u_{i2} = \tilde{u}_{i2} - t$, for random exponents $\tilde{u}_{i1}, \tilde{u}_{i2}, t$. Once the adversary $\mathcal{A}$ returns a Type-I forgery, then $\mathcal{B}$ could extract the solution for the DBP problem.

Theorem 1. *The* SynAS *scheme is* EUF-CMA *secure in the certified-key model under SXDH assumption in the random oracle model.*

Proof. To prove security, first, we define the following games:

$Game_R$: This is the original EUF-CMA game in the certified-key model. Recall that, after receiving PK^* from the challenger, adversary $\mathcal{A}$ makes q_C many certification-oracle queries and q many signing oracle queries adaptively and then returns an aggregate signature forgery on a set of new messages at some time period w^*.

$Game_0$: Same as $Game_R$ except that $\mathcal{A}$ returns a forgery from $\mathcal{V}_I$. Let E be the event that $\mathcal{A}$ returns a forgery from $\mathcal{V}_{II}$ in $Game_0$. In Lemma 1, we prove that the event E happens with negligible probability under the DBP_H assumption. Thus we deduce that $Game_R$ and $Game_0$ are computationally indistinguishable under DBP_H assumption. In particular, we have,

$$|Adv_{\mathcal{A}}^{Game_R} - Adv_{\mathcal{A}}^{Game_0}| \leq Pr[E] \leq \frac{1}{|\mathcal{W}| \cdot q_{H_1}} Adv_{\mathcal{B}}^{DBP_H} \qquad (5)$$

$Game_c$: Same as $Game_0$ except that the first c signing queries are answered using $Sign_B$, for $c \in [1, q]$, whereas the last $q - c$ queries are answered using $Sign_A$. For $c \in [1, q]$, in Lemma 2, we prove that $Game_{c-1}$ and $Game_c$ are computationally indistinguishable under DDH_G assumption. In particular, we have,

$$|Adv_{\mathcal{A}}^{Game_{c-1}} - Adv_{\mathcal{A}}^{Game_c}| \leq \frac{1}{|\mathcal{W}| \cdot q_{H_1}} Adv_{\mathcal{B}}^{DDH_G}. \qquad (6)$$

Finally in Lemma 3, we prove that $Adv_{\mathcal{A}}^{Game_q}$ is negligible under DBP_H assumption. In particular, we have,

$$Adv_{\mathcal{A}}^{Game_q} \leq \frac{1}{|\mathcal{W}| \cdot q_{H_1}} Adv_{\mathcal{B}}^{DBP_H}. \qquad (7)$$

Hence by the hybrid argument and from Eqs. (5), (6) and (7), we have,

$$
\begin{aligned}
Adv_{\mathcal{A}}^{UF} = Adv_{\mathcal{A}}^{Game_R} &= |Adv_{\mathcal{A}}^{Game_R} - Adv_{\mathcal{A}}^{Game_0} + Adv_{\mathcal{A}}^{Game_0} - Adv_{\mathcal{A}}^{Game_1} \\
&\quad + ... + Adv_{\mathcal{A}}^{Game_{c-1}} - Adv_{\mathcal{A}}^{Game_c} + ... - Adv_{\mathcal{A}}^{Game_q} + Adv_{\mathcal{A}}^{Game_q}| \\
&\leq |Adv_{\mathcal{A}}^{Game_R} - Adv_{\mathcal{A}}^{Game_0}| \\
&\quad + \sum_{c=1}^{q} |Adv_{\mathcal{A}}^{Game_{c-1}} - Adv_{\mathcal{A}}^{Game_c}| + |Adv_{\mathcal{A}}^{Game_q}| \\
&\leq \frac{1}{|\mathcal{W}| \cdot q_{H_1}} Adv_{\mathcal{B}}^{DBP_H} + \frac{q}{|\mathcal{W}| \cdot q_{H_1}} Adv_{\mathcal{B}}^{DDH_G} + \frac{1}{|\mathcal{W}| \cdot q_{H_1}} Adv_{\mathcal{B}}^{DBP_H} \\
&\leq \left(\frac{q+2}{|\mathcal{W}| \cdot q_{H_1}} \right) Adv_{\mathcal{B}}^{SXDH}.
\end{aligned}
$$

Lemma 1. *If DBP_H assumption holds in $\mathcal{P}$, then $Pr[E]$ is negligible.*

Proof. Assume that the event E happens with some non-negligible probability. Then we construct a simulator $\mathcal{B}$ to break the DBP_H assumption as follows. $\mathcal{B}$ is given $\Theta = (p, G, H, G_T, e, g, h)$ and h_r, h_s from H and her/his goal is to compute $(R, S) \neq (1, 1)$ from G^2 such that $e(R, h_r)e(S, h_s) = 1$. Now $\mathcal{B}$ defines the hash functions $H_0 : \mathcal{W} \to G$ and $H_1 : \{0,1\}^* \times \mathcal{W} \to \mathbb{Z}_p^*$ and initializes the sets $\mathcal{Q}_0 = \emptyset$, $\mathcal{Q}_1 = \emptyset$ and $\mathcal{Q}_s = \emptyset$. Let q_{H_0} (resp. q_{H_1}) be the maximum number of hash oracle H_0 (resp. H_1) queries. Then, $\mathcal{B}$ interacts with the adversary $\mathcal{A}$ in the following manner:

Setup: First $\mathcal{B}$ sets $h_z := h_r$ and $h_0 := h_s$. Then, $\mathcal{B}$ chooses (u_1, u_2, v_1, v_2) uniformly at random from $\mathbb{Z}_p$ and defines the public key as $PK^* = \big(h_1^* = h_z^{u_2} h_0^{-u_1}, h_2^* = h_z^{v_2} h_0^{-v_1} \big)$. Now $\mathcal{B}$ chooses a random value h' from $\mathbb{Z}_p$ and computes $\tilde{A} = g^{r(u_1 + h' v_1)}$, $\tilde{B} = g^r$ and $\tilde{C} = g^{r(u_2 + h' v_2)}$ for some r randomly chosen from $\mathbb{Z}_p$. $\mathcal{B}$ chooses an index k and guesses a random time period $w' \in \mathcal{W}$ ($w' \leq W$) of the forgery signature and sets $H_0(w') = g^r$. $\mathcal{B}$ stores $(\tilde{A}, \tilde{B}, \tilde{C}, w', h')$. Then $\mathcal{B}$ sets the public parameter $pp = (p, G, H, G_T, e, g, h, h_z, h_0, H_0, H_1)$ and sends pp, PK^* to $\mathcal{A}$. Also, $\mathcal{B}$ initializes the key-pair list $KeyList = \emptyset$.

Certification Query: Before $\mathcal{A}$ can use any public key PK_i, $\mathcal{A}$ uses certificate oracle to validate the key pair (PK_i, SK_i). In the process, $\mathcal{A}$ requests for the certification of key pair to $\mathcal{B}$ by providing $PK_i = \{h_{i1}, h_{i2}\}$ and its corresponding secret key $SK_i = \{u_{i1}, u_{i2}, v_{i1}, v_{i2}\}$. Now $\mathcal{B}$ checks $h_{i1} = h_z^{u_{i2}} \cdot h_0^{-u_{i1}}$ and $h_{i2} = h_z^{v_{i2}} \cdot h_0^{-v_{i1}}$ holds or not. If the above checks are valid, $\mathcal{B}$ adds the key-pair (PK_i, SK_i) to the $KeyList$. Otherwise, $\mathcal{B}$ rejects the query.

Hash Query on H_0: $\mathcal{A}$ adaptively requests a hash value for H_0 by giving $w_j \in \mathcal{W}$ as input. First $\mathcal{B}$ checks whether w_j has been queried earlier to H_0. If so, then $\mathcal{B}$ returns the corresponding value $H_0(w_j)$. Otherwise, $\mathcal{B}$ proceeds as follows. If $w_j = w'$, then $\mathcal{B}$ extracts the stored pair $(\tilde{A}, \tilde{B}, \tilde{C}, w', h')$ and returns $H_0(w') = \tilde{B}$. Otherwise, $\mathcal{B}$ chooses r_j uniformly at random from $\mathbb{Z}_p$ and returns $H_0(w_j) = g^{r_j}$. Also, $\mathcal{B}$ adds the pair (w_j, g^{r_j}) into the list $\mathcal{Q}_0$ i.e. $\mathcal{Q}_0 \leftarrow \mathcal{Q}_0 \cup \{(w_j, g^{r_j})\}$.

Hash Query on H_1: $\mathcal{A}$ adaptively requests a hash value for H_1 by giving $m_i \in \{0,1\}^*$ and $w_j \in \mathcal{W}$ as input. First $\mathcal{B}$ checks whether (m_i, w_j) has been queried earlier to H_1. If so, then $\mathcal{B}$ returns the corresponding h_{ij} retrieved from $\mathcal{Q}_1$. Otherwise, $\mathcal{B}$ proceeds as follows. If $i = k$ and $w_j = w'$, then $\mathcal{B}$ extracts the stored pair $(\tilde{A}, \tilde{B}, \tilde{C}, w', h')$ and returns $H_1(m_i \| w') = h'$. Otherwise, $\mathcal{B}$ chooses h_{ij} uniformly at random from $\mathbb{Z}_p$ and returns $H_1(m_i \| w_j) = h_{ij}$. Also, $\mathcal{B}$ adds the pair (m_i, w_j, h_{ij}) to $\mathcal{Q}_1$ i.e. $\mathcal{Q}_1 \leftarrow \mathcal{Q}_1 \cup \{(m_i, w_j, h_{ij})\}$.

Signing query: $\mathcal{A}$ adaptively requests signing oracle queries to $\mathcal{B}$ on the message m_i at time period w_j. Since $\mathcal{B}$ knows all the secret key components $SK^* = \{u_1, u_2, v_1, v_2\}$, s/he can answer the signing queries using $Sign_A$ algorithm as follows:

1. If $w_j = w'$ and $i = k$, then $\mathcal{B}$ extracts the stored pair $(\tilde{A}, \tilde{B}, \tilde{C}, w', h')$ and returns the message-signature pair $(m_k, \sigma = (\tilde{A}, \tilde{B}, \tilde{C}, w'))$. $\mathcal{B}$ also updates the list $\mathcal{Q}_s \leftarrow \mathcal{Q}_s \cup \{(m_k, w', \sigma)\}$.
2. If $w_j = w'$ and $i \neq k$, then $\mathcal{B}$ aborts the simulation, because $\mathcal{A}$ can only request at most one signature per time period.
3. If $w_j \neq w'$, then using $\mathcal{Q}_0$, $\mathcal{B}$ checks whether w_j has been queried to H_0. If so, then $\mathcal{B}$ retrieves $H_0(w_j)$ from $\mathcal{Q}_0$. Otherwise, $\mathcal{B}$ requests hash oracle H_0 and obtains $H_0(w_j)$ and updates the list $\mathcal{Q}_0$. Similarly, using $\mathcal{Q}_1$, $\mathcal{B}$ checks whether (m_i, w_j) has been queried to H_1. If so, then $\mathcal{B}$ retrieves $H_1(m_i \| w_j)$ from $\mathcal{Q}_1$. Otherwise, $\mathcal{B}$ simulates the hash oracle H_1 and obtains $H_1(m_i \| w_j) = h_{ij}$ from $\mathbb{Z}_p$ and updates the list $\mathcal{Q}_1$. Now $\mathcal{B}$ computes $A_{ij} = H_0(w_j)^{u_1 + H_1(m_i \| w_j) v_1}$, $B_{ij} = H_0(w_j)$ and $C_{ij} = H_0(w_j)^{u_2 + H_1(m_i \| w_j) v_2}$. Then, $\mathcal{B}$ returns the message-signature pair $(m_i, \sigma_{ij} = (A_{ij}, B_{ij}, C_{ij}, w_j))$ to $\mathcal{A}$ and updates the list $\mathcal{Q}_s \leftarrow \mathcal{Q}_s \cup \{(m_i, w_j, \sigma_{ij})\}$.

Output: After q many signing and hashing oracles query, $\mathcal{A}$ eventually outputs an aggregate signature forgery $(\overrightarrow{\mathbf{PK}}^*, \overrightarrow{m}^*, \sigma_{\Sigma}^*)$, where $\overrightarrow{m}^* = (m_1^*, m_2^*, \ldots, m_n^*)$ with each $m_i^* \in \{0,1\}^*$, $\overrightarrow{\mathbf{PK}}^* = (PK_1, \ldots, PK_n)$ and $\sigma_{\Sigma}^* = (A_{\Sigma}^*, B_{\Sigma}^*, C_{\Sigma}^*, w^*)$. Now $\mathcal{B}$ checks

(i) The pair $(\overrightarrow{\mathbf{PK}}^*, \overrightarrow{m}^*, \sigma_{\Sigma}^*)$ is a valid aggregate signature forgery by checking Eq. (2).
(ii) There exists $j^* \in [1, n]$ such that $PK_{j^*} = PK^*$ and $m_{j^*}^*$ is not queried to signing oracle by checking the list $\mathcal{Q}_s$.
(iii) For each $i \in [1, n] \setminus \{j^*\}$, $\mathcal{B}$ retrieves the secret key $SK_i = \{u_{i1}, u_{i2}, v_{i1}, v_{i2}\}$ corresponding to PK_i from $KeyList$.

If any of these checks fail to hold, then $\mathcal{B}$ aborts. Otherwise, $\mathcal{B}$ solves the DBP_H assumption as follows. First $\mathcal{B}$ computes S_1^* and S_2^* as in Eq. (3), i.e.

$$S_1^* = A_{\Sigma}^* (B_{\Sigma}^*)^{(-\sum_{i=1}^n u_{i1} - \sum_{i=1}^n v_{i1} H_1(m_i^* \| w^*))},$$
$$S_2^* = (C_{\Sigma}^*)^{-1} (B_{\Sigma}^*)^{(\sum_{i=1}^n u_{i2} + \sum_{i=1}^n v_{i2} H_1(m_i^* \| w^*))}.$$

Since the forgery is valid, the signature components satisfy the verification Eq. (2), i.e.

$$e(A_\Sigma^*, h_0)e(B_\Sigma^*, \prod_{i=1}^n h_{i1} \prod_{i=1}^n h_{i2}^{H_1(m_i^*\|w^*)}) = e(C_\Sigma^*, h_z)$$

$$e(C_\Sigma^*, h_r)^{-1}e(A_\Sigma^*, h_s) \cdot e(B_\Sigma^*, \prod_{i=1}^n h_r^{u_{i2}} h_s^{-u_{i1}} \prod_{i=1}^n (h_r^{v_{i2}} h_s^{-v_{i1}})^{H_1(m_i^*\|w^*)}) = 1$$

$$e((C_\Sigma^*)^{-1}, h_r)e(A_\Sigma^*, h_s) \cdot e(B_\Sigma^*, h_r^{\sum_{i=1}^n u_{i2} + \sum_{i=1}^n v_{i2} H_1(m_i^*\|w^*)}$$

$$\cdot h_s^{-\sum_{i=1}^n u_{i1} - \sum_{i=1}^n v_{i1} H_1(m_i^*\|w^*)}) = 1$$

$$e((C_\Sigma^*)^{-1}(B_\Sigma^*)^{\sum_{i=1}^n u_{i2} + \sum_{i=1}^n v_{i2} H_1(m_i^*\|w^*)}, h_r)$$

$$\cdot e(A_\Sigma^*(B_\Sigma^*)^{-\sum_{i=1}^n u_{i1} - \sum_{i=1}^n v_{i1} H_1(m_i^*\|w^*)}, h_s) = 1$$

$$e(S_2^*, h_r)e(S_1^*, h_s) = 1$$

$$e(S_1^*, h_0)e(S_2^*, h_s) = 1.$$

In order to break the DBP_H assumption, it is sufficient to argue that the pair $(S_1^*, S_2^*) \neq (1,1)$. From our contradiction assumption, $\mathcal{A}$ returns a Type-II forgery with some non-negligible probability. Then from the definition of Type-II forgery, it must satisfy $S_1^* \neq 1$ and $S_2^* \neq 1$. Hence, $\mathcal{B}$ returns (S_1^*, S_2^*) as a non-trivial solution for the DBP_H instance to her/his challenger $\mathcal{C}$.

Now, we analyze the success probability of $\mathcal{B}$ as follows. First, we observe that $\mathcal{B}$ succeeds in the simulation only if s/he does not abort in the simulation of signature queries and s/he correctly guesses the time period w^* such that $w^* = w'$ in the adversary's aggregate signature forgery. During the above security reduction, $\mathcal{B}$ aborts the simulation of the signature queries if the time period given from $\mathcal{A}$ is w', i.e., $w^* = w'$ and the message index satisfies $i \neq k$. Also, the output of H_1 is chosen uniformly at random from $\mathbb{Z}_p$, then the probability of $\mathcal{B}$ correctly guessing the message index k is $\frac{1}{q_{H_1}}$. Similarly, the time period w' of the aggregate signature forgery is chosen randomly from $\mathcal{W}$, then the probability of $\mathcal{B}$ correctly guessing $w^* = w'$ is $\frac{1}{|\mathcal{W}|}$. So the probability of $\mathcal{A}$ coming up with a forgery (without abort) is calculated using the following advantage

$$Pr[E] \leq \frac{1}{|\mathcal{W}| \cdot q_{H_1}} Adv_\mathcal{B}^{DBP_H}.$$

Lemma 2. If DDH_G assumption holds in $\mathcal{P}$, then $Game_{c-1}$ is indistinguishable from $Game_c$ in the random oracle model, for $c \in [1, q]$.

Proof. Suppose, there exists a PPT Adversary $\mathcal{A}$, who distinguishes $Game_{c-1}$ from $Game_c$ with some non-negligible probability under the condition that $\mathcal{A}$ returns a Type-II forgery. Then we construct a simulator $\mathcal{B}$ to break the DDH_G assumption as follows. $\mathcal{B}$ is given $\Theta = (p, G, H, G_T, e, g, h)$, $g^a, g^b, g^{ab+\theta}$ and her/his goal is to decide whether $\theta = 0$ mod p or not. As in Lemma 1, $\mathcal{B}$ defines H_0, H_1, q_{H_0}, q_{H_1} and initializes $\mathcal{Q}_0$, $\mathcal{Q}_1$, $\mathcal{Q}_s$ and $KeyList$ as empty. Then $\mathcal{B}$ interacts with the adversary $\mathcal{A}$ in the following manner:

Setup: $\mathcal{B}$ chooses $\delta_0, \tilde{u}_1, \tilde{u}_2, t, v_1, v_2$ uniformly at random from $\mathbb{Z}_p$ and implicitly sets $u_1 = \tilde{u}_1 + tb/\delta_0$ and $u_2 = \tilde{u}_2 + tb$. Now $\mathcal{B}$ chooses h_z uniformly at random from H and sets $h_0 = h_z^{\delta_0}$. Then $\mathcal{B}$ defines $\delta_1 = u_2 - \delta_0 u_1 = \tilde{u}_2 - \delta_0 \tilde{u}_1$, $\delta_2 = v_2 - \delta_0 v_1$ and sets $h_1 = h_z^{\tilde{u}_2 - \delta_0 \tilde{u}_1}$, $h_2 = h_z^{v_2 - \delta_0 v_1}$. Thus $\mathcal{B}$ can define the challenge public key PK^* as (h_1, h_2). $\mathcal{B}$ simulates $U_1 = g^{\tilde{u}_1}(g^b)^{t/\delta_0} = g^{\tilde{u}_1 + tb/\delta_0} = g^{u_1}$, $U_2 = g^{\tilde{u}_2}(g^b)^t = g^{\tilde{u}_2 + tb} = g^{u_2}$ and computes $V_1 = g^{v_1}$ and $V_2 = g^{v_2}$. Using U_1, U_2, V_1, V_2, $\mathcal{B}$ can answer for the signing queries of $Sign_A$ type and by additionally knowing δ_0 s/he can answer $Sign_B$ type too. Now $\mathcal{B}$ chooses a random value h' from $\mathbb{Z}_p^*$, chooses an index k and guesses a random time period $w' \in \mathcal{W}$ of the forgery signature and embeds the DDH instance as $H_0(w') = g^a$ and stores $(\tilde{A}, \tilde{B}, \tilde{C}, w', h')$, where $\tilde{A} = (g^a)^{\tilde{u}_1 + h'v_1}$, $\tilde{B} = g^a$ and $\tilde{C} = (g^a)^{\tilde{u}_2 + h'v_2}$. Then, $\mathcal{B}$ sets the public parameter $pp = (p, G, H, G_T, e, g, h, h_z, h_0, H_0, H_1)$ and sends pp, PK^* to $\mathcal{A}$.

Certification Query: As in Lemma 1, $\mathcal{B}$ answers for the certification query (PK_i, SK_i) made by the adversary $\mathcal{A}$, where $PK_i = \{h_{i1}, h_{i2}\}$ and $SK_i = \{u_{i1}, u_{i2}, v_{i1}, v_{i2}\}$.

Hash Query on H_0: Note that, for the simulation purpose, $\mathcal{B}$ needs to store the randomness r_j in $\mathcal{Q}_0$. Now, $\mathcal{A}$ adaptively requests a hash value for H_0 by giving $w_j \in \mathcal{W}$ as input. If $w_j = w'$, then $\mathcal{B}$ extracts the stored value $H_0(w') = g^a$, adds the pair $(w', \emptyset, g^a)$ to the list $\mathcal{Q}_0$ and returns g^a to $\mathcal{A}$. Otherwise, $\mathcal{B}$ chooses r_j uniformly at random from $\mathbb{Z}_p$, updates $\mathcal{Q}_0 \leftarrow \mathcal{Q}_0 \cup \{(w_j, r_j, g^{r_j})\}$ and returns $H_0(w_j) = g^{r_j}$ to $\mathcal{A}$.

Hash Query on H_1: As in Lemma 1, $\mathcal{B}$ answers for the hash oracle H_1 queries on m_i, w_j as input and updates $\mathcal{Q}_1 \leftarrow \mathcal{Q}_1 \cup \{(m_i, w_j, h_{ij})\}$ and returns $H_1(m_i \| w_j) = h_{ij}$ to $\mathcal{A}$, for h_{ij} randomly chosen from $\mathbb{Z}_p$.

Signing Query: $\mathcal{A}$ makes q many signing oracle queries to $\mathcal{B}$ on the message m_i at time period w_j.

1. If $w_j = w'$ and $i = k$, then $\mathcal{B}$ extracts the stored pair $(\tilde{A}, \tilde{B}, \tilde{C}, w', h')$ and returns the message-signature pair $\left(m_k, \sigma = (\tilde{A}, \tilde{B}, \tilde{C}, w')\right)$. $\mathcal{B}$ also updates the list $\mathcal{Q}_s \leftarrow \mathcal{Q}_s \cup \{(m_k, w', \sigma)\}$. Note that due to the relation $u_2 - \delta_0 u_1 = \tilde{u}_2 - \delta_0 \tilde{u}_1$, one can verify that the pair (m_k, σ) satisfies the verification Eq. (1). Even though, the above signature component do not use u_1 and u_2, but uses $\tilde{u}_1$ and $\tilde{u}_2$, all components of the signature σ are distributed as in $Sign_A$ algorithm.

2. If $w_j = w'$ and $i \neq k$, then $\mathcal{B}$ aborts the simulation, because $\mathcal{A}$ can only request at most one signature per time period.

3. If $w_j \neq w'$, then using $\mathcal{Q}_0$, $\mathcal{B}$ checks whether w_j has been queried to H_0. If so, $\mathcal{B}$ retrieves $H_0(w_j)$ from $\mathcal{Q}_0$. Otherwise, $\mathcal{B}$ requests hash oracle H_0 and obtains $(w_j, H_0(w_j))$ and updates the list $\mathcal{Q}_0$. Similarly, using $\mathcal{Q}_1$, $\mathcal{B}$ checks whether (m_i, w_j) has been queried to H_1. If so, $\mathcal{B}$ retrieves $H_1(m_i \| w_j)$ from $\mathcal{Q}_1$. Otherwise, $\mathcal{B}$ simulates the hash oracle H_1 and obtains h_{ij} from $\mathbb{Z}_p$ and updates the list $\mathcal{Q}_1$. Since, $\mathcal{B}$ knows U_1, U_2, V_1, V_2 and δ_0, the first $(c-1)$ queries are answered by $Sign_B$ algorithm and the last $(q-c)$ queries are answered by $Sign_A$ algorithm.

$\mathcal{B}$ computes A_{ij} as follows:

$$A_{ij} = (U_1 V_1^{H_1(m_i\|w_j)})^{r_j}(g^{r_j})^{s/\delta_0} = g^{r_j\left(u_1 + v_1 H_1(m_i\|w_j) + s/\delta_0\right)}$$

$$= H_0(w_j)^{u_1 + v_1 H_1(m_i\|w_j) + s/\delta_0}$$

Similarly,

$$C_{ij} = (U_2 V_2^{H_1(m_i\|w_j)})^{r_j}(g^{r_j})^s = H_0(w_j)^{u_2 + v_2 H_1(m_i\|w_j) + s}$$

The $Sign_A$ queries are answered by letting $s = 0$ in the above signature obtained by $Sign_B$ algorithm.

For the c-th query, $\mathcal{B}$ embeds the DDH instance to construct the signature $\sigma_c = (A_c, B_c, C_c, w_c)$. $\mathcal{B}$ extracts $H_0(w') = g^a$ from $\mathcal{Q}_0$ and $H_1(m_i\|w')$ from $\mathcal{Q}_1$. Then sets $B_c = H_0(w') = g^a$ and computes A_c as follows:

$$A_c = (g^a)^{\tilde{u}_1}(g^{ab+\theta})^{t/\delta_0}(g^a)^{H_1(m_i\|w')v_1} = (g^a)^{\tilde{u}_1 + tb/\delta_0}(g^a)^{H_1(m_i\|w')v_1}g^{t\theta/\delta_0}$$

$$= (g^a)^{u_1 + H_1(m_i\|w')v_1}(g^a)^{a^{-1}t\theta/\delta_0} \doteq H_0(w')^{u_1 + H_1(m_i\|w')v_1 + s/\delta_0},$$

where s is implicitly set as $a^{-1}t\theta \pmod{p}$. Similarly $\mathcal{B}$ computes C_c

$$C_c = (g^a)^{\tilde{u}_2}(g^{ab+\theta})^t(g^a)^{H_1(m_i\|w')v_2} = (g^a)^{\tilde{u}_2 + tb}(g^a)^{H_1(m_i\|w')v_2}g^{t\theta}$$

$$= (g^a)^{u_2 + H_1(m_i\|w')v_2}(g^a)^{a^{-1}t\theta} = H_0(w')^{u_2 + H_1(m_i\|w')v_2 + s}.$$

After each signing query, $\mathcal{B}$ updates the list as $\mathcal{Q}_s \leftarrow \mathcal{Q}_s \cup \{(m_i, w_j, \sigma_{ij})\}$. Note that the exponent a from the DDH instance is used to simulate the signature component B_c and hence the random oracle function H_0, whereas $\mathcal{B}$ implicitly sets $s = a^{-1}t\theta$ modulo p. If $\theta = 0$ modulo p, then $s = 0$ modulo p, i.e., the signature σ_c is distributed as an output of $Sign_A$. If $\theta \neq 0$ modulo p, then $s \neq 0$ modulo p, i.e., the signature σ_c is distributed as an output of $Sign_B$ with non-zero exponent $s = a^{-1}t\theta$ modulo p.

Output: After q many signing and hashing oracles query, $\mathcal{A}$ eventually outputs an aggregate signature forgery $(\overrightarrow{\mathbf{PK}}^*, \overrightarrow{m}^*, \sigma_\Sigma^*)$, where $\overrightarrow{m}^* = (m_1^*, m_2^*, \ldots, m_n^*)$ with each $m_i^* \in \{0,1\}^*$, $\overrightarrow{\mathbf{PK}}^* = (PK_1, \ldots, PK_n)$ and $\sigma_\Sigma^* = (A_\Sigma^*, B_\Sigma^*, C_\Sigma^*, w^*)$. Now $\mathcal{B}$ checks

- The pair $(\overrightarrow{\mathbf{PK}}^*, \overrightarrow{m}^*, \sigma_\Sigma^*)$ is a valid aggregate signature forgery by checking Eq. (2)
- There exists $j^* \in [1, n]$ such that $PK_{j^*} = PK^*$ and $m_{j^*}^*$ is not queried to signing oracle by checking the list $\mathcal{Q}_s$ and
- For each $i \in [1, n] \setminus \{j^*\}$, $\mathcal{B}$ retrieves the secret key $SK_i = \{u_{i1}, u_{i2}, v_{i1}, v_{i2}\}$ corresponding to PK_i from $KeyList$.

If any of the above checks fail to hold, then $\mathcal{B}$ aborts. Otherwise, $\mathcal{B}$ proceeds to solve the DDH_G assumption. Note that σ_c was generated using the DDH_G instance. As $\mathcal{B}$ knows δ_0, U_1, U_2 and all other secret key components, $\mathcal{B}$ can generate k-th signature of any type properly. However, $\mathcal{B}$ cannot on her/his own decide the type of signatures generated using the problem instance, as s/he cannot compute S_1^* and S_2^* (defined in Eq. (3)) which uses the exponents u_1 and u_2. So $\mathcal{B}$ needs to rely on the advantage of $\mathcal{A}$. From Lemma 1, under the DBP assumption, $\mathcal{A}$ only returns a Type-I forgery. Also from our initial contradiction assumption, $\mathcal{A}$ distinguishes between $Game_{c-1}$ and $Game_c$ with some non-negligible probability. So $\mathcal{B}$ leverages $\mathcal{A}$ to break the DDH$_G$ assumption. As in Lemma 1, we have

$$|Adv_{\mathcal{A}}^{Game_{c-1}} - Adv_{\mathcal{A}}^{Game_c}| \leq \frac{1}{|\mathcal{W}| \cdot q_{H_1}} Adv_{\mathcal{B}}^{DDH_G}.$$

Lemma 3. *If* DBP_H *assumption holds in* $\mathcal{P}$, *then* Adv^{Game_q} *is negligible in the random oracle model.*

Proof. Suppose, assume that there exists a PPT adversary $\mathcal{A}$, who wins in $Game_q$ and produces a Type-I forgery with some non-negligible probability. Then we construct a simulator $\mathcal{B}$ to break the DBP_H assumption as follows. $\mathcal{B}$ is given $\Theta = (p, G, H, G_T, e, g, h)$, and h_r, h_s from H and her/his goal is to compute $(R, S) \neq (1, 1)$ from G^2 such that $e(R, h_r)e(S, h_s) = 1$. As in Lemma 1, $\mathcal{B}$ defines $H_0, H_1, q_{H_0}, q_{H_1}$ and initializes $\mathcal{Q}_0, \mathcal{Q}_1, \mathcal{Q}_s$ and $KeyList$ as empty. Now, $\mathcal{B}$ interacts with the adversary $\mathcal{A}$ in the following manner:

Setup: First $\mathcal{B}$ sets $h_z := h_r$, $h_0 := h_s$ and implicitly sets $u_1 = \tilde{u}_1 - t/\delta_0$ and $u_2 = \tilde{u}_2 - t$ for some randomly chosen $\tilde{u}_1, \tilde{u}_2, t$ from $\mathbb{Z}_p$. Thus $\mathcal{B}$ can simulate $U_1' = g^{\tilde{u}_1} = g^{u_1 + t/\delta_0}$ and $U_2' = g^{\tilde{u}_2} = g^{u_2 + t}$. $\mathcal{B}$ implicitly sets $\delta_1 = u_2 - \delta_0 u_1 = \tilde{u}_2 - \delta_0 \tilde{u}_1$ and chooses v_1, v_2 uniformly at random from $\mathbb{Z}_p$ and defines $\delta_2 = v_2 - \delta_0 v_1$. Then $\mathcal{B}$ computes $h_1 = h_z^{\delta_1} = h_r^{\tilde{u}_2} h_s^{-\tilde{u}_1}$ and $h_2 = h_z^{\delta_2} = h_r^{v_2} h_s^{-v_1}$. $\mathcal{B}$ defines PK^* as (h_1, h_2). $\mathcal{B}$ chooses a random value h' from $\mathbb{Z}_p$ and computes $\tilde{A} = (U_1' V_1^{h'})^r = (g^{u_1 + t/\delta_0} \cdot g^{v_1 h'})^r = (g^r)^{u_1 + v_1 h' + t/\delta_0}$, $\tilde{B} = g^r$ and $\tilde{C} = (U_2' V_2^{h'})^r = (g^{u_2 + t} \cdot g^{v_2 h'})^r = (g^r)^{u_2 + v_2 h' + t}$ for some r randomly chosen from $\mathbb{Z}_p$. $\mathcal{B}$ chooses an index k and guesses a random time period $w' \in \mathcal{W}$ of the forgery signature and sets $H_0(w') = g^r$. $\mathcal{B}$ stores $(\tilde{A}, \tilde{B}, \tilde{C}, w', h')$. $\mathcal{B}$ sets the public parameter $pp = (p, G, H, G_T, e, g, h, h_z, h_0, H_0, H_1)$ and sends pp, PK^* to $\mathcal{A}$. Also, $\mathcal{B}$ initializes the key-pair list $KeyList = \emptyset$.

Certification Query: As in Lemma 1, $\mathcal{B}$ answers for the certification query on (SK_i, PK_i) by the adversary $\mathcal{A}$.

Hash Query on H_0 and H_1 : As in Lemma 1, $\mathcal{B}$ answers for the hash oracle queries on H_0 and H_1.

Signing Query: As in Lemma 1, $\mathcal{B}$ can answer the signing queries for the cases (i) $w_j = w'$ and $i = k$, (ii) $w_j = w'$ and $i \neq k$. While answering for the case $w_j \neq w'$, $\mathcal{B}$ answers using $Sign_B()$ algorithm as follows. As before, $\mathcal{B}$ retrieves $(w_j, H_0(w_j)) \in \mathcal{Q}_0$ and $(m_i, w_j, h_{ij}) \in \mathcal{Q}_1$.

Now $\mathcal{B}$ computes $B_{ij} = H_0(w_j)$, A_{ij} as

$$A_{ij} := (U_1' V_1^{H_1(m_i \| w_j)})^{r_j} = (g^{u_1 + t/\delta_0} \cdot g^{v_1 H_1(m_i \| w_j)})^{r_j}$$
$$= (g^{r_j})^{u_1 + v_1 H_1(m_i \| w_j) + t/\delta_0} = H_0(w_j)^{u_1 + v_1 H_1(m_i \| w_j) + t/\delta_0}.$$

Similarly $\mathcal{B}$ computes C_{ij} as follows:

$$C_{ij} := (U_2' V_2^{H_1(m_i \| w_j)})^{r_j} = (g^{u_2+t} \cdot g^{v_2 H_1(m_i \| w_j)})^{r_j}$$
$$= (g^{r_j})^{u_2 + v_2 H_1(m_i \| w_j)+t} = H_0(w_j)^{u_2 + v_2 H_1(m_i \| w_j)+t}.$$

From the above derivation, it is clear that the signature $\sigma_{ij} = (A_{ij}, B_{ij}, C_{ij}, w_j)$ is properly distributed as an output of $Sign_B$ with $s = t \bmod p$. Then, $\mathcal{B}$ returns the message-signature pair $(m_i, \sigma_{ij} = (A_{ij}, B_{ij}, C_{ij}, w_j))$ to $\mathcal{A}$ and updates the list $\mathcal{Q}_s \leftarrow \mathcal{Q}_s \cup \{(m_i, w_j, \sigma_{ij})\}$.

Output: After q many signing and hashing oracles query, $\mathcal{A}$ eventually outputs an aggregate signature forgery signature $(\overrightarrow{\mathbf{PK}}^*, \overrightarrow{m}^*, \sigma_{\Sigma}^*)$, where $\overrightarrow{m}^* = (m_1^*, m_2^*, \ldots, m_n^*)$ with each $m_i^* \in \{0,1\}^*$, $\overrightarrow{\mathbf{PK}}^* = (PK_1, \ldots, PK_n)$ and $\sigma_{\Sigma}^* = (A_{\Sigma}^*, B_{\Sigma}^*, C_{\Sigma}^*, w^*)$. Now $\mathcal{B}$ checks

- The pair $(\overrightarrow{\mathbf{PK}}^*, \overrightarrow{m}^*, \sigma_{\Sigma}^*)$ is a valid aggregate signature forgery by checking Eq. (2)
- There exists $j^* \in [1, n]$ such that $PK_{j^*} = PK^*$ and $m_{j^*}^*$ is not queried to signing oracle by checking the list $\mathcal{Q}_s$ and
- For each $i \in [1, n] \setminus \{j^*\}$, $\mathcal{B}$ retrieves the secret key $SK_i = \{u_{i1}, u_{i2}, v_{i1}, v_{i2}\}$ corresponding to PK_i from $KeyList$.

If any of these checks fail to hold, then $\mathcal{B}$ aborts. Otherwise, $\mathcal{B}$ solves the DBP_H assumption as follows. From the contradiction assumption, $\mathcal{A}$ returns a Type-I forgery with some non-negligible probability. Then, $\mathcal{B}$ can write the Type-I forgery components as, $B_{\Sigma}^* = g^r = H_0(w^*)$,

$$A_{\Sigma}^* = (g^r)^{\sum_{i=1}^{n} u_{i1} + \sum_{i=1}^{n} v_{i1} H_1(m_i^* \| w^*)} = H_0(w^*)^{\sum_{i=1}^{n} u_{i1} + \sum_{i=1}^{n} v_{i1} H_1(m_i^* \| w^*)},$$
$$C_{\Sigma}^* = (g^r)^{\sum_{i=1}^{n} u_{i2} + \sum_{i=1}^{n} v_{i2} H_1(m_i^* \| w^*)} = H_0(w^*)^{\sum_{i=1}^{n} u_{i2} + \sum_{i=1}^{n} v_{i2} H_1(m_i^* \| w^*)}$$

for some r from $\mathbb{Z}_p$.

Now $\mathcal{B}$ computes S_1^* and S_2^* as follows:

$$S_1^* = A_{\Sigma}^* (B_{\Sigma}^*)^{- \sum_{i=1, i \neq j^*}^{n} u_{i1} - \sum_{i=1}^{n} v_{i1} H_1(m_i^* \| w^*) - \tilde{u}_1}$$
$$= H_0(w^*)^{\sum_{i=1}^{n} u_{i1} + \sum_{i=1}^{n} v_{i1} H_1(m_i^* \| w^*)}$$
$$\cdot (H_0(w^*))^{- \sum_{i=1, i \neq j^*}^{n} u_{i1} - \sum_{i=1}^{n} v_{i1} H_1(m_i^* \| w^*) - \tilde{u}_1}$$
$$= H_0(w^*)^{u_{j^*1} - \tilde{u}_1} = H_0(w^*)^{u_1 - \tilde{u}_1} = H_0(w^*)^{-t/\delta_0}$$
$$S_2^* = (C_{\Sigma}^*)^{-1} (B_{\Sigma}^*)^{\sum_{i=1, i \neq j^*}^{n} u_{i2} + \sum_{i=1}^{n} v_{i2} H_1(m_i^* \| w^*) + \tilde{u}_2}$$
$$= H_0(w^*)^{- \left(\sum_{i=1}^{n} u_{i2} + \sum_{i=1}^{n} v_{i2} H_1(m_i^* \| w^*) \right)}$$
$$\cdot H_0(w^*)^{\sum_{i=1, i \neq j^*}^{n} \tilde{u}_{i2} + \sum_{i=1}^{n} v_{i2} H_1(m_i^* \| w^*) + \tilde{u}_2}$$
$$= H_0(w^*)^{-u_{j^*2} + \tilde{u}_2} = H_0(w^*)^{-u_2 + \tilde{u}_2} = H_0(w^*)^{t}.$$

Then one can verify that,

$$e(S_1^*, h_s)e(S_1^*, h_r) = e(H_0(w^*)^{-t/\delta_0}, h_z^{\delta_0})e(H_0(w^*)^t, h_z) = 1.$$

From the verification Eq. (2), $B_\Sigma^* \neq 1$ holds and hence $S_2^* \neq 1$ holds. In other words, (S_1^*, S_2^*) is a non-trivial solution of DBP_H problem instance. As in Lemma 1, we have,

$$Adv_A^{Game_q} \leq \frac{1}{|\mathcal{W}| \cdot q_{H_1}} Adv_B^{DBP_H}.$$

4 Comparison

In this section, we compare our SynAS construction with synchronized aggregate signatures of [3,25,38] and [39]. Ahn, Green, and Hohenberger [3] constructed two synchronized aggregate signature schemes based on the Hohenberger-Waters [20] short signature scheme. One is constructed in the random oracle model (denoted as AGH-Sym-ROM in symmetric pairing, AGH-Asym-ROM in asymmetric pairing) and the other is constructed in the standard model (denoted as AGH-Sym-SM in symmetric pairing, AGH-Asym-SM in the asymmetric pairing). The security of Ahn et al.'s [3] scheme is proved under the standard computational Diffie-Hellman (CDH) assumption. However, their scheme has the drawback of having limited message space. In particular, message space has Z-bits which is divided into k chunks of ℓ bits each. Lee, Lee, and Yung [25] proposed a synchronized aggregate signature scheme based on the Camenisch-Lysyanskaya signature (CL) scheme [10]. The security of this scheme relies on the static OT-LRSW assumption in the random oracle model.

Table 1. Comparison of various SynAS schemes

		|PK|	|σ|	Signing cost	Aggregation cost	Agg Verify cost	Assumption
AGH-Sym-SM [3]	Symmetric Pairing	1|G|	2|G|+ 1|W| *	$(k+5)E_G +$ $(k+3)M_G$	$(2n-2)M_G$	$(k+3)\mathbb{P}+ 2nM_G+$ $2nE_G+(k+1) M_{G_T}$	CDH
AGH-Sym-ROM [3]		1|G|	2|G| + 1|W| *	$6E_G + 4M_G$	$(2n-2)M_G$	$4\mathbb{P} + 2n\, M_G+$ $(2n+2)E_G+2M_{G_T}$	CDH+ROM
LLY13 [25]		1|G|	1|G|+ 1|W|	$2E_G +1M_G$	$(n-1)M_G$	$3\mathbb{P}+(2n-2)M_G+$ $nE_G+1M_{G_T}$	OT-LRSW + ROM
TT20 [38]		1|G|	1|G| + 1 |W|	$2E_G +1M_G$	$(n-1)M_G$	$3\mathbb{P}+(2n-2)M_G +$ $nE_G+1M_{G_T}$	1-MSDH-2 + ROM
AGH-Asym-SM [3]	Asymmetric Pairing	1|G|	2 |G| + 1|W| *	$(k+5)E_G +$ $(k+3)M_G$	$(2n-2)M_G$	$(k+3)P + nE_G +$ $(2n-2)M_G + 2E_H +$ $2M_H + (k+1)M_{G_T}$	co-CDH
AGH-Asym-ROM [3]		1|G|	2|G| + 1|W| *	$6E_G +$ $4M_G$	$(2n-2)M_G$	$4P + nE_G + (2n-$ $2)M_G + 2E_H +$ $2M_H + 2M_{G_T}$	co-CDH + ROM
PS-SynAS [39]		2|H|	2|G| + 1|W|	$1E_G$	$(n-1)M_G$	$2P + (2n-2)M_H +$ nE_H	Generalized PS + ROM
SynAS (Sect. 3.1)		2|H|	3|G| + 1|W|	$2E_G$	$(2n-2)M_G$	$3P+(2n-2)M_H +$ $nE_H+1M_{G_T}$	SXDH + ROM

* In all [3]'s construction, the time-bound size is exponential in the security parameter λ, whereas other constructions time bound size is polynomial in λ.

Recently, Tezuka-Tanaka [39] constructed (denoted as PS-SynAS) synchronized aggregate signature scheme based on efficient Pointcheval-Sanders randomizable signatures in the random-oracle model. However, the security of the scheme is proved under the Generalized PS assumption [24], which is an interactive assumption. Whereas, our scheme SynAS is inspired by the construction of Tezuka-Tanaka along with an additional signature component (corresponding to a QA-NIZK simulated proof) and proved its security under SXDH assumption. Due to the additional signature component, there is a slight deficiency of signature size, signing cost, aggregation cost, and aggregation verification cost with respect to [39]'s PS-SynAS scheme.

Table 1 describes the comparison of both symmetric and asymmetric pairing-based synchronized aggregate signature constructions in both the random oracle model (ROM) and the standard model (SM). Let $\mathcal{W}$ be the time-period space and n be the number of signers. Consider the tuple (G, H, G_T) to denote the bilinear groups from the asymmetric pairing setting and P denotes the asymmetric pairing computation cost. Similarly the tuple $(\mathbb{G}, \mathbb{G}_T)$ denotes the symmetric bilinear groups and $\mathbb{P}$ denotes the symmetric pairing computation cost. For any group $X \in \{\mathbb{G}, \mathbb{H}, \mathbb{G}_T, G, G_T\}$, $|X|$ is the bit size of X and E_X, M_X respectively denote the cost of exponentiation, multiplication in X.

References

1. Abdalla, M., Pointcheval, D.: Interactive Diffie-Hellman assumptions with applications to password-based authentication. In: Patrick, A.S., Yung, M. (eds.) FC 2005. LNCS, vol. 3570, pp. 341–356. Springer, Heidelberg (2005). https://doi.org/10.1007/11507840_31

2. Abe, M., Fuchsbauer, G., Groth, J., Haralambiev, K., Ohkubo, M.: Structure-preserving signatures and commitments to group elements. J. Cryptol. **29**(2), 363–421 (2016)

3. Ahn, J.H., Green, M., Hohenberger, S.: Synchronized aggregate signatures: new definitions, constructions and applications. In: Al-Shaer, E., Keromytis, A.D., Shmatikov, V. (eds.) Proceedings of the 17th ACM Conference on Computer and Communications Security, CCS 2010, Chicago, Illinois, USA, 4–8 October 2010, pp. 473–484. ACM (2010)

4. Bellare, M., Namprempre, C., Pointcheval, D., Semanko, M.: The one-more-RSA-inversion problems and the security of Chaum's blind signature scheme. J. Cryptol. **16**(3), 185–215 (2003)

5. Benhamouda, F., Lepoint, T., Loss, J., Orrù, M., Raykova, M.: On the (in)security of ROS. J. Cryptol. **35**(4), 25 (2022)

6. Boldyreva, A., Gentry, C., O'Neill, A., Yum, D.H.: Ordered multisignatures and identity-based sequential aggregate signatures, with applications to secure routing. In: Ning, P., De Capitani di Vimercati, S., Syverson, P.F. (eds.) Proceedings of the 2007 ACM Conference on Computer and Communications Security, CCS 2007, Alexandria, Virginia, USA, 28–31 October 2007, pp. 276–285. ACM (2007)

7. Boneh, D., Boyen, X., Goh, E.-J.: Hierarchical identity based encryption with constant size ciphertext. In: Cramer, R. (ed.) EUROCRYPT 2005. LNCS, vol. 3494, pp. 440–456. Springer, Heidelberg (2005). https://doi.org/10.1007/11426639_26

8. Bourse, F., Pointcheval, D., Sanders, O.: Divisible E-cash from constrained pseudo-random functions. In: Galbraith, S.D., Moriai, S. (eds.) ASIACRYPT 2019, Part I. LNCS, vol. 11921, pp. 679–708. Springer, Cham (2019). https://doi.org/10.1007/978-3-030-34578-5_24

9. Camenisch, J., Drijvers, M., Lehmann, A., Neven, G., Towa, P.: Short threshold dynamic group signatures. In: Galdi, C., Kolesnikov, V. (eds.) SCN 2020. LNCS, vol. 12238, pp. 401–423. Springer, Cham (2020). https://doi.org/10.1007/978-3-030-57990-6_20

10. Camenisch, J., Lysyanskaya, A.: Signature schemes and anonymous credentials from bilinear maps. In: Franklin, M. (ed.) CRYPTO 2004. LNCS, vol. 3152, pp. 56–72. Springer, Heidelberg (2004). https://doi.org/10.1007/978-3-540-28628-8_4

11. Chatterjee, S., Kabaleeshwaran, R.: From rerandomizability to sequential aggregation: efficient signature schemes based on SXDH assumption. In: Liu, J.K., Cui, H. (eds.) ACISP 2020. LNCS, vol. 12248, pp. 183–203. Springer, Cham (2020). https://doi.org/10.1007/978-3-030-55304-3_10

12. Desmoulins, N., Lescuyer, R., Sanders, O., Traoré, J.: Direct anonymous attestations with dependent basename opening. In: Gritzalis, D., Kiayias, A., Askoxylakis, I. (eds.) CANS 2014. LNCS, vol. 8813, pp. 206–221. Springer, Cham (2014). https://doi.org/10.1007/978-3-319-12280-9_14

13. Gentry, C., Ramzan, Z.: Identity-based aggregate signatures. In: Yung, M., Dodis, Y., Kiayias, A., Malkin, T. (eds.) PKC 2006. LNCS, vol. 3958, pp. 257–273. Springer, Heidelberg (2006). https://doi.org/10.1007/11745853_17

14. Gerbush, M., Lewko, A., O'Neill, A., Waters, B.: Dual form signatures: an approach for proving security from static assumptions. In: Wang, X., Sako, K. (eds.) ASIACRYPT 2012. LNCS, vol. 7658, pp. 25–42. Springer, Heidelberg (2012). https://doi.org/10.1007/978-3-642-34961-4_4

15. Ghadafi, E.: Efficient round-optimal blind signatures in the standard model. In: Kiayias, A. (ed.) FC 2017. LNCS, vol. 10322, pp. 455–473. Springer, Cham (2017). https://doi.org/10.1007/978-3-319-70972-7_26

16. Goldwasser, S., Micali, S., Rivest, R.L.: A "paradoxical" solution to the signature problem (extended abstract). In: 25th Annual Symposium on Foundations of Computer Science, West Palm Beach, Florida, USA, 24–26 October 1984, pp. 441–448. IEEE Computer Society (1984)

17. Goldwasser, S., Micali, S., Rivest, R.L.: A digital signature scheme secure against adaptive chosen-message attacks. SIAM J. Comput. 17(2), 281–308 (1988)

18. Groth, J.: Homomorphic trapdoor commitments to group elements. Cryptology ePrint Archive 2009/007 (2009)

19. Hartung, G., Kaidel, B., Koch, A., Koch, J., Rupp, A.: Fault-tolerant aggregate signatures. In: Cheng, C.-M., Chung, K.-M., Persiano, G., Yang, B.-Y. (eds.) PKC 2016, Part I. LNCS, vol. 9614, pp. 331–356. Springer, Heidelberg (2016). https://doi.org/10.1007/978-3-662-49384-7_13

20. Hohenberger, S., Waters, B.: Realizing hash-and-sign signatures under standard assumptions. In: Joux, A. (ed.) EUROCRYPT 2009. LNCS, vol. 5479, pp. 333–350. Springer, Heidelberg (2009). https://doi.org/10.1007/978-3-642-01001-9_19

21. Hwang, J.Y., Lee, D.H., Yung, M.: Universal forgery of the identity-based sequential aggregate signature scheme. In: Li, W., Susilo, W., Tupakula, U.K., Safavi-Naini, R., Varadharajan, V. (eds.) Proceedings of the 2009 ACM Symposium on Information, Computer and Communications Security, ASIACCS 2009, Sydney, Australia, 10–12 March 2009, pp. 157–160. ACM (2009)

22. Ishii, R., et al.: Aggregate signature with traceability of devices dynamically generating invalid signatures. In: Zhou, J., et al. (eds.) ACNS 2021. LNCS, vol. 12809, pp. 378–396. Springer, Cham (2021). https://doi.org/10.1007/978-3-030-81645-2_22

23. Kiltz, E., Wee, H.: Quasi-adaptive NIZK for linear subspaces revisited. In: Oswald, E., Fischlin, M. (eds.) EUROCRYPT 2015, Part II. LNCS, vol. 9057, pp. 101–128. Springer, Heidelberg (2015). https://doi.org/10.1007/978-3-662-46803-6_4

24. Kim, H., Sanders, O., Abdalla, M., Park, J.H.: Practical dynamic group signatures without knowledge extractors. IACR Cryptol. ePrint Arch. 351 (2021)

25. Lee, K., Lee, D.H., Yung, M.: Aggregating CL-signatures revisited: extended functionality and better efficiency. In: Sadeghi, A.-R. (ed.) FC 2013. LNCS, vol. 7859, pp. 171–188. Springer, Heidelberg (2013). https://doi.org/10.1007/978-3-642-39884-1_14

26. Libert, B., Mouhartem, F., Peters, T., Yung, M.: Practical "signatures with efficient protocols" from simple assumptions. In: Chen, X., Wang, X., Huang, X. (eds.) Proceedings of the 11th ACM on Asia Conference on Computer and Communications Security, AsiaCCS 2016, Xi'an, China, 30 May–3 June 2016, pp. 511–522. ACM (2016)

27. Libert, B., Peters, T., Yung, M.: Short group signatures via structure-preserving signatures: standard model security from simple assumptions. In: Gennaro, R., Robshaw, M. (eds.) CRYPTO 2015. LNCS, vol. 9216, pp. 296–316. Springer, Heidelberg (2015). https://doi.org/10.1007/978-3-662-48000-7_15

28. Lu, S., Ostrovsky, R., Sahai, A., Shacham, H., Waters, B.: Sequential aggregate signatures and multisignatures without random oracles. In: Vaudenay, S. (ed.) EUROCRYPT 2006. LNCS, vol. 4004, pp. 465–485. Springer, Heidelberg (2006). https://doi.org/10.1007/11761679_28

29. Lysyanskaya, A., Micali, S., Reyzin, L., Shacham, H.: Sequential aggregate signatures from trapdoor permutations. In: Cachin, C., Camenisch, J.L. (eds.) EUROCRYPT 2004. LNCS, vol. 3027, pp. 74–90. Springer, Heidelberg (2004). https://doi.org/10.1007/978-3-540-24676-3_5

30. Lysyanskaya, A., Rivest, R.L., Sahai, A., Wolf, S.: Pseudonym systems. In: Heys, H., Adams, C. (eds.) SAC 1999. LNCS, vol. 1758, pp. 184–199. Springer, Heidelberg (2000). https://doi.org/10.1007/3-540-46513-8_14

31. Ma, D., Tsudik, G.: A new approach to secure logging. ACM Trans. Storage 5(1), 2:1–2:21 (2009)

32. Maurer, U.: Abstract models of computation in cryptography. In: Smart, N.P. (ed.) Cryptography and Coding 2005. LNCS, vol. 3796, pp. 1–12. Springer, Heidelberg (2005). https://doi.org/10.1007/11586821_1

33. Pointcheval, D., Sanders, O.: Short randomizable signatures. In: Sako, K. (ed.) CT-RSA 2016. LNCS, vol. 9610, pp. 111–126. Springer, Cham (2016). https://doi.org/10.1007/978-3-319-29485-8_7

34. Sato, S., Shikata, J., Matsumoto, T.: Aggregate signature with detecting functionality from group testing. IACR Cryptol. ePrint Arch. 1219 (2020)

35. Shoup, V.: Lower bounds for discrete logarithms and related problems. In: Fumy, W. (ed.) EUROCRYPT 1997. LNCS, vol. 1233, pp. 256–266. Springer, Heidelberg (1997). https://doi.org/10.1007/3-540-69053-0_18

36. Sonnino, A., Al-Bassam, M., Bano, S., Meiklejohn, S., Danezis, G.: Coconut: threshold issuance selective disclosure credentials with applications to distributed ledgers. In: 26th Annual Network and Distributed System Security Symposium, NDSS 2019, San Diego, California, USA, 24–27 February 2019. The Internet Society (2019)

37. Szydlo, M.: A note on chosen-basis decisional Diffie-Hellman assumptions. In: Di Crescenzo, G., Rubin, A. (eds.) FC 2006. LNCS, vol. 4107, pp. 166–170. Springer, Heidelberg (2006). https://doi.org/10.1007/11889663_14

38. Tezuka, M., Tanaka, K.: Improved security proof for the Camenisch-Lysyanskaya signature-based synchronized aggregate signature scheme. In: Liu, J.K., Cui, H. (eds.) ACISP 2020. LNCS, vol. 12248, pp. 225–243. Springer, Cham (2020). https://doi.org/10.1007/978-3-030-55304-3_12

39. Tezuka, M., Tanaka, K.: Pointcheval-sanders signature-based synchronized aggregate signature. In: Seo, S.H., Seo, H. (eds.) ICISC 2022. LNCS, vol. 13849, pp. 317–336. Springer, Cham (2022). https://doi.org/10.1007/978-3-031-29371-9_16

Malleable Commitments from Group Actions and Zero-Knowledge Proofs for Circuits Based on Isogenies

Mingjie Chen[1], Yi-Fu Lai[2,3], Abel Laval[4(✉)], Laurane Marco[5],
and Christophe Petit[1,4]

[1] University of Birmingham, Birmingham, UK
[2] University of Auckland, Auckland, New Zealand
[3] Ruhr-University Bochum, Bochum, Germany
[4] Université Libre de Bruxelles, Bruxelles, Belgium
abel.laval@ulb.be
[5] EPFL, Lausanne, Switzerland

Abstract. Zero-knowledge proofs for NP statements are an essential tool for building various cryptographic primitives and have been extensively studied in recent years. In a seminal result from Goldreich, Micali and Wigderson [17], zero-knowledge proofs for NP statements can be built from any one-way function, but this construction leads very inefficient proofs. To yield practical constructions, one often uses the additional structure provided by homomorphic commitments.

In this paper, we introduce a relaxed notion of homomorphic commitments, called *malleable commitments*, which requires less structure to be instantiated. We provide a malleable commitment construction from the ElGamal-type isogeny-based group action from Eurocrypt'22 [5]. We show how malleable commitments with a group structure in the malleability can be used to build zero-knowledge proofs for NP statements, improving on the naive construction from one-way functions. We compare three different approaches, namely from arithmetic circuits, rank-1 constraint systems and branching programs.

Keywords: group action · isogeny-based cryptography ·
commitments · generic zero-knowledge proof of knowledge ·
post-quantum cryptography

1 Introduction

Many cryptographic applications such as multiparty computations, verifiable computations, cryptocurrencies and more, rely on zero-knowledge proofs as an underlying primitive. In 1991, Goldreich, Micali and Wigderson [17] introduced a construction of proof systems for all NP statements from one-way functions. However, such proofs remain inefficient in general and in practice, one usually relies on the structure given by *homomorphic* commitments to build efficient zero-knowledge proofs [10,11,20].

© The Author(s), under exclusive license to Springer Nature Switzerland AG 2024
A. Chattopadhyay et al. (Eds.): INDOCRYPT 2023, LNCS 14459, pp. 221–243, 2024.
https://doi.org/10.1007/978-3-031-56232-7_11

Current homomorphic commitment constructions rely on classical hardness assumptions such as the hardness of the discrete logarithm problem [23] and factorisation or pairing-based assumptions [19]. The security of these problems no longer holds in a post-quantum setting. Lattice-based cryptography provides some constructions of post-quantum commitments with some homomorphic properties [9,16,18]. However, due to the weaker algebraic structure offered by isogeny-based cryptography, having homomorphic commitments does not seem plausible. Hence, this brings us to the main question of this work:

Can we construct a zero-knowledge proof for generic NP statements from isogenies better than from one-way functions?

Our Contributions. In this work, we build a generic zero-knowledge proof of knowledge from isogeny-based assumptions. More specifically, we work under the known-order effective group action model [1] which can be instantiated from isogenies [7,13]. We consider two representations for NP statements: arithmetic circuits over a small field and matrix-based branching programs over a large field, where the latter can serve as an efficient representation for a shallow circuit. We construct proof systems for proving the addition and multiplication gates for the arithmetic circuits over a small field or $\mathbb{Z}_h$ for some small $h \in \mathbb{N}$. Then, using these two components we construct a proof system for the rank-1 constraints system over a small field or ring, which is a more efficient representation for an arithmetic circuit. Furthermore, we construct a proof system for a matrix branching relation. Using the technique given in [5], each of these constructions leads to an online-extractable zero-knowledge proof of knowledge for NP statements without trusted setup under the random oracle model.

We use the known-order effective group action (KO-EGA) model to develop our schemes, where we know the group order and have an efficient algorithm for computing the action of ng for any g within the (additive) group and $n \in \mathbb{N}$. Currently, the only two known post-quantum instantiations of KO-EGA are from isogeny-based group actions, specifically CSI-FiSh and SCALLOP [7,13], and the quantum security levels of these two instantiations are somewhat lower than the NIST-1 requirement [24]. As such, our results remain predominantly theoretical. However, it is essential to highlight that the KO-EGA model implies the existence of numerous cryptographic primitives with significant relevance in real-world applications [5,6,14,21]. The challenge of instantiating a stronger KO-EGA remains an open problem and an active area of research, which is orthogonal to the focus of our paper. Our primary focus centers on constructing schemes within this model and providing performance metrics in terms of the number of group actions, group elements, and set elements.

Another limitation of our proof systems is that their sizes grow linearly with the input length. We describe the obstacles and challenges towards obtaining succinct proofs in Sect. 6.4.

Technical Overview. In this paper, we first introduce *malleable commitments*, a generalization of homomorphic commitments. Intuitively, a malleable commitment scheme for a given relation must satisfy the following property: given a commitment com on a message m, there exists an efficient algorithm that can derive a commitment com' on m' such that m and m' are related. The idea is to settle for a middle point between the absence of structure given by commitments from one-way functions, and the constraints of homomorphic commitments.

Our construction of malleable commitment is based on an adaptation of the public key encryption scheme of [5] (Sect. 4). It is computationally hiding and perfectly binding, both classically and in a post-quantum setting. We develop ad hoc proof systems for proving the correctness of a given commitment, taking advantage of the malleability property for greater efficiency (Sect. 5).

Finally, we use these proof systems as fundamental tools to provide zero-knowledge proofs for NP statements (Sect. 6). More precisely, we derive proof systems for arithmetic circuits, rank-1 constraint systems and branching programs. The constructions for arithmetic circuits and rank-1 constraint systems impose some size restrictions on the underlying field as well as the message space, which can be lifted using the branching program approach.

2   Preliminaries

We say that a function $f : \mathbb{N} \to \mathbb{R}$ is negligible, written $\mathsf{negl}(\lambda)$ if its absolute value is asymptotically dominated by $\mathcal{O}(x^{-n})$ for all $n > 0$. We write PPT for probabilistic polynomial time algorithm. For $n \in \mathbb{N}$, we let $[n] = \{1, \cdots, n\}$.

2.1   Commitment Scheme

Definition 1 (Commitment scheme). *A commitment scheme is, Given a security parameter λ, the algorithms* Setup, Commit, Verify *are defined in the following way:*

- Setup(1^λ) $\to$ pp*: A probabilistic polynomial time algorithm that takes as input the security parameter λ and outputs the public parameters* pp*. The public parameters* pp *are implicitly given as input to the following two algorithms.*
- Commit(m, r) $\to$ com*: A probabilistic polynomial time algorithm that takes a message $m \in \mathcal{M}$ and randomness $r \in \mathcal{R}$ and outputs a commitment* com $\in \mathcal{C}$.
- Verify(com, (m, r)) $\to$ 0/1*: A deterministic algorithm that takes as input a commitment value* com *and a message-randomness pair (m, r) and returns 1 if* Commit(m, r) = com*, 0 otherwise.*

The security of a commitment scheme is characterized by two properties: it must be *hiding* and *binding*. Hiding guarantees that adversary should not be able to recover the original committed message from seeing the commitment, and binding guarantees that an adversary cannot open a commitment to two distinct values.

Definition 2 (Hiding security). *A commitment scheme is* hiding *if for any probabilistic polynomial time adversary $\mathcal{A}$ playing the* Hide$_b$ *game the advantage is negligible i.e.*

$$\mathsf{Adv}^{\mathsf{Hide}}(\mathcal{A}) = |\Pr[\mathsf{Hide}_1(\mathcal{A}) \to 1] - \Pr[\mathsf{Hide}_0(\mathcal{A}) \to 1]| = \mathsf{negl}(\lambda)$$

where the Hide$_b$ *game is defined as follows: given public parameters* pp, $\mathcal{A}$ *outputs two challenge messages* m_0, m_1 *and a state* st. *The challenger samples some randomness* r, *and then computes* com $=$ Commit(m_b, r). *On input* st *and* com, $\mathcal{A}$ *returns a bit* b' *corresponding to its guess for the value of* b.

Definition 3 (Binding security). *A commitment scheme is binding if for any probabilistic polynomial time adversary* $\mathcal{A}$ *playing the* Bind *game, the advantage is negligible, i.e.,*

$$\mathsf{Adv}^{\mathsf{Bind}}(\mathcal{A}) = \Pr[\mathsf{Bind}(\mathcal{A}) \to 1] = \mathsf{negl}(\lambda)$$

where the Bind *game is defined as follows: Given the public parameters* pp, $\mathcal{A}$ *must output two different messages* m_0, m_1 *and associated randomness* r_0, r_1 *and wins if* Commit$(m_0, r_0) =$ Commit(m_1, r_1).

2.2 Group Actions

Definition 4 (Group Action). *A group* G, *with group operation* $+$, *is said to act on a set* $\mathcal{E}$ *if there is a map* $\star : G \times \mathcal{E} \to \mathcal{E}$ *that satisfies the following properties:*

1. *(Identity.) If* 0 *is the identity element of* G, *then for any* $E \in \mathcal{E}$, *we have* $0 \star E = E$.
2. *(Compatibility.) For any* $g, h \in G$ *and any* $E \in \mathcal{E}$, *we have* $(g + h) \star E = g \star (h \star E)$.

Furthermore, this action is called regular *if*

3. *For every* $x_1, x_2 \in \mathcal{E}$ *there exists* $g \in G$ *such that* $x_2 = g \star x_1$.
4. *For* $g \in G$, g *is the identity element if and only if there exists some* $x \in \mathcal{E}$ *such that* $x = g \star x$.

In order to use group actions to build our primitives, we require some efficient (PPT) algorithms. We adopt the *known-order effective group action* (KO-EGA) framework introduced in [1]. As discussed in Sect. 1, post-quantum instantiations of KO-EGA are only known from isogenies. Obtaining a stronger parameter set from isogenies is an active research area [7,13]. In this model, there exists various constructions, such as logarithmic (linkable) ring signatures [6], threshold signatures [14], logarithmic (and tightly secure) group signatures [5], and compact blind signatures [21].

Definition 5 (Known-order Effective Group Action). *Let* $(G, \mathcal{E}, \star)$ *be a group action. For an* $E_0 \in \mathcal{E}$, *we say* $(G, \mathcal{E}, \star, E_0)$ *forms a* known-order effective group action *if the following properties are satisfied:*

1. *The group* G *is finite and both the structure of* $G \cong \bigoplus_i \mathbb{Z}/m_i\mathbb{Z}$ *and a minimal generating set* $\langle g_i \rangle = G$ *are known.*
2. *The set* $\mathcal{E}$ *is finite and there exist* PPT *algorithms for membership testing and for generating a unique bit-string representation for every element in* $\mathcal{E}$.

3. *There exists a distinguished element $E_0 \in \mathcal{E}$ for which the bit-string representation is publicly known.*
4. *There exists a PPT algorithm to evaluate $ng_i \star E$ for any natural number $n \in \{1, ..., \prod_i m_i\}$, $E \in \mathcal{E}$ and g_i from the generating set.*

Since the group order of G is given, one can sample uniformly at random from G. Similarly, by multiplying by a proper factor, one can obtain a generating set for any subgroup of G and sample uniformly at random from the subgroup. Throughout this paper, we consider a regular known-order effective group action from an abelian group. Hard problems in the context of group actions extend naturally from their classical counterpart, namely we have an analog of the discrete logarithm problem called group-action inverse problem (GAIP) as well as analogous of the decisional Diffie-Hellman problems for group actions.

Definition 6 (Group Action Inverse Problem). *Let $(G, \mathcal{E}, \star, E_0)$ be a group action with a distinguished element $E_0 \in \mathcal{E}$. Given E sampled uniformly in $\mathcal{E}$, the Group Action Inverse Problem (GAIP) consists in finding an element $g \in G$ such that $g \star E_0 = E$.*

Definition 7 (Decisional Diffie-Hellman Problem). *Let $(G, E_0, \mathcal{E}, \star)$ be a group action. The challenger generates $s_1, s_2, s_3 \in G$ and $b \in \{0,1\}$, and gives $(s_1 \star E_0, s_2 \star E_0, ((1-b)(s_1 + s_2) + b(s_3)) \star E_0)$ to the adversary $\mathcal{A}$. Then, Decisional Diffie-Hellman (DDH) problem consists in $\mathcal{A}$ returning $b' \in \{0,1\}$ to guess b. The adversary $\mathcal{A}$ wins if $b = b'$. We define the advantage of $\mathcal{A}$ to be $\mathsf{Adv}^{\mathsf{DDH}}(\mathcal{A}) := |\Pr[\mathcal{A} \text{ wins.}] - 1/2|$ where the probability is taken over the randomness of s_1, s_2, s_3, b.*

2.3 Sigma Protocols

Definition 8 (Sigma Protocol). *A sigma protocol Π_Σ is a three-move proof system for a relation R consisting of oracle-calling PPT algorithms $(P = (P_1, P_2), V = (V_1, V_2))$, where V_2 is deterministic. We assume P_1 and P_2 share states and so do V_1 and V_2. Let ChSet be the challenge set. Π_Σ proceeds as follows:*

- *The prover, on input $(\mathsf{st}, \mathsf{wt}) \in \mathcal{R}$, runs $\mathsf{com} \leftarrow P_1^\mathcal{O}(\mathsf{st}, \mathsf{wt})$ and sends a commitment com to the verifier.*
- *The verifier runs $\mathsf{chall} \leftarrow V_1^\mathcal{O}(1^\lambda)$, drawing a random challenge from ChSet, and sends it to the prover.*
- *The prover, given chall, runs $\mathsf{resp} \leftarrow P_2^\mathcal{O}(\mathsf{st}, \mathsf{wt}, \mathsf{chall})$ and returns a response resp to the verifier.*
- *The verifier runs $V_2^\mathcal{O}(\mathsf{st}, \mathsf{com}, \mathsf{chall}, \mathsf{resp})$ and outputs $\top$ (accept) or $\bot$ (reject).*

Here, $\mathcal{O}$ is modeled as a random oracle. For simplicity, we often drop $\mathcal{O}$ from the superscript when it is clear from context. We assume the statement st is always given as input to both the prover and the verifier. The protocol transcript $(\mathsf{com}, \mathsf{chall}, \mathsf{resp})$ is said to be valid in case $V_2(\mathsf{com}, \mathsf{chall}, \mathsf{resp})$ outputs $\top$.

Definition 9 (Correctness). *A sigma protocol Π_Σ is said to be correct if for all $\lambda \in \mathbb{N}$, $(\mathsf{st}, \mathsf{wt}) \in \mathcal{R}$, given that the prover and the verifier both follow the protocol specifications, then the verifier always outputs $\top$.*

Definition 10 (High Min-Entropy). *We say a sigma protocol Π_Σ has $\alpha(\lambda)$ min-entropy if for any $\lambda \in \mathbb{N}$, $(\mathsf{st}, \mathsf{wt}) \in \mathcal{R}$, and a possibly computationally-unbounded adversary $\mathcal{A}$, we have*

$$\Pr\left[\mathsf{com} = \mathsf{com}' \;\middle|\; \mathsf{com} \leftarrow P_1^{\mathcal{O}}(\mathsf{st}, \mathsf{wt}), \mathsf{com}' \leftarrow \mathcal{A}^{\mathcal{O}}(\mathsf{st}, \mathsf{wt})\right] \leq 2^{-\alpha},$$

where the probability is taken over the randomness used by P_1 and by the random oracle. We say Π_Σ has high min-entropy if $2^{-\alpha}$ is negligible in λ.

Additionally, we require a sigma protocol to be honest verifier zero-knowledge (HVZK) as well as 2-special sound.

Definition 11 (Honest Verifier Zero-Knowledge (HVZK)). *We say Π_Σ is honest-verifier-zero-knowledge for relation R if there exists a PPT simulator $\mathsf{Sim}^{\mathcal{O}}$ with access to a random oracle $\mathcal{O}$ such that for any statement-witness pair $(\mathsf{st}, \mathsf{wt}) \in \mathcal{R}$, $\mathsf{chall} \in \mathsf{ChSet}$, $\lambda \in \mathbb{N}$ and any computationally-unbounded adversary $\mathcal{A}$ that makes at most a polynomial number of queries to $\mathcal{O}$, the advantage $\mathsf{Adv}_{\Pi_\Sigma}^{\mathsf{HVZK}}$ of $\mathcal{A}$ is negligible, where*

$$\mathsf{Adv}_{\Pi_\Sigma}^{\mathsf{HVZK}}(\mathcal{A}) := \left| \Pr[\mathcal{A}^{\mathcal{O}}(P^{\mathcal{O}}(\mathsf{st}, \mathsf{wt}, \mathsf{chall})) = 1] - \Pr[\mathcal{A}^{\mathcal{O}}(\mathsf{Sim}^{\mathcal{O}}(\mathsf{st}, \mathsf{chall})) = 1] \right|,$$

where $P = (P_1, P_2)$ is a prover running on $(\mathsf{st}, \mathsf{wt})$ with a challenge fixed to chall and the probability is taken over the randomness used by (P, V) and by the random oracle.

Definition 12 (2-Special Soundness). *We say a sigma protocol Π_Σ has 2-special soundness if there exists a polynomial-time extraction algorithm such that, given a statement st and any two valid transcripts $(\mathsf{com}, \mathsf{chall}, \mathsf{resp})$ and $(\mathsf{com}, \mathsf{chall}', \mathsf{resp}')$ relative to st and such that $\mathsf{chall} \neq \mathsf{chall}'$, outputs a witness wt satisfying $(\mathsf{st}, \mathsf{wt}) \in \mathcal{R}$.*

Remark 1. In some circumstances, we can relax the relation for 2-special soundness to R' where $R \subseteq R'$. That is, we allow the extractor to output wt such that $(\mathsf{st}, \mathsf{wt}) \in \mathcal{R}'$ but $(\mathsf{st}, \mathsf{wt}) \notin R$. As long as given st to find wt such that $(\mathsf{st}, \mathsf{wt}) \in \mathcal{R}'$, the sigma protocol can still serve as a proof system for some applications. For instance, [5,6] both relaxes the relation by including the collision of the hash function. Therefore, the resulting proof system, via the Fiat-Shamir transform, is sound if the hash function is collision-resistant.

Finally, note that sigma protocols can be used to build non-interactive zero-knowledge proofs of knowledge.

2.4 Proof Systems

We consider *non-interactive zero-knowledge proof of knowledge* (NIZKPoK) over the programmable random oracle model with statistical zero-knowledge and simulation-extractability.

Formally, a NIZKPoK for a relation $\mathcal{R}$ is a tuple of PPT algorithms $\Pi =$ (Prove, Verify) with access to the random oracle $\mathcal{O}$ such that for any valid public parameter of the scheme pp and the security parameter $\lambda \in \mathbb{N}$, the scheme satisfies the following properties:

- Completeness: for every $(\mathsf{st}, \mathsf{wt}) \in \mathcal{R}$,

$$\Pr[V^{\mathcal{O}}(\mathsf{st}, \pi) = 1 \mid \pi \leftarrow P^{\mathcal{O}}(\mathsf{st}, \mathsf{wt})] = 1.$$

- Statistical zero-knowledge: there exists a simulator Sim such that for any $(\mathsf{st}, \mathsf{wt}) \in \mathcal{R}$ and any possibly unbounded adversary $\mathcal{A}$ can distinguish the distributions: $\{\pi | \pi \leftarrow \mathsf{Prove}^{\mathcal{O}}(\mathsf{st}, \mathsf{wt})\}$, $\{\pi | \pi \leftarrow \mathsf{Sim}^{\mathcal{O}}(\mathsf{st})\}$ with only negligible advantage where the probability is taken over the randomness used in the experiment and the scheme.
- Simulation-extractability: for any adversary $\mathcal{A}$ there exists an extractor $\mathcal{E}_{\mathcal{A}}$ such that

$$\Pr\left[\begin{array}{c} \mathsf{Verify}(\mathsf{st}, \mathsf{wt}, \pi) = \top \\ (\mathsf{st}, \mathsf{wt}) \notin \mathcal{R} \wedge (\mathsf{st}, \mathsf{wt}, \pi) \notin L \end{array} \middle| \begin{array}{c} (\mathsf{st}, \pi) \leftarrow \mathcal{A}^{\mathsf{SimProve}, \mathcal{O}}(\mathsf{pp}) \\ \mathsf{wt} \leftarrow \mathcal{E}(\mathsf{st}, \pi) \end{array}\right] \leq \mathsf{negl}(\lambda)$$

where $\mathsf{SimProve}(\mathsf{st}', \mathsf{wt}')$ is an oracle that returns a simulated proof using $\pi' \leftarrow \mathsf{Sim}(\mathsf{st}')$ when $(\mathsf{st}', \mathsf{wt}') \in \mathcal{R}$. Otherwise, SimProve returns $\bot$. Also, SimProve maintains the list L, which is initially empty, and records $(\mathsf{st}', \mathsf{wt}', \pi')$ for each valid query of $(\mathsf{st}', \mathsf{wt}')$.

3 Malleable Commitments

In this section we introduce the notion of malleable commitments. We will use this notion to build commitments based on group actions and isogenies from which we can derive zero-knowledge proofs for general NP statements. We first give a general definition together with the associated security notion and then restrict the notion to a specific kind of malleability which we will use later on in our constructions.

3.1 A Generic Notion of Malleability

Intuitively, we say that a commitment scheme is *malleable* if given a commitment for an unknown value m, anyone can create a second commitment for some m' related to m. We formalise this notion below.

Definition 13 (Malleable Commitment). *Given a set* $\mathcal{M}$ *on which we have a relation* R, *a commitment scheme* $(\mathcal{M}, \mathcal{R}, \mathcal{C}, \mathsf{Setup}, \mathsf{Commit}, \mathsf{Verify})$ *is said to be R-malleable if there exists a PPT algorithm* $\mathcal{A}_m^R : \mathcal{C} \to \mathcal{C}$ *with the following property: for any* $\mathsf{com}' = \mathcal{A}_m^R(\mathsf{com})$, *if there exists some* $(m, r) \in \mathcal{M} \times \mathcal{R}$ *such that* $\mathsf{com} = \mathsf{Commit}(m, r)$, *then there exists* $(m', r') \in \mathcal{M} \times \mathcal{R}$ *such that* $\mathsf{com}' = \mathsf{Commit}(m', r')$ *and* $R(m, m')$.

Remark 2. Depending on the application, r and r' might be equal. Additionally, the algorithm $\mathcal{A}_m$ does not have to be able to retrieve any of the values m, m' or r. However, in our constructions we only consider schemes for which whenever one party knows (m, r), they can retrieve the message m' corresponding to the commitment com'.

Remark 3. One can easily check that homomorphic commitments are a special case of malleable commitments for which $\mathcal{M}, \mathcal{R}$ and $\mathcal{C}$ are all groups.

As is, malleable commitments *a priori* do not impose any structure on $\mathcal{M}, \mathcal{R}$ or $\mathcal{C}$ and provide a very generic framework. Within this framework, we instantiate a specific type of malleable commitments exploiting the structure provided by isogeny-based cryptography. Namely, we enforce a group structure on the message and randomness spaces $\mathcal{M}$ and $\mathcal{R}$, while the commitment space $\mathcal{C}$ is only assumed to be a generic set. This is motivated by the fact that messages and randomness will be encoded as isogenies, which form a group under composition, while commitments will be seen as (isomorphism classes of) elliptic curves, which possesses no built-in group structure. The malleability will then be obtained by defining a group action of $\mathcal{M} \times \mathcal{R}$ on $\mathcal{C}$. In particular, since $\mathcal{C}$ is not a group, what we obtain is *not* an homomorphic commitment.

We formalize this additional structure required through the notion of *Admissible Group-Action-based Malleable Commitment* (AGAMC).

Definition 14 (Admissible group-action-based malleable commitment (AGAMC)). *A commitment scheme* $\Pi_{\mathsf{Commit}} = (\mathsf{Setup}, \mathsf{Commit}, \mathsf{Verify})$ *is said to be an* admissible group-action based malleable commitment scheme *if the following requirements are satisfied:*

1. *The public parameters instantiated by* Setup *are* $\mathsf{pp} = (\mathcal{M}, \mathcal{R}, \mathcal{C}, X, \star)$ *where* $\mathcal{M}$ *and* $\mathcal{R}$ *are (additive) abelian groups—with respective identities* $0_{\mathcal{M}}$ *and* $0_{\mathcal{R}}$, *and where* $X \in \mathcal{C}$ *is a distinguished element defined as* $X := \mathsf{Commit}(0_{\mathcal{M}}, 0_{\mathcal{R}})$.
2. *The action* $\star : (\mathcal{M} \times \mathcal{R}) \times \mathcal{C} \to \mathcal{C}$ *is regular and can be computed in polynomial-time. Any commitment is computed as* $\mathsf{Commit}(m, r) := (m, r) \star X$.
3. *The group action gives rise to the following malleability: for any* $(m, r), (m', r') \in \mathcal{M} \times \mathcal{R}$, *we have* $(m', r') \star \mathsf{Commit}(m, r) = \mathsf{Commit}(m + m', r + r')$.
4. *There are efficient uniform sampling methods for any subgroups of* $\mathcal{M}, \mathcal{R}$.
5. *There is an efficient algorithm to determine the membership of* $\mathcal{C}$.
6. *Every element in* $\mathcal{M}, \mathcal{R}, \mathcal{C}$ *has a unique representation.*

Remark 4. We do not require the algorithm Commit specified by pp to be computed efficiently itself. Instead, one uses malleability to produce a commitment $\mathsf{Commit}(m, r) = (m, r) \star \mathsf{Commit}(0_{\mathcal{M}}, 0_{\mathcal{R}})$. We abuse the notation 0 to represent $0_{\mathcal{M}}, 0_{\mathcal{R}}$. Also, we simplify the commitment scheme notion to be $\Pi_{\mathsf{Commit}} = (\mathsf{Setup}, \mathsf{Commit}, \mathsf{Verify})$ and pp is implicitly used when the context is clear.

4 Malleable Commitments from Group Actions

We present a construction of an AGAMC scheme from a known-order effective group action (Definition 5), and prove its security.

Construction. $\Pi = (\mathsf{Setup}, \mathsf{Commit}, \mathsf{Verify})$ proceeds as follows.

- $\mathsf{Setup}(1^{\lambda}) \to \mathsf{pp}$: on input λ, output $\mathsf{pp} = (G, G, \mathcal{E} \times \mathcal{E}, (E_0, E), \star')$ i.e. $\mathcal{M} = \mathcal{R} = G$, $\mathcal{C} = \mathcal{E} \times \mathcal{E}$ and $X := \mathsf{Commit}(0_{\mathcal{M}}, 0_{\mathcal{R}}) = (E_0, E)$, where $E = s \star E_0$ for some $s \in G$ and $s \neq 0$. The action $\star'$ is then defined as:

$$(m, r) \star' (Y_0, Y) := (r \star Y_0, (r + m) \star Y),$$

  for $(G, \mathcal{E}, E_0, \star)$ a known-order effective group action where $N = |G|$ is given. The group size is implicitly parameterized by the security parameter λ.
- $\mathsf{Commit}(m, r) \to,$: on input $(m, r) \in G^2$, returns $\mathsf{com} = (m, r) \star' X$.
- $\mathsf{Verify}(\mathsf{com}, (m, r)) \to 0/1$: on input $(\mathsf{com}, (m, r)) \in \mathcal{C} \times G^2$, checks whether $(m, r) \star' X = \mathsf{com}$. If yes, then outputs 1 and outputs 0 otherwise.

The requirements of an AGAMC follow naturally from the known-order effective group action. The uniqueness and membership test of $\mathcal{C}$ and the feasibility of $\star$ come from Definition 5. The uniform sampling methods of any subgroups of $\mathcal{M}, \mathcal{R}$ come from the known order and minimal generating set of G. Since we know minimal generating sets, we have a unique representation of $\mathcal{M}, \mathcal{R}$.

Theorem 1. *The commitment scheme described above is (computationally) hiding assuming DDH problem over $(G, \mathcal{E}, E_0, \star)$ is hard.*

Proof. Consider a DDH setup $(G, \mathcal{E}, E_0, \star)$. Given a DDH challenge (E_1, E_2, E_3) and access to an hiding adversary $\mathcal{A}$, we build an adversary $\mathcal{B}$ against DDH as follows.

1. Set $X = (E_0, E_1)$ and $\mathsf{pp} = (G, G, \mathcal{E} \times \mathcal{E}, X, \star')$. This gives a valid setup for our commitment construction.
2. Invoke the adversary $\mathcal{A}$ on the the public parameter pp.
3. Upon receiving $m_0, m_1 \in \mathcal{M}$ from $\mathcal{A}$, reply with $(E_2, m_b \star E_3)$ where $b \leftarrow \{0, 1\}$.
4. Output $(b = b')$ where b' is returned by $\mathcal{A}$.

We claim that this adversary $\mathcal{B}$ has half of the advantage of $\mathcal{A}$. Write $E_1 = s_1 \star E_0$ and $E_2 = s_2 \star E_0$ where s_1, s_2 are sampled uniformly from G by the DDH challenger. In the case that $E_3 = (s_1 + s_2) \star E_0$, i.e. $b = 1$, it is clear that $\mathcal{B}$ returns a well-formed commitment of m_b with respect to pp.

In the case that $E_3 \leftarrow \mathcal{E}$ by the challenger, the commitment is potentially malformed. Since the input $(E_2, m_i \star E_3)$ follows a uniform distribution over $\mathcal{E}^2$ regardless of m_b and b in the view of $\mathcal{A}$, the random bit b equals b' returned by $\mathcal{A}$ with probability exactly $1/2$ in this case. Hence, the reduction $\mathcal{B}$ can win with probability exactly $1/2$ in this case. Therefore,

$$
\begin{aligned}
|\Pr[\mathcal{B} \text{ wins}] - 1/2| &= |1/2 \Pr[\mathcal{B} \text{ wins} \mid E_3 = (s_1 + s_2) \star E_0] \\
&\quad + 1/2 \Pr[\mathcal{B} \text{ wins} \mid E_3 \neq (s_1 + s_2) \star E_0] - 1/2| \\
&= 1/2 |\Pr[\mathcal{B} \text{ wins} \mid E_3 = (s_1 + s_2) \star E_0] - 1/2| \\
&= 1/2 \, \mathsf{Adv}_{\Pi}^{\mathsf{Hide}}(\mathcal{A}).
\end{aligned}
$$

Hence, we have $\mathsf{Adv}^{\mathsf{DDH}}(\mathcal{B}) = \frac{1}{2}\mathsf{Adv}_{\Pi}^{\mathsf{Hide}}(\mathcal{A})$.

Theorem 2. *The commitment scheme described above is (perfectly) binding.*

Proof. Given $(m_0, r_0) \star' X = (m_1, r_1) \star' X$, we have $(r_0 \star Y_0, (r_0 + m_0) \star Y) = (r_1 \star Y_0, (r_1 + m_1) \star Y)$ for some $Y_0, Y \in \mathcal{E}$. Since the action is regular, we have $r_0 = r_1$ and $m_0 = m_1$, and the result follows.

4.1 Commitment Products

We show that given two AGAMCs we can derive a new one via a direct product. Concretely, given two AGAMCs $\Pi_1 = (\mathsf{Setup}_1, \mathsf{Commit}_1, \mathsf{Verify}_1)$ and $\Pi_2 = (\mathsf{Setup}_2, \mathsf{Commit}_2, \mathsf{Verify}_2)$ we can define $\Pi = (\mathsf{Setup}, \mathsf{Commit}, \mathsf{Verify})$ in the following way: Given $\mathsf{Setup}_1 \rightarrow (\mathcal{M}_1, \mathcal{R}_1, \mathcal{C}_1, X_1, \star_1)$, $\mathsf{Setup}_2 \rightarrow (\mathcal{M}_2, \mathcal{R}_2, \mathcal{C}_2, X_2, \star_2)$ then $\mathsf{Setup} \rightarrow (\mathcal{M}, \mathcal{R}, \mathcal{C}, X, \star)$ with $\mathcal{M} = (\mathcal{M}_1, \mathcal{M}_2)$, $\mathcal{R} = (\mathcal{R}_1, \mathcal{R}_2)$, $\mathcal{C} = (\mathcal{C}_1, \mathcal{C}_2)$, $X = (X_1, X_2)$ and $((m_1, m_2), (r_1, r_2)) \star X = ((m_1, r_1) \star_1 X_1, (m_2, r_2) \star_2 X_2)$. The definitions of Commit and Verify follow accordingly.

Proposition 1 (Commitment Product). *Given two AGAMCs Π_1 and Π_2, then their product Π defined above gives an AGAMC. Moreover, Π is perfectly (resp. computationally) hiding (resp. binding) when both Π_1, Π_2 are perfectly (resp. computationally) hiding (resp. binding).*

Proof. The uniqueness and membership test of $\mathcal{M}, \mathcal{R}, \mathcal{C}$, the feasibility of $\star$, and the uniform sampling immediately follow from the properties of Π_1 and Π_2. Regarding the hiding property, by using a standard hybrid argument, for any hiding adversary $\mathcal{A}$ against Π, there exists B_1 against Π_1 and B_2 against Π_2 such that $\mathsf{Adv}_{\Pi}^{\mathsf{Hide}}(\mathcal{A}) \leq \mathsf{Adv}_{\Pi_1}^{\mathsf{Hide}}(B_1) + \mathsf{Adv}_{\Pi_2}^{\mathsf{Hide}}(B_2)$. Regarding the binding property, suppose $(m, r) \star X = (m', r') \star X$ where $m = (m_1, m_2), m' = (m_1', m_2'), r = (r_1, r_2), r' = (r_1', r_2')$. Then, we have $(m_1, r_1) \star_1 X_1 = (m_1', r_1') \star_1 X_1$ and $(m_2, r_2) \star_2 X_2 = (m_2', r_2') \star_2 X_2$. Therefore, Π is binding when both Π_1, Π_2 are binding.

Remark 5. Proposition 1 implies that the AGAMC is closed under the product. Alternatively, given an AGAMC with message and randomness space $\mathcal{M}$ and $\mathcal{R}$, we can extend both to vector spaces or matrix spaces $\mathcal{M}^{m \times n}, \mathcal{R}^{m \times n}$. As we will show in Sect. 6.2, this algebraic structure is sufficient to efficiently encode the information of circuits.

5   Proof Systems for an Admissible Group-Action Based Commitment

In this section, we build two proof systems for AGAMC that will serve as essential tools for constructing a generic proof system from this special commitment scheme. Both of them are proving the knowledge of the input of a commitment with a restricted message space. The nuance between the two protocols lies in the structure of the message space being a subgroup or not.

The first system (Sect. 5.1 and Fig. 1) will be needed to commit on the multiplicative gates of an arithmetic circuit (as $\mathcal{M}$ is not a *multiplicative* group) while the second (Sect. 5.2 and Fig. 2) will be used for the addition gates (and thus we can exploit the additive group structure of $\mathcal{M}$). The additional structure available in the second makes it more efficient : it is therefore simpler to commit on additions than on multiplications.

Formally, let $\mathsf{pp} = (\mathcal{M}, \mathcal{R}, \mathcal{C}, X, \star)$, the parameters of an AGAMC and M' a subset of $\mathcal{M}$. Consider the following relation $\mathcal{R}_{\mathsf{pp},M'} \subseteq \mathcal{C} \times \mathcal{M} \times \mathcal{R}$ where

$$\mathcal{R}_{\mathsf{pp},M'} = \{(\mathsf{st} = c, \mathsf{wt} = (m,r)) \mid c = (m,r) \star X \wedge m \in M'\}. \quad (1)$$

Equivalently, $\mathcal{R}_{\mathsf{pp},M'} = \{(\mathsf{st} = c, \mathsf{wt} = (m,r)) \mid c = \mathsf{Commit}(m,r) \wedge m \in M'\}$ due to the malleability. Therefore, to have a proof system for the relation $\mathcal{R}_{\mathsf{pp},M'}$ implies to have a proof system for the commitment without revealing m or r.

We then turn it into non-interactive zero-knowledge proof (NIZK) using parallel repetition and the Fiat-Shamir transform ([15]). Looking ahead, in Sects. 6.1 and 6.2 we will use this tool to build NIZK to prove satisfiability of arithmetic circuits and for rank-1 constraint systems (R1CS).

As mentioned earlier, commitment schemes from one-way functions can indeed be used to build generic zero-knowledge proof, but they result in highly inefficient ones. Homomorphic commitments (e.g. Pedersen commitments) have additional structure that makes them useful in many pre-quantum succinct zero-knowledge proofs, such as [10,25,26]. Here, we demonstrate that a malleable commitment (AGAMC) can also provide a structure suitable for use in post-quantum generic proof systems for arithmetic circuits when the underlying ring is small.

5.1   Proof System for Small Message Space

We start with a fundamental proof system of AGAMC for proving the knowledge of the input of a commitment with an arbitrary small message space.

```
round 1: P₁ᴼ((pp, M′, u), (m = m_I, r))
 1: seed ← {0,1}^λ
 2: (r′, bits₁, ···, bits_t) ← H(EXP‖seed)          ▷ Sample r′ ∈ 𝓡 and bits_i ∈ {0,1}^λ
 3: for i ∈ [t] do
 4:     com_i ← (−m_i, r′) ⋆ u
 5:     C_i ← H(COM‖com_i‖bits_i)                    ▷ Create commitments C_i ∈ {0,1}^{2λ}
 6: (root, tree) ← MerkleTree(C₁, ···, C_N)
 7: P sends comm ← root to V.

round 2: V₁(comm)                          Verification: V₂ᴼ(comm, chall, resp)
 1: c ← {0,1}                                1: (root, c) ← (comm, chall)
 2: V sends chall ← c to P.                  2: if c = 1 then
                                             3:     (r″, path, bits) ← resp
round 3: P₂((m = m_I, r), chall)            4:     c̄om ← (0, r″) ⋆ X
 1: c ← chall                                5:     C̃ ← H(COM‖c̄om‖bits)
 2: if c = 1 then                            6:     r̃oot ← ReconstructRoot(C̃, path)
 3:     r″ ← r′ + r                          7: else
 4:     path ← getMerklePath(tree, I)        8:     Repeat round 1 with seed ← resp.
 5:     resp ← (r″, path, bits_I)            9: Output accept if the computation results in
 6: else                                        root, and reject otherwise.
 7:     resp ← seed
 8: P sends resp to V.
```

Fig. 1. Our construction for the relation $\mathcal{R}_{\mathsf{pp},M'}$ (Eq. (1)) using an AGAMC with $\mathsf{pp} = (\mathcal{M}, \mathcal{R}, \mathcal{C}, X, \star)$. The set M' is of cardinality t and can be publicly enumerated as $M' = \{m_1, \cdots, m_t\}$. The hash function H is modeled as a random oracle with prefixes EXP (for expand) and COM, for a pseudo-random number generator (PRNG) and a commitment scheme respectively and is also implicitly used in MerkleTree($\cdot$).

We first consider the case where M' is polynomially small. A concrete construction is given in Fig. 1 using Merkle trees to make the proof size logarithmic in $|M'|$. The use of Merkle trees is standard except for one modification: two leaves are concatenated in alphabetical order, resulting in an index-hiding Merkle tree [6].

To be more precise, the modification to the alphabetical order serves as a random (in the ROM) permutation of the input, ensuring that revealing a path in the tree does not reveal the location of the leaf. The construction is inspired by [5]. The high-level idea is that the prover uses the malleability property to create multiple new commitments $(-m', r')$ for each $m' \in M'$, and generates a Merkle root. The reason M' is required to be small is to make this possible. The prover is then challenged to either reveal all randomnesses used in the malleability, including the randomness used to generate the whole tree, or to reveal $(0, r + r')$ and the path of the tree that results in the same root. Intuitively, knowing the whole tree and the secret path allows for the determination of m along with r. Zero-knowledge follows from the uniform distribution of $r + r'$ (remember $\mathcal{R}$ has a group structure), and the use of an index-hiding Merkle tree to protect m.

Theorem 3. *The proof system described in Fig. 1 is correct, 2-special sound (assuming collision resistance of the hash function H), and statistical honest-verifier zero-knowledge (HVZK) (in the random oracle model).*

---

round 1: $P_1^{\mathcal{O}}((\mathsf{pp}, M', u), (m, r))$

1: $\mathsf{seed} \leftarrow \{0,1\}^\lambda$
2: $(m', r') \leftarrow H(\mathsf{EXP}\|\mathsf{seed})$ $\triangleright$ Sample $(m', r') \in M' \times \mathcal{R}$
3: $\mathsf{C} \leftarrow H(\mathsf{COM}\|(m', r') \star u)$ $\triangleright$ Create commitments $\mathsf{C} \in \{0,1\}^{2\lambda}$
4: P sends $\mathsf{comm} \leftarrow \mathsf{C}$ to V.

round 2: $V_1(\mathsf{comm})$ **Verification:** $V_2^{\mathcal{O}}(\mathsf{comm}, \mathsf{chall}, \mathsf{resp})$

1: $c \leftarrow \{0,1\}$ 1: $(\mathsf{C}, c) \leftarrow (\mathsf{comm}, \mathsf{chall})$
2: V sends $\mathsf{chall} \leftarrow c$ to P.   2: **if** $c = 1$ **then**
                                   3:     $(m'', r'',) \leftarrow \mathsf{resp}$
round 3: $P_2((m, r), \mathsf{chall})$ 4: $\widetilde{\mathsf{C}} \leftarrow H(\mathsf{COM}\|(m'', r'') \star X)$

1: $c \leftarrow \mathsf{chall}$ 5: **else**
2: **if** $c = 1$ **then**   6:   Repeat **round 1** with $\mathsf{seed} \leftarrow \mathsf{resp}$.
3:     $(m'', r'') \leftarrow (m' + m, r' + r)$ 7: Output **accept** if $\widetilde{\mathsf{C}} = \mathsf{C}$ or **round 1** with
4:     $\mathsf{resp} \leftarrow (m'', r'')$ $\mathsf{seed} \leftarrow \mathsf{resp}$ leads to C
5: **else**
6:     $\mathsf{resp} \leftarrow \mathsf{seed}$
7: P sends resp to V.

Fig. 2. Our construction for the relation $\mathcal{R}_{\mathsf{pp}, M'}$ (Eq. (1)) using an AGAMC with $\mathsf{pp} = (\mathcal{M}, \mathcal{R}, \mathcal{C}, X, \star)$ where M' is a subgroup of $\mathcal{M}$. The hash function H is modeled as a random oracle with prefixes EXP and COM as a PRNG and a commitment scheme respectively.

Correctness. For the challenge $\mathsf{chall} = 0$, the correctness holds naturally. For the challenge $\mathsf{chall} = 1$, for the response $\mathsf{resp} = (r'', \mathsf{path}, \mathsf{bits})$, we have $r'' = r' + r$. Therefore, $\widetilde{\mathsf{com}} = (0, r'') \star X = (0, r' + r) \star X$. Recall that $\mathsf{com}_I = (-m_I, r') \star u$. Since $\mathsf{bits}_I = \mathsf{bits}$, we have $\widetilde{\mathsf{com}} = \mathsf{com}_I$ so that $\widetilde{\mathsf{C}} = \mathsf{C}_I$. ReconstructRoot($\widetilde{\mathsf{C}}, \mathsf{path}$) will result in the same root and the correctness follows.

2-Special Soundness. Let $(\mathsf{comm} = \mathsf{root}, 0, \mathsf{resp}_0 = \mathsf{seed}), (\mathsf{comm} = \mathsf{root}, 1, \mathsf{resp}_1 = (r'', \mathsf{path}, \mathsf{bits}))$ be two valid transcripts. We can either extract a collision of the hash function H or a witness for the statement (pp, M', u). Following **round 1** of Fig. 1, say seed generates r', $\mathsf{com}_i = (-m_i, r') \star u$ for each $i \in [t]$. Let $\widetilde{\mathsf{com}} = (0, r'') \star X$. Since both $\mathsf{resp}_0, \mathsf{resp}_1$ result in the same root root, we assume $\widetilde{\mathsf{com}} = \mathsf{com}_{\widetilde{I}}$ for some $\widetilde{I} \in [t]$; otherwise, a collision of the hash function H can be found in the process of generating the root (tree, ReconstructRoot) due to the binding property of the Merkle tree [6, Lemma 2.9]. We claim that $(m_{\widetilde{I}}, r'' - r')$ is the witness. This is because we have $(0, r'') \star X = (-m_{\widetilde{I}}, r') \star u$. Therefore, $(m_{\widetilde{I}}, r'' - r') \star X = u$. The result follows.

Statistical HVZK. We show the scheme is statistical HVZK by modeling the hash function H as a random oracle, which is essential to achieve statistical zero-knowledge. For the challenge $\mathsf{chall} = 0$, the simulator follows the first round of Fig. 1 and generates $(\mathsf{seed}, 0, \mathsf{comm})$ as the transcript. For the challenge $\mathsf{chall} = 1$, the simulator generates r'' from $\mathcal{R}$ uniformly at random and computes $\widetilde{\mathsf{com}}_1 = (0, r'') \star X$. Then, the simulator generates uniformly r' from $\mathcal{R}$ and $\mathsf{bits}_1, \cdots, \mathsf{bits}_t$ from $\{0,1\}^\lambda$, computes $\widetilde{\mathsf{C}}_1(\mathsf{COM}\|\widetilde{\mathsf{com}}_1\|\mathsf{bits}_1)$ and $\mathsf{C}_i = H(\mathsf{COM}\|(-m_i, r') \star u\|\mathsf{bits}_i)$ for $i \in \{2, \cdots, t\}$, and generates $(\mathsf{root}, \mathsf{path}) \leftarrow \mathsf{MerkleTree}(\widetilde{\mathsf{C}}_1, \mathsf{C}_2, \cdots, \mathsf{C}_N)$. Finally, the simulator outputs the transcript $(\mathsf{root}, 1, (r'', \mathsf{path}, \mathsf{bits}_1))$. Even though the simulator simulates for $I = 1$, the distribution is independent of this particular choice due to the hiding property of the Merkle tree [6, Lemma 2.10]. The only difference is that $r', \text{bits}_1, \cdots, \text{bits}_t$ are not generated by $H(\text{EXP}||\cdot)$. Since seed has λ bits of min-entropy and is information-theoretically hidden from the distinguisher, for any unbounded distinguisher with Q queries to H of the form $H(\text{EXP}||\cdot)$ the statistical difference between the distributions is $Q_H/2^\lambda$.

Efficiency. Suppose it requires x bits to represent an element over the message space M' and y bits for the randomness space $\mathcal{R}$. The scheme in Fig. 1 has proof size $\frac{\lambda+(y+\log\lambda|M'|+\lambda)}{2}$ bits on average and requires $O(|M'|)$ malleability actions.

5.2   Proof System for Message Spaces with a Subgroup Structure

In Sect. 5.1, we require the message space to be small since we do not have any assumptions on its structure. Here we show that we can remove this restriction when the message space M' is a subgroup of $\mathcal{M}$, we obtain a more efficient NIZK based on the proof system shown in Fig. 2.

Theorem 4. *The proof system described in Fig. 2 is correct, 2-special sound (assuming collision resistance of the hash function H) and statistical honest verifier zero-knowledge.*

Proof. See the appendix in the full version of the paper.

Efficiency. Say it requires x bits to represent an element over the message space M' and y bits for the randomness space $\mathcal{R}$. The scheme in Fig. 2 has proof size $\frac{\lambda+(x+y)}{2}$ bits on average and requires 2 malleability actions.

5.3   NIZK via the Fiat-Shamir Transform

By applying the Fiat-Shamir transform [15], we obtain NIZKs for the relation $\mathcal{R}_{\text{pp},M'}$ as shown in Fig. 3. Using Theorems 3 and 4 respectively, and the properties of the Fiat-Shamir transform we may prove that the NIZKs are complete and statistical zero-knowledge with standard proofs.

In order to demonstrate simulation-extractability, we can reuse the proof of Thm. 6.1 in [5] and show that our NIZKs are also (multi-proof) online-extractable (see Def. 2.10 of [5] for the definition), which implies the simulation extractability and has online-extractability. From a high level, the scheme is online-extractable because we use the PRNG to generate the randomness in the commitment in the sigma protocol. In the security proof, the PRNG is modeled as a random oracle and the challenge is binary. Alternatively, a simulator can extract the response for the challenge 0 by checking the random oracle queries regardless of the challenge $c \in \{0, 1\}$. In other words, since we use the random oracle as PRNG to generate the randomness for the commitment and the challenge is binary, the proof of Theorem 6.1 in [5] is agnostic to the underlying sigma protocol by replacing Step 2.(b) with the 2-special soundness extractor. We, therefore, omit the proof for simplicity.

Theorem 5. *The NIZKs of Figs. 1 and 2 compiled by Fig. 3 is complete, simulation-extractable and statistically zero-knowledge in the random oracle model.*

```
Prove^O(st, wt)                              Verify^O(st, π)
1: for i ∈ [λ] do                            1: (com     =     (com_1, ⋯ , com_λ),
2:     com_i ← P_1^O(st, wt)                     chall   =     (c_1, ⋯ , c_λ), resp   =
3: com ← (com_1, ⋯ , com_λ)                      (resp_1, ⋯ , resp_λ)) ← π
4: (c_1, ⋯ , c_λ) ← O(FS||st||com)           2: output = 1
5: chall ← (c_1, ⋯ , c_λ)                    3: for i ∈ [λ] do
6: for i ∈ [λ] do                            4:     r ← V_2(com_i, c_i, resp_i)
7:     resp_i ← P_2^O(st, com_i, c_i)         5:     output ← output · r
8: resp ← (resp_1, ⋯ , resp_λ)               6: output   ←   output · (chall   ==
9: return π ← (com, chall, resp)                 O(FS||st||com))
                                             7: return output
```

Fig. 3. NIZK for the relation by applying the Fiat-Shamir transform to $\Pi = (P = (P_1, P_2), V = (V_1, V_2))$ from either Fig. 1 or Fig. 2 with λ repetitions.

6   Proof Systems for NP Statements

This section presents proof systems for an arithmetic circuit over a finite field or ring constructed from an AGAMC. Firstly, we consider two cases: the arithmetic circuit in general with the addition and multiplication gates in Sect. 6.1, and the representation using rank-1 constraint system in Sect. 6.2. In both cases, the underlying finite field or ring must be small or embedded into a small subgroup of the message space. In Sect. 6.3, we construct a proof system for the branching program representation without size restrictions.

6.1   Arithmetic Circuits over a Small Ring

This subsection gives a NIZK for the satisfiability of an arithmetic circuit over a small ring. We encode this ring as $M' \subseteq M$ of a given AGAMC $\Pi = (M, R, C, X, \star)$ (e.g. via a dictionary map).

The idea is fairly straightforward. For each gate of the circuit, the prover uses an AGAMC to make a commitment to the inputs and the output. Then, the prover uses Fig. 1 to generate a proof based on the commitment product Prop. 1. Concretely, for the input (m_1, m_2) and the output m_3 for a certain gate. The prover computes $\mathsf{Commit}(m_1, r_1)$, $\mathsf{Commit}(m_2, r_2)$, and $\mathsf{Commit}(m_3, r_3)$. Say the ring of the circuit is encoded as $M' \subseteq M$. The prover uses Fig. 1 for the triple commitment $\Pi \times \Pi \times \Pi$ with the message space $M'' \subseteq M'^3$ by collating the valid input-output tuple corresponding to the gate. For example, consider an addition gate with inputs m_1, m_2 and output m_3. We need to prove that $\mathsf{Commit}(m_1, r_1)$, $\mathsf{Commit}(m_2, r_2)$, and $\mathsf{Commit}(m_3, r_3)$ are all in Π and that $m_1 + m_2 = m_3$. Regarding a multiplication gate with inputs m_1, m_2 and output m_3. We need to prove that $\mathsf{Commit}(m_1, r_1)$, $\mathsf{Commit}(m_2, r_2)$, and $\mathsf{Commit}(m_3, r_3)$ are all in Π and that $m_1 * m_2 = m_3$.

Efficiency. Say it requires x bits to represent an element over the message space M' and y bits for the randomness space $\mathcal{R}$. Let the number of parallel repetitions of the base schemes Figs. 1 and 2 be λ in Fig. 3. The scheme in Fig. 1 has proof size $\frac{\lambda^2 + \lambda(3y + x \log \lambda + 3\lambda)}{2} x$ bits on average and requires $O(|M'|^3)$ malleability actions. The equality test can also be done in the same way. Say the prover commits to $\mathsf{Commit}(m_1, r_1)$, $\mathsf{Commit}(m_1, r_2)$. Then, the prover uses Fig. 1 for the commitment product $\Pi \times \Pi$ with the message space $M'' \subseteq M'^2$ by collecting $(m, m) \in M'^2$.

Optimization for a Group M'. When the ring of the circuit can be embedded as a subgroup $M' \leq M$, the proofs for a variety of gates can be significantly improved, both with respect to proof size and efficiency.

Concretely, we can improve the proofs of the gates of equalities, additions, addition-by-a-constant, or multiply-by-a-constant. This is because the tuple of valid input-output tuples will form a subgroup of $\mathcal{M}^3$ by Proposition 2. In this case we can use Fig. 2 to generate the proofs. In contrast, the proof size is improved by a factor of $\log_2(M'^2)$ and the number of actions is reduced by a factor of M'.

However, the main bottleneck of our technique is rooted in the multiplication gate where the input-output tuples cannot form a subgroup over $\mathcal{M}^3$. Therefore, it still requires Fig. 1 to generate a proof for a multiplication gate and the proof size is of $O(\log(M'))$. This is also the reason that our result cannot be extended to a larger ring in a computationally feasible manner.

6.2 Proof System for Rank-1 Constraint System over a Small Ring

Let $\mathbb{F}$ be a finite field (or a ring), the rank-1 Constraint System (R1CS) relation consists of statement-witness pairs $((A, B, C, v), w)$ where $A, B, C \in \mathbb{F}^{m \times (n+1)}$ and v, w are matrices such that $(Az) \circ (Bz) = Cz$ for $z := (1, v, w) \in \mathbb{F}^{n+1}$ and $\circ$ denotes the entry-wise product (i.e. the Hadamard product). R1CS capture arithmetic circuit satisfaction : A, B, C encode the circuit's gates, the circuit's public input v, and w is the circuit's private input and wire values.

Recall from Remark 5 that we can assume the message space of an AGAMC to be a vector. Therefore, by using an AGAMC, a prover generates proofs for the rowcheck and lincheck problems as follows.

1. Commit to $z := (1, v, w)$, Az, Bz, Cz using Commit. Let $\mathsf{C}_z, \mathsf{C}_A, \mathsf{C}_B, \mathsf{C}_C$ be the resulting commitment vectors.
2. **Rowcheck:** Generate the proof for $\mathsf{C}_z, \mathsf{C}_A, \mathsf{C}_B, \mathsf{C}_C$ using Fig. 2 for the relation

$$
\left\{
\begin{array}{l}
(\ \mathsf{st} = (A, B, C, \mathsf{C}_z, \mathsf{C}_A, \mathsf{C}_B, \mathsf{C}_C), \\
\quad \mathsf{wt} = (z, r_z, r_A, r_B, r_C)\ )
\end{array}
\middle|
\begin{array}{l}
\mathsf{C}_z = \mathsf{Commit}(z; r_z), \\
\mathsf{C}_A = \mathsf{Commit}(Az; r_A), \\
\mathsf{C}_B = \mathsf{Commit}(Bz; r_B), \\
\mathsf{C}_C = \mathsf{Commit}(Cz; r_C).
\end{array}
\right\}.
$$

3. **Lincheck:** Generate the proof for C_A, C_B, C_C using Fig. 1 for the relation

$$\left\{ \begin{array}{l} (\ \mathsf{st} = (C_A, C_B, C_C), \\ \quad \mathsf{wt} = (y_A, y_B, y_C, r_A, r_B, r_C)\ ) \end{array} \left| \begin{array}{l} C_A = \mathsf{Commit}(y_A; r_A), \\ C_B = \mathsf{Commit}(y_B; r_B), \\ C_C = \mathsf{Commit}(y_C; r_C), \\ y_C = y_A \circ y_B. \end{array} \right. \right\}.$$

4. Output C_z, C_A, C_B, C_C and the proofs.

Note that we are allowed to apply Fig. 2 to the rowcheck because if M' is a subgroup of $\mathcal{M}$, where $\mathcal{M}$ is of dimensional $n+1$, then $\{(m, Am, Bm, Cm) | m \in M'\}$ is also a subgroup of $\mathcal{M}^4$, which can be formalized as Proposition 2. In contrast, Fig. 2 is not applicable to the lincheck due to the non-subgroup structure of the restriction $y_C = y_A \circ y_B$.

Proposition 2. *Let M_1, M_2 be subgroups of a finite abelian group $(\mathcal{M}, +)$. Then,*

(i) $M_1 \times M_2$ *is also a subgroup of* $\mathcal{M}^2 := \mathcal{M} \times \mathcal{M}$.
(ii) *the set* $\{(x, x) \subseteq M_1^2\}$ *is a subgroup of* $\mathcal{M}^2$.
(iii) *for any integer $a, b, c \in \mathbb{Z}$, the set* $\{(x, y, z) \subseteq M_1 \times M_2 \times \langle M_1, M_2 \rangle \mid ax + by = cz\}$ *is a subgroup of* $\mathcal{M}^3$.

For a given instantiation of an AGAMC scheme, the estimation is given in Table 1 with $A, B, C \in \mathbb{F}^{m \times (n+1)}$ and $z \in \mathbb{F}^{1+n}$ for an arithmetic circuit over a finite field $\mathbb{F}$ which can be embedded as a subgroup into $\mathcal{M}$. For example, when $\lambda = 128$ with an instantiation of AGAMC using CSIDH-2048 (assuming the existence), we have $\eta \approx \gamma^2 \approx 2048$.

Table 1. The proof (both rowcheck and lincheck proofs) size estimation of R1CS over $\mathbb{F}$ where $\mathbb{F}$ can be embedded as a subgroup using Figs. 1 and 2. The parameters η, γ are the numbers of bits to represent the elements of $\mathcal{C}$ and $\max\{|\mathcal{M}|, |\mathcal{R}|\}$, respectively. The density $\rho \in (0, 1)$ is the Hamming weight of A, B, C divided by $3m(n+1)$. The computational cost is evaluated for both parties in terms of the malleability operations.

Item	Cost
pp	$\frac{\eta}{4}$ B
Commitment	$(3m + n + 1)\frac{\eta}{8}$ B
Rowcheck Proof	$\frac{\lambda\rho}{16}(\lambda + (3m + n + 1)\log_2(|\mathbb{F}|) + \gamma(3m + n + 1)$ B
Lincheck Proof	$\frac{\lambda\rho}{16}(\lambda + \eta(m + \lambda \log_2(|\mathbb{F}|)))$ B
Computational Cost	$\mathcal{O}(mn\lambda|\mathbb{F}^2|)$

6.3 Zero-Knowledge Proofs for Branching Programs

Barrington's theorem is a fundamental result in computational complexity theory that has important implications for circuit design. It states that any circuit of depth d and fan-in 2 can be represented by a matrix branching program of length at most 4^d and width 5 [3].

In more formal terms, a matrix branching program over a field $\mathbb{F}$ of depth d and width w for ℓ-bit inputs is defined by a sequence $\mathsf{BP} = (i, A_{i,0}, A_{i,1})_{i\in[d]}$, where $A_{i,b}$ is a square matrix of $\mathbb{F}$ of dimension w for $i \in [d]$ and $b \in \{0,1\}$. The evaluation function e maps each index $i \in [d]$ to a bit position $e(i) \in [\ell]$.

Given an input $x \in \{0,1\}^\ell$, the matrix branching program evaluates by computing the product of matrices $\Pi_{i=1}^d A_{i,x_{e(i)}}$. The program outputs 1 if and only if the resulting product equals the identity matrix.

In this section, we show how to prove the satisfiability of a circuit in the form of a branching program (over a ring) using AGAMCs. We start by introducing the high-level idea. Let N be a natural number and $(i, A_{j,0}, A_{j,1})_{j\in[d]}$ represent a matrix branching program defined over $\mathbb{Z}_N$, with depth d and width w and let square matrices $M_0 = I_w, (M_j)_{j\in[d]}$ be a transcript of the program where $M_1 = A_{1,0}M_0$ or $= A_{1,1}M_0$, depending on the input x and the evaluation function e, and M_d is the final product.

Our proof system of the matrix branching program will proceed as follows.

1. Commit to M_j via $u_j = \mathsf{Commit}(M_j, R_j)$.
2. Compute $M_{j+1} = A_{j,b}M_j$ depending on the secret $b \in \{0,1\}$.
3. Commit to M_{j+1} via $u_{j+1} = \mathsf{Commit}(M_{j+1}, R_{j+1})$.
4. Generate a proof that the values committed in u_{j+1} and u_j are related by either $A_{j,0}$ or $A_{j,1}$ (i.e. $M_{j+1} = A_{j,0}M_j$ or $A_{j,1}M_j$).

It suffices to build a proof system for Item 4, as described in the following.

Base Proof System. This subsection introduces a fundamental tool using AGAMC for our proof system of a circuit in the form of a branching program. Concretely, for a given parameter $\mathsf{pp} = (\mathcal{M}'^2, \mathcal{R}'^2, \mathcal{C}'^2, X, \star)$, where $\mathcal{M}', \mathcal{R}', \mathcal{C}'$ correspond to the message space, the randomness space and the commitment space in the square matrix variant as explained in Rem. 5, we consider the following relation

$$\mathcal{R}_{\mathsf{pp},A_0,A_1} = \left\{ \begin{array}{l} (\ \mathsf{st} = (u_1, u_2), \\ \mathsf{wt} = ((M_1, M_2),(R_1, R_2))) \\ \in \mathcal{C}'^2 \times \mathcal{M}'^2 \times \mathcal{R}'^2 \end{array} \middle| \begin{array}{l} M_1, M_2 \in \mathcal{M}' \\ u_1 = (M_1, R_1) \star X \wedge \\ u_2 = (M_2, R_2) \star X \wedge \\ A_0 M_1 = M_2 \vee A_1 M_1 = M_2 \end{array} \right\}.$$

Note that a standard OR-proof [12] does not seem feasible here. An OR-proof requires statements for both relations and proves the knowledge of a witness for one of the two statements. Therefore, using OR-proof for a branching program will induce a statement and the proof size growing exponentially in the length because the prover needs to provide all possible statements in advances.

In contrast, we construct a proof system with linear growth as in Fig. 4. The high-level idea is that the prover generates and commits to $(M', A_0 M', R'),(M', A_1 M', R')$, then reveals either $(M_b + M', R + R')$ or (M', R').

However, this will disclose the information of b when the response is the $(M_b + M', R + R')$ and the witness is given. To this end, the prover additionally commits to $(M_1 + M', A_{1-b}(M_1 + M'), R + R')$ and shuffles with the previous commitments in a specific way to hide the information of b.

Notations. When the context is clear, we abuse notation by writing $(M'_1, M'_2, R'_1, R'_2) \star (x_1, x_2)$ to represent $((M'_1, R'_1) \star x_1, (M'_2, R'_2) \star x_2)$.

---

round 1: $P_1^{\mathcal{O}}((\mathsf{pp}, A_0, A_1, u = (u_1, u_2)), (M_1, M_2, R = (R_1, R_2), b))$

1: Parse $\mathsf{pp} = (\mathcal{M}'^2, \mathcal{R}'^2, \mathcal{C}'^2, X, \star)$
2: $\mathsf{seed}, \mathsf{bits}_0, \mathsf{bits}_1, \mathsf{bits}_2 \leftarrow_\$ \{0,1\}^\lambda$
3: $(M', R') \leftarrow H(\mathtt{EXP}\|\mathsf{seed})$ ▷ Sample $(M', R') \in \mathcal{M}' \times \mathcal{R}'^2$ and $\mathsf{bits}_i \in \{0,1\}^\lambda$
4: $\mathsf{C}_b \leftarrow H(\mathtt{COM}\|(M', A_0 M', R') \star u\|\mathsf{bits}_b)$ ▷ Commitments $\mathsf{C}_i \in \{0,1\}^{2\lambda}$
5: $\mathsf{C}_{b+1} \leftarrow H(\mathtt{COM}\|(M', A_1 M', R') \star u\|\mathsf{bits}_{b+1})$
6: $\mathsf{C}_{2-2b} \leftarrow H(\mathtt{COM}\|(M_1 + M', A_{1-b}(M_1 + M'), R + R') \star (X, X)\|\mathsf{bits}_{2-2b})$
7: $\mathsf{root} \leftarrow \mathsf{MerkleTree}((\mathsf{C}_0\|\mathsf{C}_1\|\mathsf{C}_2), (\mathsf{C}_2\|\mathsf{C}_0\|\mathsf{C}_1))$ ▷ Index-hiding Merkle Tree
8: Prover sends $\mathsf{comm} \leftarrow \mathsf{root}$ to Verifier.

---

round 2: $V_1(\mathsf{comm})$

1: $c \leftarrow_\$ \{0,1\}$
2: Verifier sends $\mathsf{chall} \leftarrow c$ to Prover.

round 3: $P_2((M, R, b), \mathsf{chall})$

1: $c \leftarrow \mathsf{chall}$
2: **if** $c = 1$ **then**
3:     $(M'', R'') \leftarrow (M_1 + M', R + R')$
4:     $\mathsf{resp} \leftarrow (M'', R'', \mathsf{bits}_0, \mathsf{bits}_2, \mathsf{C}_1)$
5: **else**
6:     $\mathsf{resp} \leftarrow (\mathsf{seed}, \mathsf{bits}_b, \mathsf{bits}_{b+1}, \mathsf{C}_{2-2b})$
7: Prover sends resp to Verifier

Verification: $V_2^{\mathcal{O}}(\mathsf{comm}, \mathsf{chall}, \mathsf{resp})$

1: $(\mathsf{root}, c) \leftarrow (\mathsf{comm}, \mathsf{chall})$
2: **if** $c = 1$ **then**
3:     $(M'', R'', \mathsf{bits}_0, \mathsf{bits}_2, \mathsf{C}_1) \leftarrow \mathsf{resp}$
4:     $\widetilde{\mathsf{C}} \leftarrow H(\mathtt{COM}\|(M'', A_0 M'', R'') \star (X, X)\|\mathsf{bits}_0)$
5:     $\widetilde{\mathsf{C}}' \leftarrow H(\mathtt{COM}\|(M'', A_1 M'', R'') \star (X, X)\|\mathsf{bits}_2)$
6:     $\widetilde{\mathsf{root}} \leftarrow \mathsf{MerkleTree}((\widetilde{\mathsf{C}}\|\mathsf{C}_1\|\widetilde{\mathsf{C}}'), (\widetilde{\mathsf{C}}'\|\widetilde{\mathsf{C}}\|\mathsf{C}_1))$
7: **else**
8:     $(\mathsf{seed}, \mathsf{bits}_b, \mathsf{bits}_{b+1}, \mathsf{C}_{2-2b}) \leftarrow \mathsf{resp}$
9:     $(M', R') \leftarrow H(\mathtt{EXP}\|\mathsf{seed})$
10:    $\widetilde{\mathsf{C}} \leftarrow H(\mathtt{COM}\|(M', A_0 M', R') \star u\|\mathsf{bits}_b)$
11:    $\widetilde{\mathsf{C}}' \leftarrow H(\mathtt{COM}\|(M', A_1 M', R') \star u\|\mathsf{bits}_{b+1})$
12:    $\widetilde{\mathsf{root}} \leftarrow \mathsf{MerkleTree}((\widetilde{\mathsf{C}}\|\widetilde{\mathsf{C}}'\|\mathsf{C}_{2-2b}), (\mathsf{C}_{2-2b}\|\widetilde{\mathsf{C}}\|\widetilde{\mathsf{C}}'))$
13: Output accept if $\widetilde{\mathsf{root}} = \mathsf{root}$, and reject otherwise.

Fig. 4. A three-move interactive protocol $(P^{\mathcal{O}} = (P_1^{\mathcal{O}}, P_2), V = (V_1, V_2^{\mathcal{O}}))$ to prove the possession of a witness (M_1, M_2, R_1, R_2, b) for a statement $u = (u_1, u_2)$ such that $u = (M_1, M_2, R_1, R_2) \star (X, X)$ and $M_2 = A_b M_1$.

Theorem 6. *In the random oracle model, by assuming* $\mathsf{MerkleTree}$ *to be collision-resistant and* $H(\mathtt{COM}\|\cdot)$ *is binding, the construction* $(P^{\mathcal{O}} = (P_1^{\mathcal{O}}, P_2), V = (V_1, V_2^{\mathcal{O}}))$ *in Fig. 4 has correctness, 2-special soundness, and HVZK for the relation* $\mathcal{R}_{\mathsf{pp}, A_0, A_1}$ *with* Q_H *queries to* H.

Proof. The correctness holds clearly when the challenge bit is 0. When the challenge is 1 the verifier have $\mathsf{resp} = (M'', R'', \mathsf{bits}_0, \mathsf{bits}_2, \mathsf{C}_1)$ where $M'' = M_1 + M'$ and $R'' = R + R'$. Then, we have either $\{\widetilde{\mathsf{C}}, \widetilde{\mathsf{C}}'\} = \{\mathsf{C}_0, \mathsf{C}_2\}$ where $\widetilde{\mathsf{C}} = H(\mathtt{COM}\|(M'', A_0 M'', R'') \star (X, X)\|\mathsf{bits}_0)$ and $\widetilde{\mathsf{C}}' \leftarrow H(\mathtt{COM}\|(M'', A_1 M'', R'') \star (X, X)\|\mathsf{bits}_2)$. Therefore, the Merkle tree on input $(\widetilde{\mathsf{C}}\|\mathsf{C}_1\|\widetilde{\mathsf{C}}'), (\widetilde{\mathsf{C}}'\|\widetilde{\mathsf{C}}\|\mathsf{C}_1)$ and

$(C_0||C_1||C_2), (C_2||C_0||C_1)$ will result in the same root since the entries of the input will be ordered alphabetically. We leave the proof for 2-special soundness and honest-verifier zero-knowledge for the appendix of the full version of this paper.

Remark 6. The Merkle tree only has two leaves in **round 1** so the prover does not need to store additional information for the tree to generate the response.

Proof Size and Computational Cost. We turn Fig. 4 into a NIZK via Fig. 3, giving an essential tool for our proof system for a branching program.

Let N be a natural number and $(i, A_{j,0}, A_{j,1})_{j \in [d]}$ represent a matrix branching program defined over $\mathbb{Z}_N$, with depth d and width w. For a given instantiation of an AGAMC scheme, the estimation is given in Table 2 using as described in Remark 5 to produce the commitment scheme in the matrix form. For example, regarding the proof size, when $\lambda = 128$ with an instantiation of AGAMC using CSIDH-2048 (assuming the existence), we have $\eta \approx \gamma^2 \approx 2048$.

Regarding the computational complexity, as shown in Table 2, we measure the cost in terms of the malleability operation, which is the group action operation that gives malleability. Within each layer $j \in [d]$, both the prover and verifier need to perform $O(\lambda)$ malleability operation. Therefore, the proof takes both parties $O(d\lambda w^2)$ CSIDH group actions regardless of the size of the underlying ring. This cost estimate applies to each layer of the proof and takes into account the size of the matrices used in the computation.

Table 2. An evaluation of the commitment size and the proof size Fig. 4 for a matrix branching problem over $\mathbb{Z}_N$ and of length d and width w, where N is also the number of bits to represent $\mathcal{M}$. The parameters η, γ are the numbers of bits to represent the elements of $\mathcal{C}$ and $\mathcal{R}$, respectively. Note that $(\mathcal{C}', \mathcal{M}', \mathcal{R}') = (\mathcal{C}^{w \times w}, \mathcal{M}^{w \times w}, \mathcal{R}^{w \times w})$. The computational cost is evaluated for both parties given in terms of the malleability operations.

Item	Cost (Bytes)
pp	2η B
Commitment	$dw^2 \frac{\eta}{8}$ B
Proof	$\frac{\lambda d}{16}((N + 2\gamma)w^2 + 9\lambda)$ B
Computational Cost	$\mathcal{O}(\lambda dw^2)$

6.4   Discussion and Further Work

Compared to the classical setting, there are three major challenges in our current context that require further attention.

The first bottleneck, concerning our proof systems for generic arithmetic circuits over a ring (Sects. 6.1 and 6.2) is that to prove the multiplication relation of the committed messages, we employ an OR-proof-type proof system. Although the proof remains logarithmic in the ring size (of the arithmetic circuit), both the prover and the verifier must compute the actions of all possible combinations of $(M_1, M_2, M_1 M_2)$. This limits the ring size as well as the message space, as they cannot both be large. Developing a more efficient proof system for the multiplication relation of the committed messages would improve the capability of the proof systems in Sects. 6.1 and 6.2. Note however, that this restriction *does not affect* the proof system for branching program, which relies on an affine relation, as discussed in Sect. 6.3.

Another restriction is the lack of an efficient commitment scheme that allows a prover to commit to a vector of messages using only one element. In classical cryptography, the generalized Pedersen commitment provides such a functionality. In our setting, a naive approach is to use randomly generated elements $y, g_1, \cdots, g_n$, and Y_0, Y from the group, and commit to $(y^r \star Y_0, (r \prod_i^n g_i^{m_i}) \star Y)$ for the messages $m_1, \cdots, m_n$ (using multiplicative notation for the group operation). However, this commitment is not quantumly binding since the relations between the g_i's can be recovered using a quantum period-finding algorithm. This is a challenge that is worth investigating in future research.

Finally, the folding technique, which enables proof compression [10,11] does not extend from homomorphic to malleable commitment or the Naor-Reingold-type VRF [22], resulting in an argument with square root communication complexity. A variant of that technique making use of the malleability rather than the homomorphic property is an interesting question to explore.

However, our construction does offer several advantages. First, we do not require any form of trusted set-up. Second, we achieve statistical, rather than computational, zero-knowledge, as well as simulation extractability, meaning that an adversary cannot simulate a new proof unless they know a witness, regardless of the number of simulated proofs they have seen. This is a stronger notion than soundness or even knowledge soundness, and highly desirable in situations where the non-malleability of the proofs is required. Finally, we obtain linear efficiency in the size of circuit. Despite being less efficient than state of the art post-quantum zero-knowledge proofs, such as [2,4,8] whose proofs sizes are sublinear or even polylogarithmic in the size of the witness, our approach still improves on the naive construction from one-way functions and provides a first general purpose zero-knowledge proof from isogenies.

7   Conclusion

In this paper, we introduce and formalize the notion of malleable commitment, which extends the notion of homomorphic commitments but enforces less structure on the underlying commitment scheme. We show how to instantiate such malleable commitments from an isogeny-based group action. Subsequently, we build different flavors of generic proofs for NP statements. Our first two approaches, based on arithmetic circuits and rank-1 constraint systems impose restrictions on the size of the underlying ring and message space. However, this restriction is lifted by our third approach based on branching programs.

Regardless, these three approaches offer the first feasibility results towards post-quantum zero-knowledge proofs based on isogeny assumptions, improving on the naive construction from one-way functions.

Acknowledgements. Mingjie Chen and Christophe Petit are partly supported by EPSRC through grant number EP/V011324/1. Yi-Fu Lai thanks the New Zealand Ministry for Business and Employment for financial support.

References

1. Alamati, N., De Feo, L., Montgomery, H., Patranabis, S.: Cryptographic group actions and applications. In: Moriai, S., Wang, H. (eds.) ASIACRYPT 2020, Part II. LNCS, vol. 12492, pp. 411–439. Springer, Cham (2020). https://doi.org/10.1007/978-3-030-64834-3_14
2. Ames, S., Hazay, C., Ishai, Y., Venkitasubramaniam, M.: Ligero: lightweight sub-linear arguments without a trusted setup. In: ACM CCS 2017, pp. 2087–2104 (2017)
3. Barrington, D.A.: Bounded-width polynomial-size branching programs recognize exactly those languages in NC. In: Proceedings of the Eighteenth Annual ACM Symposium on Theory of Computing, pp. 1–5 (1986)
4. Ben-Sasson, E., Chiesa, A., Riabzev, M., Spooner, N., Virza, M., Ward, N.P.: Aurora: transparent succinct arguments for R1CS. In: Ishai, Y., Rijmen, V. (eds.) EUROCRYPT 2019, Part I. LNCS, vol. 11476, pp. 103–128. Springer, Cham (2019). https://doi.org/10.1007/978-3-030-17653-2_4
5. Beullens, W., Dobson, S., Katsumata, S., Lai, Y.-F., Pintore, F.: Group signatures and more from isogenies and lattices: generic, simple, and efficient. In: Dunkelman, O., Dziembowski, S. (eds.) EUROCRYPT 2022, Part II. LNCS, vol. 13276, pp. 95–126. Springer, Cham (2023). https://doi.org/10.1007/978-3-031-07085-3_4
6. Beullens, W., Katsumata, S., Pintore, F.: Calamari and Falafl: logarithmic (linkable) ring signatures from isogenies and lattices. In: Moriai, S., Wang, H. (eds.) ASIACRYPT 2020, Part II. LNCS, vol. 12492, pp. 464–492. Springer, Cham (2020). https://doi.org/10.1007/978-3-030-64834-3_16
7. Beullens, W., Kleinjung, T., Vercauteren, F.: CSI-FiSh: efficient isogeny based signatures through class group computations. In: Galbraith, S.D., Moriai, S. (eds.) ASIACRYPT 2019, Part I. LNCS, vol. 11921, pp. 227–247. Springer, Cham (2019). https://doi.org/10.1007/978-3-030-34578-5_9
8. Beullens, W., Seiler, G.: LABRADOR: compact proofs for R1CS from module-SIS. Cryptology ePrint Archive, Paper 2022/1341 (2023)
9. Boneh, D., Gentry, C., Gorbunov, S., Halevi, S., Nikolaenko, V., Segev, G., Vaikuntanathan, V., Vinayagamurthy, D.: Fully key-homomorphic encryption, arithmetic circuit ABE and compact garbled circuits. In: Nguyen, P.Q., Oswald, E. (eds.) EUROCRYPT 2014. LNCS, vol. 8441, pp. 533–556. Springer, Heidelberg (2014). https://doi.org/10.1007/978-3-642-55220-5_30
10. Bootle, J., Cerulli, A., Chaidos, P., Groth, J., Petit, C.: Efficient zero-knowledge arguments for arithmetic circuits in the discrete log setting. In: Fischlin, M., Coron, J.-S. (eds.) EUROCRYPT 2016, Part II. LNCS, vol. 9666, pp. 327–357. Springer, Heidelberg (2016). https://doi.org/10.1007/978-3-662-49896-5_12
11. Bünz, B., Bootle, J., Boneh, D., Poelstra, A., Wuille, P., Maxwell, G.: Bulletproofs: Short proofs for confidential transactions and more. In: 2018 IEEE Symposium on Security and Privacy, pp. 315–334 (2018)

12. Cramer, R., Damgård, I., Schoenmakers, B.: Proofs of partial knowledge and simplified design of witness hiding protocols. In: Desmedt, Y.G. (ed.) CRYPTO 1994. LNCS, vol. 839, pp. 174–187. Springer, Heidelberg (1994). https://doi.org/10.1007/3-540-48658-5_19
13. De Feo, L., et al.: SCALLOP: scaling the CSI-FiSh. In: Boldyreva, A., Kolesnikov, V. (eds.) PKC 2023, Part I. LNCS, vol. 13940, pp. 345–375. Springer, Cham (2023)
14. De Feo, L., Meyer, M.: Threshold schemes from isogeny assumptions. In: Kiayias, A., Kohlweiss, M., Wallden, P., Zikas, V. (eds.) PKC 2020, Part II. LNCS, vol. 12111, pp. 187–212. Springer, Cham (2020). https://doi.org/10.1007/978-3-030-45388-6_7
15. Fiat, A., Shamir, A.: How To prove yourself: practical solutions to identification and signature problems. In: Odlyzko, A.M. (ed.) CRYPTO 1986. LNCS, vol. 263, pp. 186–194. Springer, Heidelberg (1987). https://doi.org/10.1007/3-540-47721-7_12
16. Gentry, C., Sahai, A., Waters, B.: Homomorphic encryption from learning with errors: conceptually-simpler, asymptotically-faster, attribute-based. In: Canetti, R., Garay, J.A. (eds.) CRYPTO 2013. LNCS, vol. 8042, pp. 75–92. Springer, Heidelberg (2013). https://doi.org/10.1007/978-3-642-40041-4_5
17. Goldreich, O., Micali, S., Wigderson, A.: Proofs that yield nothing but their validity or all languages in np have zero-knowledge proof systems. J. ACM 38, 691–729 (1991)
18. Gorbunov, S., Vaikuntanathan, V., Wichs, D.: Leveled fully homomorphic signatures from standard lattices. In: Proceedings of the Forty-Seventh Annual ACM Symposium on Theory of Computing, New York, NY, USA, pp. 469–477 (2015)
19. Groth, J.: Homomorphic trapdoor commitments to group elements. Cryptology ePrint Archive, Paper 2009/007 (2009)
20. Groth, J.: Linear algebra with sub-linear zero-knowledge arguments. In: Halevi, S. (ed.) CRYPTO 2009. LNCS, vol. 5677, pp. 192–208. Springer, Heidelberg (2009). https://doi.org/10.1007/978-3-642-03356-8_12
21. Katsumata, S., Lai, Y., LeGrow, J.T., Qin, L.: CSI-otter: isogeny-based (partially) blind signatures from the class group action with a twist. In: Handschuh, H., Lysyanskaya, A. (eds.) CRYPTO 2023. LNCS, vol. 14083, pp. 729–761. Springer, Cham (2023). https://doi.org/10.1007/978-3-031-38548-3_24
22. Lai, Y.-F.: CAPYBARA and TSUBAKI: verifiable random functions from group actions and isogenies. Cryptology ePrint Archive, Report 2023/182 (2023)
23. Pedersen, T.P.: Non-interactive and information-theoretic secure verifiable secret sharing. In: Feigenbaum, J. (ed.) CRYPTO 1991. LNCS, vol. 576, pp. 129–140. Springer, Heidelberg (1992). https://doi.org/10.1007/3-540-46766-1_9
24. Peikert, Chris: He gives C-sieves on the CSIDH. In: Canteaut, Anne, Ishai, Yuval (eds.) EUROCRYPT 2020, Part II. LNCS, vol. 12106, pp. 463–492. Springer, Cham (2020). https://doi.org/10.1007/978-3-030-45724-2_16
25. Wahby, R.S., Tzialla, I., shelat, A., Thaler, J., Walfish, M.: Doubly-efficient zkSNARKs without trusted setup. In: 2018 IEEE Symposium on Security and Privacy, pp. 926–943 (2018)
26. Zhang, Y., Genkin, D., Katz, J., Papadopoulos, D., Papamanthou, C.: vSQL: verifying arbitrary SQL queries over dynamic outsourced databases. In: 2017 IEEE Symposium on Security and Privacy, pp. 863–880 (2017)

Attacks

A CP-Based Automatic Tool
for Instantiating Truncated Differential
Characteristics

François Delobel[1], Patrick Derbez[2], Arthur Gontier[2], Loïc Rouquette[3,4(✉)],
and Christine Solnon[5]

[1] Université Clermont-Auvergne, CNRS, Mines de Saint-Étienne,
LIMOS, Clermont-Ferrand, France
`francois.delobel@uca.fr`
[2] Univ Rennes, CNRS, IRISA, Rennes, France
{`patrick.derbez,arthur.gontier`}`@irisa.fr`
[3] EPITA Research Laboratory (LRE), 14/16 rue Voltaire,
94270 Le Kremlin-Bicêtre, France
`loic.rouquette@epita.fr, publications@loicrouquette.fr`
[4] LORIA, Université de Lorraine, 54000 Nancy, France
[5] INSA Lyon, CITI, INRIA CHROMA, 69621 Villeurbanne, France
`christine.solnon@insa-lyon.fr`

Abstract. An important criteria to assert the security of a cryptographic primitive is its resistance against differential cryptanalysis. For word-oriented primitives, a common technique to determine the number of rounds required to ensure the immunity against differential distinguishers is to consider truncated differential characteristics and to count the number of active S-boxes. Doing so allows one to provide an upper bound on the probability of the best differential characteristic with a reduced computational cost. However, in order to design very efficient primitives, it might be needed to evaluate the probability more accurately. This is usually done in a second step, during which one tries to instantiate truncated differential characteristics with actual values and computes its corresponding probability. This step is usually done either with ad-hoc algorithms or with CP, SAT or MILP models that are solved by generic solvers. In this paper, we present a generic tool for automatically generating these models to handle all word-oriented ciphers. Furthermore the running times to solve these models are very competitive with all the previous dedicated approaches.

Keywords: Differential cryptanalysis · Constraint Programming · Automatic tool

The work presented in this article was funded by the French National Research Agency as part of the DeCrypt project (ANR-18-CE39-0007).

1   Introduction

The security of a symmetric primitive mostly relies on applying all known crypt-analysis techniques to show that none of them is close to endanger it and to ensure that there is enough security margin against all known attacks. Among others, this evaluation process typically involves identifying the differential and linear characteristic which allows one to distinguish the primitive from a random permutation on as many rounds as possible. Indeed, several decades after their introduction, both differential [3] and linear [22] cryptanalysis are still two very powerful techniques receiving a lot of attention from the community.

Differential cryptanalysis aims at searching for differential distinguishers, which are formed by both an input and output difference such that the transition occurs with a probability higher than expected for a random permutation. The main issue with this cryptanalysis technique lies in the inherent difficulty of finding such differentials due to the extremely large search space. To overcome this difficulty it was proposed to look for differential characteristics, meaning that, instead of defining only the input and output differences, all differences in internal states are specified as well. However, searching for the best differential characteristics is not easier than searching for differential and thus, especially for word-oriented primitives, it was proposed to search for truncated differential characteristics. In a truncated differential characteristic, the exact difference on each word is omitted and replaced by a Boolean variable indicating whether there is a non-zero difference on the word or not. Associated with the best probability of a transition through the non-linear function involved in the primitive, truncated differential analysis provides an upper bound on the probability of any differential characteristic.

However, the bounds computed from truncated differential cryptanalysis are most often not tight, since it might be impossible to instantiate a truncated differential characteristic such that all its internal transitions occur with the highest possible probability. To evaluate more accurately the resistance of a primitive against differential cryptanalysis, an interesting problem is to find the best possible instantiation of a given truncated differential characteristic, *i.e.* , find the actual difference values maximizing the overall probability. Due to the huge search space, and because difference distribution tables (DDT) of non-linear cryptographic components are hard to model by means of linear or Boolean constraints, searching for the best instanciation of a given truncated differential characteristic most often relies on Constraint Programming (CP) solvers. Constraint programming is a flexible declarative language that comes with a large collection of constraints. A CP model is composed of variables, their corresponding domains and a set of constraint on these variables. A constraint is a relation between the variables and comes with an algorithm to check if the constraint is satisfied and an algorithm to remove the values that do not respect the relation from the domains of the variables. This algorithm is called a *filtering algorithm*. The CP solving method uses both a Depth-First Search algorithm and all the filtering algorithms of each variable to solve a problem. There are many different types of constraints and new ones can be added to CP solvers if we can provide a corresponding filtering algorithm.

Regarding the specific problem of differential cryptanalysis, we can cite the collection of works from Gerault, Minier and their co-authors [9–11,23,30], in which the ultimate goal is to provide the probability of the best differential characteristic for each round-reduced version of all versions of the Rijndael cipher. It is also worth mentioning the recent work of Delaune et al. [5], who proposed a dedicated CP model to instantiate truncated differential characteristics on the cipher SKINNY.

Our Contribution. The common point of the previous CP models is that they are all dedicated to a specific cipher. Designing these dedicated CP models is time-consuming and error prone. In this paper, we propose a generic approach for automatically generating these models to handle all word-oriented ciphers, providing running times close enough to the dedicated models on both Rijndael and SKINNY while being fully generic. This CP model generator is a part of the TAGADA project, a tool which aims at providing a simple and easy-to-use API to describe ciphers as well as the automatic generation of CP models to search for differential characteristics. However, our CP model generator can also be used as a standalone object, taking as input the description of the cipher and a truncated differential characteristic and outputting its best possible instantiation.

Organisation of the Paper. Section 2 describes the TAGADA tool, it defines the notion of Differential Cryptanalysis as well as the way in which TAGADA works. Section 3 describes the first contribution of this article, namely a CP model generator for computing the best differential characteristic given a truncated characteristic. Section 4 explains the integration of this model generator within the TAGADA library. Section 5 presents some optimizations and Sect. 6 presents the results obtained. Lastly, Sect. 7 recalls the tools introduced by this article and proposes new areas of research allowing the extension of automated differential cryptanalysis.

2   Tagada

TAGADA (Tool for Automatic Generation of Abstraction-based Differential Attack) is a tool proposed in [20] that can generate models for computing truncated differential characteristics for any word-oriented cipher. It relies on a graph representation of the cipher and uses several techniques to optimize the models. In this background section, we first recall how differential is modelled and then explain the graph representation of TAGADA. Moreover, we describe some of the optimizations TAGADA proposed for the first step of this problem.

2.1   Differential Cryptanalysis

Differential analysis is a method to analyze the effect of differences in plaintext pairs on the differences of the resultant ciphertexts. The difference is usually obtained with the bit-wise XOR; we will note it $+$. For a cipher function F and two plaintexts x and y where y is created by injecting an input difference δ_{in}, $i.e.$ $y = x + \delta_{in}$, the output difference δ_{out} is computed with $\delta_{out} = F(x) + F(y)$. There is a differential distinguisher in F if the probability that $\delta_{out} = F(x) + F(x + \delta_{in})$ is high $i.e.$ the input difference δ_{in} has a good probability of ending up in the output difference δ_{out}. Symmetric ciphers are iterated functions like $F(x) = f(f(\ldots f(f(x))\ldots))$. To see if the difference δ_{in} can end up in a difference δ_{out}, we study the propagation of the difference through all the rounds and all the ciphers operators. We usually note δx_i the difference of an intermediate variable of the cipher x_i. The tracking of differences from δ_{in} to δ_{out} through the complete cipher is called a *differential trail* or *differential characteristic*.

Symmetric ciphers are generally composed of two types of operators: linear operators like XORs or permutations and non-linear operators like S-Boxes and multiplications:

- *Linear operators* will always propagate differences with probability 1. Indeed a permutation will simply reorganize the differences. Another common linear operator is the XOR of three variables $y = x_2 + x_3$ were the differences will be propagated with another XOR: $\delta y = \delta x_2 + \delta x_3$. Note that if $\delta x_2 = \delta x_3$ then the output difference is cancelled $\delta y = 0$.
- However, *non-linear operators*, called S-Boxes, will propagate a difference with a probability that may be lower than one. For each S-Box, the propagation probabilities of the pair of input-output differences can be computed with their *Difference Distribution Table* (DDT). Since it is necessary to enumerate all the possible pairs of inputs to generate the DDTs they are cached and it is not possible to compute them if the input size of the associated operator is too big. Usually it is possible to compute DDTs for word oriented ciphers as they are working with 4-bit (nibbles) and 8-bit (bytes) words. TAGADA is designed for word oriented ciphers, for which the propagation of differences in non-linear operators can be computed with DDTs. TAGADA will have very poor performances on bit-oriented ciphers (*e.g.* [1]), or on word-oriented ciphers that operate on words larger than bytes.

If a differential trail contains two or more S-Boxes, the probability of the trail is the multiplication of the probabilities given by the DDTs because we assume that the probabilities are independent [15]. Therefore, the probability of a trail will be lower if we add more S-Boxes to it, and low-probability trails may not be useful for standard differential attacks (differential trail with zero probability can be used in impossible attacks for example). To make ciphers more resistant, we could want to search for S-Boxes with perfectly balanced probabilities. However, this seems really hard to achieve among all the other required properties of S-Boxes [12].

As the search of differential trails is hard, it is usually split into two steps in the most recent works [4,6,10].

First Step: Find Truncated Trails. The first step is the search for *truncated differential trails* [13]. A truncated differential trail is an abstraction of a differential trail in which we only retain whether a difference exists or not, *i.e.* each difference variable δx_i, associated to cipher's intermediate word x_i of size n, is abstracted by a Boolean variable Δx_i where

$$\Delta x_i = \begin{cases} 0 & \text{if} \quad \delta x_i = 0 \\ 1 & \text{if} \quad \delta x_i \in [1; 2^n - 1] \end{cases}$$

Because each Boolean variable Δx_i encodes the existence of the difference without tracking its value, several aspects of the problem change.

- For the probability of propagation through the S-Boxes, we cannot know which probability to pick in the DDT except when the Boolean variable associated with the input difference is equal to 0, in which case we know for sure that the Boolean variable associated with the output difference is also equal to 0. When the Boolean variable associated with the input variable is equal to 1, we only know that the S-Box is *active i.e.* involved in the trail. Therefore, we will use the highest probability to have an upper bound on the probability of the differential trail.
- The second change is that we cannot capture the difference cancellations of the XOR operations. Therefore, there are usually a lot of false positive trails *i.e.* trails that cannot be instantiated.

Truncated differential analysis can be enough to say that a cipher is secure in the case where the best truncated trail (the one with the fewest active S-Boxes) has a low enough probability. On the other hand, when a truncated trail has a high probability, we must successfully instantiate it with real difference values to have a differential trail.

Second Step: Instantiate the Trails. In the second step, we enumerate all possible differential trails (starting from active S-boxes) to find the best one. The two steps can be done separately (find all truncated trails, then try to instantiate them all) or together like in Algorithm 3 that will be explained more in Sect. 4.

In [5], the authors compare a dedicated implementation (based on dynamic programming) with SAT, MILP, and CP models that are solved by generic solvers on the two steps of the differential cryptanalysis of the cipher SKINNY. In conclusion, they found that their hand-made algorithm is the fastest on the first step, followed by SAT and MILP. For the second step, they conclude that the CP solver is the most efficient solver by far. This work points out a key issue of cryptanalysis problems. It has to be done on every cipher, and the development time is not the same for a hand-made algorithm or a model for tools like SAT, MILP and CP.

2.2 How Tagada Works

In [20], a generic graph representation of ciphers was proposed. This graph is encoded in a text format to be able to simplify the communication of the cipher definition. TAGADA uses this graph to generate CP,SAT of MILP models to solve the first step of the differential analysis, and we will use the same graph to generate the second step too.

DAG: Unifying Description of Ciphers. The input graph is a Directed Acyclic graph where the nodes correspond to all the cipher parameters (inputs, outputs, constants...) and operators (XOR, S-Box, permutations...). The edges of the DAG link the operators to the parameters. Hence the DAG is a bipartite graph. Figure 1 shows the DAG of a 2-round toy example Feistel cipher.

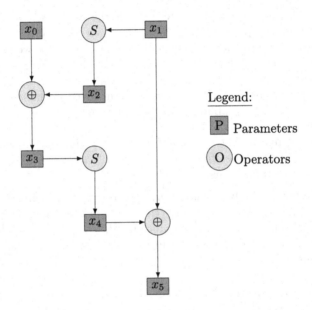

Fig. 1. DAG of a simple 2-round toy example Feistel cipher.

The text format used to define the DAG is a JSON composed of a list of three types of objects.

- The variables with their domain ranges.
- The functions (the operators) with input domains, output domains and specificities. For example, an S-Box will declare its lookup table, a LFSR will declare its length, shift direction and feedback polynomial...
- The transitions are triplets composed of a list of variables, a function and another list of variables. They describe the link between the operators and their input and output variables.

The advantage of a text format like JSON is that it is easy to generate and parse for any language. Moreover, the DAG has to be made only once for each cipher, thus saving a lot of time compared to the development of a hand-made algorithm or a solver-specific model. The difficulty of the DAG representation is that the DAG must be able to represent all the operators in order to be able to model any cipher. TAGADA currently handles the following operators: equality, bit-wise XOR, Galois field multiplication, LFSR, left shift register, right shift register, permutation, concatenation, split and S-Box. TAGADA also proposes to describe new operators by means of tables describing all the possible in/out tuples. However, this option is possible only when the table is not too large.

Truncated Differential Graph and Optimizations. To solve the first step of the differential analysis, TAGADA first builds a truncated version of the graph of the input cipher. Then, this graph is optimized with a simplification of the useless parts of the graph. Indeed, differential analysis does not use constants nodes, and equality operators can be removed from the graph by merging the equals nodes.

A second optimization is done to detect some inconsistent differential trails by adding constraints, as illustrated in Example 1.

Example 1. Let $\delta_1, \delta_2, \delta_3$ be three differential variables, and $\Delta_1, \Delta_2, \Delta_3$ be their corresponding truncated differential variables. Let $\Delta_1 + \Delta_2 = 0$ and $\Delta_1 + \Delta_2 + \Delta_3 = 0$ be two equations generated from the DAG.

If we look at the first equation, it is satisfied if $\Delta_1 = 0$ and $\Delta_2 = 0$. However, the difference can also be cancelled if they have the same value ($\delta_1 = \delta_2$). Therefore, the equation is also satisfied if $\Delta_1 = 1$ and $\Delta_2 = 1$.

For the same reason, the second equation accepts the solution $\Delta_1 = 1, \Delta_2 = 1, \Delta_3 = 1$. In the truncated model, there is no problem. However, in the second step model, the first equation implies that $\delta_1 = \delta_2$ and that they are equals to a non-zero difference, so $\delta_3 \neq 0$ would never be a valid assignment.

To detect this kind of inconsistencies, TAGADA combines XOR equations to generate new equations. Generating new equations is a key point for an efficient model. These equations were hand-made in [10,11], whereas they are automatically derived from the DAG in TAGADA. In [28], an abstract-XOR constraint has been designed to better propagate XOR constraint in CP solvers.

Once optimized, a mathematical model is automatically generated from the truncated graph. This model is expressed using the MiniZinc language, which is a high-level language for defining constraint satisfaction problems [24]. Many different solvers are able to solve problems defined in MiniZinc such as, for example, Choco, Chuffed or Picat. We may also use Picat to automatically generate SAT or MILP models from a MiniZinc model, thus allowing one to use SAT or MILP solvers. In many cases, the best solver is Picat-SAT.

2.3   First Step Results

In [20] the TAGADA models were able to recover the state-of-the-art truncated trails of the ciphers AES, Midori, SKINNY, and Craft in either single-key or related-key scenarios.

However, truncated trails may not lead to valid differential trails. To be able to make a strong statement on the differential characteristics of a cipher, we must try to instantiate these trails with the second step. Therefore, another program or model has to be made to solve the second step, and we propose to generate it from the DAG representation of the cipher like TAGADA generated the first step models.

3   Model Generation for the Second Step

In this section, we present our contribution to the TAGADA project, focusing on the modelling of the second step of the differential analysis, the instantiation of truncated characteristics. Contrary to the first step, we rely only on a CP solver, namely Choco [25]. There are two reasons for that. Firstly, in both [11] and [5], the CP solver was said to perform very well for this particular problem. Secondly, we wanted to develop dedicated filtering algorithms for operators like the bit-wise XOR in a CP solver to improve the overall efficiency of the solving process.

In the second step of the differential analysis, we use the solutions of the first step and we try to instantiate them. More precisely, we make a complete model of the cipher, and we constrain the S-Boxes variables according to the truncated trail *i.e.* the active S-Boxes in the truncated trail have a full domain, and the inactive S-Boxes be set to zero. To generate models for the second step, we need constraints for each operator and the most important one is the S-Box.

3.1   Modelling DDT with Table Constraints

The probability of the propagation of a differences through an S-Box is described in a *Difference Distribution Table* (DDT). For example, the DDT of an S-Box of the F function of DES is given in Table 1. This S-Box has six input bits and four output bits. Therefore, the DDT is a table with $2^6 \times 2^4$ entries. In this table, we can see that the first difference (0) always propagates to 0 (64 times over the 64 possible input pairs of difference 0). In the second line, we can see that the input difference 1 propagates to the output difference 3 six times over 64 possible input pairs of difference 1. Therefore, the probability of this propagation is 6×2^{-6}. In the truncated model, we would use only the best probability of this table which is 16×2^{-6}, but this probability holds only for one transition ($34 \rightarrow 2$). We use the best transition probability in order to use the probability of the truncated differential trail as an upper bound approximation of the probability of the differential trail - the best possible differential trail is the differential trail that uses only optimal transitions.

Table 1. DDT of one S-Box of DES

	0	1	2	3	4	5	6	7	8	9	A	B	C	D	E	F
0	64	0	0	0	0	0	0	0	0	0	0	0	0	0	0	0
1	0	0	0	6	0	2	4	4	0	10	12	4	10	6	2	4
2	0	0	0	8	0	4	4	4	0	6	8	6	12	6	4	2
3	14	4	2	2	10	6	4	2	6	4	4	0	2	2	2	0
4	0	0	0	6	0	10	10	6	0	4	6	4	2	8	6	2
5	4	8	6	2	2	4	4	0	4	0	12	2	4	6	2	4
6	0	4	2	4	8	2	6	2	8	4	4	2	4	2	0	12
7	2	4	10	4	0	4	8	4	2	4	8	2	2	2	4	4
8	0	0	0	12	0	8	8	4	0	6	2	8	8	2	2	4
9	10	2	4	0	2	4	6	0	2	2	8	0	10	0	2	12
A	0	8	6	2	2	8	6	0	6	4	6	0	4	0	2	10
B	2	4	0	10	2	2	4	0	2	6	2	6	6	4	2	12
C	0	0	0	8	0	6	6	0	0	6	6	4	6	6	14	2
D	6	6	4	8	4	8	2	6	0	6	4	6	0	2	0	2
E	0	4	8	8	6	6	4	0	6	6	4	0	0	4	0	8
F	2	0	2	4	4	6	4	2	4	8	2	2	2	6	8	8
10	0	0	0	0	0	0	2	14	0	6	6	12	4	6	8	6
11	6	8	2	4	6	4	8	6	4	0	6	6	0	4	0	0
12	0	8	4	2	6	6	4	6	6	4	2	6	6	0	4	0
13	2	4	4	6	2	0	4	6	2	0	6	8	4	6	4	6
14	0	8	8	0	10	0	4	2	8	2	2	4	4	8	4	0
15	0	4	6	4	2	2	4	10	6	2	0	10	0	4	6	4
16	0	8	10	8	0	2	2	6	10	2	0	2	0	6	2	6
17	4	4	6	0	10	6	0	2	4	4	4	6	6	6	2	0
18	0	6	6	0	8	4	2	2	2	4	6	8	6	6	2	2
19	2	6	2	4	0	8	4	6	10	4	0	4	2	8	4	0
1A	0	6	4	0	4	6	6	6	2	2	0	4	4	6	8	6
1B	4	4	2	4	10	6	6	4	6	2	2	4	2	2	4	2
1C	0	10	10	6	6	0	0	12	6	4	0	0	2	4	4	0
1D	4	2	4	0	8	0	0	2	10	0	2	6	6	6	14	0
1E	0	2	6	0	14	2	0	0	6	4	10	8	2	2	6	2
1F	2	4	10	6	2	2	2	8	6	8	0	0	0	4	6	4
20	0	0	0	10	0	12	8	2	0	6	4	4	4	2	0	12
21	0	4	2	4	4	8	10	0	4	4	10	0	4	0	2	8
22	10	4	6	2	2	8	2	2	2	2	6	0	4	0	4	10
23	0	4	4	8	0	2	6	0	6	6	2	10	2	4	0	10
24	12	0	0	2	2	2	2	0	14	14	2	0	2	6	2	4
25	6	4	4	12	4	4	4	10	2	2	2	0	4	2	2	2
26	0	0	4	10	10	10	2	4	0	4	6	4	4	4	2	0
27	10	4	2	0	2	4	2	0	4	8	0	4	8	8	4	4
28	12	2	2	8	2	6	12	0	0	2	6	0	4	0	6	2
29	4	2	2	10	0	2	4	0	0	14	10	2	4	6	0	4
2A	4	2	4	6	0	2	8	2	2	14	2	6	2	6	2	2
2B	12	2	2	2	4	6	6	2	0	2	6	2	6	0	8	4
2C	4	2	2	4	0	2	10	4	2	2	4	8	8	4	2	6
2D	6	2	6	2	8	4	4	4	2	4	6	0	8	2	0	6
2E	6	6	2	2	0	2	4	6	4	0	6	2	12	2	6	4
2F	2	2	2	2	2	6	8	8	2	4	4	6	8	2	4	2
30	0	4	6	0	12	6	2	2	8	2	4	4	6	2	2	4
31	4	8	2	10	2	2	2	6	0	0	2	2	4	2	10	8
32	4	2	6	4	4	2	2	4	6	6	4	8	2	2	8	0
33	4	4	6	2	10	8	4	2	4	0	2	2	4	6	2	4
34	0	8	16	6	2	0	0	12	6	0	0	0	0	8	0	6
35	2	2	4	0	8	0	0	0	14	4	6	8	0	2	14	0
36	2	6	2	2	8	0	2	2	4	2	6	8	6	4	10	0
37	2	2	12	4	2	4	4	10	4	4	2	6	0	2	2	4
38	0	6	2	2	2	0	2	2	4	6	4	4	4	6	10	10
39	6	2	2	4	12	6	4	8	4	0	2	4	2	4	4	0
3A	6	4	6	4	6	8	0	6	2	2	6	2	2	6	4	0
3B	2	6	4	0	0	2	4	6	4	6	8	6	4	4	6	2
3C	0	10	4	0	12	0	4	2	6	0	4	12	4	4	2	0
3D	0	8	6	2	2	6	0	8	4	4	0	4	0	12	4	4
3E	4	8	2	2	2	4	4	14	4	2	0	2	0	8	4	4
3F	4	8	4	2	4	0	2	4	4	2	4	8	8	6	2	2

DDT to Table Constraint. The main advantage of the CP solver is that we can directly model the DDT with a table constraint. A table constraint constrains a list of n variables to be instantiated to an n-tuple chosen within a collection of all valid n-tuples. Any constraint can be declared in *extension i.e.* declared as a table constraint. However, this might not be the best way to model a constraint, especially if there are a lot of tuples. For example, to constrain three Boolean variables a, b, and c to have a sum equal to 2, we may define the table constraint $(a, b, c) \in \{(1, 1, 0), (1, 0, 1), (0, 1, 1)\}$ but the same sum with integer variables (with domains like $[\![-1000, 1000]\!]$) would require too many tuples. This is why constraints are often best declared in *intention i.e.* with a dedicated filtering algorithm. However, an efficient filtering algorithm may not always be available.

Table Filtering. The table constraint not only requires a filtering algorithm but also needs to be cautious of the data structure used to store and manipulate the table. This is because the memory used by the table depends on the number of tuples. In the literature [7, 16–18, 21, 33], a lot of work has been done to optimize this algorithm for various situations (positive tables, binary tables, big or small tables...). The general idea is to pre-compute and try to maintain a set of indexes of valid tuples. When a tuple is no longer valid, an efficient search method will search for another valid tuple in the table. If there are no more valid tuples, the constraint is violated. There are a lot of variations depending on the situation.

The DDT is an extensive definition of all the possible transitions of a difference through an S-Box with its probability and it is usually not possible to define these transitions in intention, by means of a small number of arithmetic constraints. Therefore, the table constraint is the most suited way to model it. In CP, this takes the form of a table T composed of a list of tuples $(\delta x_{in}, \delta x_{out}, p) \in T$ where δx_{in} and δx_{out} are the input and output variables of the S-Box, respectively, p is a variable which corresponds to the probability of observing the output difference δx_{out} given the input difference δx_{in}. To avoid rounding errors, the probability is replaced by the negation of its base 2 logarithm (and probability multiplications are replaced with additions).

For MILP and SAT, this table would require a lot more intermediate variables and constraints [32].

3.2   Modelling Other Operators

Unfortunately, the other operators are not available in CP solvers except for some exceptions, like the modular addition. To model the other operators, we could also use table constraints. However, tables are often too large to be efficient. Therefore, we will develop new filtering algorithms. In particular for the bit-wise XOR operator because it is used everywhere.

Bit-Wise XOR. We first consider the bit-wise XOR in the case with three variables: $a + b = c$. Note that if there is only one variable, the XOR is a constant. If there are two variables, the XOR can be replaced with an equality operator. For the three variable case, we used some previous work from [5]. The filtering algorithm is not very smart. To filter the values of D_c, the domain of c, the algorithm computes a set that contains all the possible XORs between the values of the domains of a and b. This set is then used to remove the inconsistent values from the domain of c. The algorithm is given in Algorithm 1.

This algorithm has two weaknesses. First, the computing of the set is time-consuming but most of all, the set can reach the maximum size very fast. Indeed, if a, b and c have the same domain sizes, for example, 8 bits, then this algorithm will compute $2^8 \times 2^8$ XORs but will fill a set of maximum 2^8 values. In practice, this filtering algorithm takes a lot of time to build the set that does not filter anything in most cases. Therefore, this filtering is only performed in some cases decided by a chosen condition. In [5], this condition is that the sum of the domains of a and b is lower than the maximal domain size of c. We will see later that this condition is not the best one.

Bit-Wise XOR of Arbitrary Arity. To model a XOR of higher arity with only three variable XOR constraints, we need to introduce intermediate variables and declare additional XORs. For TAGADA, we wanted to avoid the introduction of new variables, so we extended the idea of this algorithm to a XOR of arbitrary

---

Algorithm 1: 3-variable XOR filtering algorithm

Input: IntVar a, IntVar b, IntVar c: the target domain to filter

1 set $\leftarrow \emptyset$;
   // Loop through possible values
2 **for all** $v1 \in D_a$ **do**
3   **for all** $v2 \in D_b$ **do**
4   set $\leftarrow$ set $\cup \{v1 \oplus v2\}$;
5   **if** set *contains all possible values in variable domains* **then**
6   **return**;

7 $D_c \leftarrow D_c \cap$ set;

---

arity. The algorithm uses a recursive loop to compute the set of all the possible XORs between $n - 1$ variable domain values to filter the target domain. The algorithm is depicted in Algorithm 2. This algorithm is less efficient to constrain a 3-variable XOR than the previous propagator but the more variables we have, the more efficient this algorithm becomes compared to a decomposition with 3-variable XORs constraints and new intermediate variables (more than five variables XORs decompositions gives slower models). However, as we will see at the end of this section, the chosen condition to activate the filtering was also a problem.

Operations in the Galois Field. In cryptography, we sometimes use addition and multiplication in some fields. For CP modelling, operations like the Mix-Columns of AES have often been modelled with table constraints [23]. For the modular addition, the modulo constraint exists in CP solvers. To model the modular addition $a \boxplus b = c \mod m$, we need one intermediate variable x and the two constraints $a + b = x$ and $x \mod m = c$. Unfortunately, modelling the multiplication is not possible with existing constraints. Like the XOR, we must make a new filtering algorithm for multiplication and division in a finite field. The filtering algorithm we made follows the same idea of Algorithm 1. We create a set of all the possible values of c from the domains of a and b except that in the line "set $\leftarrow$ set $\cup \{v1 \oplus v2\}$" the $\oplus$ is replaced by ProdGF($v1$,$v2$) or DivGF($v1$,$v2$) where ProdGF and DivGF are algorithms to perform the modular product and division depending on the context. In real ciphers, the product is usually between a variable and a constant, so the algorithm is more likely to filter properly.

LFSR. Another operator we added is the LFSR. Similarly to the bit-wise XOR, we replace the line 4 of Algorithm 1 with a function that can compute the next step of the LFSR.

Algorithm 2: n-variable XOR filtering algorithm

Input: int *target*: index in vars table of the domain to filter

```
1 Function combiXor(target, current, xor):
     // skip target
2    if current == target then
3    |   combiXor(target, current + 1, xor);
4    else
        // add the value xor to the set of values
5       if current == vars.length - 1
6          or (current + 1 == target
7          and current + 1 == vars.length - 1) then
8          for all v ∈ D_current do
9          |   set ← set ∪ {xor ⊕ v};
10      else
           // Loop through domain with recursion
11         for all v ∈ D_current do
12         |   combiXor(target, current + 1, xor ⊕ v);

13 set ← ∅;
14 combiXor(target, 0, 0);
15 D_target ← D_target ∩ set;
```

Concat and Split. Bit concatenation and bit splitting are modelled by constraint tables.

Filtering Efficiency. As stated before, the filtering of bit-wise XOR constraints needs a parameter to avoid useless computations. During our tests, we observed that the trade-off between the time gained from filtering and the time taken by the filtering algorithm is in favour of less filtering. To be efficient, we must find at which domain size we would have a good chance to filter. For the XOR with two variables, we can write it as follows. Let a, b be two variables, D_a and D_b, their domain of max sizes n. Let c be a constant in the constraint $a + b = c$. In this case, we can filter the values in D_b if this set of values is not contained in D_a. For two variables, this condition is very probable. However, if now c is also a variable with a domain D_c of max size n, then the filtering is possible if the XORs of all the possible values of D_a and D_b is a set that does not include D_c. The number of values from the XORs is $\#D_a \times \#D_b$, and each value is in the domain range of D_c. In the end, it is like if we pick at random $\#D_a \times \#D_b$ values in the range of $[\![0, n]\!]$ and hope that we did not pick all the values of D_c. For the XOR with more variables ($\bigoplus_i x_i$), the number of values is $\prod_{i \neq j} \#D_{x_i}$. As a consequence, the probability of being able to filter some values of D_j is very low. To test the filtering efficiency of our XOR constraints, we implemented a *forward-checking* version of the filtering algorithms. The forward-checking (FC) method only filters when all the variables are fixed except one. In some early tests, we saw that the FC version was performing nearly as well as the full filtering algorithms.

This means that the filtering algorithms for the 3-variable and n-variable XOR constraints are not helping the solving process that much. Therefore, we deduce that the CP model's strength is the DDT's table constraint. Moreover, the new dedicated filtering algorithms are still helpful because we would have to use table constraints instead, so they at least reduce the memory used by the model. A list of the operators we added to the model generator is depicted in Table 2.

Table 2. List of supported operators in TAGADA (both first and second steps). For exemple, Rijndael uses the operators: $\oplus$, $=$, $\odot_K$ and DDT, while Skinny uses the operators $\oplus$, $=$, $\odot_K$, **LFSR** and DDT.

Operator	Name	First step support	Second step Implementation
Linear Operators			
$=$	Equal	✓	Native support
LFSR	Linear Feedback Shift Register	✓	Custom filtering algorithm
$AB \rightarrow (A, B)$	Split	✓	Constraint table (native)
$(A, B) \rightarrow AB$	Concat	✓	Constraint table (native)
$\ll$ or $\gg$	Left (Right) Shift	✓	Custom filtering algorithm
$\lll$ or $\ggg$	Left (Right) Circular Shift	✓	Custom filtering algorithm
$\&_K$	Bitwise AND with Constant	✓	Constraint table (native)
$\|_K$	Bitwise OR with Constant	✓	Constraint table (native)
$\oplus$	N-ary Bitwise XOR	✓	Custom filtering algorithm or decomposition (for n-ary equations)
$\otimes_K$	Galois Field Multiplication with Constant	✓	Custom filtering algorithm
$\odot_K$	Galois Field Matrix Multiplication with Constant Matrix	✓	Decomposition and delegation to the $\otimes_K$ and $\oplus$ operators
T	Linear Lookup Table	✓	Constraint table (native)
Non-linear Operators			
DDT	Differential Distribution Table	✓	Constraint table (native)

4 Connect the Two Steps

At the start the first step of TAGADA was not designed to work with the second step but only to find the best truncated differential trail. While it can be possible to only use truncated differential trails optimizing, the whole process can be more efficient than only optimizing the two steps separately. To do so we improve the linking algorithm of [27] by splitting the first step search in three parts.

Step1-opt. The aim of Step1-opt is to find the optimal truncated differential trail. We define its signature with Signature 1.

Step1-next. Instead of looking for an optimal solution Step1-next (Signature 2) is designed to find one truncated differential trail with a given upper bound (UB). While Step1-opt solves an optimization problem, Step1-next solves a satisfaction problem.

Signature 1. STEP1-OPT

Input:
G_Δ: the Differential Graph of the cipher
seen: the set of all the already found solutions
UB: the current upper bound
Output:
sol: the Truncated Differential Trail with the highest probability such as $P(sol) \leq UB$ and *sol* not in *seen*. If no such solution exists, returns `null`.

Signature 2. STEP1-NEXT

Input:
G_Δ: the Differential Graph of the cipher
seen: the set of all the already found solutions
UB: the current upper bound
Output:
sol: a Truncated Differential Trail such as $P(sol) \leq UB$ and *sol* not in *seen*. If no such solution exists, returns `null`.

Usually solving optimization problems is more complicated than solving a satisfaction problem. Indeed for both optimization and satisfaction problems we have to find a solution but for optimization problems we also need to find the best one which is generally done by finding a sequence of solutions of increasing quality and proving that there is no better solution than the last found one.

To improve the overall time of the two steps algorithm (Algorithm 3) we try to use the Step1-next method as much as possible instead of the Step1-opt method. Step1-opt is only called at the beginning of the function when we have to compute the upper bound. The second call of Step1-opt is done when we have iterated over all the solutions that reach the current upper bound.

Step1-next-possible-UB (Signature 3). As said previously, step1-opt is the function that consumes the most calculation time. When we have iterated over all the solutions of a current upper bound we need to find the next upper bound. The search of the next upper bound is performed by another call to Step1-opt. However it is possible in some cases to bypass the function by using a new function Step1-next-possible-UB. The purpose of Step1-next-possible-UB is to find very quickly a lower approximation of the next upper bound. If this approximation is equal or lower than the current lower bound then we can stop the search without doing any more computation.

Example 2. Let us take the example of the 4-round `Rijndael-128-128` instance. The step1-opt finds a truncated differential trail with an upper bound probability of 2^{-72}. For this trail the second step will find a valid differential trail of probability 2^{-75} and set it as the current lower bound. As $2^{-75} < 2^{-72}$ we need to find another truncated differential trail that matches 2^{-72}. For this cipher we only have one truncated differential trail of this probability. As we have a gap between the two probabilities we need to tighten the bounds which is done by decreasing the upper bound. For `Rijndael` all the S-Boxes are the same

Algorithm 3: TWOSTEP(G_Δ, G_δ)

Input:
  G_Δ: step1 model
  G_δ: step2 model
1 LB ← 0
2 UB ← 1
3 best ← null
4 sol_1 ← STEP1-OPT($G_\Delta, seen, UB$)
5 seen ← {}
6 UB ← $P(sol_1)$
7 **while** $LB < UB$ **do**
8   seen ← seen ∪ {sol_1}
9   sol_2 ← STEP2(G_δ, sol_1,LB)
10   LB ← $P(sol_2)$
11   **if** $LB < UB$ **then**
12   sol_1 ← STEP1-NEXT(G_Δ, seen, UB)
13   **if** sol_1 is null **then**
14   UB ← STEP1-NEXT-POSSIBLE-UB(G_Δ, seen, UB)
15   **if** $LB \geq UB$ **then**
16   break
17   sol_1 ← STEP1-OPT($G_\Delta, seen, UB$)
18   UB ← $P(sol_1)$

19 **return** best

and their maximum probability is 2^{-6} (a trail of 2^{-72} is composed of 12×2^{-6} S-Boxes). In our case, the only way to find a new trail is to activate another S-Box, in that case the probability will be at least of $13 \times 2^{-6} = 2^{-78}$ which is lower than 2^{-75}. This simple computation saves one call of Step1-opt.

Signature 3. STEP1-NEXT-POSSIBLE-UB

Input:
  G_Δ: the Differential Graph of the cipher
  UB: the current upper bound
Output:
  UB': an approximation of the next reachable UB.

5 Second Step Optimizations

To gain some generic solving efficiency, we propose two optimizations.

5.1 Heuristics

Another way to improve the search speed is to use heuristics. Constraint Programming solvers usually use two kind of heuristics, a value heuristic and a variable heuristic. As their names suggest, the value heuristic is responsible of selecting the next value to test for a variable and the variable heuristic is used to select the next variable on which to branch. By default, generic solvers propose known general purpose search heuristics for both value and variable heuristics, but when we have extra information on a specific problem we can help the solver by designing custom heuristics.

We propose a custom value heuristic adapted to our second step. In this step, we want to maximize the probability of the differential trail. As the probability only depends on non-linear operators we can focus our heuristics on those operators. For each non-linear operator we have three available variables:

- δx which is the input differential variable
- δsx which is the output differential variable
- p which is the probability of the transition $\delta x \to \delta sx$

If the solver branches on a p variable, we only have to select the highest probability available. When the solver branches on a δx variable, if its δsx variable is instantiated, *i.e.* it has only one possible value, then we can select the value that maximize the transition to δsx, more formally:

$$\texttt{next-value}(\delta x) = \operatorname*{argmax}_{v \in \texttt{dom}(\delta x)} P(v \to \texttt{value}(\delta sx))$$

When the solver branches on a δsx we can performs the same computation with its corresponding δx variable.

5.2   Competitive Parallel Solving

In the previous CP model for the second step on SKINNY [5], a parallel method was used. Each model from the first step was launched on a separate thread, and the best solution found was shared with all the other threads. This new bound is added to the models that remain to be solved, and this information can be used to cut a large part of their search space.

In TAGADA, we added a similar parallel competition between the models to solve the second step models from a list of truncated trails. In this setup, TAGADA was able to recover the same results with similar solving times than the dedicated model of [5]. The generated models were only two to three times slower than the ad-hoc models.

An interesting future work would be to study the feasibility of parallel computing in the two-step method (Algorithm 3).

6   Results

We have implemented the model generator in Java and Kotlin to communicate more easily with the Choco solver. This generator can parse the JSON file of the TAGADA DAGs and an associated file for the truncated trails. The generator then builds a CP model and calls the Choco solver to find the differential characteristics. The second step model generator can be used with the first step of TAGADA in the two-step solving algorithm depicted in Algorithm 3, or it can also be used alone if we can give a truncated trails list as input.

To compare the generated models, we reproduced the results of ad-hoc models with TAGADA on the two-step differential analysis. We set a time limit of one day. The computation was done on a Debian GNU/Linux 11 (bullseye) x86_64 sever with two Intel Xeon Gold 6254 (3.10 to 4.00 GHz) processors and 64MB of RAM. Each instance was launch on a single core. We were able to reproduce the results shown in Table 3.

The code will be available at: https://gitlab.com/tagada-framework/tagada

Table 3. Best differential trails recovered with TAGADA (time limit of one day). Detailed results will be available in an extended version of the current paper on eprint.

Cipher	Max Round	Probability	Reference
Midori-64	16	2^{-16}	[8]
Midori-128	20	2^{-40}	[8]
Warp	41	2^{-40}	[31]
Twine-80	18	2^{-64}	[29]
Twine-128	16	2^{-52}	[29]
Skinny-64-TK1	11	2^{-64}	[5]
Skinny-128-TK1	11	2^{-74}	[5]
Rijndael-128-128	5	2^{-105}	[10]
Rijndael-128-160	7	2^{-120}	[27]
Rijndael-128-192	9	2^{-146}	[10]
Rijndael-128-224	12	2^{-212}	[27]
Rijndael-128-256	14	2^{-146}	[10]
Rijndael-160-128	4	2^{-112}	[27]
Rijndael-160-160	6	2^{-138}	[27]
Rijndael-160-192	8	2^{-141}	[27]
Rijndael-160-224	9	2^{-190}	[27]
Rijndael-160-256	11	2^{-204}	[27]
Rijndael-192-128	3	2^{-54}	[27]
Rijndael-192-160	5	2^{-118}	[27]
Rijndael-192-192	7	2^{-153}	[27]
Rijndael-192-224	8	2^{-205}	[27]
Rijndael-192-256	9	2^{-179}	[27]
Rijndael-224-128	3	2^{-54}	[27]
Rijndael-224-160	4	2^{-122}	[27]
Rijndael-224-192	5	2^{-124}	[27]
Rijndael-224-224	7	2^{-196}	[27]
Rijndael-224-256	8	2^{-182}	[27]
Rijndael-256-128	3	2^{-54}	[27]
Rijndael-256-160	4	2^{-130}	[27]
Rijndael-256-192	5	2^{-148}	[27]
Rijndael-256-224	4	2^{-115}	[27]
Rijndael-256-256	6	2^{-128}	[27]

7   Conclusion

We have presented a model generator for the second step of the differential analysis that relies on the DAG representation of the model generator of the first step TAGADA. We have shown that these generated models can recover the results of state-of-the-art ad-hoc models in reasonable times.

7.1   Next Optimization: DAG Simplification

To make the model more efficient, we would like to add a graph simplification algorithm. From the first step solution, we know which S-Boxes are active and which are not. Therefore, we can simplify the model by removing the variables in the inactive part of the graph and all the related constraints. For example, let G be a graph composed of two S-Boxes $S1$ and $S2$, their output variables out_1 and out_2 and one XOR operation $out_1 \oplus out_2 = out_3$ (see Fig. 2). If the truncated trail says that only the first S-Box is active, then instead of fixing the domain of out_2 to 0, we can remove the $S2$ and out_2 nodes from the graph. After this, we can further simplify the graph by removing the XOR node and the out_1 variable. In the end, we only need to keep $S1$ and out_3.

The simplification is split into two parts. The first part propagates from the active S-Boxes, all the nodes that can or will have a difference. In some cases, we can be more precise. In particular, we know that the difference always propagates through unary operators, so we can propagate this information in the graph. This information can then be used in the CP model. When we know that a variable necessarily has a difference, we can remove the value 0 from the variable domain when we declare it. We start with a graph containing all nodes with an "uncertain" marker. For each input and output variable of the S-Boxes, we use the information of the truncated trail to fix them to "active" or "inactive". For the linear operators, we can deduce the following *status propagation rules*:

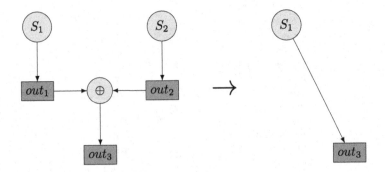

Fig. 2. Step2 graph shaving example

- If there is only one "uncertain" variable in its input or output variables and all the other variables are inactive, it can be set to "inactive".
- If there is only one variable "active" and all the others are inactive, this is an error. This is not possible if the truncated trail is correct.
- If there is only one "active" variable and one "uncertain" variable, they both must be "active".

"active" and "inactive" information are iteratively propagated in the graph until reaching a fixpoint where no more propagations are possible.

After these statuses are propagated, we can reduce the graph by removing all the "inactive" variables and all the operators only linked to them. We can also replace the XOR operators that have one inactive variable with equality operators, and finally, we can remove the equality operators and merge their input and output nodes. The simplified graph is then transformed into the CP model.

If we give the complete graph to the CP solver, it would eventually reach the same conclusion and set all the inactive variables to 0. However, this simplification is easy to do in advance and removing useless variables and constraints in a model is always a good idea.

7.2   Future Work

The idea of a simple tool to perform differential analysis is interesting, and other recent works are also working in this direction [2,26]. Moreover, we think that a unified cipher format would help the comparison and development of these tools. TAGADA could be improved by integrating previous solving methods dedicated to ARX ciphers [19] and bit-based ciphers [14]. In these cases, more work is needed to determine if the CP solver is still the best solving tool. In the end, TAGADA could be extended to search for the variants of differential analysis (boomerangs, impossible differentials...).

Acknowledgements. The authors would like to express their very great appreciation to Charles Prud'homme, Ph.D. from IMT for his valuable and constructive expertise of Choco during the development of this research work.

References

1. Beaulieu, R., Shors, D., Smith, J., Treatman-Clark, S., Weeks, B., Wingers, L.: The SIMON and SPECK families of lightweight block ciphers. Cryptology ePrint Archive, Report 2013/404 (2013). https://eprint.iacr.org/2013/404
2. Bellini, E., et al.: CLAASP: a cryptographic library for the automated analysis of symmetric primitives. IACR Cryptol. ePrint Arch., p. 622 (2023). https://eprint.iacr.org/2023/622
3. Biham, E., Shamir, A.: Differential cryptanalysis of des-like cryptosystems. In: Menezes, A., Vanstone, S.A. (eds.) Advances in Cryptology – CRYPTO '90, 10th Annual International Cryptology Conference, Santa Barbara, California, USA, 11–15 August 1990, Proceedings. LNCS, vol. 537, pp. 2–21. Springer, Cham (1990). https://doi.org/10.1007/3-540-38424-3_1

4. Biryukov, A., Nikolic, I.: Automatic search for related-key differential characteristics in byte-oriented block ciphers: application to AES, camellia, Khazad and others. In: Gilbert, H. (eds.) Advances in Cryptology – EUROCRYPT 2010. EUROCRYPT 2010. LNCS, vol. 6110, pp. 322–344. Springer, Berlin, Heidelberg (2010). https://doi.org/10.1007/978-3-642-13190-5_17

5. Delaune, S., Derbez, P., Huynh, P., Minier, M., Mollimard, V., Prud'homme, C.: Efficient methods to search for best differential characteristics on SKINNY. In: Sako, K., Tippenhauer, N.O. (eds.) Applied Cryptography and Network Security. ACNS 2021. LNCS, vol. 12727, pp. 184–207. Springer, Cham (2021). https://doi.org/10.1007/978-3-030-78375-4_8

6. Fouque, P., Jean, J., Peyrin, T.: Structural evaluation of AES and chosen-key distinguisher of 9-round AES-128. In: Canetti, R., Garay, J.A. (eds.) Advances in Cryptology – CRYPTO 2013. CRYPTO 2013. LNCS, vol. 8042, pp. 183–203. Springer, Berlin, Heidelberg (2013). https://doi.org/10.1007/978-3-642-40041-4_11

7. Gent, I.P., Jefferson, C., Miguel, I., Nightingale, P.: Data structures for generalised arc consistency for extensional constraints. In: Proceedings of the Twenty-Second AAAI Conference on Artificial Intelligence, 22–26 July 2007, Vancouver, British Columbia, Canada, pp. 191–197. AAAI Press (2007). http://www.aaai.org/Library/AAAI/2007/aaai07-029.php

8. Gérault, D.: Security analysis of contactless communication protocols. (Analyse de sécurité des protocoles de communication sans contact). Ph.D. thesis, University of Clermont Auvergne, Clermont-Ferrand, France (2018). https://tel.archives-ouvertes.fr/tel-02536478

9. Gérault, D., Lafourcade, P.: Related-key cryptanalysis of Midori. In: Dunkelman, O., Sanadhya, S. (eds.) Progress in Cryptology – INDOCRYPT 2016. INDOCRYPT 2016. LNCS, vol. 10095, pp. 287–304. Springer, Cham (2016). https://doi.org/10.1007/978-3-319-49890-4_16

10. Gérault, D., Lafourcade, P., Minier, M., Solnon, C.: Computing AES related-key differential characteristics with constraint programming. Artif. Intell. **278** (2020)

11. Gérault, D., Minier, M., Solnon, C.: Constraint programming models for chosen key differential cryptanalysis. In: Rueher, M. (eds.) Principles and Practice of Constraint Programming. CP 2016. LNCS, vol. 9892, pp. 584–601. Springer, Cham (2016). https://doi.org/10.1007/978-3-319-44953-1_37

12. Heys, H.M.: A tutorial on linear and differential cryptanalysis. Cryptologia **26**(3), 189–221 (2002). https://doi.org/10.1080/0161-110291890885

13. Knudsen, L.R.: Truncated and higher order differentials. In: Preneel, B. (ed.) Fast Software Encryption. FSE 1994. LNCS, vol. 1008, pp. 196–211. Springer, Berlin, Heidelberg (1994). https://doi.org/10.1007/3-540-60590-8_16

14. Kölbl, S.: Cryptosmt: an easy to use tool for cryptanalysis of symmetric primitives (2015). https://github.com/kste/cryptosmt

15. Lai, X., Massey, J.L., Murphy, S.: Markov ciphers and differential cryptanalysis. In: Davies, D.W. (ed.) Advances in Cryptology – EUROCRYPT'91. EUROCRYPT 1991. LNCS, vol. 547, pp. 17–38. Springer, Berlin, Heidelberg (1991). https://doi.org/10.1007/3-540-46416-6_2

16. Lecoutre, C.: STR2: optimized simple tabular reduction for table constraints. Constraints Int. J. **16**(4), 341–371 (2011). https://doi.org/10.1007/s10601-011-9107-6

17. Lecoutre, C., Likitvivatanavong, C., Yap, R.H.C.: A path-optimal GAC algorithm for table constraints. In: Raedt, L.D., et al. (eds.) ECAI 2012–20th European Conference on Artificial Intelligence. Including Prestigious Applications of Artificial

Intelligence (PAIS-2012) System Demonstrations Track, Montpellier, France, 27–31 August 2012. Frontiers in Artificial Intelligence and Applications, vol. 242, pp. 510–515. IOS Press (2012). https://doi.org/10.3233/978-1-61499-098-7-510

18. Lecoutre, C., Szymanek, R.: Generalized arc consistency for positive table constraints. In: Benhamou, F. (ed.) Principles and Practice of Constraint Programming – CP 2006. CP 2006. LNCS, vol. 4204, pp. 284–298. Springer, Berlin, Heidelberg (2006). https://doi.org/10.1007/11889205_22

19. Leurent, G.: Analysis of differential attacks in ARX constructions. In: Wang, X., Sako, K. (eds.) Advances in Cryptology – ASIACRYPT 2012. ASIACRYPT 2012. LNCS, vol. 7658, pp. 226–243. Springer, Berlin, Heidelberg (2012). https://doi.org/10.1007/978-3-642-34961-4_15

20. Libralesso, L., Delobel, F., Lafourcade, P., Solnon, C.: Automatic generation of declarative models for differential cryptanalysis. In: Michel, L.D. (ed.) 27th International Conference on Principles and Practice of Constraint Programming, CP 2021, Montpellier, France (Virtual Conference), 25–29 October 2021. LIPIcs, vol. 210, pp. 40:1–40:18. Schloss Dagstuhl - Leibniz-Zentrum für Informatik (2021). https://doi.org/10.4230/LIPIcs.CP.2021.40

21. Mairy, J., Hentenryck, P.V., Deville, Y.: Optimal and efficient filtering algorithms for table constraints. Constraints Int. J. 19(1), 77–120 (2014). https://doi.org/10.1007/s10601-013-9156-0

22. Matsui, M.: Linear cryptanalysis method for DES cipher. In: Helleseth, T. (ed.) Advances in Cryptology – EUROCRYPT '93. EUROCRYPT 1993. LNCS, vol. 765, pp. 386–397. Springer, Berlin, Heidelberg (1993). https://doi.org/10.1007/3-540-48285-7_33

23. Minier, M., Solnon, C., Reboul, J.: Solving a symmetric key cryptographic problem with constraint programming. In: ModRef 2014, Workshop of the CP 2014 Conference, p. 13 (2014)

24. Nethercote, N., Stuckey, P.J., Becket, R., Brand, S., Duck, G.J., Tack, G.: Minizinc: towards a standard CP modelling language. In: Bessiere, C. (ed.) Principles and Practice of Constraint Programming – CP 2007. CP 2007. LNCS, vol. 4741, pp. 529–543. Springer, Berlin, Heidelberg (2007). https://doi.org/10.1007/978-3-540-74970-7_38

25. Prud'homme, C., Fages, J.G.: Choco-solver: a java library for constraint programming. J. Open Source Softw. 7(78), 4708 (2022). https://doi.org/10.21105/joss.04708

26. Ranea, A., Rijmen, V.: Characteristic automated search of cryptographic algorithms for distinguishing attacks (CASCADA). IET Inf. Secur. 16(6), 470–481 (2022). https://doi.org/10.1049/ise2.12077

27. Rouquette, L., Gérault, D., Minier, M., Solnon, C.: And rijndael? Automatic related-key differential analysis of rijndael. In: Batina, L., Daemen, J. (eds.) Progress in Cryptology – AFRICACRYPT 2022. AFRICACRYPT 2022. LNCS, vol. 13503, pp. 150–175. Springer, Cham (2022). https://doi.org/10.1007/978-3-031-17433-9_7

28. Rouquette, L., Solnon, C.: abstractXOR: a global constraint dedicated to differential cryptanalysis. In: Simonis, H. (ed.) Principles and Practice of Constraint Programming. CP 2020. LNCS, vol. 12333, pp. 566–584. Springer, Cham (2020). https://doi.org/10.1007/978-3-030-58475-7_33

29. Sakamoto, K., et al.: Security of related-key differential attacks on twine, revisited. IEICE Trans. Fundam. Electron. Commun. Comput. Sci. 103-A(1), 212–214 (2020). https://doi.org/10.1587/transfun.2019CIL0004, http://search.ieice.org/bin/summary.php?id=e103-a_1_212

30. Sun, S., et al.: Analysis of AES, SKINNY, and others with constraint programming. IACR Trans. Symmetric Cryptol. **2017**(1), 281–306 (2017)
31. Teh, J.S., Biryukov, A.: Differential cryptanalysis of WARP. J. Inf. Secur. Appl. **70**, 103316 (2022). https://doi.org/10.1016/j.jisa.2022.103316
32. Udovenko, A.: MILP modeling of Boolean functions by minimum number of inequalities. IACR Cryptol. ePrint Arch., p. 1099 (2021). https://eprint.iacr.org/2021/1099
33. Ullmann, J.R.: Partition search for non-binary constraint satisfaction. Inf. Sci. **177**(18), 3639–3678 (2007). https://doi.org/10.1016/j.ins.2007.03.030

Falling into Bytes and Pieces – Cryptanalysis of an Apple Patent Application

Gregor Leander[(✉)], Lukas Stennes[(✉)], and Jan Vorloeper[(✉)]

Ruhr University Bochum, Bochum, Germany
{gregor.leander,lukas.stennes,jan.vorloeper}@rub.de

Abstract. In this paper we take a look at a cipher that has escaped public cryptanalysis so far. It is a block cipher published by Apple in a patent application describing its functionality and, as usual for a patent, its innovation. Here, we mainly focus on its security and, as we will see, there are very serious flaws that allow us to break the cipher completely. Moreover, breaking it does not need as much innovation as the patent claims for the cipher itself: Our observations leading to efficient attacks are either generic or tool based.

Keywords: Block Cipher · Apple · Patent · Tool based cryptanalysis

1 Introduction

Block ciphers, and in modern modes of operations, tweakable block ciphers are omnipresent in our daily life. Indeed, the AES is responsible for securing a large fraction of the data we produce daily. The AES is well established and no serious attacks have been proposed ever since its invention more than 25 years ago. However, there are (for good and not good reasons) many alternative and more recent proposals of block ciphers in the literature. Nowadays, any new cipher proposal is expected to be accompanied with security arguments against the most important attack vectors to gain initial trust in its design.

Ciphers where the proposal does not give such arguments are often ciphers that are proprietary and have only been disclosed due to an reverse engineering effort [3,6]. Other examples include ciphers in (military) standards [8,9] or ciphers proposed by the NSA [2]. In this paper, we exemplary look at a cipher from yet another source of proprietary designs: patent applications.

Searching Patents for Ciphers

Searching for interesting patents is surprisingly simple. A search engine like Google Patents[1] makes it easy to search patents containing a specific term like *block cipher* or a specific assignee like *Apple*. Once an interesting patent is identified, we can even look-up the correspondence of the inventors and the patent

---

[1] https://patents.google.com/.

A. Chattopadhyay et al. (Eds.): INDOCRYPT 2023, LNCS 14459, pp. 269–286, 2024.
https://doi.org/10.1007/978-3-031-56232-7_13

office. For instance, for the patent application [7] which we study in this work, the USPTO publicly provides[2] documents and transaction regarding, e.g., fees or discussions about the claims of the patent. These surely might be interesting, entertaining or even irritating. E.g., the patent application [7] which dates back to 2008, claims that it is innovative to evaluate an S-box by using Boolean functions instead of a lookup table. For context, the AES finalist SERPENT, which is well-known for an efficient bitsliced implementation, dates back to the 90 s. However, in this work, we focus only on the security of the cipher proposed in [7].

ABC

The cipher we found interesting, and that we describe and analyze in detail here, does actually not have a name. In our work, we refer to it as ABC[3]. ABC is a Feistel-like cipher, operating on plaintext blocks of 128 bits, split into two equal parts. While the number of rounds is not fixed entirely, eight rounds are recommended as a reasonable choice. The cipher uses a 128-bit master key to produce, again in a Feistel-like construction, 64-bit round keys.

We already mentioned above that any decent block cipher should come with good arguments for its security against attack vectors like differential and linear attacks and their variants and combinations. However, one even more basic property is diffusion. Actually diffusion is among the two (quintessential) properties which every good cipher should fulfill, already described by Shannon in [13]. For a binary block cipher, diffusion, in a nutshell, is the property that any output bit of the cipher depends on every input and every key bit. This property alone certainly does not ensure a secure cipher, but violating this property ensures a weak cipher.

As it turns out, the 128-bit ABC actually decomposes into five subciphers, as depicted in Fig. 1, where two subciphers operate on two bytes each and the remaining three subciphers on four bytes each. While this already demonstrates that ABC is far away from being a good cipher, there is more. We show how to efficiently recover all the round keys in the subciphers working on two bytes. First, we use a generic meet-in-the-middle [10] approach, based on the fact that very limited key-material is used here, to recovery the round keys of a subcipher in *two minutes* on a laptop. Alternatively, we can use an (all-in-one) differential attack following the framework of [1] based on the weak differential properties together with the fact that small round keys allow to partially decrypt more than one round easily. This approach recovers the round keys of a subcipher in *less than a second.*

Given that, the last hope one could have is that the other parts, operating on four bytes are more secure. However, as a generic black box approach reveals, the key schedule allows to efficiently lift the knowledge of the round keys for the two-byte subciphers to extracting the entire master key with a complexity of roughly 2^{49}. Thus, it is not even necessary to analyze how to break the larger subciphers directly, their security complete vanishes with the weakness of the smaller parts.

---

[2] See https://patentcenter.uspto.gov/applications/12055244/ifw/docs?application=.

[3] ABC for **A**pple's **N**ot So Good **C**ipher.

Outline of the Paper

We start with a detailed description of ABC in Sect. 2 followed by our cryptanalytic results and observations in Sect. 3. There, we start by generic observations,
mainly driven by the above mentioned fact that ABC decomposes into parts, and
then continue to less generic observations and in particular describe a devastating differential attack. In Sect. 4, we analyse the key schedule and show how we
can recover the entire master key from the easy to recover parts of the rounds
keys. We conclude the paper in Sect. 5.

Throughout our work, we quote the patent application to highlight important
parts. The paragraphs in the patent application are numbered. When quoting
from it we signal that using type writer font and we also state the corresponding
paragraph. For example, the patent application states:

[0033]: One of ordinary skill in the art will understand the
deciphering process there from.

Which we would like to complement by saying that, even more, one of ordinary
skill in the art will actually see that the key is not needed for this process.

2   Description of ABC

In this section, we describe ABC in detail. ABC is a 128-bit block cipher which
uses a Feistel-like structure. The 128-bit plaintext is divided into equal-sized left
and right blocks $L_0, R_0 \in \mathbb{F}_2^{64}$, which will be transformed into L_i, R_i by applying
the round function i times. In each round, a 64-bit round key KR_i is XORed
into the state, which is derived by a 128-bit master key through its key schedule
algorithm. In total, r rounds are performed. Finally, the 128-bit ciphertext is
formed by simply concatenating $L_r \| R_r$.

The patent application defines the number of rounds r like this:

[0033]: [...] The number of needed rounds is, e.g., 8 to 10
typically, but it may be more for greater security or less for
greater efficiency.

Given this vague description, we assumed $r = 8$ for our work. However, some
attacks are generic and break this cipher regardless of the value of r.

Further, the specification of the cipher does *not* include any test vectors.
Hence, there is some ambiguity left, especially regarding the bit order. Some of
our generic attacks are not impacted by this and for the others we would only
expect minor changes. To avoid any ambiguity within our work, we provide our
implementation of ABC and our attacks at https://github.com/rub-hgi/attack-
on-abc.

In the following, we start by describing the details of the round function
followed by the details of the key schedule algorithm. Notationwise, we will
denote by $L_i[j]$ the j.th byte of the left part of the state after i rounds. For the
right part the notation is $R_i[j]$ respectively.

2.1   The Round Function of ABC

The ABC round function is summarized in the figure below:

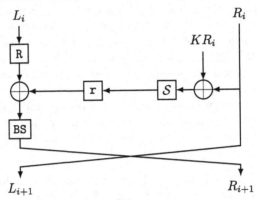

Each round of ABC uses the following operations:

The S-box Layer $\mathcal{S}(x)$. The $\mathcal{S}$-Layer uses a 4-bit S-box S 16 times in parallel to output

$$\mathcal{S}(x) \ : \mathbb{F}_2^{64} \rightarrow \mathbb{F}_2^{64}$$
$$\mathcal{S}(x) = S^{\otimes 16}(x) = (S(x_1), \ldots, S(x_{16})).$$

The S-box S is defined through the following lookup table using hexadecimal notation:

x	0	1	2	3	4	5	6	7	8	9	a	b	c	d	e	f
$S(x)$	4	c	0	8	6	e	1	b	9	d	2	5	a	f	3	7

Note that in the patent application the S-box is not a bijection, as $S(8) = S(c) = 9$, which we assumed to be a typo, as the given Boolean expressions for S translate to our corrected lookup table. As we will see later, the differential (and linear) properties of this S box are rather weak.

The BS-Layer. The BS-Layer uses an 8-bit function B eight times in parallel.

$$\text{BS}(x) \ : \mathbb{F}_2^{64} \rightarrow \mathbb{F}_2^{64}$$
$$\text{BS}(x) = B^{\otimes 8}(x) = (B(x_1), \ldots, B(x_8)).$$

For this, each byte of the 64-bit input is viewed as an element $x \in \mathbb{Z}_{256}$ and is used to compute the output

$$B(x) = 3 \cdot x + 155 \bmod 256.$$

The r-Layer. This function rotates the i.th byte by i bits. In order to describe the r-layer more precisely, i.e., on bit-level, we cite directly from the patent application.

[0021] [...] Function r applied to a 64-bit word (W=w0||...||w7)
is expressed as follows: r(W) = r(w0,0)||r(w1,1)||...||r(w7,7)
[...] A concrete example would be r(0x82, 1) = 0x05 since 0x82 =
01000001 and if one rotates this 1 bit to the right then the
result is 10100000 = 0x05.

The given example explains the endianness used in this patent application. As 0x82 = 0b10000010 and 0x05 = 0b00000101, the example being called a rotateRight by 1 bit indicates that the LSB is on the left.

The R-Layer. This layer consists of a byte permutation. Here, the 64-bit input is again considered as 8 bytes, counting from 0 to 7. Those bytes are permuted by the self-inverse permutation denoted in cycle-notation as

$$(0\ 6)(1\ 5)(2)(3\ 7)(4).$$

More explicitly, we have

$$\mathrm{R} : (\mathbb{F}_2^8)^8 \ \to \ (\mathbb{F}_2^8)^8$$
$$\mathrm{R}(x_0, x_1, x_2, x_3, x_4, x_5, x_6, x_7) = (x_6, x_5, x_2, x_7, x_4, x_1, x_0, x_3).$$

The main weakness, as we will see soon, are the two fixed-points, i.e., the fact that byte 2 and 4 are not permuted.

2.2   The Key Schedule

The key schedule also has a Feistel shape and is performed for one iteration less than the number r of encryption rounds as KR_0 is used as the first round key. The right part KR_i of each state is used as round key, while the left part KL_i is not used directly. The round function of the key schedule is described in the figure below:

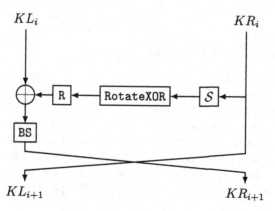

As we can see, the key schedule round function resembles the round function of the cipher to a large extend. The only new function used is `RotateXOR`, which is defined in the following.

The `RotateXOR`-Layer. The 64-bit input is divided into two 32-bit parts, which are transformed independently of each other. Each of the blocks is XORed with a rotated version of itself. The left part is rotated by $13 + i$ bits in the i.th round and the right block is rotated $29 + i$ bits in the i.th round. This procedure is depicted below.

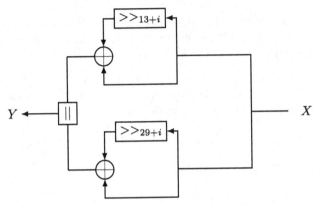

As a side note, the patent application described the rotation-coefficients to be

[0035]: (13+1) modulo 32 and (29+i) modulo 32

which we assumed to be a typo, with $(13 + i)$ and $(29 + i)$, each modulo 32 being the intended rotation-coefficients.

3   Cryptanalysis of ABC

This section contains our security analysis of ABC. We structure the observations by starting with generic attacks, based solely on the lack of diffusion and then later discuss the more specific attacks, in particular differential cryptanalysis.

3.1   Exploiting Lack of Diffusion

When analysing (8-round) ABC, it is noticeable that most of the ABC components operate on small blocks independently. BS and r both operate on each byte, while S consists of 4-bit S-boxes and acts solely as a layer of confusion. The only function which diffuses the state on a byte level is R. However, as R is a permutation which has a maximum cycle length of two (while also having two fixed points), the diffusion generated by R is not sufficient, making it possible to split up the cipher into multiple subciphers. More precisely, ABC splits into five subciphers, two of them, denoted by $\mathcal{B}_2$ and $\mathcal{B}_4$ operating on two bytes, and

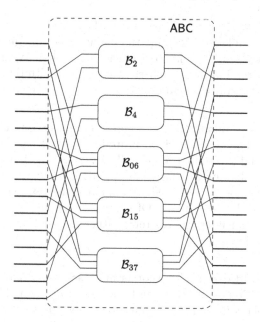

Fig. 1. Decomposition of ABC into smaller subciphers.

three, denoted by $\mathcal{B}_{06}, \mathcal{B}_{15}$ and $\mathcal{B}_{37}$, on four bytes each. This is depicted in Fig. 1. We chose to call those subciphers according to the bytes they represent in the ABC-state.

$$\mathcal{B}_2, \mathcal{B}_4 : \mathbb{F}_2^{16} \times \mathbb{F}_2^{64} \mapsto \mathbb{F}_2^{16}$$
$$\mathcal{B}_{06}, \ \mathcal{B}_{15}, \mathcal{B}_{37} : \mathbb{F}_2^{32} \times \mathbb{F}_2^{128} \mapsto \mathbb{F}_2^{32}$$

Here, the encryption is denoted as $c = \mathcal{B}_j(p_{\mathcal{B}_j}, k_{\mathcal{B}_j})$ with

$$p_{\mathcal{B}_j} = L_0[j] \ || \ R_0[j], \ c_{\mathcal{B}_j} = L_8[j] \ || \ R_8[j], \ k_{\mathcal{B}_j} = KR_0[j] \ || \ldots || \ KR_7[j].$$

3.2 Generic Attacks

As we show in this section, the fact that ABC decomposes into subciphers already makes it easy to break it with much less effort than a naive brute-force attack on the 128-bit master key. We now present multiple approaches that take advantage of ABC's weakness.

Codebook-Attack. The first approach we discuss is a naive codebook attack, as there are (for a fixed key) at most 2^{32} plaintext-ciphertext pairs for each subcipher. An attacker simply needs to determine the ciphertext to every possible plaintext of each subcipher and store them inside a lookup table, which requires

2^{32} encryption queries in total. After sorting the entries by their corresponding ciphertext, it is possible to determine the plaintext to any given ciphertext by looking up the corresponding plaintext of each subcipher. With this attack, the key remains unknown, but every ciphertext encrypted with that same key is not protected by the encryption anymore.

Brute-Force. For r rounds of ABC, there are only $8r$ bits of round keys for the subcipher $\mathcal{B}_2$. Therefore, a trivial brute-force attack on $r = 8$ rounds that recovers those round keys has time complexity 2^{64}. This is rather unpractical but already shows that ABC is far from providing a security level of 128 bits. In terms of data complexity, we need 4 plaintext-ciphertext pairs to expect only one additional false key candidate. Of course, the same applies so $\mathcal{B}_4$. As we explain in more detail later, the thereby gathered round keys are sufficient to recover the master key.

Meet-In-The-Middle-Attack. Another approach that takes advantage of the possible decomposition of ABC is to perform a meet-in-the-middle-attack, where each of the subciphers $\mathcal{B}_2$ and $\mathcal{B}_4$ are split up into two parts, the first four rounds and the last four rounds. To obtain the round keys $k_{\mathcal{B}_2}$ (or $k_{\mathcal{B}_4}$, respectively), an attacker can do the following:

Given a pair $(p^{(0)}, c^{(0)}) \in \mathbb{F}_2^{16} \times \mathbb{F}_2^{16}$, the attacker first uses every possible value of the first four round key bytes $k_{\mathcal{B}_2,0:3}$ to encrypt the first four rounds of $\mathcal{B}_2$, which yields $u_{k_{\mathcal{B}_2,0:3}}^{(0)} \in \mathbb{F}_2^{16}$. This, together with the corresponding guess for $k_{\mathcal{B}_2,0:3}$, will be stored in a lookup table. After all 2^{32} guesses are stored and the table is sorted by intermediate value $u_{k_{\mathcal{B}_2,0:3}}^{(0)}$, the attacker now proceeds to guess the key for the last four rounds, $k_{\mathcal{B}_2,4:7}$ and computes the last four inverse rounds of $\mathcal{B}_2$, which yields $v_{k_{\mathcal{B}_2,4:7}}^{(0)}$. Now, if

$$u_{k_{\mathcal{B}_2,0:3}}^{(0)} = v_{k_{\mathcal{B}_2,4:7}}^{(0)}$$

for any value in the lookup table, then $k'_{\mathcal{B}_2} = (k_{\mathcal{B}_2,0:3} \parallel k_{\mathcal{B}_2,4:7})$ is a potential key candidate. We depict this below, where $\mathcal{R}$ represents one round of the subcipher (Fig. 2).

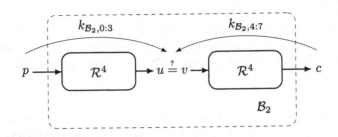

Fig. 2. Meet-in-the-middle approach on $\mathcal{B}_2$ and $\mathcal{B}_4$

As there are 2^{64} possible keys and 2^{16} possible values for u, the attacker expects to get $2^{64-16} = 2^{48}$ key candidates. To reduce the number of false positives, more pairs $(p^{(i)}, c^{(i)})$ can be used to check if $u^{(i)}_{k_{B_2},0:3} = v^{(i)}_{k_{B_2},4:7}$ continues to hold. The more pairs are used, the less false positives are to be expected, as the wrong key candidates fail to pass this check. Using four pairs already reduces the number of expected key candidates down to $2^{64-4\cdot16} = 1$.

We implemented a slightly improved version of this attack. That is, we can omit the last round key of the forward direction and only use one half of the state for the check. With this, our implementation takes about two minutes to restore the round keys on a laptop.

Performing this meet-in-the-middle attack does not recover the master key of ABC, but rather the round keys of the two small subciphers, i.e., k_{B_2} and k_{B_4}. However, as we present later, the knowledge of these keys is sufficient to recover the master key.

3.3   A Closer Look at S and BS

For now, we only considered attacks that are independent of the concrete instantiation of S and BS. In the next sections we change this and therefore now have a closer look at them. What we see is surprisingly weak. The S-box S is chosen to be efficient to implement in a bit-sliced manner. The patent application states that

> [0011]: [...] the S-boxes here are not the same as those previously known and are structured to be relatively easy to design in hardware (logic gates).

While this might be true, it comes at a high price. Indeed, consider the algebraic normal form of the S-Box S:

$$\begin{pmatrix} x_0x_1x_3 + x_0x_2x_3 + x_1x_2x_3 + x_1x_2 + x_1x_3 + x_2x_3 + x_3 \\ x_0x_1x_2 + x_0x_1x_3 + x_1x_2 + x_1x_3 + x_2 \\ x_0x_3 + x_1x_3 + x_1 + x_3 + 1 \\ x_0x_3 + x_0 + x_1x_3 + x_3 \end{pmatrix}$$

One sharp glance, or computing its linear approximation table with a tool like SageMath [14], is all it takes to realize that the sum of the last two components is $x_0 + x_1 + 1$, i.e., an affine function. This, of course, means that S has the maximal linearity 16. In other words, to make this crystal clear, S is a *terrible* S-Box unless special care is taken to avoid those linear approximations from propagating for multiple rounds.

Next, consider the first four components of the algebraic normal form of B:

$$\begin{pmatrix} x_0 + 1 \\ x_1 + 1 \\ x_0 + x_2 \\ x_0x_2 + x_1 + x_2 + x_3 + 1 \end{pmatrix}$$

This is even worse. Not only are there three linear or rather affine components but all four depend only on the first four inputs.

To sum up, the only two *non-linear* building blocks of ABC both exhibit *linear* components. This opens the door for statistical attacks, in particular linear and differential attacks. In the following we focus on differential attacks, and as they already allow to break the cipher entirely, we do not bother investigating linear attacks.

3.4   Differential Cryptanalysis of $\mathcal{B}_2$ and $\mathcal{B}_4$

Differential cryptanalysis, introduced in [5], aims to trace the difference of message pairs through the encryption process. In its initial form, it focuses on the most likely output difference. Later many variants have been proposed such as truncated differentials [11] and impossible differentials [4,12]. One framework to cover all of those variants jointly was presented in [1], however the applicability of this work is limited to very small block sizes as it requires to compute the expected probabilities for all output differences, i.e., 2^n values for a block cipher of n bits. However, as ABC splits into very small parts, it is one of the rare cases where this becomes practical and we can exploit differential cryptanalysis in an (under reasonable assumptions) optimal way.

The Basics. We start by taking a closer look at the difference distribution of $\mathcal{B}_2$ and $\mathcal{B}_4$. As the approaches for both subciphers are identical, we will use $\mathcal{B}_2$ to explain our approach. For a (round-reduced) cipher $E : \mathbb{F}_2^n \to \mathbb{F}_2^n$ we consider an input difference $\alpha \in \mathbb{F}_2^n$ and an output difference $\beta \in \mathbb{F}_2^n$ and are interested in

$$\delta_E(\alpha, \beta) = |\{x \in \mathbb{F}_2^n \mid E(x) \oplus E(x \oplus \alpha) = \beta\}|$$

which counts how often a pair of plaintexts with input difference α result in a difference of β when being encrypted with the same key. Assuming uniform plaintexts, we thus get the probability of the *differential* as

$$\tilde{P}_E(\alpha, \beta) = \mathbf{Pr}[E(X) \oplus E(X \oplus \alpha) = \beta] = 2^{-n} \cdot \delta_E(\alpha, \beta).$$

To collect information about all possible $\delta_E(\alpha, \beta)$ at once, it is common to use the *Differential Distribution Table* $\mathcal{D}$, where the values of α make up the columns and the values of β make up the rows, respectively. Analogously, $\tilde{\mathcal{D}}$ contains the values of $\tilde{P}_E(\alpha, \beta)$ indexed by α and β respectively.

In order to compute the probabilities involved for iterated constructions, we usually assume that the individual rounds behave independently, i.e., the cipher is a so called Markov cipher. Moreover, for keyed primitives, the often used hypothesis of stochastic equivalence assumes that those probabilities are very similar for all keys. This allows to compute the (normalized, expected) DDT $\tilde{\mathcal{D}}_r$ for r rounds of the cipher by observing that

$$\tilde{P}_r(\alpha, \beta) = \sum_{\gamma \in \mathbb{F}_2^{16}} \tilde{P}_{r-1}(\alpha, \gamma) \cdot \tilde{P}_1(\gamma, \beta).$$

This exactly corresponds to computing

$$\tilde{\mathcal{D}}_r = \tilde{\mathcal{D}}_{r-1} \cdot \tilde{\mathcal{D}}_1 = \tilde{\mathcal{D}}_1^r$$

as $\tilde{\mathcal{D}}_r$ contains $\tilde{P}_r(\alpha, \beta)$ for all α, β and each entry is computed accordingly through matrix multiplication.

Application to $\mathcal{B}_2$ and $\mathcal{B}_4$. As mentioned above, while the computation of the entire DDT is infeasible in most cases, it is rather easy in the case of the smallest subciphers of ABC. Indeed, given the block length of just 16 bits, computing $\tilde{\mathcal{D}}_i$ with $i \in \{2, 4, 6\}$ for $\mathcal{B}_2$ and $\mathcal{B}_4$, took 12 hours in total on a standard desktop PC and, using 32 bit floating point values, can be stored in 16 GiB per DDT.

In the following table, we summarize the maximal entries for both subciphers for rounds 2, 4 and 6, where the differences α and β are denoted in hex.

r	$\max_{\alpha,\beta}(\mathcal{D}_{r,\mathcal{B}_2})$	α_2	β_2	$\max_{\alpha,\beta}(\mathcal{D}_{r,\mathcal{B}_4})$	α_4	β_4
2	0.265625	c0c0	40	0.265625	3c0	40
4	0.0664062	c000	80	0.0247269	8003	400a
6	0.0277326	8040	4080	0.0030473	e007	400a

Even though $\mathcal{B}_4$ seems (relatively) more resistant against differential attacks, those values describe distributions which are far from being uniform. For reference, for a random permutation on n bits, we expect a uniformity of approximately $2n$, thus for $n = 16$ we would expect a maximal probability of 0.000488, i.e., at least one order of magnitude smaller. Moreover, the (expected, normalized) DDT contains a significant amount of zero values.

This clearly opens up the possibility of standard differential attacks based on the most probable output difference, zero-correlation attacks based on the impossible differentials and, finally, the all-in-one differential attack. While we implemented all three approaches, as we were interested in how much we actually gain from using all differentials together, we describe only the all-in-one approach in a bit more detail. As all our attacks are practical, we do not bother with theoretical estimates of the complexities and success probabilities, but provide experimental data instead.

All-in-One Differential Attack on $\mathcal{B}_2$. We will show the attack on $\mathcal{B}_2$, as it works analogously on $\mathcal{B}_4$. Given a message pair $(p, p \oplus \alpha)$ along with their ciphertexts (c, c') after eight rounds, the attacker first guesses the last two round keys $(k_{\mathcal{B}_2,6}, k_{\mathcal{B}_2,7}) \in \mathbb{F}_2^{16}$ and trial decrypts the last two rounds $\mathcal{B}_{2,2}^{-1}(k_{\mathcal{B}_2,6} || k_{\mathcal{B}_2,7})$ with those keys, getting y and y' as shown in the picture below.

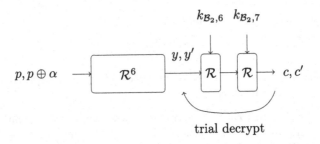

trial decrypt

To simplify, we assumed that a wrong key-guess would yield a uniform difference of $y \oplus y'$ and a correct key guess would result in a difference $\beta = y \oplus y'$ with probability $\tilde{\mathcal{D}}_6(\alpha, \beta)$ that we precomputed. According to [1], we increased the counter of the key-guess in turn by $\log(2^{16}\tilde{\mathcal{D}}_6(\alpha, \beta))$. That is, using multiple pairs (y, y') the counter of a key guess k is defined as

$$c(k) = \sum_{(y,'y)} \log\left(\tilde{\mathcal{D}}_6(\alpha, y \oplus y')\right)$$

where we skip the factor 2^{16} as we are only interested in the order of the counters. The expectation is that the correct key guess ends up having the largest counter.

Experimental Data. We performed these attacks with an increasing amount of plaintext pairs N and computed the average rank of the correct round key over multiple tries each. The results are summarized in Fig. 3a for $\mathcal{B}_2$ and in Fig. 3b for $\mathcal{B}_4$.

As can be seen, for $\mathcal{B}_2$, for the all-in-one approach 40 pairs are enough to recover the correct key as the one with the top rank in most cases. An impossible differential attack would still be left with rather many possible keys and require 200 pairs to have only a few candidates left with good probability. A traditional differential attack on $\mathcal{B}_2$ would require about 1700 plaintext pairs for a good probability to discover the right key correctly.

For $\mathcal{B}_4$, the situation is slightly worse. A traditional differential attack on $\mathcal{B}_4$ focusing on the most probable differential would not be effective at all, as the average rank hovers around 10.000 even when thousands of plaintext pairs are used. This is somewhat surprising. However, we do not bother to investigate this any further simply because the all-in-one approach still works.

After recovering the round keys for rounds 6 and 7, the attacker assumes the key guess to be correct and decrypts the last 2 rounds, which leads to him being able to do the exact same attack for the rounds 4 and 5 (and rounds 2 and 3 afterwards) in order to determine $(k_{\mathcal{B}_2,4}, k_{\mathcal{B}_2,5}) \in \mathbb{F}_2^{16}$, or $(k_{\mathcal{B}_2,2}, k_{\mathcal{B}_2,3})$ respectively. However, when the attack on rounds 6 and 7 lead to a wrong key guess k', the decryption is incorrect and leads to another difference distribution (namely over $\mathcal{B}_{2,2}^{-1}(k') \circ \mathcal{B}_{2,8}(k)$) which yields differentials that are impossible over $\mathcal{B}_{2,6}$ and thus have a high probability to get filtered out by the attack when enough pairs are used; The scoreboard would then be empty. In the case where the correct $(k_{\mathcal{B}_2,6}, k_{\mathcal{B}_2,7})$ was recovered, the next iterations of the attack are easier to perform, as the difference distribution becomes even less uniform,

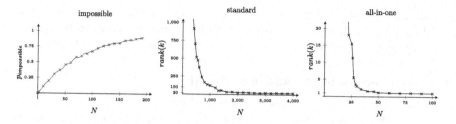

(a) Efficiency of each differential attack on $\mathcal{B}_2$.

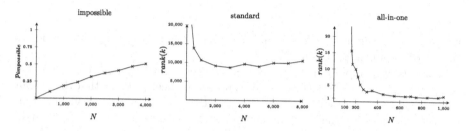

(b) Efficiency of each differential attack on $\mathcal{B}_4$.

Fig. 3. Efficiency of our impossible, standard and all-in-one differential attacks on $\mathcal{B}_2$ and $\mathcal{B}_4$. N is the number of used differential pairs. $p_{impossible}$ is the probability that a wrong key is identified as impossible. $rank(k)$ is the ranking of the key, i.e., $rank(k) = 1$ means that the correct key is identified as most probable.

the less rounds are performed. Thus, the attack complexity is dominated by the cost of recovering the last two round keys.

Again, just like the meet-in-the-middle attack, this differential attack only recovers $k_{\mathcal{B}_2}$ and $k_{\mathcal{B}_4}$ but is sufficient to recover the master key as we will explain in the next section.

4   Key Recovery

In this section, we assume that we are given $k_{\mathcal{B}_2}$ and $k_{\mathcal{B}_4}$, that could either be due to a successful meet-in-the-middle attack or a differential attack as in both cases we end up with all the round keys of those two subciphers. Now we want to recover the master key. For this, we have to study the properties of the mapping that maps the 128 bit master key k to the round keys $k_{\mathcal{B}_2}$ and $k_{\mathcal{B}_4}$, which correspond again to 128 bits in total. We denote this mapping by

$$\phi : \mathbb{F}_2^{128} \mapsto \mathbb{F}_2^{128}.$$

Recall that the key schedule resembles a lot the round function itself, so the first idea is to check again if this function decomposes again into parts. This time, luckily for us, it does not. If it would decompose, we would have been unlikely to recover the master keys from parts of the round keys. However, while it does not decompose, it is still far from having full dependency as we will explain next.

Note that full dependency is not a strict requirement for a key schedule algorithm. Indeed, there are examples of ciphers that, for now, withstand all cryptanalytic attempts and have (up to a constant) identical round keys.

4.1 Dependencies in the ABC Key Schedule

Instead of investigating the properties of the function ϕ by hand, we take a simple algorithmic approach to discover the dependencies. We are interested to determine if the i.th bit of the master key k has an influence on the j.th output bit of $\phi(k)$. This is the case if and only if there exists an input k such that

$$\phi_j(k) \neq \phi_j(k \oplus e_i),$$

where e_i is the vector that has a one exactly at position i and ϕ_j corresponds to the j.th component function of ϕ.

For this, we simply iterate over a few hundred randomly selected keys k and compute $\phi_j(k) \oplus \phi_j(k \oplus e_i)$. If the result is non-zero once, we know that the output depends on the i.th input bit. If we always get zero as a result, we conclude that the i.th bit does not have an influence on the output. This procedure might obviously overlook actual dependencies, but this theoretical drawback of this black-box approach does not have any influence in practice. We collected all those dependencies in a big binary matrix T of size 128×128 where a 1 entry at column i and row j corresponds to the case where bit i in the input influences bit j in the output.

$$T = (T_{j,i})_{j,i}, \quad T_{ji} = \begin{cases} 1 & \text{if } \phi_j \text{ depends on } k_i \\ 0 & \text{else.} \end{cases}$$

We show a (condensed) version of T in Fig. 4. This description is not meant to be readable entirely, but rather give a visual intuition that an attack as described next is reasonable.

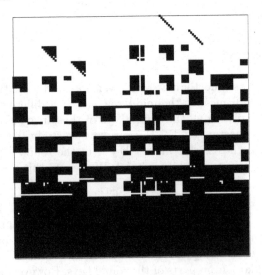

Fig. 4. Dependency matrix T.

4.2   Recovering the Master Key from $k_{\mathcal{B}_2}$ and $k_{\mathcal{B}_4}$

Given the dependency matrix T, we aim at designing an efficient guess-and-determine approach for recovering the master key k. The idea is rather obvious. We start with a row of T that contains only very few ones. The position of the one entries correspond to master key bits, we have to guess. Every row furthermore corresponds to a one bit condition on the guesses, as we know the output value $\phi(x)$, i.e., the round keys.

After guessing the first (hopefully few) bits of the master key and filtering with the one bit condition, i.e., verifying that the guess is consistent with the round keys obtained, we have to continue guessing the next few bits and verifying the next condition. Of course, in every following step, we already have guessed some bits and that potentially reduces the new guesses to make.

Thus, what we are searching for is actually an order of the output bits, i.e., the rows of the matrix T, such that the additional guesses to make remains small at any step. While finding the optimal solution to this problem seems hard in general, a greedy approach is easy to implement and test. We simply start with one of the rows with a minimal number of ones and at each following step select one of the rows with the minimal number of additional guesses. The derived ordering directly results in a guess-and-determine strategy and comes with an easy to estimate complexity. We summarize the strategy found by this greedy approach in Table 1. There are some interesting observations to make. First, the overall complexity is dominated by the large number of expected key candidates at step 78 where 2^{49} possible candidates are estimated. We depict this in the picture below. Moreover, it cannot be expected that we end up with a single candidate for the master key, but rather with 2^4 candidates. This is due to the fact that the first 23 steps depend only on 19 master key bits.

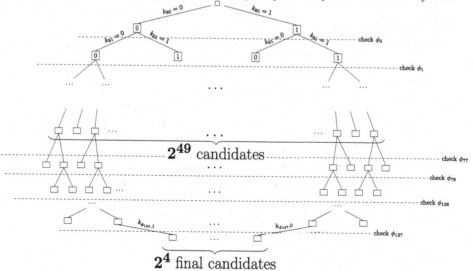

Table 1. The key-recovery strategy.

Step	Bit ϕ	guess	cands	Step	Bit ϕ	guess	cands	Step	Bit ϕ	guess	cands
0	0	1	0	43	74	0	7	86	92	0	41
1	1	1	0	44	75	0	6	87	93	0	40
2	2	1	0	45	76	0	5	88	95	0	39
3	3	1	0	46	77	0	4	89	80	4	42
4	4	1	0	47	94	0	3	90	81	0	41
5	5	1	0	48	40	11	13	91	82	0	40
6	6	1	0	49	41	0	12	92	83	0	39
7	7	1	0	50	57	0	11	93	84	0	38
8	8	1	0	51	58	0	10	94	85	0	37
9	9	1	0	52	59	0	9	95	87	0	36
10	10	1	0	53	60	0	8	96	96	0	35
11	11	1	0	54	72	0	7	97	97	0	34
12	12	1	0	55	73	0	6	98	98	0	33
13	13	1	0	56	38	12	17	99	99	0	32
14	14	1	0	57	39	0	16	100	100	0	31
15	15	1	0	58	34	8	23	101	101	0	30
16	30	1	0	59	35	0	22	102	102	0	29
17	31	1	0	60	36	0	21	103	103	0	28
18	29	1	0	61	37	0	20	104	104	0	27
19	28	1	0	62	32	8	27	105	105	0	26
20	46	0	0	63	33	0	26	106	106	0	25
21	47	0	0	64	53	10	35	107	107	0	24
22	78	0	0	65	54	0	34	108	108	0	23
23	79	0	0	66	55	0	33	109	109	0	22
24	27	1	0	67	49	8	40	110	110	0	21
25	23	4	3	68	50	0	39	111	111	0	20
26	21	5	7	69	51	0	38	112	112	0	19
27	22	1	7	70	52	0	37	113	113	0	18
28	20	1	7	71	48	4	40	114	114	0	17
29	19	1	7	72	68	8	47	115	115	0	16
30	18	5	11	73	69	0	46	116	116	0	15
31	17	1	11	74	70	0	45	117	117	0	14
32	16	1	11	75	71	0	44	118	118	0	13
33	26	5	15	76	86	0	43	119	119	0	12
34	25	1	15	77	56	4	46	120	120	0	11
35	24	1	15	78	64	4	**49**	121	121	0	10
36	42	0	14	79	65	0	48	122	122	0	9
37	43	0	13	80	66	0	47	123	123	0	8
38	44	0	12	81	67	0	46	124	124	0	7
39	45	0	11	82	88	0	45	125	125	0	6
40	61	0	10	83	89	0	44	126	126	0	5
41	62	0	9	84	90	0	43	127	127	0	4
42	63	0	8	85	91	0	42				

5   Conclusion

In conclusion, we presented standard techniques to recover parts of the round keys of ABC and a very generic black box approach to deal with the task of recovering the master key. As it turns out, even though the strategy we discovered might be non-optimal, it is efficient enough to break ABC.

As the shown attacks on the roundkeys are only possible because of missing diffusion inbetween the state bytes, the first intuition to address this issue is to use a fix-point free permutation R' instead of the provided R. However, the weak properties of S and BS still pose threats to the security level of ABC.

This strikingly shows why serious cipher proposals must contain a security analysis which argues the resistance against standard attack vectors. Furthermore, it shows that patents, which appear to have been a blind spot so far, might be an interesting source for potential future works.

From an experimental perspective, ABC provides a nice target and we use that to experimentally investigate the question on how much better an all-in-one differential attack is compared to impossible differential and standard differential attacks.

References

1. Albrecht, M.R., Leander, G.: An all-in-one approach to differential cryptanalysis for small block ciphers. In: Knudsen, L.R., Wu, H. (eds.) Selected Areas in Cryptography. SAC 2012. LNCS, vol. 7707, pp. 1–15. Springer, Berlin, Heidelberg (2012). https://doi.org/10.1007/978-3-642-35999-6_1
2. Beaulieu, R., Shors, D., Smith, J., Treatman-Clark, S., Weeks, B., Wingers, L.: The SIMON and SPECK lightweight block ciphers. In: Proceedings of the 52nd Annual Design Automation Conference, San Francisco, CA, USA, 7–11 June 2015, pp. 175:1–175:6. ACM (2015). https://doi.org/10.1145/2744769.2747946
3. Beierle, C., et al.: Cryptanalysis of the GPRS encryption algorithms GEA-1 and GEA-2. In: Canteaut, A., Standaert, F.X. (eds.) Advances in Cryptology – EUROCRYPT 2021. EUROCRYPT 2021. LNCS, vol. 12697, pp. 155–183. Springer, Cham (2021). https://doi.org/10.1007/978-3-030-77886-6_6
4. Biham, E., Biryukov, A., Shamir, A.: Cryptanalysis of skipjack reduced to 31 rounds using impossible differentials. In: Stern, J. (ed.) Advances in Cryptology – EUROCRYPT '99. EUROCRYPT 1999. LNCS, vol. 1592, pp. 12–23. Springer, Berlin, Heidelberg (1999). https://doi.org/10.1007/3-540-48910-X_2
5. Biham, E., Shamir, A.: Differential cryptanalysis of des-like cryptosystems. In: Menezes, A., Vanstone, S.A. (eds.) Advances in Cryptology – CRYPTO '90, 10th Annual International Cryptology Conference, Santa Barbara, California, USA, 11–15 August 1990, Proceedings. LNCS, vol. 537, pp. 2–21. Springer, Berlin, Heidelberg (1990). https://doi.org/10.1007/3-540-38424-3_1
6. Briceno, M., Goldberg, I., Wagner, D.: A pedagogical implementation of A5/1 (1998)
7. Ciet, M., Farrugia, A.J., Fasoli, G., Paun, F.: Block cipher with security intrinsic aspects. US20090245510A1 (2008). Apple Inc

8. Dansarie, M.: Cryptanalysis of the sodark cipher for HF radio automatic link establishment. IACR Trans. Symmetric Cryptol. **2021**(3), 36–53 (2021). https://doi.org/10.46586/tosc.v2021.i3.36-53
9. Dansarie, M., Derbez, P., Leander, G., Stennes, L.: Breaking HALFLOOP-24. IACR Trans. Symmetric Cryptol. **2022**(3), 217–238 (2022). https://doi.org/10.46586/tosc.v2022.i3.217-238
10. Diffie, W., Hellman, M.E.: Special feature exhaustive cryptanalysis of the NBS data encryption standard. Computer **10**(6), 74–84 (1977). https://doi.org/10.1109/C-M.1977.217750
11. Knudsen, L.R.: Truncated and higher order differentials. In: Preneel, B. (ed.) Fast Software Encryption. FSE 1994. LNCS, vol. 1008, pp. 196–211. Springer, Berlin, Heidelberg (1994). https://doi.org/10.1007/3-540-60590-8_16
12. Knudsen, L.: Deal-a 128-bit block cipher. department of informatics, university of bergen. Tech. rep., Norway. Technical report (1998)
13. Shannon, C.: A mathematical theory of cryptography (1945)
14. The Sage Developers: SageMath, the Sage Mathematics Software System (2022). https://doi.org/10.5281/zenodo.6259615, https://www.sagemath.org

Grover on Chosen IV Related Key Attack Against GRAIN-128a

Arpita Maitra$^{1(\boxtimes)}$, Asmita Samanta^2, Subha Kar^2, Hirendra Kumar Garai^3,
Mintu Mandal^3, and Sabyasachi Dey^3

^{1} TCG Centres for Research and Education in Science and Technology,
Kolkata 700091, West Bengal, India
arpita76b@gmail.com
^{2} Indian Statistical Institute, Kolkata 700108, West Bengal, India
^{3} BITS Pilani Hyderabad Campus, Hyderabad 500078, Telangana, India

Abstract. In this paper, we present a chosen IV related key attack on Grain-128a, that exploits Grover's algorithm as a tool. Earlier a classical version of such a chosen IV related key attack was considered by Banik et al. in ACISP 2013. They showed that using around $\gamma \cdot 2^{32}$ related keys (where γ is an experimentally determined constant and is estimated as 2^8), and $\gamma \cdot 2^{64}$ chosen IVs one can mount the attack in the classical domain. This is because for each related key on an average 2^{32} chosen IVs need to be examined. Thus, the query complexity becomes $O(2^{32} \cdot 2^{32})$, i.e., $O(2^{64})$. Contrary to this, thanks to the quantum paradigm, we use the superposition of all these 2^{64} queries at a time and feed them to the oracle. As a result, we could manage to decrease the complexity of the related key search to the order of 2^{16}, consequently reducing the number of required IVs to 2^{32} through the exploitation of the Grover search algorithm. Simulation of the attack against a reduced version of Grain-128a like cipher in the IBMQ simulator has also been presented as proof of the concept. Resource estimation for hardware implementation of the attack is presented and analyzed under NIST MAXDEPTH limit.

Keywords: Grain 128a · Chosen IV · Related keys · Grover's Algorithm

1 Introduction

The devastating impact of the Shor's [1] algorithm on classical public key cryptosystems led the researchers to design public key cryptosystems that are resistant against any quantum adversary. As a consequence, a standardization process has been initiated by NIST and fourth round candidates [2] have been announced. This is broadly known as Post Quantum Cryptography (PQC).

On the other hand, Grover's algorithm, as described in [3], poses significant challenges to the security of symmetric ciphers. In the context of an ideal cipher with a secret key size of k bits, the sole classical attack method is an exhaustive search, which demands $O(2^k)$ complexity. However, when Grover's search algorithm is applied generically, it reduces this complexity to $O(\sqrt{2^k})$. Consequently, the effective security level of the secret key length is halved.

This led to the quantum cryptanalysis exploiting Grover's search against symmetric ciphers, especially on block ciphers. In this direction, key recovery attacks on AES [4-7], lightweight cipher Simon [8], and Speck [9] have been explored. In [10], the author showed the speed-up of Quantum information-set-decoding attacks on McEliece over non-quantum information-set-decoding attacks. It is also exhibited that classically secure ciphers can be broken with quantum algorithms [11-15]. Additionally, certain studies demonstrate that quantum algorithms can accelerate classical attack methods [16-18].

Unlike block ciphers, stream ciphers are less explored. A stream cipher is a type of encryption algorithm that operates on individual bits of plaintext and produces a stream of pseudo-random bits, which are then combined with the plaintext using a bitwise XOR operation to generate the ciphertext. Unlike block ciphers, which process fixed-size blocks of data, stream ciphers encrypt data in real time on a per-bit or per-byte basis.

One notable stream cipher is Grain-128a. This stream cipher was first proposed at the Symmetric Key Encryption Workshop (SKEW) in 2011 [19]. It served as an advancement over its precursor, Grain 128, incorporating security enhancements and offering optional message authentication through the Encrypt and MAC approach.

An important characteristic of the Grain family is its ability to boost the throughput at the expense of additional hardware. Grain 128a was sketched by Martin Ågren, Martin Hell, Thomas Johansson, and Willi Meier [19]. It is a synchronous stream cipher that generates a keystream based on a nonlinear feedback shift register (NFSR) and a linear feedback shift register (LFSR) structure combined with a non-linear filtering function. The filtering function uses a combination of shift and XOR operations to produce the keystream output, which is then used to encrypt the plaintext.

Grain-128a achieves its popularity over time because of its small hardware footprint and low power consumption. These make the cipher suitable for implementation in constrained environments such as wireless sensors, RFID tags, and other embedded systems. It has been extensively analyzed and has been shown to provide a high level of security against various attacks. However, it is worth noting that stream ciphers need to be used with caution to ensure the security of the encryption process, as they are vulnerable to specific types of attacks, such as keystream reuse.

Very recently, Anand et al. [20] considered some popular Feedback Shift Register (FSR) based ciphers and explored quantum cryptanalysis against those ciphers. The list contains Grain-128-AEAD, TinyJAMBU, LIZARD, and Grain-v1. In this direction, the implementation of the ciphers in different quantum environments has been studied and Grover search for recovery of the key has been explored. The resources required for mounting the attack have also been estimated with respect to NIST's depth restriction (MAXDEPTH). However, the other existing classical attacks on Grain-128a remain untouched. In this backdrop, we explore the Grover's search algorithm in the chosen IV related key attack against the cipher.

Our Contributions:

I. Banik et al. in [21], mounted a chosen IV related key attack on Grain-128a. They showed that a key recovery attack on Grain-128a, in a chosen IV related key setting, can be done using around $\gamma \cdot 2^{32}$ related keys, where γ is an experimentally determined constant and it is sufficient to estimate it as 2^8, and $\gamma \cdot 2^{64}$ chosen IVs. They also proved that it is possible to obtain $32 \cdot \gamma$ simple nonlinear equations and solve them to recover the secret key. Hence, the computational complexity of the attack is $O(2^{64})$.
Contrary to this, we explore Grover's algorithm to find the related keys for a chosen IV. IVs are chosen noting the 32-bit shift in the keystreams. Details of the attack are presented in Sect. 2.3. In our approach, the complexity is reduced to $O(2^{16})$ from $O(2^{32})$ for finding the related keys. This signifies the quantum advances for the attack model over the classical paradigm.

II. In [20], the authors considered Grover's search for key recovery on Grain-128a. As expected, the key is found in complexity $O(2^{64})$. In this context, we exploit Grover's search for chosen IV related key recovery. In this case, the computational complexity is reduced to $O(2^{32})$ which is a significant improvement over Grover's search for key recovery.

III. Due to the restriction of the resources in the IBMQ interface, we could not implement the full version of the attack on Grain-128a. This is because, for the full implementation of the attack against Grain-128a, we require more than 128 qubits (128 qubits for the key, 128 qubits for the IV (including 32 qubits for padding), and some ancillary qubits (at least 64 qubits) for mounting the Grover search). However, we can access only 32 qubits simulator for free of cost. So, we designed a toy version of a Grain-like cipher and mounted the attack as a validation of the concept of attack. Simulation details are presented in Sect. 5. However, we provide the resource estimation for the hardware implementation of the full attack.

IV. While estimating the resources under NIST MAXDEPTH limit, we found that for key recovery using Grover's algorithm for MAXDEPTH $= 2^{40}$, one requires $O(2^{121})$ many parallelized Grover [20], whereas, in our case, it reduces to $O(2^{59})$. We interpret the result as follows. Our attack overperforms the attack of Anand et al. where Grover search is exploited for key recovery against Grain-128a. Related key attack seems more powerful in quantum domain than exhaustive search. One may think that the related key attack is not a serious attack. However, under MAXDEPTH limit, it requires much less resources than Grover search for key recovery against Grain-128a.

Organization of the Paper:

We organize the paper as follows.

i. In Sect. 2, we define the preliminaries needed for the readers.
ii. Sect. 3 describes the classical model of the attack.
iii. The quantum chosen IV related key attack on Grain-128a using Grover's algorithm is discussed in Sect. 4.
iv. We present the simulation of the attack on a reduced version of Grain-like cipher in the IBMQ interface in Sect. 5.
v. In the Sect. 6, we provide the resource estimation for the hardware implementation of the attack. The cost of the attack is analyzed according to the NIST MAXDEPTH limit.
vi. Section 7 concludes the paper.

2   Preliminaries

In this section, we discuss the design of Grain-128a, Chosen IV Attacks and Chosen IV Related Key Attack, and the Grover's algorithm.

2.1   Design of GRAIN-128a

Grain-128a, an enhanced version of the original Grain-v0 cipher, incorporates several fundamental principles to ensure its security and efficiency.

Pre-output Function: The pre-output function of Grain-128a uses two 128-bit feedback shift registers, one is linear (LFSR) and the other is non-linear (NFSR) along with a non-linear Boolean function h. The LFSR is updated as

$$S_{t+n} = f(S_t),$$

where $S_t = [s_t, s_{t+1}, \ldots, s_{t+n-1}]$ is the n-bit LFSR state at cycle t and f is the linear feedback function

$$f(x) = 1 + x^{32} + x^{47} + x^{58} + x^{90} + x^{121} + x^{128}$$

over $GF(2)$. The NFSR state is updated by

$$B_{t+n} = S_t \oplus g(B_t),$$

where $B_t = [b_t, b_{t+1}, \ldots, b_{t+n-1}]$ is the n-bit NFSR state at time t and g is the non-linear function

$$
\begin{aligned}
g(x) = {} & 1 + x^{32} + x^{37} + x^{72} + x^{102} + x^{128} + x^{44}x^{60} + x^{61}x^{125} + x^{63}x^{67} + x^{69}x^{101} \\
& + x^{80}x^{88} + x^{110}x^{111} + x^{115}x^{117} + x^{46}x^{50}x^{58} + x^{103}x^{104}x^{106} + x^{33}x^{35}x^{36}x^{40}
\end{aligned}
\tag{1}
$$

in $GF(2)[x]$. The function h takes two bits from the NFSR (x_0 and x_4) and seven bits from the LFSR ($x_1, x_2, x_3, x_5, x_6, x_7$ and x_8) and gives the pre-output keystream. It is defined as

$$h(x_0, x_1, \ldots, x_8) = x_0x_1 \oplus x_2x_3 \oplus x_4x_5 \oplus x_6x_7 \oplus x_0x_4x_8.$$

Finally, the pre-output keystream (y_t) at any cycle t is defined as

$$y_t = h(x) \oplus S_{t+93} \oplus B_{t+2} \oplus B_{t+15} \oplus B_{t+36} \oplus B_{t+45} \oplus B_{t+64} \oplus B_{t+73} \oplus B_{t+89}.$$

NFSR and LFSR State Initialization: Grain-128a uses a 128-bit key (K) that is copied into the NFSR, i.e.

$$B_i = K_i, \text{ for } 0 \leqslant i \leqslant 127.$$

The LFSR is initialized with a 96-bit initial vector (IV) along with a 32-bit padding (P) in the ending (0xffff fffe), i.e.

$$\begin{aligned} S_i &= IV_i, \text{ for } 0 \leqslant i \leqslant 95 \\ &= 1, \text{ for } 96 \leqslant i \leqslant 126 \\ &= 0, \text{ for } i = 127. \end{aligned} \tag{2}$$

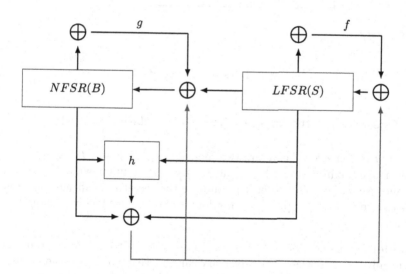

Fig. 1. Key Initialisation

Startup Clocking: After the initialization, the pre-output function is clocked 256 times before it begins to produce its pre-output stream. At this step, the pre-output stream is fed into the feedback polynomials f and g.

Keystream: The keystream (z) and MAC in Grain-128a both use the pre-output function (y). The keystream output is given by

$$z_t = y_{64+2t}, \ t = 0, 1, 2, \dots$$

That means every other bit of the output is considered as the keystream bit after the first 64 bits (those are used for authentication).

The circuits for *Key Initialization* and *Keystream Generation* are presented in Fig. 1 and 2 respectively.

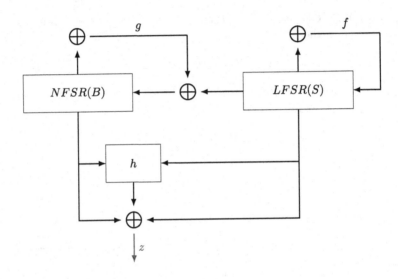

Fig. 2. Keystream Generation

2.2   Chosen *IV* Attacks and Chosen *IV* Related Key Attack

Chosen *IV* Attacks. There are two models used in chosen *IV* attacks, where an adversary can interact with an oracle possessing an unknown quantity (secret key) and use various *IV*s to either compute the secret key efficiently or distinguish the keystream output from random streams. The following are the two models:

■ **Model 1:** The adversary has access to an oracle with an unknown quantity, typically the secret key. The adversary can choose a public parameter, usually the *IV* (Initialization Vector), and request the oracle to encrypt a message of their choice. This allows the adversary to obtain keystream bits by querying the oracle with different *IV*s of their choice. This is shown in Fig. 3.

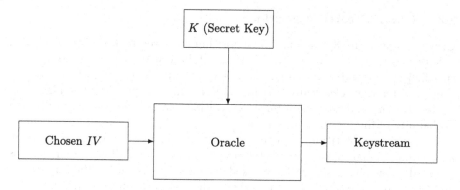

Fig. 3. Chosen IV attack model

The adversary's objective is to efficiently compute the secret key. This model has been used in cube attacks on stream ciphers.

■ **Model 2:** The adversary again has access to an oracle with an unknown quantity, typically the secret key. He/She can choose an IV and ask the oracle to encrypt a message. The goal of the adversary is to distinguish the keystream output from any random stream. It is used particularly in reduced round variants of stream and block ciphers.

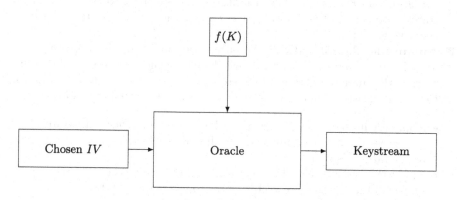

Fig. 4. Chosen IV-related key attack model

Chosen IV Related Key Attack: This is a relaxed version of the chosen IV attack model. In this model, the adversary can obtain keystream bits corresponding to a specific key-IV pair $[f(K), IV]$, where $K \in \mathcal{K}$ is the secret key and f is a function from the Key-space $\mathcal{K}$ onto itself, i.e. $f : \mathcal{K} \to \mathcal{K}$. The adversary's objective remains the same, which is to recover the value of the secret key K. The attack is depicted in Fig. 4.

2.3   Grover's Search Algorithm

Grover's Algorithm [3] is a quantum algorithm proposed by Lov Grover in 1996. It is primarily used to search an unsorted database with N items, looking for a specific item marked as the target with $O(\sqrt{N})$ time complexity, which is significantly faster than classical algorithms' $O(N)$ time complexity. The algorithm utilizes the principles of quantum superposition and interference to amplify the probability of finding the target item.

In Grover's Algorithm, we can represent the quantum state as a vector in a two-dimensional plane known as 'Grover's search space'. The search space is spanned by two basis states: the marked state ($|s\rangle$) representing the target item and the unmarked states ($|u\rangle$) representing the rest of the items in the database.

- ■ **Initialization:** At the beginning of the algorithm, all states are equally probable, and we can represent this as the state vector $|\psi\rangle$, which is the equal superposition of all possible states:

$$|\psi\rangle = \frac{1}{\sqrt{N}} \sum |u\rangle,$$

  where N is the total number of states ($N = 2^n$, where n is the number of qubits).
- ■ **Oracle (Marking the target state):** The oracle is a quantum operation that marks the target state ($|s\rangle$) by flipping its phase. Geometrically, this operation can be visualized as a reflection with respect to the origin of the state vector $|\psi\rangle$. After applying the oracle, the marked state will have a negative amplitude.
- ■ **Amplitude Amplification:** Amplitude amplification is a process that amplifies the amplitude of the marked state while suppressing the amplitude of the unmarked states. This step can be represented as a rotation in the two-dimensional plane. Geometrically, it can be visualized as follows:

  - • Apply the inversion about the mean: This operation reflects the state vector $|\psi\rangle$ across the average amplitude of all states, effectively rotating it closer to the marked state.
  - • Apply the oracle again: This operation marks the target state ($|s\rangle$) again, further rotating the state vector closer to the marked state.

- ■ **Measurement:** After a certain number of iterations (about $\sqrt{N}$ times), we perform a measurement on the state vector $|\psi\rangle$. The probability of measuring the marked state $|s\rangle$ is significantly increased compared to other states. Thus, we have a high probability of finding the target item in the database.

Geometrically, Grover's Algorithm can be visualized as a rotation of the initial state vector $|\psi\rangle$ toward the marked state $|s\rangle$, with successive rotations bringing the state vector closer to the target state, making it more likely to be measured. The amplitude amplification process concentrates the probability around the marked state, leading to an efficient search for the target item in an unsorted database.

Grover's Algorithm provides a quadratic speedup over classical algorithms for unstructured search problems. It has potential applications in cryptanalysis to find secret keys used in an encryption scheme. However, Grover's Algorithm does not provide a super-polynomial speedup like some other quantum algorithms e.g. Shor's algorithm for factorization.

3   Classical Chosen IV Related Key Attack on Grain-128a

In [21], Banik et al. showed that it is possible to find a 32-bit shifted keystream by searching among 2^{32} key-IV pairs for Grain-128a. That means there exists two key-IV pairs (K, IV) and (K', IV') such that the keystream K'_s generated by (K', IV') is a 32-bit shift of the keystream K_s generated by (K, IV). They exhibited that there are some particular cases when we can get these types of pairs and the required conditions are as follows:

- The 96 bit IV will be of the form $(v_0 \, v_1 \ldots v_{63} \, 11 \ldots 10)$. Here the last 32-bits have been taken the same as the 32-bits padding $P = 11...10$.
- $(K', IV') = (KSA^{-1})^{32}(K, IV)$ i.e. $(K', IV') = KSA^{-32}(K, IV)$.
- The first 32 bits of the keystream generated from (K', IV'), after applying 256 round KSA followed by the $PRGA$, will be all zero.

They have also shown that by using these key-IV pairs one can perform a key recovery attack on Grain-128a, in a chosen IV-related key setting. The time complexity of this attack is $O(2^{64})$. An important point of this attack is that KSA^{-1} can be expressed as a combination of two functions f and g such that

$$KSA^{-32}(k_0||k_1||k_2||k_3, \beta_0||\beta_1||111...10||111...10)$$
$$= (f_{\beta_0,\beta_1}(k_0, k_1, k_2, k_3)||k_0||k_1||k_2, g_{\beta_0,\beta_1}(k_0, k_1, k_2, k_3)||\beta_0||\beta_1||111...10),$$

where the key is $k_0||k_1||k_2||k_3$ and IV is of the form $\beta_0||\beta_1||111...10$ (β_0, β_1 each is of 32 bits).

To implement the attack they have used an algorithm to find the pairs (K, IV) and (K', IV'), for a fixed IV of the particular form $\beta_0||\beta_1||P$ (and IV' will be of the form $\eta||\beta_0||\beta_1$). The algorithm is described below:

Algorithm 1: Finding η

Input: β_0, β_1

Output: η or $\perp$

1 **for** *all the values of* $\eta \in \{0,1\}^{32}$ **do**

2 $\quad$ $K' \leftarrow f_{\beta_0,\beta_1}(k_0, k_1, k_2, k_3)||k_0||k_1||k_2$, where $K = k_0||k_1||k_2||k_3$.

3 $\quad$ Calculate the two key streams $(K_S, K_{S'})$ produced by the states
$\quad$ $S = K||\beta_0||\beta_1||P||P$ and $S' = K'||\eta||\beta_0||\beta_1||P$ respectively.

4 $\quad$ **if** K_S *and* $K_{S'}$ *maintain a 32-bit shift after 64-th clock cycle* **then**

5 $\quad\quad$ | **return** η.

6 $\quad$ **end**

7 **end**

8 **return** $\perp$

The strategy they have followed for the key recovery attack is as follows:

i. First assume that the secret key is $(k_0||k_1||k_2||k_3)$, where each k_i is of 32 bits and they all are variables.
ii. Fix β_0, β_1.
iii. Then we call Algorithm 1 with the input (β_0, β_1) to get the output η for which we can get a 32-bits shifted keystream.
iv. If we do not get such η, then change the fixed β_0, β_1 and repeat the above process.

If we get some η for which we are getting the 32-bit shifted keystreams, then we can surely say the following:

■ First 32 bits of the keystream generated by $f_{\beta_0,\beta_1}(k_0, k_1, k_2, k_3)||k_0||k_1||k_2$, $\eta||\beta_0||\beta_1$ are all zeros.
■ $g_{\beta_0,\beta_1}(k_0, k_1, k_2, k_3) = \eta$.

So we get 32 equations from the fact that $g_{\beta_0,\beta_1}(k_0, k_1, k_2, k_3) = \eta$.

This whole process is repeated several times with a shifted key in the motivation to collect more equations. Lastly, the equations are solved to get the secret key $k_0||k_1||k_2||k_3$. For more details, one can refer to [21].

4   Quantum Chosen *IV* Related Key Attack on Grain-128a Using Grover's Algorithm

Our quantum attack has the same kind of strategy as the work in [21]. First, we present the idea intuitively and then state it formally. In our attack strategy, we have used Grover's algorithm to find η, 32 bits of β_0, and β_1 for which we can generate 32-bit shifted keystream. Our assumptions are as follows.

i. We are given two keystreams: K_s and K'_s. K'_s is the 32-bits shifted keystream of K_s.
ii. K_s is generated from actual Grain-128a whereas K'_s is generated from 32 times KSA^{-1} followed by Grain-128a.
iii. For these two keystreams we know from [21] that $IV = \beta_0||\beta_1||P$ and $IV' = \eta||\beta_0||\beta_1$.
iv. Additionally, we know some fixed 32-bit positions of β_0 and β_1. For example, let us consider 8-bits for each of β_0 and β_1. Suppose, we know the bit values of the first, third, fifth, and seventh position of β_0, i.e., $\beta_0 = 1*1*1*1*$ and the bit values of the second, third, sixth and eighth position of β_1, i.e., $\beta_1 = *01**1*0$.
v. We do not have any information regarding η.

In this backdrop, we need to recover the unknown 32 bits of β_0 and β_1 along with the 32-bit values for η which provide the given 32-bit shifted keystream K'_s.

In the case of an exhaustive search, in the worst case scenario, the query complexity will be $O(2^{64})$. Due to the application of Grover's search, we can reduce it to $O(2^{32})$.

After getting the values of η, β_0 and β_1, the key recovery remains same as [21], i.e., generating $\gamma \cdot 32$ non-linear equations from the rotated key and solving those equations, we can recover the secret key. In [21], the authors commented that the experimentally estimated value for γ is 2^8. Hence, in our case, the number of related keys would be $\gamma \cdot 2^{16}$. The number of chosen IVs would be $\gamma \cdot 2^{32}$. The computational effort is therefore $O(2^{32})$. We now describe the attack.

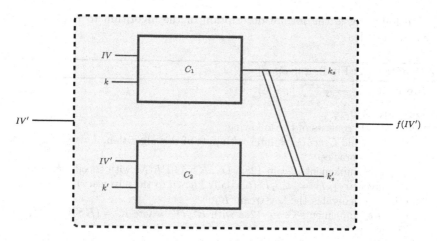

Fig. 5. Classical Boolean function (Here \\ implies 32-bits shifted XOR, i.e. we are computing bit-wise XOR of first 32-bits of K_s with the last 32-bits of the K'_s)

i. We make an equal superposition of all possible 2^{32} η and 2^{32} β_0 and β_1. We name it as $|\psi_1\rangle$. Thus,

$$|\psi_1\rangle = \frac{1}{\sqrt{2^{64}}} \sum_{\eta, \beta_0, \beta_1} |\eta||\beta_0||\beta_1\rangle$$

ii. We construct the Boolean function f for the Grover oracle to mount the attack which follows as:

$$f(IV') = 1, \quad \text{iff } K'_s \text{ is the 32-bit shifted keystream of } K_s$$
$$= 0, \quad \text{otherwise}$$

where K_s is the keystream generated from the actual Grain-128a ($KSA + PRGA$) and K'_s is the keystream generated from 32 times KSA^{-1} followed by Grain-128a. The classical oracle corresponding to this is presented in Fig. 5.

iii. $|\psi_1\rangle$ is fed to the Grover oracle.

iv. The oracle is iterated for $\frac{\pi}{4}\sqrt{2^{64}}$ times and amplifies the amplitude of the IV' for which the shifted keystream is generated. We call it as IV_0'. Grover oracle amplifies the amplitude of IV_0'.
v. After that, IV' is measured to get IV_0'. The corresponding quantum circuit is presented in Fig. 6
vi. We thus get η, β_0 and β_1.
vii. After getting η, β_0, and β_1, we follow the same methodology presented in [21] for recovering the secret key. That is, generating $\gamma \cdot 32$ many non-linear equations by considering i-bit cyclically left rotated key, where $i = 0, 1, 2, \ldots, \gamma - 1$ and solve those all together to recover the key.

The following is the formal description of the algorithm to implement the above attack idea.

Algorithm 2: Finding η, β_0, β_1

Input: $|\psi_1\rangle = \frac{1}{\sqrt{2^{64}}} \sum_{\eta\beta_0\beta_1} |\eta\beta_0\beta_1\rangle$
Output: β_0, β_1, η
2 Grover oracle consists of the following:
   i. $Circuit_1$ and $Circuit_2$ as internal inputs of this algorithm. This algorithm can call these two oracles.
   ii. $Circuit_1$ implements Grain-128a, i.e., $KSA+PRGA$ with specific K (unknown to the adversary), $IV = \beta_0\beta_1 P$ (partially known to the adversary).
   iii. $Circuit_1$ generates the keystream K_s.
   iv. $Circuit_2$ implements Grain-128a with K', IV' where $K' = (KSA^{-1})^{\otimes 32}(K)$, and $IV' = \eta\beta_0\beta_1$.
   v. $Circuit_2$ generates the 32-bit shifted keystream K_s'.
3 Now, the Grover oracle is fed by the equal superposition of all possible 2^{64} IV', i.e., by $|\psi_1\rangle$, where $|\psi_1\rangle = \frac{1}{\sqrt{2^{64}}} \sum_{\eta,\beta_0,\beta_1} |\eta||\beta_0||\beta_1\rangle$.
4 The oracle is iterated for $\frac{\pi}{4}.2^{32}$ times.
5 Measure IV'.
return η, β_0, β_1

Note that for the recovery of the secret key, we follow the same procedure as followed in [21]. This adds $\gamma \cdot 32$ to the overall complexity. Hence, the total query complexity of the attack remains $O(2^{32})$. And, we are getting a quadratic speed-up over the classical attack.

5   Simulation of the Attack in IBMQ Interface

Due to the limitations of available resources within the IBMQ interface, we were unable to implement the attack on the full version of Grain-128a. This is primarily because the full execution of the attack against Grain-128a requires the use of more than 128 qubits. Specifically, we need 128 qubits for the key, 128 qubits for the initialization vector (which includes 32 qubits for padding), and a minimum of 64 ancillary qubits for the Grover search procedure. Unfortunately, our access

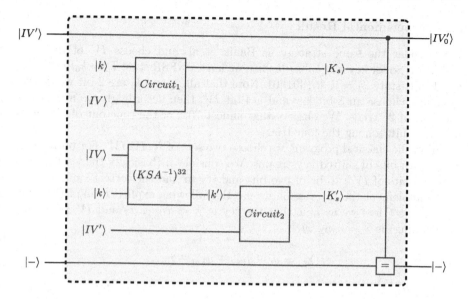

Fig. 6. Quantum Circuit to implement the attack

is restricted to a complimentary simulator with only 32 qubits. Consequently, we are unable to carry out the implementation of the attack on the actual full version of Grain-128a. Instead, we have opted to implement a scaled-down variant of the Grain-like cipher and mount an attack on that version.

Here, we consider two bits shift in the keystream. We verify our reduced version classically and check that a two-bit shift in the keystream is indeed possible for some key-IV pairs; (K, IV) and (K', IV'). The code for the classical implementation and the code for Grover search on the chosen IV related key, i.e., the quantum code for the attack are provided here [22].

5.1   Structure of Toy-Grain

We design a toy version of Grain-128a. The toy-Grain takes an 8-bit key K and a 6-bit IV with 2-bit padding 01 at the end of the IV. It has 8-bit non-linear and linear feedback shift registers. With the arrangements of K and IV, the updation process mimics the original design except for the feedback polynomials. Here

$$g(x) = 1 + x^8 + x^6 + x^5 + x^5 x^3 + x^7 x^6 + x^4 x^2 \text{ and } f(x) = 1 + x^8 + x^7 + x^2 + x.$$

The non-linear boolean function h is defined as

$$h(x_0, x_1, x_2, x_3) = x_0 x_1 \oplus x_2 x_3,$$

where $(x_0, x_1, x_2, x_3) = (S_4, B_2, S_7, B_6)$. The final keystream at cycle t is generated by $z_t = B_{2+t} + B_{4+t} + B_{7+t} + S_{7+t} + h$. The KSA is run 16 times before producing the first keystream bit.

5.2   Experimental Result

We consider the same strategy as Banik et al. and choose IV of the form $\beta_0||\beta_1||01$. So for key $k = k_0||k_1||k_2||k_3$, we have a NFSR state $T_n = k_0||k_1||k_2||k_3$ and LFSR state $T_l = \beta_0||\beta_1||01||01$. Note that all β_i and k_i are 2-bit words. We randomly choose an 8-bit key and a 4-bit IV, then accordingly we must get a shift out of 2^2 trials. We observe that almost 70% of the time out of 2^{18} times we get a shift among the four trials.

From the classical program, we choose two sets of K, IV, IV'. For both cases, we get two bits of shifted keystreams. We consider η (basically the least signifi-cant two bits of IV') to be of two bits and try to find the correct value for η.

To implement the toy version we use the following explicit form of the func-tions f, g, z (Here we assume that the key is $X = x_0x_1...x_7$ and IV (including the padding) is $Y = y_0y_1...y_7$):

$$f(Y_t) = y_t + y_{1+t} + y_{6+t} + y_{7+t}.$$

$$g(X_t) = x_t + x_{2+t} + x_{3+t} + x_{3+t}x_{5+t} + x_{1+t}x_{2+t} + x_{4+t}x_{6+t}.$$

$$z_t = x_{2+t} + x_{4+t} + x_{7+t} + y_{5+t} + h(x_{4+t}, y_{2+t}, x_{7+t}, y_{6+t}).$$

For the state updation in KSA, we use $x_{t+8} = y_t + z_t + g(X_t)$ and $y_{t+8} = z_t + f(Y_t)$ and for the state updation in $PRGA$, we use $x_{t+8} = y_t + g(X_t)$ and $y_{t+8} = f(Y_t)$.

In the current implementation, we assume that we know the values of β_0 and β_1 (if we have varied selected 2-bits of β_0 and β_1 as well, then to avoid unnecessary collision we have to generate two keystreams of 4-bits and 6-bits, but we can't do this because of the limited number of qubits availability in the free version of IBMQ simulator). Thus, the Grover oracle is fed by an equal superposition of all possible 2^2 many IV's (we have varied the 2-bit η of IV'). The oracle iterates $\lfloor \frac{\pi}{4}.\sqrt{2^2} \rfloor$ many times. After that, we measure the IV', and we get the two bits of η.

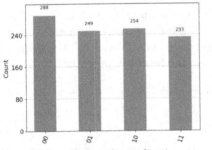

(a) State before the application of Grover's Search Algorithm

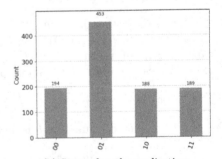

(b) State after the application of Grover's Search Algorithm

Fig. 7. First IBMQ Simulation

The histograms (Fig. 7 and 8) show the results. In the figures, the possible values of η have been presented along the X-axis, the left-most bit of the value of η is the 2nd least significant bit of IV', and the right-most bit of the value of η is the lsb of IV'.

To realize the shifting, we apply CNOTs (equivalent to classical XOR) considering the first two bits of the keystream (KS) generated from the cipher as controls and the last two bits of the keystream $(KS1)$ generated from KSA^{-1} followed by the cipher as targets. If the control bits are equal to the target bits, the target bits will be all zeros.

In Fig. 7, we get the value of $\eta = 01$. In this case, the secret key was 00010110, and the IV was 10100011. The actual value of IV' was 10001101 and its first two bits from the right side represent the value of η. That means our program is giving the correct value of η.

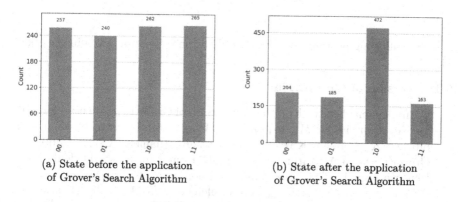

(a) State before the application of Grover's Search Algorithm

(b) State after the application of Grover's Search Algorithm

Fig. 8. Second IBMQ Simulations

In Fig. 8, we get the value of $\eta = 10$. Here, the secret key was 11110001, and the IV was 10100001. The actual value of IV' was 10000110 and its first two bits from the right side represent the value of η. That means in this case also our program is giving the correct value of η.

6   Resource Estimation for Hardware Implementation

In this section, we provide the resource estimation for the hardware implementation of the attack. That is, we calculate the number of qubits, CNOT gates, and T gates required to implement the attack. The resource estimation is given both for the toy as well as the full version of Grain-128a in Table 1. To count the number of T gates, we follow [23]. The calculation is done based on the following estimation.

– we need at most 7 T gates to implement a Toffoli gate,
– at most 22 T gates for a 3-fold NOT gate,

- at most 52 T gates for a 4-fold NOT gate
- and at most $(32 \cdot k - 84)$ T gates for any k-fold NOT gate where $k \geq 5$.

Table 1. Resource estimation of the attack on Grain-128a

Version of Grain-128a	Implementation of Module	Number of qubits	Number of CNOT-gates	Number of T-gates
Toy-version	KSA	17	12	35
	$PRGA$	21	11	35
	Full Attack	32	920	2723
Original Version	KSA	257	20	195
	$PRGA$	353	19	195
	Full Attack	546	26464	268524

In the following portion, we have described the estimation given in the above Table 1.

■ **Toy-version of Grain-128a:**

- To implement the attack on the Toy-version of Grain-128a, we need 8-qubits for Key, 8-qubits for IV, 8-qubits for IV', 6-qubits for two keystreams (2-bits for K_s (namely KS) and 4-bits for K'_s (namely KS1)), 1-qubit for Grover's Search Algorithm (for $|-\rangle$) and 1-qubit for internal calculation.
- Here we need 12 CNOT gates and 5 Tofolli gates to implement KSA; 11 CNOT gates and 5 Tofolli gates to implement $PRGA$; 12 CNOT gates and 5 Tofolli gates to implement KSA^{-1}; and 6 CNOT gates and 3 Tofolli gates to implement $PRGA^{-1}$.
- To implement the full attack we need to run KSA for a total of 34 times, KSA^{-1} for a total of 34 times, $PRGA$ for a total of 6 times, and $PRGA^{-1}$ for a total of 6 times.
- We also need 2 extra CNOT gates and 1 Tofolli gate to implement the full attack.

■ **Original Version of Grain-128a:**

- To implement the attack on the original Grain-128a, we need 128-qubits for Key, 128-qubits for IV, 128-qubits for IV', 160-qubits for two keystreams (64-bits for K_s and 96-bits for K'_s), 1-qubit for Grover's Search Algorithm (for $|-\rangle$) and 1-qubit for internal calculation.
- Here we need 20 CNOT gates, 11 Toffoli gates, three 3-fold NOT gates, and one 4-fold NOT gate to implement KSA; 19 CNOT gates, 11 Toffoli gates, three 3-fold NOT gates, and one 4-fold NOT gate to implement $PRGA$; 20 CNOT gates, 11 Toffoli gates, three 3-fold NOT gates, and one 4-fold NOT gate to implement KSA^{-1}; and 10 CNOT gates, 7 Toffoli gates, two 3-fold NOT gates, and one 4-fold NOT gate to implement $PRGA^{-1}$.

- To implement the full attack we need to run KSA for a total of 544 times, KSA^{-1} for a total of 544 times, $PRGA$ for a total of 160 times, and $PRGA^{-1}$ for a total of 160 times.
- We also need 64 extra CNOT gates and one 64-fold NOT gate to implement the full attack.

For Grover iteration, we need to multiply each column of Table-1, except the column assigned for the number of qubits, by $\lfloor \frac{\pi}{4} \cdot 2^{n/2} \rfloor$. That is for the toy version, we need not multiply at all as we consider 2-bits for η (and we did not vary any bits of β_0 or β_1), whereas for the original version, we need to multiply the figures by $\lfloor \frac{\pi}{4} \cdot 2^{32} \rfloor$.

In our attack model, we have applied Grover's search algorithm to find (K, IV) pairs for which the shifted keystream was generated and then the secret key was recovered following the method presented in [21]. Contrary to this, in [20], Grover search is applied for key recovery on Grain-128a. In the following Table-2, we compare the resources for these two types of attacks.

Table 2. Comparison of Required resources for our attack model and an attack similar as in [20]

	Our Attack Model	Grover's Search for key recovery [20]
Number of qubits	546	396
Number of CNOT gates for one iteration of Grover's Search algorithm	26464	15484
Number of T gates for one iteration of Grover's Search algorithm	268524	151092
Total number of Grover's iteration required for attack implementation	2^{32}	2^{64}

Interestingly, from the above table, one may notice that we need more gates for one iteration of our attack than the attack model presented in [20]. However, the attack reported in [20] needs 2^{64} iterations of Grover's search whereas we need only 2^{32} iterations of Grover's search algorithm. Hence, the overall gate count in our attack implementation is much less than the attack described in [20]. We analyze the cost of our attack model under the NIST MAXDEPTH limit.

6.1   Cost of the Attack Under NIST MAXDEPTH Limit

NIST provides an estimation of the resources for quantum attacks restricted to a fixed running time, or circuit depth. They call it MAXDEPTH [24]. The values considered for MAXDEPTH are 2^{40}, 2^{64} and 2^{96} logical gates. NIST also provides a rough estimation of time complexity for the circuits with these many gates [24]. As an example, considering the results from [4], the resource estimations are presented for AES-128, -192, and -256 under these MAXDEPTH limit.

In the current initiative, we also report the gate counts for our quantum chosen IV related key attack under the MAXDEPTH limit in Table 3. Here, G implies the total number of gates and D represents the depth of the full circuit. In our model $G = 294988 \cdot 2^{32} \approx 1.125 \cdot 2^{50}$ and $D = 161292 \cdot 2^{32} \approx 1.230 \cdot 2^{49}$.

Table 3. Cost of Grover search for the related key attack on Grain-128a under MAXDEPTH limit.

k		MAXDEPTH $= 2^{40}$	MAXDEPTH $= 2^{64}$	MAXDEPTH $= 2^{96}$	$G * D$
128	NIST [24]	2^{130}	2^{106}	2^{74}	2^{170}
	Grain-128a (Model of [20])	$1.795 \cdot 2^{121}$	$1.795 \cdot 2^{97}$	$1.795 \cdot 2^{65}$	$1.795 \cdot 2^{161}$
	AES [5]	$1.07 \cdot 2^{117}$	$1.07 \cdot 2^{93}$	$1.07 \cdot 2^{61}$	$\approx 2^{157}$
	Grain-128a (Our Model)	$1.385 \cdot 2^{59}$	$1.385 \cdot 2^{35}$	$1.385 \cdot 2^{3}$	$1.385 \cdot 2^{99}$

Considering MAXDEPTH $= 2^{40}$, one may interpret the results as follows.

- Our attack overperforms the attack of Anand et al. [20] where Grover Search is exploited for key recovery against Grain-128a.
- For key recovery using Grover's algorithm, one requires $O(2^{121})$ many parallelized Grover, whereas, in our case, it reduces to $O(2^{59})$.
- Hence, the related key attack seems more powerful in the quantum domain than an exhaustive search. One may think that a related key attack is not a serious attack. However, under the MAXDEPTH limit, it requires much less resources than Grover's search for key recovery against Grain-128a.

7   Conclusion

The chosen IV-related key attack against Grain-128a is sped up by a quadratic factor in this work. In the classical paradigm, Banik et al. [21] discovered that an attack required around $\gamma \cdot 2^{32}$ related keys (with γ being a constant approximated as 2^8) and $\gamma \cdot 2^{64}$ chosen IVs. However, the quantum paradigm lets us use the power of superposition, which permits the oracle to respond to every potential query simultaneously. This improvement brings the total complexity down to $O(2^{32})$, which is equal to 2^{16} for related keys and the next 2^{16} for the chosen IVs. We use the same strategy to improve the key recovery complexity of [21] as well. In the paper, we have also implemented the attack on a toy version of Grain-128a to show the correctness of the model. We have also estimated the required resources to implement the attack on the original version of Grain-128a and analyzed it under NIST MAXDEPTH limit.

References

1. Shor, P.W.: Algorithms for quantum computation: discrete logarithms and factoring. In: Proceedings 35th Annual Symposium on Foundations of Computer Science, pp. 124–134 (1994)
2. NIST. Post-Quantum Cryptography: Round 4 Submissions. NIST (2022)
3. Grover, L.K.: A Fast Quantum Mechanical Algorithm for Database Search. In: Proceedings of the Twenty-Eighth Annual ACM Symposium on Theory of Computing, STOC 1996, New York, NY, USA, pp. 212–219 (1996)
4. Grassl, M., Langenberg, B., Roetteler, M., Steinwandt, R.: Applying Grover's Algorithm to AES: quantum resource estimates. In: Takagi, T. (ed.) PQCrypto 2016. LNCS, vol. 9606, pp. 29–43. Springer, Cham (2016). https://doi.org/10.1007/978-3-319-29360-8_3
5. Jaques, S., Naehrig, M., Roetteler, M., Virdia, F.: Implementing Grover oracles for quantum key search on AES and LowMC. In: Canteaut, A., Ishai, Y. (eds.) EUROCRYPT 2020. LNCS, vol. 12106, pp. 280–310. Springer, Cham (2020). https://doi.org/10.1007/978-3-030-45724-2_10
6. Langenberg, B., Pham, H., Steinwandt, R.: Reducing the cost of implementing AES as a quantum circuit. IACR Cryptol. ePrint Arch. **2019**, 854 (2019)
7. Bonnetain, X., Naya-Plasencia, M., Schrottenloher, A.: Quantum security analysis of AES. IACR Trans. Symmetric Cryptol. **55–93**, 2019 (2019)
8. Anand, R., Maitra, A., Mukhopadhyay, S.: Grover on SIMON. Quantum Inf. Process **19** (2020)
9. Anand, R., Maitra, A., Mukhopadhyay, S.: Evaluation of quantum cryptanalysis on SPECK. In: Bhargavan, K., Oswald, E., Prabhakaran, M. (eds.) INDOCRYPT 2020. LNCS, vol. 12578, pp. 395–413. Springer, Cham (2020). https://doi.org/10.1007/978-3-030-65277-7_18
10. Bernstein, D.J.: Grover vs. McEliece. In: Sendrier, N. (ed.) PQCrypto 2010. LNCS, vol. 6061, pp. 73–80. Springer, Heidelberg (2010). https://doi.org/10.1007/978-3-642-12929-2_6
11. Kuwakado, H., Morii, M.: Security on the quantum-type Even-Mansour cipher. In: Proceedings of the International Symposium on Information Theory and its Applications, ISITA 2012, Honolulu, HI, USA, 28–31 October 2012, pp. 312–316 (2012)
12. Kaplan, M.: Quantum attacks against iterated block ciphers (2015)
13. Kaplan, M., Leurent, G., Leverrier, A., Naya-Plasencia, M.: Breaking symmetric cryptosystems using quantum period finding. In: Robshaw, M., Katz, J. (eds.) CRYPTO 2016, Part II. LNCS, vol. 9815, pp. 207–237. Springer, Heidelberg (2016). https://doi.org/10.1007/978-3-662-53008-5_8
14. Hosoyamada, A., Sasaki, Yu.: Quantum Demiric-Selçuk meet-in-the-middle attacks: applications to 6-round generic feistel constructions. In: Catalano, D., De Prisco, R. (eds.) SCN 2018. LNCS, vol. 11035, pp. 386–403. Springer, Cham (2018). https://doi.org/10.1007/978-3-319-98113-0_21
15. Leander, G., May, A.: Grover meets Simon – quantumly attacking the FX-construction. In: Takagi, T., Peyrin, T. (eds.) ASIACRYPT 2017, Part II. LNCS, vol. 10625, pp. 161–178. Springer, Cham (2017). https://doi.org/10.1007/978-3-319-70697-9_6
16. Kaplan, M., Leurent, G., Leverrier, A., Naya-Plasencia, M.: Quantum differential and linear cryptanalysis. IACR Trans. Symmetric Cryptol. 71–94 (2016)

17. Hosoyamada, A., Sasaki, Yu.: Cryptanalysis against symmetric-key schemes with online classical queries and offline quantum computations. In: Smart, N.P. (ed.) CT-RSA 2018. LNCS, vol. 10808, pp. 198–218. Springer, Cham (2018). https://doi.org/10.1007/978-3-319-76953-0_11

18. Santoli, T., Schaffner, C.: Using Simon's algorithm to attack symmetric-key cryptographic primitives. Quantum Inf. Comput. **7**, 65–78 (2017)

19. Ågren, M., Hell, M., Johansson, T., Meier, W.: Grain-128a: a new version of Grain-128 with optional authentication. Int. J. Wireless Mobile Comput. **5**(1), 48–59 (2011). https://doi.org/10.1504/IJWMC.2011.044106

20. Anand, R., Maitra, A., Maitra, S., Mukherjee, C.S., Mukhopadhyay, S.: Quantum resource estimation for FSR based symmetric ciphers and related Grover's attacks. In: Adhikari, A., Küsters, R., Preneel, B. (eds.) INDOCRYPT 2021. LNCS, vol. 13143, pp. 179–198. Springer, Cham (2021). https://doi.org/10.1007/978-3-030-92518-5_9

21. Banik, S., Maitra, S., Sarkar, S., Meltem Sönmez, T.: A chosen IV related key attack on grain-128a. In: Boyd, C., Simpson, L. (eds.) ACISP 2013. LNCS, vol. 7959, pp. 13–26. Springer, Heidelberg (2013). https://doi.org/10.1007/978-3-642-39059-3_2

22. ToyGrain. GitHub repository (2023). https://github.com/namenotpublished/ToyGrain.git

23. Wiebe, N., Roetteler, M.: Quantum arithmetic and numerical analysis using repeat-until-success circuits. Quantum Inf. Comput. **16**, 134–178 (2016)

24. Submission Requirements and Evaluation Criteria for the Post-Quantum Cryptography Standardization Process (2016). https://csrc.nist.gov/CSRC/media/Projects/Post-Quantum-Cryptography/documents/call-for-proposals-final-dec-2016.pdf

Concrete Time/Memory Trade-Offs in Generalised Stern's ISD Algorithm

Sreyosi Bhattacharyya and Palash Sarkar$^{(\boxtimes)}$

Indian Statistical Institute, 203, B.T. Road, Kolkata 700108, India
palash@isical.ac.in

Abstract. The first contribution of this work is a generalisation of Stern's information set decoding (ISD) algorithm. Stern's algorithm, a variant of Stern's algorithm due to Dumer, as well as a recent generalisation of Stern's algorithm due to Bernstein and Chou are obtained as special cases of our generalisation. Our second contribution is to introduce the notion of a set of effective time/memory trade-off (TMTO) points for any ISD algorithm for given ranges of values of parameters of the algorithm. Such a set succinctly and uniquely captures the entire landscape of TMTO points with only a minor loss in precision. We further describe a method to compute a set of effective TMTO points. As an application, we compute sets of effective TMTO points for the five variants of the Classic McEliece cryptosystem corresponding to the new algorithm as well as for Stern's, Dumer's and Bernstein and Chou's algorithms. The results show that while Dumer's and Bernstein and Chou's algorithms do not provide any interesting TMTO points beyond what is achieved by Stern's algorithm, the new generalisation that we propose provide about twice the number of effective TMTO points that is obtained from Stern's algorithm. Consequences of the obtained TMTO points to the classification of the variants of Classic McEliece in appropriate NIST categories are discussed.

Keywords: information set decoding · code based cryptography · Classic McEliece cryptosystem · concrete security · time/memory trade-off

1 Introduction

Code based cryptography [20] is one of the approaches to building post-quantum secure cryptosystems. There are three code-based cryptosystems, namely BIKE [1], Classic McEliece [5] and HQC [21], in Round 4 of the NIST PQC competition [24].

The best known attack against general code-based cryptosystems is based on information set decoding (ISD). The first ISD algorithm was proposed by Prange [25] and since then there have been many important developments. The early ISD algorithms by Prange [25], Lee and Brickell [16] and Leon [17] do not require any memory beyond what is needed to store the input.

© The Author(s), under exclusive license to Springer Nature Switzerland AG 2024
A. Chattopadhyay et al. (Eds.): INDOCRYPT 2023, LNCS 14459, pp. 307–328, 2024.
https://doi.org/10.1007/978-3-031-56232-7_15

A cornerstone of ISD algorithms is the algorithm by Stern [27] which introduced a meet-in-the-middle approach to ISD algorithms. The minimum time complexity of Stern's algorithm, however, is achieved by using a large amount of memory. In fact, Section 8.2 of the specification of Classic McEliece (available from [5]) mentions the following. "A closer look shows that the attack in [11] is bottlenecked by random access to a huge array (much larger than the public key being attacked)." The reference [11] in the quotation is [6] which introduced certain important practical efficiency improvements to Stern's algorithm. A key point to note in the above criticism of Stern's algorithm is that the memory requirement is much larger than the size of the public key of the cryptosystem being attacked.

Our first contribution is to introduce a generalisation of Stern's algorithm. Stern's algorithm as well as a variant of Stern's algorithm due to Dumer [9] (and later by Finiasz and Sendrier [13]) are obtained as special cases of the generalisation. Further, a recent generalisation of Stern's algorithm due to Bernstein and Chou [4] is also obtained as a special case. A justification for focusing on Stern's algorithm is that it is a watershed in the literature on ISD algorithms. While later work led to more advanced algorithms, the community still continues to use Stern's algorithm as the baseline ISD algorithm to evaluate code based cryptosystems. This is exemplified by the discussion on the PQC forum available at [14]. We also briefly sketch how our generalisation can be further extended to obtain a unified algorithm from which Prange's, Lee and Brickell's, Leon's as well as Stern's, Dumer's, and Bernstein and Chou's algorithms are obtained as special cases.

The basic idea of our generalisation of Stern's algorithm is the following. In Stern's algorithm, there is an enumeration step which builds a list. Storing this list requires a large amount of memory. Our generalisation allows the list to be significantly smaller by ignoring some elements which would otherwise be stored. This reduces the storage requirement of the algorithm. On the flip side, it also reduces the success probability, thus requiring more iterations. So the generalisation does not improve the runtime of Stern's algorithm. Rather it provides a larger number of time/memory trade-off (TMTO) points than what would be achieved by Stern's algorithm.

Varying the parameters of an ISD algorithm provides a large number of TMTO points. It is quite difficult to directly analyse the entire set of all TMTO points. Our second contribution tackles this issue. We introduce the notion of a set of effective TMTO points of an ISD algorithm with respect to a range of values of the parameters of the algorithm. Such a set succinctly and uniquely captures the entire TMTO landscape at only a minor loss in precision. Further, we describe a method to compute a set of effective TMTO points for any ISD algorithm.

As an application, we have obtained sets of effective TMTO points corresponding to the five variants of the Classic McEliece cryptosystem for the generalisation of Stern's algorithm that we propose as well as for Stern's, Dumer's and Bernstein and Chou's algorithms. The results show that Dumer's and Bernstein and Chou's algorithms do not provide any interesting TMTO points beyond what is achieved by Stern's algorithm. On the other hand, the sets of effective TMTO points obtained from the generalisation that we introduce are about twice the sizes of the corresponding sets of effective TMTO points obtained from Stern's algorithm. Further, in each case, the set of effective TMTO points obtained from the new generalised algorithm essentially subsumes the corresponding set obtained from Stern's algorithm. In particular, there are certain TMTO points which are achieved by the new generalised algorithm, but not by either of Stern's, Dumer's and Bernstein and Chou's algorithms. The TMTO points themselves are quite interesting. For example, these points show that in certain cases, by letting the time estimates increase by factors which are at most 2, it is possible to reduce the memory requirements by factors of about 2^8 to 2^{10}.

By obtaining sets of effective TMTO points, we are able to address the question of what time complexity can be achieved without requiring memory much larger than the public key size. For mceliece-4608-096, the size of the public key is about 2^{22} and the NIST target is 2^{207} gates. For this variant, it is possible to obtain TMTO points having time complexities less than 2^{207} with memory requirement only about 1.5 times the size of the public key for both the cases of constant memory access cost and logarithmic memory access cost. For mceliece-6688-128 and mceliece-6960-119, the sizes of the public key are also about 2^{22} bits and the NIST target for both is 2^{272} gates. With constant memory access cost, time complexities of about 2^{265} can be obtained for both these variants. The corresponding memory complexities are $2^{57.97}$ and $2^{67.37}$ respectively which are much larger than the size of the public key. By choosing values of the parameters appropriately, we find that it is possible to obtain time complexities slightly less than 2^{272} while keeping the memory complexities to be less than 2^{30}. While the memory complexities are still larger than the size 2^{22} of the public key, they are not too large. If, on the other hand, logarithmic cost of memory access is incorporated into the time estimates, then with memory restricted to be less than 2^{30} bits, the minimum time estimates for both the variants turn out to be more than the target value of 2^{272} by factors which are less than 8. For the other two variants of the Classic McEliece cryptosystem, we do not obtain any TMTO point, even with constant memory access cost, whose time complexity is less than what is required for NIST classification.

1.1   Previous and Related Works

A number of works [3,7,8,11,18,19] described advanced ISD algorithms with improved asymptotic time complexities. The advanced ISD algorithms require much more memory compared to Stern's algorithm. It has been proved in [28] that for codes which are relevant to the McEliece cryptosystem, all known ISD algorithms have essentially the same asymptotic complexity as that of Prange's algorithm.

TMTO for ISD algorithms has not received much attention in the literature. A recent work on this topic is [12] which mentions that "there has been very limited work on time-memory trade-offs for ISD algorithms." The work [12] considered TMTO with respect to the MMT [18] algorithm and for Classic McEliece, reported time complexities with an upper bound (2^{60} or 2^{80}) on the memory complexities. Considering the time estimates for memory at most 2^{60} bits and comparing with the time estimates of the (generalised) Stern's algorithm also with memory at most 2^{60} bits, we find that the MMT algorithm is faster than Stern's algorithm by factors which are less than 8. We note, though, that the gap could be somewhat wider if certain techniques for improving practical efficiency are incorporated into the method used in [12] for obtaining the time estimates.

A systematic work to assess the concrete security of various code based cryptosystems against a number of important ISD algorithm was reported in [2]. A work along the same line and the corresponding code was provided in [10]. A recent work [4] provides a rigorous approach to obtaining time complexity estimates by comprehensively automating the effects of various practical improvements as well as hidden costs. The works [2,4,10] focus on obtaining the minimum time complexities achievable by the different ISD algorithms. Unlike our approach, they do not provide any method to analyse the entire landscape of all TMTO points.

2 Preliminaries

Let $\mathbb{F}_2$ denote the finite field of two elements. The cardinality of a finite set S will be denoted as $\#S$. For a positive integer k, the set $\{1, \ldots, k\}$ will be denoted as $[k]$.

Vectors will be considered to be row vectors and will be denoted by bold lower case letters. Matrices will be denoted by bold upper case letters. By $\mathbf{M}^\top$ we will denote the transpose of the matrix $\mathbf{M}$. The identity matrix of order m will be denoted as $\mathbf{I}_m$. If $\mathbf{M}_1$ and $\mathbf{M}_2$ are two matrices having the same number of rows, then by $[\mathbf{M}_1|\mathbf{M}_2]$ we will denote the matrix obtained by juxtaposing $\mathbf{M}_1$ and $\mathbf{M}_2$. For a positive integer k, $\mathbf{0}_k$ and $\mathbf{1}_k$ will denote the all-zero and all-one vectors of length k respectively.

Vectors with elements from $\mathbb{F}_2$ will also be called binary strings. For a binary string $\mathbf{x}$, by $\mathsf{wt}(\mathbf{x})$ we will denote the number of positions where $\mathbf{x}$ is 1, i.e. for $\mathbf{x} = (x_1, \ldots, x_m)$, $\mathsf{wt}(\mathbf{x}) = \#\{i : x_i = 1\}$. For a subset S of $[k]$, by $\chi(S)$ we will denote the k-bit string $\mathbf{x} = (x_1, \ldots, x_k)$ such that $x_i = 1$ if and only if $i \in S$.

Let n and k be positive integers such that $k < n$. We will consider codes over $\mathbb{F}_2$. A linear code $\mathcal{C}$ is a subspace of dimension k of $\mathbb{F}_2^n$. Elements of $\mathcal{C}$ are called codewords. A basis for $\mathcal{C}$ is given by a matrix $\mathbf{G}_0$ of order $k \times n$ and so $\mathcal{C} = \{\mathbf{x}\mathbf{G}_0 : \mathbf{x} \in \mathbb{F}_2^k\}$. Such a matrix $\mathbf{G}_0$ is called a generator matrix for $\mathcal{C}$. The null space of $\mathbf{G}_0$ has dimension $r = n - k$. Let $\mathbf{H}_0$ be a matrix of order $r \times n$ such that the rows of $\mathbf{H}_0^\top$ form a basis for the null space of $\mathbf{G}_0$, i.e. $\mathbf{G}_0\mathbf{H}_0^\top = \mathbf{0}$. Such a matrix $\mathbf{H}_0$ is called a parity check matrix for $\mathcal{C}$. For any codeword $\mathbf{y} \in \mathcal{C}$, we have $\mathbf{H}_0\mathbf{y}^\top = \mathbf{0}$.

Let $w < n$ be a positive integer and $\mathbf{e} \in \mathbb{F}_2^n$ be a vector of weight w. Let $\mathbf{y} \in \mathcal{C}$ be a codeword and define $\mathbf{z} = \mathbf{y} + \mathbf{e}$. The vector $\mathbf{e}$ is called an error vector. Note that $\mathbf{s}^\top = \mathbf{H}_0\mathbf{z}^\top = \mathbf{H}_0(\mathbf{y}^\top + \mathbf{e}^\top) = \mathbf{H}_0\mathbf{e}^\top$, since $\mathbf{H}_0\mathbf{y}^\top = \mathbf{0}$. The vector $\mathbf{s} \in \mathbb{F}_2^r$ is called a syndrome.

The computational problem to be solved is the following.

Definition 1 (Syndrome decoding problem (SDP)). *Let n, k, r and w be positive integers such that $k, w < n$ and $r = n - k$. Let $\mathbf{H}_0$ be a parity check matrix for a linear code $\mathcal{C} \subseteq \mathbb{F}_2^n$ of dimension k. Given a syndrome $\mathbf{s}_0 \in \mathbb{F}_2^r$, the goal is to find an error vector $\mathbf{e} \in \mathbb{F}_2^n$ of weight w such that $\mathbf{H}_0 \mathbf{e}^\top = \mathbf{s}_0^\top$. The triplet $(\mathbf{H}_0, \mathbf{s}_0, w)$ is an instance of the SDP.*

2.1 ISD Algorithms from Prange to Stern

The first ISD algorithm which can be used to solve SDP was proposed by Prange [25]. Let $(\mathbf{H}_0, \mathbf{s}_0, w)$ be an instance of SDP. The requirement is to obtain a vector $\mathbf{e}_0 \in \mathbb{F}_2^n$ of weight w such that $\mathbf{H}_0 \mathbf{e}_0^\top = \mathbf{s}_0^\top$. The algorithm proceeds as follows. Choose a random permutation matrix $\mathbf{P}$ of order $n \times n$ and apply Gaussian elimination using row operations to the augmented matrix $[\mathbf{H}_0 \mathbf{P} \mid \mathbf{s}_0^\top]$ to obtain a matrix $[\mathbf{H} \mid \mathbf{s}^\top]$, where $\mathbf{H} = \mathbf{U}\mathbf{H}_0\mathbf{P}$ is such that $\mathbf{H}$ can be written as $\mathbf{H} = [\mathbf{A} \mid \mathbf{I}_r]$ and $\mathbf{s}^\top = \mathbf{U}\mathbf{s}^\top$. Here $\mathbf{U}$ is the invertible matrix of order $r \times r$ which corresponds to the sequence of row operations. (This may not always be possible; see Remark 2 in Sect. 3.) Note that $\mathbf{A}$ is a matrix of order $r \times k$.

To solve SDP on instance $(\mathbf{H}_0, \mathbf{s}_0, w)$, it is sufficient to obtain a vector $\mathbf{e} \in \mathbb{F}_2^n$ of weight w such that $\mathbf{H}\mathbf{e}^\top = \mathbf{s}^\top$, since in this case $\mathbf{e}_0^\top = \mathbf{P}\mathbf{e}^\top$ satisfies the relation $\mathbf{H}_0 \mathbf{e}_0^\top = \mathbf{s}_0^\top$.

The structure of the algorithm is the following. In each iteration, it chooses an independent and uniform random permutation matrix $\mathbf{P}$ and obtains $\mathbf{H}$ and $\mathbf{s}$ as mentioned above. Then it checks the weight of $\mathbf{s}$ and returns $\mathbf{e} = [\mathbf{0}_k \mid \mathbf{s}]$ if the weight is equal to w. (Note: $\mathbf{H}\mathbf{e}^\top = \mathbf{s}^\top$.) Iterations are repeated until a solution is obtained.

Conceptually, each iteration has two steps, namely the linear algebra step and the search step. The linear algebra step yields $[\mathbf{H} \mid \mathbf{s}^\top]$, while the search step looks for a solution using $\mathbf{H}$ and $\mathbf{s}$. The search step in Prange's algorithm is trivial and amounts to checking the weight of $\mathbf{s}$. Subsequent algorithms introduced more sophisticated ideas to perform the search step.

Lee and Brickell [16] described an algorithm which can be considered to be a modification of Prange's algorithm. In each iteration, the linear algebra step remains the same as in Prange's algorithm, i.e. $[\mathbf{H} \mid \mathbf{s}^\top]$, where $\mathbf{H} = [\mathbf{A} \mid \mathbf{I}_r]$ is obtained as in Prange's algorithm. The search step in the Lee-Brickell algorithm uses a parameter p which is a positive integer satisfying $p \leq w$. This step proceeds over all possible subsets of $[k]$ of size $p/2$. For each such p-element subset S, let $\mathbf{e}_1 = \chi(S)$ and $\mathbf{x} = \mathbf{A}\mathbf{e}_1^\top$. If $\mathrm{wt}(\mathbf{x} + \mathbf{s}) = w - p$, then $\mathbf{e} = [\mathbf{e}_1 \mid \mathbf{s}]$ satisfies $\mathbf{H}\mathbf{e}^\top = \mathbf{s}^\top$.

Leon [17] put forward an algorithm which can be considered to be a further modification of Prange's algorithm. Apart from the parameter p in Lee and Brickell's algorithm, Leon's algorithm introduced an additional parameter ℓ with $1 \leq \ell \leq r$. The linear algebra step yielding $[\mathbf{H} \mid \mathbf{s}^\top]$ remains the same as in Prange's algorithm. In Leon's algorithm, $\mathbf{H}$ and $\mathbf{s}$ are written as

$$
\mathbf{H} = \begin{array}{|c|c|c|}
\hline
\mathbf{A}_{\ell \times k} & \mathbf{I}_\ell & \mathbf{0}_{\ell \times (r-\ell)} \\
\hline
\mathbf{B}_{(r-\ell) \times k} & \mathbf{0}_{(r-\ell) \times \ell} & \mathbf{I}_{r-\ell} \\
\hline
\end{array}, \qquad
\mathbf{s}^\top = \begin{array}{|c|}
\hline
\mathbf{u}^\top \\
\hline
\mathbf{v}^\top \\
\hline
\end{array}
\tag{1}
$$

The search step uses the parameter p as in Lee and Brickell's algorithm and proceeds over all possible subsets of size of $[k]$ of size p. For each p-element subset S, let $\mathbf{e}_1 = \chi(S)$ and $\mathbf{x} = \mathbf{A}\mathbf{e}_1^\top$. (Note that in this case $\mathbf{A}$ is an $\ell \times k$ matrix and not an $r \times k$ matrix as in Lee and Brickell's algorithm.) If $\mathbf{x} = \mathbf{u}$, then let $\mathbf{y} = \mathbf{B}_1 \mathbf{e}_1^\top$; if $\mathrm{wt}(\mathbf{y}+\mathbf{v}) = w - p$, then $\mathbf{e} = [\mathbf{e}_1\,|\,\mathbf{0}^\ell\,|\,\mathbf{v}]$ satisfies $\mathbf{H}\mathbf{e}^\top = \mathbf{s}^\top$.

Stern [27] introduced the idea of using the meet-in-the-middle technique to perform the search step. The algorithm uses the two parameters ℓ and p. For simplicity of the basic description, assume that p and also k are even. The output $[\mathbf{H}\,|\,\mathbf{s}^\top]$ of the linear algebra step is written as in (1). Next write

$$
\mathbf{A} = [\mathbf{A}_1|\mathbf{A}_2] \text{ and } \mathbf{B} = [\mathbf{B}_1|\mathbf{B}_2],
\tag{2}
$$

where $\mathbf{A}_1$, $\mathbf{A}_2$ are $\ell \times k/2$ matrices and $\mathbf{B}_1$, $\mathbf{B}_2$ are $(r-\ell) \times k/2$ matrices.

The search step has two phases. In the first phase, all possible subsets of $[k/2]$ of size $p/2$ are considered. For each such subset S, let $\mathbf{e}_1 = \chi(S)$ and $\mathbf{a}_1 = \mathbf{A}_1 \mathbf{e}_1^\top$. All such pairs $(\mathbf{a}_1, S)$ are stored in a list $\mathcal{L}$ and $\mathcal{L}$ is indexed on the first components of the entries. In the second phase, again all possible subsets of $[k/2]$ of size $p/2$ are considered. For each such subset T, let $\mathbf{e}_2 = \chi(T)$ and $\mathbf{a}_2 = \mathbf{A}_2 \mathbf{e}_2^\top$. Next, for each such pair $(\mathbf{a}_2, T)$, find all entries $(\mathbf{a}_1, S)$ in $\mathcal{L}$ such that $\mathbf{a}_1 + \mathbf{u} = \mathbf{a}_2$. Let $\mathbf{b}_1 = \mathbf{B}_1 \mathbf{e}_1^\top$ and $\mathbf{b}_2 = \mathbf{B}_2 \mathbf{e}_2^\top$. If $\mathrm{wt}(\mathbf{y}_1 + \mathbf{y}_2 + \mathbf{v}) = w - p$, then $\mathbf{e} = [\mathbf{e}_1\,|\,\mathbf{e}_2\,|\,\mathbf{0}_\ell\,|\,\mathbf{v}]$ satisfies $\mathbf{H}\mathbf{e}^\top = \mathbf{s}^\top$.

3   A Generalisation of Stern's ISD Algorithm

We introduce a generalisation of Stern's algorithm. The generalisation is obtained through the use of two parameters λ and δ in addition to the parameters ℓ and p used in Stern's algorithm. The maximum possible ranges of values of the parameters are as follows.

$$
0 \le \ell \le r, \quad -(k-2) \le \lambda \le \ell, \quad 0 \le p \le \min(k+\lambda, w), \quad \delta \in (0,1].
\tag{3}
$$

Remark 1. A generalisation of Stern's algorithm (called isd1) has been given by Bernstein and Chou [4, Section 4.8] through the use of a parameter z, which is related to the parameter λ that we use by $z + \lambda = \ell$. The only values of z mentioned in [4] are $z = 0$ and $z = \ell$. It is not clear whether [4] allows z to be greater than ℓ (corresponding to λ less than 0). Note that due to the use of the additional parameter δ, our generalisation of Stern's algorithm subsumes the generalisation in [4]. As we will see later (Remark 7 in Sect. 5), it is the parameter δ that turns out to determine the non-triviality of the generalisation that we propose.

The basic structure of the algorithm is shown in Algorithm 1. It consists of a linear algebra step and a search step as in Stern's algorithm. These two steps are explained below. The input to the algorithm is $(\mathbf{H}_0, \mathbf{s}_0, w)$ and the requirement is to find a vector $\mathbf{e}_0 \in \mathbb{F}_2^n$ such that $\mathbf{H}_0 \mathbf{e}_0^\top = \mathbf{s}_0^\top$.

Algorithm 1: A general formulation of Stern's ISD algorithm.

Input: $(\mathbf{H}_0, \mathbf{s}_0, w)$
Output: $\mathbf{e}_0$ such that $\mathbf{H}_0 \mathbf{e}_0^\top = \mathbf{s}_0^\top$ and $\mathrm{wt}(\mathbf{e}_0) = w$.

1 **while** *true* **do**
2   Choose a random permutation matrix $\mathbf{P}$ of order $n \times n$
3      $(\mathbf{A}_1, \mathbf{A}_2, \mathbf{B}_1, \mathbf{B}_2, \mathbf{u}, \mathbf{v}) \leftarrow \mathsf{LinAlg}(\mathbf{H}_0, \mathbf{P}, \mathbf{s}_0, \ell, \lambda)$
4      $(\mathsf{flg}, \mathbf{e}) \leftarrow \mathsf{Srch}(\mathbf{A}_1, \mathbf{A}_2, \mathbf{B}_1, \mathbf{B}_2, \mathbf{u}, \mathbf{v}, w, p, \delta)$
5   **if** flg = yes **then**
6   return $\mathbf{P}\mathbf{e}^\top$

Linear Algebra. Apply Gaussian elimination using row operations to the augmented matrix $[\mathbf{H}_0\mathbf{P} \,|\, \mathbf{s}_0^\top]$ to obtain a matrix $[\mathbf{H} \,|\, \mathbf{s}^\top]$, where $\mathbf{H} = \mathbf{U}\mathbf{H}_0\mathbf{P}$ and $\mathbf{s}^\top = \mathbf{U}\mathbf{s}^\top$ are of the following forms

$$
\mathbf{H} = \left[
\begin{array}{c|c|c}
\mathbf{A}_{\ell \times (k+\lambda)} & \mathbf{C}_{\ell \times (\ell-\lambda)} & \mathbf{0}_{\ell \times (r-\ell)} \\
\hline
\mathbf{B}_{(r-\ell) \times (k+\lambda)} & \mathbf{D}_{(r-\ell) \times (\ell-\lambda)} & \mathbf{I}_{r-\ell}
\end{array}
\right], \quad
\mathbf{s}^\top = \begin{bmatrix} \mathbf{u}^\top \\ \mathbf{v}^\top \end{bmatrix}
\tag{4}
$$

Here $\mathbf{U}$ is the invertible matrix of order $r \times r$ which corresponds to the sequence of row operations, $\mathbf{u} \in \mathbb{F}_2^\ell$ and $\mathbf{v} \in \mathbb{F}_2^{r-\ell}$.

Remark 2. Consider the event E_1 that $\mathbf{H}_0\mathbf{P}$ can be reduced to the form shown in (4), i.e. the bottom right sub-matrix of $\mathbf{H}_0\mathbf{P}$ of order $(r - \ell) \times (r - \ell)$ is invertible. So E_1 is the event that the linear algebra step succeeds, and let π_1 be the probability of E_1. Under the heuristic assumption that the entries of $\mathbf{H}_0$ are independent and uniform random bits, $\pi_1 = \prod_{i=1}^{r-\ell}(1 - 2^{-i})$, which is about 0.288 for large enough values of $r - \ell$.

Write

$$
\mathbf{A} = [\mathbf{A}_1 | \mathbf{A}_2] \text{ and } \mathbf{B} = [\mathbf{B}_1 | \mathbf{B}_2],
\tag{5}
$$

where $\mathbf{A}_1$ (resp. $\mathbf{A}_2$) is an $\ell \times \lfloor (k+\lambda)/2 \rfloor$ (resp. $\ell \times \lceil (k+\lambda)/2 \rceil$) matrix, and $\mathbf{B}_1$ (resp. $\mathbf{B}_2$) is an $(r-\ell) \times \lfloor (k+\lambda)/2 \rfloor$ (resp. $(r-\ell) \times \lceil (k+\lambda)/2 \rceil$) matrix. The call $\mathsf{LinAlg}(\mathbf{H}_0, \mathbf{P}, \mathbf{s}_0, \ell, \lambda)$ in Algorithm 1 returns $(\mathbf{A}_1, \mathbf{A}_2, \mathbf{B}_1, \mathbf{B}_2, \mathbf{u}, \mathbf{v})$.

Search. As explained in Sect. 2.1, it is sufficient to obtain a vector $\mathbf{e} \in \mathbb{F}_2^n$ such that $\mathbf{He}^\top = \mathbf{s}^\top$. The search step looks for such an $\mathbf{e}$. The parameter δ is used in the search step. As in Stern's algorithm, the search step is done in two parts. The first part prepares a list $\mathcal{L}$ and the second part searches for a collision. The difference with Stern's algorithm is that instead of storing all the $\binom{\lfloor (k+\lambda)/2 \rfloor}{\lfloor p/2 \rfloor}$ ℓ-bit strings arising from considering all $\lfloor p/2 \rfloor$ possible combinations of the $\lfloor (k+\lambda)/2 \rfloor$ columns of $\mathbf{A}_1$, only $\left(\binom{\lfloor (k+\lambda)/2 \rfloor}{\lfloor p/2 \rfloor} \right)^\delta$ of such combinations are considered and the corresponding ℓ-bit strings stored in $\mathcal{L}$. The second part of the search step proceeds more or less in the same manner as that of Stern's algorithm. The complete search algorithm is shown in Algorithm 2. Note that the list $\mathcal{L}$ is stored indexed on the first component of its entries.

Remark 3. From the description of the algorithm, we note that for the error vector $\mathbf{e}$ returned by the algorithm, the matrix vector product $\mathbf{He}^\top$ is of the following form.

$$
\begin{bmatrix} \mathbf{A}_1 & \mathbf{A}_2 & \mathbf{C} & \mathbf{0}_{\ell \times (r-\ell)} \\ \mathbf{B}_1 & \mathbf{B}_2 & \mathbf{D} & \mathbf{I}_{(r-\ell)} \end{bmatrix}
\begin{bmatrix} \mathbf{e}_1^\top \\ \mathbf{e}_2^\top \\ \mathbf{0}_{\ell-\lambda}^\top \\ \mathbf{e}_3^\top \end{bmatrix}.
\tag{6}
$$

Note that there are $\ell - \lambda$ positions where $\mathbf{e}$ has the value 0.

Algorithm 2: The Srch procedure.

Input: $(\mathbf{A}_1, \mathbf{A}_2, \mathbf{B}_1, \mathbf{B}_2, \mathbf{u}, \mathbf{v}, w, p, \delta)$ (see (4) and (5))
Output: (yes, $\mathbf{e}$) such that $\mathbf{He}^\top = \mathbf{s}^\top$ and $\mathsf{wt}(\mathbf{e}) = w$, or (no, $\perp$).
1 $c \leftarrow k + \lambda$; $L_1 \leftarrow \binom{\lfloor c/2 \rfloor}{\lfloor p/2 \rfloor}$
2 $\mathcal{L} \leftarrow ()$; $i = 0$
3 **for** $S \subseteq [\lfloor c/2 \rfloor]$ *with* $\#S = \lfloor p/2 \rfloor$ **do**
4   **if** $i < \lceil L_1^\delta \rceil$ **then**
5          $\mathbf{e}_1 = \chi(S)$; $\mathbf{a}_1^\top = \mathbf{A}_1 \mathbf{e}_1^\top$; $\mathbf{b}_1^\top = \mathbf{B}_1 \mathbf{e}_1^\top$
6          $\mathcal{L} \leftarrow \mathcal{L} \cup \{(\mathbf{a}_1, \mathbf{b}_1, S)\}$
7          $i \leftarrow i + 1$
8   **else**
9   break
10 **for** $T \subseteq [\lceil c/2 \rceil]$ *with* $\#T = \lceil p/2 \rceil$ **do**
11     $\mathbf{e}_2 = \chi(T)$; $\mathbf{a}_2^\top = \mathbf{A}_2 \mathbf{e}_2^\top$; $\mathbf{b}_2^\top = \mathbf{B}_2 \mathbf{e}_2^\top$
12   **for** *all* $(\mathbf{a}_1, \mathbf{b}_1, S) \in \mathcal{L}$ *such that* $\mathbf{a}_1 + \mathbf{u} = \mathbf{a}_2$ **do**
13   **if** $\mathsf{wt}(\mathbf{b}_1 + \mathbf{b}_2 + \mathbf{v}) = w - p$ **then**
14             $\mathbf{e}_1 = \chi(S)$; $\mathbf{e}_3 = \mathbf{b}_1 + \mathbf{b}_2 + \mathbf{v}$
15   Return (yes, $[\mathbf{e}_1 | \mathbf{e}_2 | \mathbf{0}_{\ell-\lambda} | \mathbf{e}_3]$)

16 Return (no, $\perp$)

Correctness. It is easy to check that any solution returned by Algorithm 1 is correct.

Special Cases. If we take $\delta = 1$, then we obtain the generalisation of Stern's algorithm by Bernstein and Chou [4] (see Remark 1). If we take $\delta = 1$ and $\lambda = 0$, then we essentially get back Stern's algorithm, and if we take $\delta = 1$ and $\lambda = \ell$, then we essentially obtain Dumer's algorithm.

3.1 Further Generalisation

The matrix $\mathbf{A}$ has been divided into $\mathbf{A}_1$ and $\mathbf{A}_2$, where $\mathbf{A}_1$ (resp. $\mathbf{A}_2$) has $\lfloor (k+\lambda)/2 \rfloor$ (resp. $\lceil (k+\lambda)/2 \rceil$) columns. We may instead let $\mathbf{A}_1$ and $\mathbf{A}_2$ to have κ_1 and κ_2 columns respectively, where κ_1 and κ_2 are two parameters which are non-negative integers satisfying $\kappa_1 + \kappa_2 = k + \lambda$. Correspondingly, we divide $\mathbf{B}$ into matrices $\mathbf{B}_1$ and $\mathbf{B}_2$ having κ_1 and κ_2 columns respectively. Let p_1 and p_2 be two additional parameters which are non-negative integers such that $p_1 + p_2 = p$.

The generalisation is the following. The first part of the search step will consider column combinations of $\mathbf{A}_1$ taken p_1 at a time and the second part will consider column combinations of $\mathbf{A}_2$ taken p_2 at a time. The description of the search step given in Algorithm 2 can be easily modified to give effect to this generalisation: replace $\lfloor c/2 \rfloor$ by κ_1, $\lceil c/2 \rceil$ by κ_2, $\lfloor p/2 \rfloor$ by p_1 and $\lceil p/2 \rceil$ by p_2.

With the above generalisation, we obtain a *unified* algorithm which provides as special cases all the algorithms from Prange's to Dumer's. It is clear that the generalised Stern's algorithm is obtained by taking $\kappa_1 = \lfloor (k+\lambda)/2 \rfloor$, $\kappa_2 = \lceil (k+\lambda)/2 \rceil$, $p_1 = \lfloor p/2 \rfloor$ and $p_2 = \lceil p/2 \rceil$. Prange's algorithm is obtained by setting $\ell = \lambda = \kappa_1 = p = p_1 = p_2 = 0$, $\delta = 1$ and $\kappa_2 = k$; Lee and Brickell's algorithm is obtained by setting $\ell = \lambda = \kappa_1 = p_1 = 0$, $\delta = 1$, $\kappa_2 = k$ and $p_2 = p$; and Leon's algorithm is obtained by setting $\lambda = \kappa_1 = p_1 = 0$, $\delta = 1$, $\kappa_2 = k$ and $p_2 = p$.

3.2 Practical Efficiency Improvements

There are known techniques for improving the practical efficiency of Stern's algorithms. Below we briefly describe two of these techniques which we will use in estimating the expected number of bit operations.

Chase's Sequence: For each choice S, the Srch procedure obtains $\mathbf{e}_1 = \chi(S)$ and computes $\mathbf{a}_1^\top = \mathbf{A}_1 \mathbf{e}^\top$. The latter computation involves adding together $\lfloor p/2 \rfloor$ columns of $\mathbf{A}_1$. This requires a total of $\lfloor p/2 \rfloor - 1$ additions of ℓ-bit vectors. Since L_1 subsets S are considered, the total number of ℓ-bit vector additions is $L_1 \cdot (\lfloor p/2 \rfloor - 1)$. An alternative way to perform the entire computation is to use Chase's sequence (see Section 7.2.1.3 of [15]) as was done in [29,30]. In this technique, the subsets S are generated incrementally, where the next subset is obtained from the present subset by removing one element and including a new one. So from the vector $\mathbf{a}_1$ corresponding to the present subset, the vector $\mathbf{a}_1$ corresponding to the next subset can be obtained using exactly two ℓ-bit vector additions. As a result, the total number of ℓ-bit vector additions required for all the subsets becomes $2 \cdot L_1$. This is an improvement over the naive method if $p > 5$. Similar efficiency improvements are obtained for the computations of $\mathbf{b}_1$, $\mathbf{a}_2$ and $\mathbf{b}_2$. Even though the use of Chase's sequence is advantageous only for $p > 5$, for the sake of obtaining a single expression for the time complexity, we will assume $2 \cdot L_1$ vector additions are required even for $p < 5$, and similarly for the computations of $\mathbf{b}_1$, $\mathbf{a}_2$ and $\mathbf{b}_2$. This makes the time complexity estimates slightly worse for $2 \le p \le 5$.

Early Abort: The final check for a solution is to compare the weight of $\mathbf{x} = \mathbf{b}_1 + \mathbf{b}_2 + \mathbf{v}$ with $w - p$. Note that the length of $\mathbf{x}$ is $r - \ell$, which in general is substantially greater than $w - p$. This observation forms the basis of the technique of early abort in [6]. Instead of first computing $\mathbf{x}$ and then comparing its weight to $w - p$, it is faster to compute $\mathbf{x}$ incrementally and abort once the weight of the partially computed $\mathbf{x}$ exceeds $w - p$. If $\mathbf{x}$ does not correspond to a solution, it is reasonable to assume that it will behave like a random binary string of length $r - \ell$. So the first $2(w - p + 1)$ positions is likely to have weight $w - p + 1$ and such a vector $\mathbf{x}$ can be discarded without computing the other bits. For most vectors $\mathbf{x}$, this brings down the cost of checking $\mathsf{wt}(\mathbf{x}) = w - p$ from $r - \ell$ bit operations to an expected number $2(w - p + 1)$ of bit operations.

There are several other sophisticated techniques to improve both the linear algebra step and the search step of Stern's algorithm [4,6]. All of these techniques also apply to the generalised Stern's algorithm. For the sake of simplicity we do not include the effect of these techniques in the present analysis. Nonetheless, time estimates for the variants of Classic McEliece obtained from our simple model are quite close to the time estimates obtained using the more detailed techniques of [4].

3.3   Time Complexity

We estimate the number of bit operations required by Algorithm 1. The quantity L_1 is defined in Algorithm Srch. Recall that $c = k + \lambda$ and let

$$L_2 = \binom{\lceil c/2 \rceil}{\lceil p/2 \rceil}. \tag{7}$$

Since the algorithm is probabilistic, first we calculate the success probability of the algorithm in a single iteration. We assume that there is one solution $\mathbf{e}_0$ (of weight w) to the ISD instance $(\mathbf{H}_0, \mathbf{s}_0, w)$. By success probability we mean the probability of the event that the algorithm returns this solution. (If there are more than one solutions, then the success probability will be higher.)

We assume that the permutation matrix $\mathbf{P}$ is chosen uniformly and independently in each iteration. So in each iteration, a uniform random permutation is applied to the columns of the parity check matrix $\mathbf{H}_0$. The total number of such permutations is $n!$. Let $\mathcal{P}$ be the set of 'good' permutations, i.e. if any permutation from $\mathcal{P}$ is applied to the columns of $\mathbf{H}_0$, then a solution is obtained in the search step. So the success probability π of the search step of a single iteration is

$$\pi = \frac{\#\mathcal{P}}{n!}. \tag{8}$$

The set $\mathcal{P}$ can be constructed in the following manner. Let $i_1, \ldots, i_w$ be the one-positions of $\mathbf{e}_0$ (i.e. the positions where $\mathbf{e}_0$ is 1). Call the other positions of $\mathbf{e}_0$ to be zero-positions. It is helpful to visualise the construction of the permutations in $\mathcal{P}$ as distributing the one-positions and the zero-positions to the cells of an array of length n. Distribute the first $\lfloor p/2 \rfloor$ one-positions to a subset of the cells $1, \ldots, \lfloor c/2 \rfloor$ of size $\lfloor p/2 \rfloor$ in a manner such that these cells form some subset S in $\mathcal{L}$ (there are $\#\mathcal{L}$ ways to make this distribution); distribute the next $\lceil p/2 \rceil$ one-positions to some subset of the cells $(\lfloor c/2 \rfloor +1), \ldots, c$ of size $\lceil p/2 \rceil$ (there are L_2 ways to make this distribution); distribute the remaining $w - p$ one-positions to some subset of the cells $(c + 1), \ldots, n$ of size $w - p$ (there are $\binom{n-k-\ell}{w-p}$ ways to make this distribution); and then fill the remaining $n - w$ cells with the zero-positions in some particular order. This fixes the positions for the one-positions and the zero-positions in the array. This fixing can be done in

$$\#\mathcal{L} \cdot L_2 \cdot \binom{n - k - \ell}{w - p}$$

ways. Now permute the cells filled with the one-positions among themselves and also permute the cells filled with the zero-positions among themselves, which can be done in $w!(n - w)!$ ways. So the size of $\mathcal{P}$ is

$$\#\mathcal{P} = \#\mathcal{L} \cdot L_2 \cdot \binom{n - k - \ell}{w - p} \cdot w!(n - w)!. \tag{9}$$

Algorithm Srch ensures that the size of $\mathcal{L}$ is $\lceil L_1^\delta \rceil$. Using (8) and (9), we have

$$\pi = \frac{\lceil L_1^\delta \rceil \cdot L_2 \cdot \binom{n-k-\ell}{w-p}}{\binom{n}{w}}. \tag{10}$$

Let N be the number of iterations required to obtain success. Then N follows a geometric distribution with parameter π. Consequently,

$$\mathbb{E}[N] = \frac{1}{\pi}. \tag{11}$$

Let X_i be the random variable whose value is given by the number of bit operations performed in the i-th iteration. The total number of bit operations is $X_1 + X_2 + \cdots + X_N$. Let T be the expected value of the total number of bit operations. Under the heuristic assumption that the X_i's are independent and identically distributed, and N is independent of the X_i's, by Wald's equation (see Page 300 of [22]),

$$T = \mathbb{E}[X_1 + X_2 + \cdots + X_N] = \mathbb{E}[X_1] \cdot \mathbb{E}[N]. \tag{12}$$

Next we obtain an estimate for $\mathbb{E}[X_1]$, i.e. the expected number of bit operations in each iteration. This has two components, the number of bit operations due to linear algebra step and the number of bit operations due to the search step. Denoting by T_{LA} and T_{SR} the expected number of bit operations required for linear algebra and search steps respectively, we have

$$\mathbb{E}[X_1] = T_{\mathrm{LA}} + T_{\mathrm{SR}}. \tag{13}$$

Using (11), (12) and (13), we obtain

$$T = \frac{1}{\pi} \cdot (T_{\mathrm{LA}} + T_{\mathrm{SR}}) = \frac{\binom{n}{w}}{\lceil L_1^\delta \rceil \cdot L_2 \cdot \binom{n-k-\ell}{w-p}} \cdot (T_{\mathrm{LA}} + T_{\mathrm{SR}}). \tag{14}$$

Remark 4. The above analysis ignores the effect of E_1, i.e. the event that the linear algebra step succeeds (see Remark 2). We briefly consider this effect. Let E_2 be the event that the search step succeeds and so $\pi = \Pr[E_2]$. Let M be a random variable whose value is the number of iterations required by Algorithm 1 to achieve success. For $i = 1, \ldots, M$, let T_i be the binary valued random variable which takes the value 1 if and only if E_1 occurs in the i-th step. So $\mathbb{E}[T_i] = \pi_1$. Let $N = \sum_{i=1}^{M} T_i$, i.e. N is the number of times the search step is repeated until success is obtained (i.e. E_2 occurs), and so $\mathbb{E}[N] = 1/\pi$. Hence, by an application of Wald's equation, $\mathbb{E}[M] = (\pi \cdot \pi_1)^{-1}$. Let Y_i and Z_i be random variables whose values are the numbers of bit operations required for the linear algebra and the search step respectively. As above, let X_i be the number of bit operations required in the i-th step, and so $X_i = Y_i + T_i Z_i$. Note that $\mathbb{E}[Y_i] = T_{\mathrm{LA}}$ and $\mathbb{E}[Z_i] = T_{\mathrm{SR}}$. Letting T denote the expected value of the total number of bit operations, we have $T = \mathbb{E}[\sum_{i=1}^{M} X_i] = \mathbb{E}[\sum_{i=1}^{M} Y_i] + \mathbb{E}[\sum_{i=1}^{M} T_i Z_i]$. Applying Wald's equation separately to the two terms and heuristically assuming that T_i and Z_i are independent, we obtain $T = \pi^{-1} \cdot (T_{\mathrm{LA}} \cdot \pi_1^{-1} + T_{\mathrm{SR}})$. Note the difference with the expression for T given by (14). This difference, however, does not cause noticeable difference in the concrete time estimates and so for simplicity, we consider T to be given by (14).

Next we describe estimates of T_{LA} and T_{SR}. If $\lambda \leq 0$, then the last ℓ columns of the matrices $\mathbf{C}$ and $\mathbf{D}$ in (4) are $\mathbf{I}_\ell$ and the all-zero matrices respectively. This corresponds to the row operations required in Stern's algorithm. For $0 < \lambda \leq \ell$, the matrices $\mathbf{C}$ and $\mathbf{D}$ are smaller and the linear algebra step possibly requires less number of bit operations. For simplicity of analysis, we assume that the number of bit operations required for the linear algebra step is equal to that of Stern's algorithm. Following [6], we estimate the expected number of bit operations required for the linear algebra step in Stern's algorithm to be

$$T_{\mathrm{LA}} = \frac{k^2 (n-k)(n-k-1)(3n-k)}{4n^2}. \tag{15}$$

In the search step, bit operations are required to compute the quantities $\mathbf{a}_1$, $\mathbf{a}_2$, $\mathbf{b}_1$, $\mathbf{b}_2$ and $\mathbf{b}_1 + \mathbf{b}_2 + \mathbf{v}$. Using the early abort technique (see Sect. 3.2) requires computing only the first $2(w - p + 1)$ bits of the last three quantities. Using Chase's sequence (again see Sect. 3.2) the computation of all the $\lceil L_1^\delta \rceil$ $\mathbf{a}_1$'s and $\mathbf{b}_1$'s require a total of $(2\ell + 4(w - p + 1))\lceil L_1^\delta \rceil$ bit operations. Similarly, using Chase's sequence the computation of all the L_2 $\mathbf{a}_2$'s and $\mathbf{b}_2$'s require a total of $(2\ell + 4(w - p + 1))L_2$ bit operations. Finally, we consider the number of bit operations in the collision phase. We make the heuristic assumption that the $\mathbf{a}_1$'s that arise due to the different choices of S are independent and uniformly distributed ℓ-bit strings. So for any fixed ℓ-bit string $\mathbf{x}$, the probability it arises as $\mathbf{a}_1$ due to any particular choice of S is equal to $1/2^\ell$; consequently, the total number of times $\mathbf{x}$ occurs as a first component in the list $\mathcal{L}$ follows the binomial distribution with parameters $1/2^\ell$ and $\#\mathcal{L}$. So the expected number of times $\mathbf{x}$ occurs as a first component in the list $\mathcal{L}$ is $\#\mathcal{L}/2^\ell = \lceil L_1^\delta \rceil / 2^\ell$. It then follows that for each of the L_2 $\mathbf{a}_2$'s (which are also ℓ-bit strings) generated in the collision phase, on an average there will be about $\lceil L_1^\delta \rceil / 2^\ell$ $\mathbf{a}_1$'s in $\mathcal{L}$ such that the condition $\mathbf{a}_1 + \mathbf{u} = \mathbf{a}_2$ holds. So the computation of $\mathbf{b}_1 + \mathbf{b}_2 + \mathbf{v}$ has to be done a total of about $L_2 \cdot (\lceil L_1^\delta \rceil / 2^\ell)$ times. Using the early abort technique, this requires about $L_2 \cdot (\lceil L_1^\delta \rceil / 2^\ell) \cdot (2(w - p + 1))$ bit operations. Putting the calculations together, we have

$$
\begin{aligned}
T_{\mathrm{SR}} \approx\; & (2\ell + 4(w - p + 1))\lceil L_1^\delta \rceil \\
& + (2\ell + 4(w - p + 1))L_2 + L_2 \cdot (\lceil L_1^\delta \rceil / 2^\ell) \cdot (2(w - p + 1)).
\end{aligned}
\tag{16}
$$

3.4 Memory Complexity

An instance of the SDP is a triplet $(\mathbf{H}_0, \mathbf{s}_0, w)$. So any ISD algorithm has to store the matrix $\mathbf{H}_0$ and the syndrome $\mathbf{s}_0$. Let M_{mat} be the number of bits required to store $\mathbf{H}_0$ and $\mathbf{s}_0$. Then

$$
M_{\mathrm{mat}} = (n - k)(n + 1). \tag{17}
$$

The additional memory requirement of Algorithm 1 arises from the memory requirement of Algorithm Srch, which needs to store the list $\mathcal{L}$. The size of $\mathcal{L}$ is $\lceil L_1^\delta \rceil$. Each entry of $\mathcal{L}$ is of the form $(\mathbf{a}_1, \mathbf{b}_1, S)$, where $\mathbf{a}_1$ is an ℓ-bit vector, $\mathbf{b}_1$ is an $(r - \ell)$-bit vector and M is a subset of $[\lfloor (k + \lambda)/2 \rfloor]$ of size $\lfloor p/2 \rfloor$. Since we consider the early abort technique, only the first $2(w - p + 1)$ bits of $\mathbf{b}_1$ are stored. Let M_{lst} be the number of bits required to store $\mathcal{L}$. Then

$$
M_{\mathrm{lst}} = \lceil L_1^\delta \rceil \cdot (\ell + 2(w - p + 1) + \lfloor p/2 \rfloor \cdot \lceil \log_2 \lfloor (k + \lambda)/2 \rfloor \rceil). \tag{18}
$$

The total memory required by Algorithm 1 is M, where

$$
M = M_{\mathrm{mat}} + M_{\mathrm{lst}}. \tag{19}
$$

3.5 Cost of Memory Access

If M is large, then accessing memory will require non-negligible time. The logarithmic memory access cost model has been suggested in previous works [2,11]. In this model, the time estimate is increased by a factor which is equal to the logarithm to the base two of the memory estimate. So the time estimate T_{ma} taking logarithmic memory access cost into consideration is given by

$$T_{\text{ma}} = T \cdot \log_2 M. \tag{20}$$

Remark 5. Estimating the cost of memory access using (20) is rather adhoc. For one thing it assumes that each bit operation requires a memory access which is not the case. Secondly, a large fraction of the bit operations are on the matrix **H** and not on the list $\mathcal{L}$. Compared to **H**, the list $\mathcal{L}$ can require much more memory to store. So assigning the same cost of memory access to memory operations on **H** and $\mathcal{L}$ may not be justified.

Remark 6. A Boolean circuit model based analysis of time estimates of ISD algorithms has been performed in [4]. Such an analysis inherently incorporates cost of memory access which would otherwise be ignored in an analysis assuming constant time for memory access. Note, however, that incorporating logarithmic memory access cost as in (20) captures memory access cost in a manner which is different from that in [4].

4   A Set of Effective TMTO Points

We describe what we mean by a set of effective TMTO points and a procedure to compute such a set. Our initial description is based on Algorithm 1. Later we mention how the procedure can be applied to any ISD algorithm.

For fixed values of n, k and w, it is possible to evaluate the expressions for T and M given by (14) and (19) respectively for every possible choice of the values of the parameters ℓ, p, λ and δ. This leads to a large number of TMTO points (T, M). For example, using the range of parameters given by (23) in Sect. 5 leads to more than 3 million TMTO points. There are, however, large clusters of values of T and M. As an example, for the variant `mceliece-3488-064` of Classic McEliece, we have $n = 3488$, $k = 2720$ and $w = 64$. In this case, the minimum value of $\log_2 T$ required by Algorithm 1 is 147.70988 and there are 819 values which are less than 148, the largest of which is 147.99983. Similar clustering also occurs for $\log_2 M$. Based on these observations, we define two time complexities T and T' to be equivalent if $\lceil \log_2 T \rceil = \lceil \log_2 T' \rceil$, and two memory complexities M and M' to be equivalent if $\lceil \log_2 M \rceil = \lceil \log_2 M' \rceil$. Extending to TMTO points, we define two TMTO points (T, M) and (T', M') to be equivalent if $\lceil \log_2 T \rceil = \lceil \log_2 T' \rceil$ and $\lceil \log_2 M \rceil = \lceil \log_2 M' \rceil$. So the time and memory complexities of a TMTO point differ from the respective complexities of an equivalent TMTO point by factors which are less than 2.

For fixed values of n, k and w, let $\mathcal{T}_0$ be the set of all tuples

$$(\lceil \log_2 T \rceil, \lceil \log_2 M \rceil, \ell, p, \lambda, \delta, \log_2 T, \log_2 M) \tag{21}$$

corresponding to a pre-determined range of values of the parameters ℓ, p, λ and δ. The list $\mathcal{T}_0$ captures all the TMTO points arising from the chosen range of values of the parameters. We say that a TMTO point (T, M) is represented in $\mathcal{T}_0$ if $(\log_2 T, \log_2 M)$ occurs as the last two components of some tuple in $\mathcal{T}_0$. Let $\mathcal{V}$ be a non-empty set of TMTO points represented in $\mathcal{T}_0$ satisfying the following two properties.

1. *Minimality:* If $(T, M) \neq (T', M')$ are in $\mathcal{V}$, then $\lceil \log_2 T \rceil \neq \lceil \log_2 T' \rceil$, $\lceil \log_2 M \rceil \neq \lceil \log_2 M' \rceil$, and either $\lceil \log_2 T \rceil > \lceil \log_2 T' \rceil$ or $\lceil \log_2 M \rceil > \lceil \log_2 M' \rceil$.
2. *Completeness:* If (T', M') is represented in $\mathcal{T}_0$, then there is a point (T, M) in $\mathcal{V}$ such that $\lceil \log_2 T \rceil \leq \lceil \log_2 T' \rceil$ and $\lceil \log_2 M \rceil \leq \lceil \log_2 M' \rceil$.

We say that the set $\mathcal{V}$ is a *set of effective TMTO points* with respect to the chosen range of parameters. A consequence of the first condition is that two distinct points in $\mathcal{V}$ are inequivalent. Let

$$\mathsf{CL}(\mathcal{V}) = \{(\lceil \log_2 T \rceil, \lceil \log_2 M \rceil) : (T, M) \in \mathcal{V}\}. \tag{22}$$

Proposition 1. *If $\mathcal{V}$ and $\mathcal{V}'$ are two sets of effective TMTO points for the same ranges of values of the parameters, then $\mathsf{CL}(\mathcal{V}) = \mathsf{CL}(\mathcal{V}')$.*

Proposition 1 shows that a set of effective TMTO points uniquely captures the entire landscape of all TMTO points up to loss of precision by factors which are less than 2. Later we will see examples which show that a set of effective TMTO points can be much smaller than the size of all TMTO points.

We describe a method to compute a set of effective TMTO points. The idea is to progressively process a list of tuples $\mathcal{T}$ which is initially set to be equal to $\mathcal{T}_0$. The following steps are then performed successively on $\mathcal{T}$.

1. *Sorting:* Perform an ascending order sort of the tuples in $\mathcal{T}$, where the usual lexicographic ordering of tuples is assumed, i.e. a tuple is considered to be less than another if for some $i \geq 1$, the first $i - 1$ components of the two tuples are equal and the i-th component of the first tuple is less than that of the second.
2. *First filtering:* Perform a filtering on $\mathcal{T}$ to ensure that for any particular value of $\lceil \log_2 T \rceil$, only the first tuple with the given value is retained while all other tuples with the same value of $\lceil \log_2 T \rceil$ are dropped.
3. *Second filtering:* Perform a filtering on $\mathcal{T}$ to ensure that for any particular value of $\lceil \log_2 M \rceil$, only the first tuple with the given value is retained while all other tuples with the same value of $\lceil \log_2 M \rceil$ are dropped.
4. *Pruning:* Discard all tuples from $\mathcal{T}$ starting from the point where $\lceil \log_2 T \rceil$ increases but $\lceil \log_2 M \rceil$ does not decrease.

Let $\mathcal{T}_1$ be the final state of $\mathcal{T}$ and let $\mathcal{U}$ be the set of all TMTO points (T, M) which are represented in $\mathcal{T}_1$. We have the following result.

Proposition 2. *$\mathcal{U}$ is a set of effective TMTO points.*

We used a simple SAGE [26] code to compute a set of effective TMTO points for Algorithm 1 for the range of parameters in (23). Setting $\delta = 1$ in the code provides a set of effective TMTO points for the Bernstein-Chou generalisation of Stern's algorithm; setting $\lambda = 0$ and $\delta = 1$ in the code provides a set of effective TMTO points for Stern's algorithm; setting $\lambda = \ell$ and $\delta = 1$ in the code provides a set of effective TMTO points for Dumer's algorithm.

The discussion above for obtaining a set of effective TMTO points was with reference to Algorithm 1. Changing the procedure to any other ISD algorithm is simply to change the parameters and to use the appropriate expressions (or algorithms) to compute the time and memory complexities of the algorithm. This leads to changing (21) and affects the construction of the initial state $\mathcal{T}_0$ of $\mathcal{T}$. The sorting, filtering and pruning steps are applied to $\mathcal{T}$ as described above. A set $\mathcal{U}$ of effective TMTO points (with respect to the chosen range of values of the parameters) can then be defined from the final state $\mathcal{T}_1$ of $\mathcal{T}$ in the same manner as described above.

4.1   Comparison to Previous TMTO Analysis of ISD Algorithms

The TMTO analysis considered in [12] is of the following type. Fix an upper bound M on the memory complexity and then obtain the minimum time required by the algorithm utilising at most M bits of memory. Let us call this a memory bounded approach to TMTO analysis. Such an approach provides less information in comparison to the analysis using the notion of an effective set of TMTO points. We illustrate this using an example. Consider Table 2a which provides certain TMTO points for mceliece-3488-064 achieved by Algorithm 1 using the notion of effective set of TMTO points. Now consider the memory bounded analysis. Suppose we fix the memory bound M to be 2^{50} and ask what is the minimum time that can be achieved by Algorithm 1 utilising at most 2^{50} bits of memory? From Table 2a the answer is $2^{147.93}$. However, from the table we also find that the time $2^{147.93}$ can be achieved using $2^{44.55}$ bits of memory. Similarly, if we fix M to be 2^{40}, then the memory bounded analysis will provide minimum time estimate of $2^{148.93}$, while the analysis based on effective set of TMTO points will provide the additional information that the time estimate of $2^{148.93}$ can be achieved using $2^{34.63}$ bits of memory. In both the above cases, the actual memory requirements $2^{44.55}$ or $2^{34.63}$ are lower than the upper bounds 2^{50} and 2^{40} respectively. So while the memory bounded analysis obtains the minimum time subject to an upper bound on the memory, it does not yield the actual memory that is required to achieve the minimum time. In contrast, the analysis based on the notion of effective set of TMTO points provides for each time estimate the minimum memory required to achieve the time estimate. More generally, due to the completeness property an effective set of TMTO points captures (up to a minor loss in precision) the entire landscape of TMTO points. In particular, for any TMTO point (T', M') obtained using a memory bounded analysis, the completeness property assures us that there is a TMTO point (T, M) in an effective set of TMTO points such that $\lceil \log_2 T \rceil \leq \lceil \log_2 T' \rceil$ and $\lceil \log_2 M \rceil \leq \lceil \log_2 M' \rceil$.

5   Application to Classic McEliece

The NIST call for proposals [23] for post-quantum cryptosystems outlines five categories. Of these, Categories 1, 3 and 5 require cryptosystems to be secure under attacks using 2^{143}, 2^{207} and 2^{272} classical gates respectively.

The expected security categories of the five variants are given in Section 7 of the Classic McEliece specification [5]. For the five variants, our abbreviations of the names, the values of n, k and w for the five variants, and their categories are shown in Table 1. Section 2.2.3 of the Classic McEliece specification [5], states that the public key is an $r \times k$ binary matrix. Table 1 shows $\log_2 P$, where $P = r \cdot k$ is the size of the public key in bits. An ISD instance is given by $(\mathbf{H}_0, \mathbf{s}_0, w)$. So M_{mat} bits are required to store $\mathbf{H}_0$ and $\mathbf{s}_0$. The values of $\log_2 M_{\text{mat}}$ for the variants of the Classic McEliece are shown in Table 1.

Table 1. Parameters for Classic McEliece and the corresponding values of $\log_2 P$ and $\log_2 M_{\text{mat}}$.

category	name	n	k	w	$\log_2 P$	$\log_2 M_{\text{mat}}$
1	mceliece-3488-064 (m3488)	3488	2720	64	20.99	21.35
3	mceliece-4608-096 (m4608)	4608	3360	96	22.00	22.46
5	mceliece-6688-128 (m6688)	6688	5024	128	22.00	23.41
	mceliece-6960-119 (m6960)	6960	5413	119	22.00	23.36
	mceliece-8192-128 (m8192)	8192	6528	128	23.37	23.70

For Algorithm 1, instead of using the maximum possible ranges of values of parameters given by (3), we have restricted to the following ranges of values of the parameters. The concrete results obtained using these choices indicate that there are no interesting TMTO points for values of parameters outside these ranges.

$$0 \leq \ell \leq 100, \quad 2 \leq p \leq 30, \quad -\ell \leq \lambda \leq \ell, \quad \delta = 1 - 0.025 \cdot i, \; i = 0, \dots, 12. \;(23)$$

The total number of choices of values for the parameters ℓ, p, λ and δ in the above ranges is 3712800. We have used a simple SAGE code to compute the sets of effective TMTO points for the variants of Classic McEliece corresponding to Algorithm 1 for both the cases where memory access cost is taken to be constant and the logarithmic memory access cost model is assumed. We have also computed similar sets of effective TMTO points corresponding to Stern's (i.e. with $\delta = 1$ and $\lambda = 0$) and Dumer's (i.e. with $\delta = 1$ and $\lambda = \ell$) algorithms for the ranges of ℓ and p given by (23). For every case, the sizes of the sets of effective TMTO points corresponding to Stern's and Dumer's are the same and the corresponding time complexities are equivalent. The memory complexities are also mostly equivalent, though in a few cases they vary a little. We have similarly computed the sets of effective TMTO points corresponding to the

Bernstein and Chou's (i.e. with $\delta = 1$) algorithm for the ranges of ℓ, p and λ given by (23). Again the sizes of the sets of effective TMTO points are the same as those obtained for Stern's algorithm and the corresponding time complexities are equivalent, while the memory complexities are mostly equivalent and vary a little for a small number of cases.

Remark 7. The generalisation of Algorithm 1 over Stern's algorithm arises from the use of two parameter λ and δ. From the above discussion, we see that it is the parameter δ which provides the non-triviality of the generalisation. If we set $\delta = 1$ (obtaining Bernstein and Chou's algorithm isd1), then we do not obtain any interesting TMTO points beyond what is achieved by Stern's algorithm. It is necessary to allow δ to be less than 1 to explore the TMTO landscape not covered by Stern's algorithm.

The sets of effective TMTO points for the variants of Classic McEliece obtained from Algorithm 1 are shown in Tables 2 and 3. The tables also provide the values of the parameters which achieve the corresponding TMTO points. In these tables, rows marked with (*) indicate that Stern's algorithm achieves TMTO points which are equivalent to the corresponding TMTO points achieved by Algorithm 1. Further, Stern's algorithm does not achieve any TMTO point whose time complexity is equivalent to the time complexity of any row not marked with (*). The values in the tables show that the maximum and minimum values of $\lceil \log_2 T \rceil$ remain the same for Algorithm 1 and Stern's algorithm. For Algorithm 1, $\lceil \log_2 T \rceil$ achieves every value between the maximum and the minimum, while for Stern's algorithm only about half of these values are achieved. So Algorithm 1 provides a finer time/memory trade-off compared to Stern's algorithm. Note that for Algorithm 1 sets of 3712800 TMTO points reduce to sets of 6 to 13 effective TMTO points. This underlines the usefulness of the notion of a set of effective TMTO points.

The minimum memory TMTO points for the five Classic McEliece variants have $\log_2 M$ to be equal to 21.45, 22.54, 23.49, 23.45 and 23.79. Comparing with the values of $\log_2 M_{\text{mat}}$ in Table 1, one may observe that the minimum memory is marginally greater than $\log_2 M_{\text{mat}}$.

It is interesting to observe that there are sharp drops in memory requirement at only a moderate increase in the time complexity. As an example, if we consider the estimates which take memory access cost to be constant, increasing $\log_2 T$ by the amount indicated below leads to $\log_2 M$ dropping by the stated amount.

m3488: $\log_2 T$ from 147.93 to 148.93; $\log_2 M$ from 44.55 to 34.63;

m4608: $\log_2 T$ from 189.98 to 190.80; $\log_2 M$ from 46.22 to 37.70;

m6688: $\log_2 T$ from 265.00 to 265.95; $\log_2 M$ from 57.97 to 48.89;

m6960: $\log_2 T$ from 265.00 to 265.96; $\log_2 M$ from 67.37 to 58.38;

m8192: $\log_2 T$ from 300.00 to 300.96; $\log_2 M$ from 77.99 to 68.96.

This shows that allowing the time complexity to increase by factors which are at most 2 result in the memory requirement dropping by factors varying from 2^8 to 2^{10}. In general, one may observe that the drops in the memory requirement are sharper for smaller values of T and become less sharp for larger values of T. Similar observations hold when we consider the estimates which include logarithmic memory access cost.

We consider the implications of the new TMTO points to the classification in NIST categories of the variants of Classic McEliece. Time estimates of Stern's algorithm from [2] have been used in a discussion [14] regarding the classifications of the five variants. A criticism forwarded against these time estimates was that the corresponding memory requirements are much larger than the size of the public key being attacked.

Since we have obtained the sets of effective TMTO points, we may consider points where the memory size is not much larger than the size of the public key.

Let us first consider m4608. In this case the gate count requirement is 2^{207} and the size of the public key is about 2^{22} bits. The time estimates of all the TMTO points for this variant given by Tables 2a and 2b are below the target 2^{207}. The TMTO point with highest of these time estimates is equal to $(2^{197.99}, 2^{22.52})$ when memory access cost is taken to be constant, and is equal to $(2^{202.00}, 2^{22.54})$ when logarithmic memory access cost is considered. So for m4608, it is possible to achieve time complexity less than 2^{207} by a factor of about 2^9 (for constant memory access cost) and about 2^5 (for logarithmic memory access cost) while requiring memory which is only about 1.5 times the size of the public key.

Let us now consider m6688 and m6960. The NIST gate count requirement is 2^{272} and the size of the public key is about 2^{22} bits. From Table 3a, we see that for m6688, Algorithm 1 has the TMTO points $(2^{269.99}, 2^{29.25})$ and $(2^{271.00}, 2^{27.69})$; and for m6960, Algorithm 1 has the TMTO points $(2^{270.00}, 2^{29.93})$ and $(2^{270.85}, 2^{28.83})$. All of these time estimates are slightly smaller than 2^{272}. The corresponding memory requirements, while still being larger than the size of the public key, are not too large. On the other hand, if we consider logarithmic memory access cost, then with memory requirement less than 2^{30}, we obtain time complexities $2^{274.97}$ and $2^{274.93}$ respectively, both of which are greater than the NIST requirement of 2^{272} by factors which are less than 8.

In contrast to the above, for both m3488 and m8192, there is no TMTO point in Tables 2 and 3, even with constant memory access cost, which has time estimate less than the NIST requirement for classification in the respective categories.

Remark 8. For m4608 and m6688, the minimum gate count estimates obtained in [4] for isd1 (which subsumes Stern's and Dumer's algorithms) are $2^{198.93}$ and $2^{275.41}$ respectively. For m4608, the count of $2^{198.93}$ is lower than the NIST target of 2^{207} while for m6688, the count of $2^{275.41}$ is about 10 times the NIST target of 2^{272}. The methodology used in [4] for obtaining gate count estimates incorporates cost of memory access though in a manner which is different from the logarithmic memory access cost considered here (see Remark 6). From Tables 2b and 2b, the minimum bit complexity estimates for m4608 and m6688 obtained in the present work are not too far from those obtained in [4]. The main difference with the results reported in [4] is that, as discussed above, the bit complexity estimate for m4608 falls below the NIST target and the bit complexity estimate for m6688 (and also m6960) is a little above the NIST target *even* if we restrict the memory requirement to be not too larger than the size of the public key.

Table 2. TMTO points and the corresponding values of the parameters achieved by Algorithm 1 for m3488 and m4608. Stern's algorithm provides equivalent TMTO points for only the rows marked with (*).

		$(\log_2 T, \log_2 M)$	ℓ	p	λ	δ
	(*)	(147.93, 44.55)	36	8	-36	1.000
	(*)	(148.93, 34.63)	27	6	-27	0.950
	(*)	(149.89, 27.15)	18	4	-18	1.000
m3488		(150.97, 25.71)	17	4	-17	0.925
		(151.96, 24.77)	15	4	-14	0.875
		(152.81, 23.90)	14	4	-14	0.825
		(153.99, 22.80)	13	4	-13	0.750
	(*)	(154.98, 21.44)	13	4	-13	0.975
	(*)	(189.98, 46.22)	38	8	-38	1.000
	(*)	(190.88, 37.70)	28	6	-28	0.975
		(191.81, 35.95)	26	6	-26	0.950
	(*)	(192.95, 27.75)	18	4	-18	0.975
m4608		(193.99, 26.76)	16	4	-16	0.925
		(194.85, 25.81)	15	4	-15	0.875
		(195.74, 24.92)	14	4	-14	0.825
		(196.95, 23.81)	13	4	-13	0.750
	(*)	(197.99, 22.52)	7	3	-7	0.975

(a) Constant memory access cost.

		$(\log_2 T_{\mathrm{ma}}, \log_2 M)$	ℓ	p	λ	δ
	(*)	(154.00, 34.69)	27	6	16	0.950
	(*)	(154.94, 26.67)	18	4	-18	0.975
m3488		(155.81, 25.70)	16	4	-16	0.925
		(156.91, 24.76)	14	4	-14	0.875
		(157.94, 23.50)	13	4	-13	0.800
	(*)	(158.93, 21.45)	8	3	-8	1.000
	(*)	(195.94, 36.70)	29	6	-29	0.975
	(*)	(196.98, 35.95)	26	6	-26	0.950
	(*)	(197.94, 27.75)	17	4	-17	0.975
m4608		(198.73, 26.76)	16	4	-16	0.925
		(199.92, 25.80)	14	4	-14	0.875
		(200.97, 24.51)	13	4	-13	0.800
	(*)	(202.00, 22.54)	8	3	-2	1.000

(b) Logarithmic memory access cost.

Table 3. TMTO points and the corresponding values of the parameters achieved by Algorithm 1 for m6688, m6960 and m8192. Stern's algorithm provides equivalent TMTO points for only the rows marked with (*).

		$(\log_2 T, \log_2 M)$	ℓ	p	λ	δ
	(*)	(265.00, 57.97)	50	10	-16	1.000
	(*)	(265.95, 48.89)	39	8	-39	1.000
	(*)	(267.00, 38.82)	31	6	30	0.975
		(268.80, 36.41)	28	6	-28	0.900
m6688	(*)	(269.99, 29.25)	18	4	-18	0.975
		(271.00, 27.69)	18	4	7	0.900
		(271.93, 26.68)	17	4	-17	0.850
		(273.00, 25.76)	15	4	15	0.800
		(273.95, 24.95)	14	4	-14	0.750
	(*)	(274.89, 23.48)	9	3	-9	0.975
	(*)	(265.00, 67.37)	59	12	19	1.000
	(*)	(265.96, 58.38)	48	10	-48	1.000
	(*)	(267.00, 48.29)	39	8	29	0.975
	(*)	(267.91, 39.00)	30	6	-30	0.975
		(268.92, 37.41)	29	6	-29	0.925
m6960	(*)	(270.00, 29.93)	20	4	20	1.000
		(270.85, 28.83)	19	4	-19	0.950
		(271.96, 27.77)	17	4	-17	0.900
		(272.87, 26.75)	16	4	-16	0.850
		(273.80, 25.80)	15	4	-15	0.800
		(274.74, 24.98)	14	4	-14	0.750
	(*)	(275.87, 23.45)	8	3	-8	1.000
	(*)	(300.00, 77.99)	68	14	16	1.000
	(*)	(300.96, 68.96)	58	12	-58	1.000
	(*)	(301.90, 58.55)	50	10	-50	0.975
	(*)	(302.90, 49.38)	40	8	-40	0.975
	(*)	(303.96, 39.88)	31	6	-31	0.975
		(305.00, 38.26)	30	6	-30	0.925
m8192		(305.95, 37.44)	28	6	-28	0.900
	(*)	(306.92, 29.99)	19	4	-19	0.975
		(307.80, 28.89)	18	4	-18	0.925
		(308.72, 27.81)	17	4	-17	0.875
		(309.66, 26.79)	16	4	-16	0.825
		(310.96, 25.85)	14	4	-14	0.775
	(*)	(311.85, 23.79)	10	3	-10	1.000

(a) Constant memory access cost.

		$(\log_2 T_{\mathrm{ma}}, \log_2 M)$	ℓ	p	λ	δ
	(*)	(270.96, 57.91)	49	10	-49	1.000
	(*)	(271.99, 39.55)	31	6	-31	1.000
		(272.96, 37.98)	29	6	-29	0.950
		(273.99, 36.41)	28	6	-28	0.900
m6688	(*)	(274.97, 28.72)	19	4	-19	0.950
		(275.97, 27.68)	17	4	-17	0.900
		(276.83, 26.67)	16	4	-16	0.850
		(278.00, 25.75)	14	4	-7	0.800
		(278.88, 24.95)	13	4	-13	0.750
	(*)	(279.86, 23.45)	8	3	-8	0.950
	(*)	(271.00, 75.99)	68	14	-4	1.000
	(*)	(272.00, 49.33)	40	8	32	1.000
	(*)	(272.94, 39.79)	30	6	-30	1.000
		(274.00, 37.45)	30	6	13	0.925
m6960	(*)	(274.93, 29.91)	20	4	-20	1.000
		(275.91, 28.83)	18	4	-18	0.950
		(276.75, 27.77)	17	4	-17	0.900
		(277.97, 26.75)	15	4	-15	0.850
		(278.84, 25.80)	14	4	-14	0.800
	(*)	(279.89, 23.45)	10	3	-10	1.000
	(*)	(306.00, 86.78)	77	16	46	1.000
	(*)	(307.00, 59.88)	52	10	-9	1.000
	(*)	(307.93, 50.43)	41	8	-41	1.000
	(*)	(309.00, 40.70)	31	6	-31	1.000
		(309.97, 39.88)	29	6	-29	0.975
		(310.90, 37.45)	29	6	-29	0.900
m8192	(*)	(311.82, 29.99)	19	4	-19	0.975
		(312.95, 28.35)	18	4	-18	0.900
		(313.90, 27.81)	16	4	-16	0.875
		(314.77, 26.78)	15	4	-15	0.825
		(315.90, 25.44)	14	4	-14	0.750
	(*)	(316.84, 23.78)	9	3	-9	0.975

(b) Logarithmic memory access cost.

Acknowledgement. We thank the reviewers for their kind comments which have helped in improving the discussion at several points in the paper.

References

1. Aragon, N., et al.: BIKE, Round 4 submission (2022). https://csrc.nist.gov/csrc/media/Projects/post-quantum-cryptography/documents/round-4/submissions/BIKE-Round4.zip. Accessed 9 Aug 2023
2. Baldi, M., Barenghi, A., Chiaraluce, F., Pelosi, G., Santini, P.: A finite regime analysis of information set decoding algorithms. Algorithms **12**(10), 209 (2019)
3. Becker, A., Joux, A., May, A., Meurer, A.: Decoding random binary linear codes in $2^{n/20}$: how $1 + 1 = 0$ improves information set decoding. In: Pointcheval, D., Johansson, T. (eds.) EUROCRYPT 2012. LNCS, vol. 7237, pp. 520–536. Springer, Cham (2012). https://doi.org/10.1007/978-3-642-29011-4_31
4. Bernstein, D.J., Chou, T.: CryptAttackTester: formalizing attack analyses (2023). https://cat.cr.yp.to/papers.html#cryptattacktester. Accessed 9 Aug 2023
5. Bernstein, D.J., et al.: Classic McEliece, Round 4 submission (2022). https://csrc.nist.gov/csrc/media/Projects/post-quantum-cryptography/documents/round-4/submissions/mceliece-Round4.tar.gz. Accessed 9 Aug 2023
6. Bernstein, D.J., Lange, T., Peters, C.: Attacking and defending the McEliece cryptosystem. In: Buchmann, J., Ding, J. (eds.) PQCrypto 2008. LNCS, vol. 5299, pp. 31–46. Springer, Heidelberg (2008). https://doi.org/10.1007/978-3-540-88403-3_3
7. Bernstein, D.J., Lange, T., Peters, C.: Smaller decoding exponents: ball-collision decoding. In: Rogaway, P. (ed.) CRYPTO 2011. LNCS, vol. 6841, pp. 743–760. Springer, Heidelberg (2011). https://doi.org/10.1007/978-3-642-22792-9_42
8. Both, L., May, A.: Decoding linear codes with high error rate and its impact for LPN security. In: Lange, T., Steinwandt, R. (eds.) PQCrypto 2018. LNCS, vol. 10786, pp. 25–46. Springer, Cham (2018). https://doi.org/10.1007/978-3-319-79063-3_2
9. Dumer, I.: On minimum distance decoding of linear codes. In: Proceedings of the 5th Joint Soviet-Swedish International Workshop on Information Theory, pp. 50–52 (1991)
10. Esser, A.: Syndrome decoding estimator. In: Hanaoka, G., Shikata, J., Watanabe, Y. (eds.) PKC 2022, Part I. LNCS, vol. 13177, pp. 112–141. Springer, Cham (2022). https://doi.org/10.1007/978-3-030-97121-2_5
11. Esser, A., May, A., Zweydinger, F.: Mceliece needs a break - solving McEliece-1284 and quasi-cyclic-2918 with modern ISD. In: Dunkelman, O., Dziembowski, S. (eds.) EUROCRYPT 2022, Part III, vol. 13277, pp. 433–457. Springer, Cham (2022). https://doi.org/10.1007/978-3-031-07082-2_16
12. Esser, A., Zweydinger, F.: New time-memory trade-offs for subset sum - improving ISD in theory and practice. In: Hazay, C., Stam, M. (eds.) EUROCRYPT 2023, Part V. LNCS, vol. 14008, pp. 360–390. Springer, Cham (2023). https://doi.org/10.1007/978-3-031-30589-4_13
13. Finiasz, M., Sendrier, N.: Security bounds for the design of code-based cryptosystems. In: Matsui, M. (ed.) ASIACRYPT 2009. LNCS, vol. 5912, pp. 88–105. Springer, Heidelberg (2009). https://doi.org/10.1007/978-3-642-10366-7_6
14. Fleming, K.: ROUND 3 OFFICIAL COMMENT: Classic McEliece, started on November 10, 2020, 7:05:28 AM (2020). https://groups.google.com/a/list.nist.gov/g/pqc-forum/c/EiwxGnfQgec?pli=1. Accessed 9 Aug 2023

15. Knuth, D.E.: Art of Computer Programming, Volume 4A, The Combinatorial Algorithms, Part 1
16. Lee, P.J., Brickell, E.F.: An observation on the security of McEliece's public-key cryptosystem. In: Günther, C.G. (ed.) EUROCRYPT 1988. LNCS, vol. 330, pp. 275–280. Springer, Heidelberg (1988). https://doi.org/10.1007/3-540-45961-8_25
17. Leon, J.S.: A probabilistic algorithm for computing minimum weights of large error-correcting codes. IEEE Trans. Inf. Theory **34**(5), 1354–1359 (1988)
18. May, A., Meurer, A., Thomae, E.: Decoding random linear codes in $\tilde{O}(2^{0.054n})$. In: Lee, D.H., Wang, X. (eds.) ASIACRYPT 2011. LNCS, vol. 7073, pp. 107–124. Springer, Heidelberg (2011). https://doi.org/10.1007/978-3-642-25385-0_6
19. May, A., Ozerov, I.: On computing nearest neighbors with applications to decoding of binary linear codes. In: Oswald, E., Fischlin, M. (eds.) EUROCRYPT 2015, Part I. LNCS, vol. 9056, pp. 203–228. Springer, Heidelberg (2015). https://doi.org/10.1007/978-3-662-46800-5_9
20. McEliece, R.J.: A public-key cryptosystem based on algebraic coding theory. Jet Propulsion Laboratory DSN Progress Report 42-44 (1978). http://ipnpr.jpl.nasa.gov/progressreport2/42-44/44N.PDF
21. Melchor, C.A., et al.: HQC, Round 4 submission (2022). https://csrc.nist.gov/csrc/media/Projects/post-quantum-cryptography/documents/round-4/submissions/HQC-Round4.zip. Accessed 9 Aug 2023
22. Mitzenmacher, M., Upfal, E.: Probability and Computing: Randomized Algorithms and Probabilistic Analysis. Cambridge University Press, Cambridge (2005)
23. NIST. Call for proposals (2016). https://csrc.nist.gov/CSRC/media/Projects/Post-Quantum-Cryptography/documents/call-for-proposals-final-dec-2016.pdf. Accessed 9 Aug 2023
24. NIST. Round 4 submissions (2022). https://csrc.nist.gov/Projects/post-quantum-cryptography/round-4-submissions. Accessed 9 Aug 2023
25. Prange, E.: The use of information sets in decoding cyclic codes. IRE Trans. Inf. Theory **8**(5), 5–9 (1962)
26. Stein, W.A., et al.: Sage Mathematics Software (Version 8.1), The Sage Development Team (2017). http://www.sagemath.org. Accessed 9 Aug 2023
27. Stern, J.: A method for finding codewords of small weight. In: Cohen, G.D., Wolfmann, J. (eds.) Coding Theory and Applications. LNCS, vol. 388, pp. 106–113. Springer, Heidelberg (1988). https://doi.org/10.1007/BFb0019850
28. Torres, R.C., Sendrier, N.: Analysis of information set decoding for a sub-linear error weight. In: Takagi, T. (ed.) PQCrypto 2016. LNCS, vol. 9606, pp. 144–161. Springer, Cham (2016). https://doi.org/10.1007/978-3-319-29360-8_10
29. Vasseur, V.: Information set decoding implementation (2019). https://github.com/vvasseur/isd/blob/master/src/dumer.c. Accessed 9 Aug 2023
30. Zweydinger, F.: Decoding (2022). https://github.com/FloydZ/decoding/blob/master/src/dumer.h. Accessed 9 Aug 2023

Practical Aspects of Vertical Side-Channel Analyses on HMAC-SHA-2

Lukas Vlasak[1]([✉]) [iD], Antoine Bouvet[1] [iD], and Sylvain Guilley[2,3] [iD]

[1] Secure-IC S.A.S., Cesson-Sévigné, France
{lukas.vlasak,antoine.bouvet}@secure-ic.com
[2] Secure-IC S.A.S., Paris, France
sylvain.guilley@secure-ic.com
[3] Télécom Paris, Paris, France
sylvain.guilley@telecom-paristech.fr

Abstract. Cryptographic hashing with secret key is widely used for message authentication, *e.g.*, in the popular HMAC protocol. This algorithm is a suitable target for vertical Side-Channel Analyses. In practice, such analyses involve various attacks on intermediate computations to recover security critical information. If the hash function is SHA-2, an attacker compares an expected leakage model of modular addition and logical AND operations, with the side-channel activity. In both cases, there are practical constraints that lead to errors in correlation-based analyses and make security assessment difficult. This becomes even more apparent when the target of evaluation is not theoretical values but a real device. We discuss these practical aspects and show a real evaluation on an embedded implementation.

Keywords: Side-Channel Analysis · SHA-2 · HMAC · Leakage model · Correlation Power Analysis · Normalized Inter-Class Variance

1 Introduction

Side-Channel Analyses (SCAs) have been extensively discussed on block ciphers. In this context, the Advanced Encryption Standard (AES) is considered a paragon representing symmetrical cryptography. Nevertheless, other algorithms also use secret keys and should therefore be equally protected. The Keyed-Hash Message Authentication Code (HMAC) (RFC 2104) is one of them. This protocol has multiple applications such as authentication or key derivation. It is a symmetric primitive that uses the secret key in combination with a cryptographic hash function to ensure authenticity and integrity of the messages. Interestingly, owing to the nature of the hash functions, the exploitation turns out to be non-trivial. Several attack steps are required to extract the secret key on a word-by-word basis. However, the existing literature does not describe the necessary details for this analysis to succeed in practice. For example, SHA-256 involves the non-bijective AND function and linearly dependent modular additions. The importance of Points of Interest (PoIs) detection in this context has not sufficiently been reported.

© The Author(s), under exclusive license to Springer Nature Switzerland AG 2024
A. Chattopadhyay et al. (Eds.): INDOCRYPT 2023, LNCS 14459, pp. 329–349, 2024.
https://doi.org/10.1007/978-3-031-56232-7_16

Related Works. The side-channel attack paths on HMAC have been discussed in the state-of-the-art. McEvoy *et al.* [13] present a complete theoretical attack on this scheme using Differential Power Analysis (DPA) in a Hamming Distance leakage model. This approach is generalized for a Hamming Weight leakage model and developed by Belaïd *et al.* [1]. Schuhmacher [16] further explores attacking the last round of SHA-2. Belenky *et al.* [2] present a template attack on a parallel hardware design. Benoit and Peyrin [3] present analyses for multiple candidates for the successor *SHA-3*. Oswald *et al.* [15] present attacking possibilities for *SHA-1*. PoI-detection has been widely used to improve SCA. Kim *et al.* [10] describe the usage of PoI to practically perform vertical SCA on *RSA*. Zheng *et al.* [18] stress the importance of PoI detection in practical SCA. Jungk and Bhasin [9] use Normalized Inter-Class Variance (NICV) to improve their SCA of the *ChaCha20* stream cipher. They also use a leakage model based on prior results to analyze modular additions. Furthermore, Do *et al.* [7] present two new techniques to extract PoI.

Contributions. In this paper, we propose some adjustments to the existing methodology for vertical SCAs on HMAC-SHA-2, and extend it to an applicable level in a leakage assessment context. We explore some practical constraints of SCAs on modular additions and logical AND operations and use PoI detection to make them exploitable. Indeed, we successfully apply our analysis methodology to electro-magnetic (EM) traces acquired on an STM32 Nucleo device, and show how the security evaluation of a target implementing HMAC-SHA-2 can be conducted. Besides, we establish that classical vertical SCAs as DPA [11] or Correlation Power Analysis (CPA) [5] are realistic threats to hashing-based cryptography.

Outline. Section 2 recalls the HMAC-SHA-2 protocol, and describes the attack path. Then, we expose the problems inherent to SCAs on modular addition and logical AND operations. In Sect. 3 we show real experiments conducted on an embedded SHA-2 software implementation. All SCAs required to break the secret key have been performed, providing examples of both, analyses targeting a modular addition and a logical AND operation. A comparison of our analysis and latest template analyses with respect to Common Criteria exigences is given in Sect. 4. We discuss the results and improvement possibilities in Sect. 5. Conclusions are given in Sect. 6.

2   Side-Channel Evaluation of HMAC-SHA-2

2.1   Preliminaries

Notations. Let $x, y \in \{0,1\}^\infty$ be sequences of bits and $n \in \mathbb{N}$ an integer. We write $x \parallel y$ the concatenation of x and y. The rotation by n bits to the right on x is denoted $x \ggg n$. For a bitwise logical AND (resp. OR) between the words x and y, we write $x \wedge y$ (resp. $x \vee y$). The $\oplus$ symbol denotes a bitwise eXclusive OR (XOR). The $\boxplus$ (*resp.* $\boxminus$) symbol denotes an addition (*resp.* subtraction) in $\mathbb{Z}/2^{32}\mathbb{Z}$ or $(\mathbb{Z}/2^{32}\mathbb{Z})^n$.

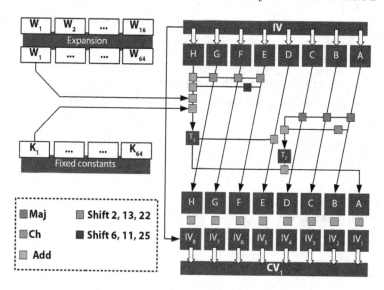

Fig. 1. Diagram of the SHA-256 compression function.

SHA-256. The Secure Hash Algorithm (SHA)-256 hash function [6] takes an input message M of arbitrary length and returns a 256-bit sequence. For that purpose, the message is split into N blocks of 512-bit size $M = M_0 \parallel \ldots \parallel M_{N-1}$. A compression function then operates on each block one by one.

The compression function $f_h : \{0,1\}^{(512)} \times \{0,1\}^{(256)} \rightarrow \{0,1\}^{(256)}$ takes two inputs: the message block M_n and a recursively defined chaining-value CV_n, where CV_0 is a fixed initial vector IV, and $\forall n \in \{1,...,N\}$, we have $CV_n = f_h(M_{n-1}, CV_{n-1})$. This function is called repeatedly until the final digest $d = CV_N = f_h(M_{N-1}, CV_{N-1})$ has been reached. During the compression, the algorithm operates on 32-bit words. The intermediate hash values are 8 words that we call $A^{(r)}, B^{(r)}, C^{(r)}, D^{(r)}, E^{(r)}, F^{(r)}, G^{(r)}, H^{(r)}$, for each round $r \in \{0,...,64\}$. Also, each message block is extended to 64 32-bit words, and since we are only analyzing one block, we write $M_n = W_1 \parallel \ldots \parallel W_{16}$. These words are expanded with the help of an `ExpansionFunction` to obtain one word per round. There are also 64 constants $K_1,...,K_{64}$ which are used at each round.

In addition, $A, B, C \in \{0,1\}^{32}$ being words, SHA-2 uses the following basic functions:

- $\Sigma_0(A) = (A \ggg 2) \oplus (A \ggg 13) \oplus (A \ggg 22)$,
- $\Sigma_1(A) = (A \ggg 6) \oplus (A \ggg 11) \oplus (A \ggg 25)$,
- $\mathrm{Ch}(A, B, C) = (A \wedge B) \vee (\neg A \wedge C)$,
- $\mathrm{Maj}(A, B, C) = (A \wedge B) \vee (A \wedge C) \vee (B \wedge C)$.

The compression function of SHA-256 is given in Algorithm 1, which is also illustrated in the schematic Fig. 1.

Input   : Message block $W_1, ..., W_{16}$,
            Chaining value $A^{(0)}, B^{(0)}, C^{(0)}, D^{(0)}, E^{(0)}, F^{(0)}, G^{(0)}, H^{(0)}$
Output : Chaining value
            $A^{(0)} \boxplus A^{(64)}, B^{(0)} \boxplus B^{(64)}, C^{(0)} \boxplus C^{(64)}, D^{(0)} \boxplus D^{(64)},$
            $E^{(0)} \boxplus E^{(64)}, F^{(0)} \boxplus F^{(64)}, G^{(0)} \boxplus G^{(64)}, H^{(0)} \boxplus H^{(64)}$

1    $(W_1, ..., W_{64}) = \texttt{ExpansionFunction}(W_1, ..., W_{16})$ // Precomputation of the 64 round words
2   for $r = 0$ to 63 do
3        $T_1 = H^{(r)} \boxplus \sum_1 (E^{(r)}) \boxplus \texttt{Ch}(E^{(r)}, F^{(r)}, G^{(r)}) \boxplus K_{r+1} \boxplus W_{r+1}$
4        $T_2 = \sum_0 (A^{(r)}) \boxplus \texttt{Maj}(A^{(r)}, B^{(r)}, C^{(r)})$
5        $H^{(r+1)} = G^{(r)}$
6        $G^{(r+1)} = F^{(r)}$
7        $F^{(r+1)} = E^{(r)}$
8        $E^{(r+1)} = D^{(r)} \boxplus T_1$
9        $D^{(r+1)} = C^{(r)}$
10       $C^{(r+1)} = B^{(r)}$
11       $B^{(r+1)} = A^{(r)}$
12       $A^{(r+1)} = T_1 \boxplus T_2$
13   return $(A^{(0)}, ..., H^{(0)}) \boxplus (A^{(64)}, ..., H^{(64)})$

Algorithm 1: SHA-256 compression function.

HMAC. In the HMAC algorithm [12], a secret key k^* and a hash function h are used to authenticate messages of arbitrary size. First, the key is padded to the correct block size with an inner padding $ipad = \texttt{0x36}, ..., \texttt{0x36}$. The messages are concatenated with this padded key and hashed. Then, the same key is padded again with an outer padding $opad = \texttt{0x5C}, ..., \texttt{0x5C}$, which is concatenated again with the result of the previous computation, and hashed again.

$$HMAC_h(M, k^*) = h((k^* \oplus opad) \;||\; h((k^* \oplus ipad) \;||\; M)) \qquad (1)$$

We define the hashed padded keys k_1 and k_2 as follows:

$$\begin{cases} k_1 = CV_1^{(1)} = h(k^* \oplus ipad) \\ k_2 = CV_1^{(2)} = h(k^* \oplus opad) \end{cases}$$

The aim of the analysis is to recover k_1 and k_2, for finally being able to forge Message Authentication Codes (MACs) without knowing the key.

To obtain all parts of k_1 (and k_2), one must perform a series of SCAs on different intermediate values of the second hash-block of each call of the hash function. Though two keys must be recovered, we analyze only the situation for k_1 without loss of generality, since the analysis of k_2 itself is identical [1].

2.2   Analysis Strategy

In the considered situation, the hash function h is SHA-256, but the general analysis path can possibly be adapted for any hash function of the SHA-2 family.

Table 1. The vertical SCAs that must be performed for this attack, with alternatives (6b and 7b) for some target values (resp. 6 and 7).

Attack	CSP	Targeted operation	Known parameter
SCA 1	$\delta^{(0)}$	$T_1^{(1)} = \delta^{(0)} \boxplus W_1$	W_1
SCA 2	$D^{(0)}$	$E^{(1)} = T_1^{(1)} \boxplus D^{(0)}$	$T_1^{(1)}$
SCA 3	$T_2^{(1)}$	$A^{(1)} = T_1^{(1)} \boxplus T_2^{(1)}$	$T_1^{(1)}$
SCA 4	$F^{(1)}$	$E^{(1)} \wedge F^{(1)}$ in Ch	$E^{(1)}$
SCA 5	$G^{(1)}$	$\neg E^{(1)} \wedge G^{(1)}$ in Ch	$\neg E^{(1)}$
SCA 6	$B^{(1)}$	$A^{(1)} \wedge B^{(1)}$ in Maj	$A^{(1)}$
SCA 7	$C^{(1)}$	$A^{(1)} \wedge C^{(1)}$ in Maj	$A^{(1)}$
SCA 8	$H^{(1)}$	$T_1^{(2)} = H^{(1)} \boxplus \epsilon^{(1)}$	$\epsilon^{(1)}$
SCA 9	$D^{(1)}$	$E^{(2)} = T_1^{(2)} \boxplus D^{(1)}$	$T_1^{(2)}$
SCA 6b	$D^{(3)}$	$E^{(4)} = T_1^{(4)} \boxplus D^{(3)}$	$T_1^{(4)}$
SCA 7b	$D^{(2)}$	$E^{(3)} = T_1^{(3)} \boxplus D^{(2)}$	$T_1^{(3)}$

The list of the necessary CPAs is given in Table 1. The attacks are enumerated in the first column. The second one shows the attacked Critical Security Parameter (CSP). In the third column we show the attacked operation and in the last one the variable parameter known by the attacker. The CPAs 1 to 8 are the ones described by Belaïd et $al.$ [1]. When all these CPAs are successful, the evaluator knows $A^{(0)} = B^{(1)}$, $B^{(0)} = C^{(1)}$, $D^{(0)}$, $E^{(0)} = F^{(1)}$, $F^{(0)} = G^{(1)}$ and $G^{(0)} = H^{(1)}$. For the sake of clarity, we define $\delta^{(i)}$ and $\epsilon^{(i)}$ as follows:

$$\delta^{(i)} = H^{(i)} \boxplus \Sigma_1(E^{(i)}) \boxplus \text{Ch}(E^{(i)}, F^{(i)}, G^{(i)}) \boxplus K_{i+1}$$

$$\epsilon^{(i)} = \Sigma_1(E^{(i)}) \boxplus \text{Ch}(E^{(i)}, F^{(i)}, G^{(i)}) \boxplus K_{i+1} \boxplus W_{i+1}$$

Note that there is a misleading formulation in the original paper. Belaïd et $al.$ [1] say that $C^{(0)}$ and $H^{(0)}$ could be computed through the already recovered values, which is untrue. While this is valid for

$$H^{(0)} = T_1^{(1)} \boxminus (\Sigma_1(E^{(0)}) \boxplus \text{Ch}(E^{(0)}, F^{(0)}, G^{(0)}) \boxplus K_1 \boxplus W_1),$$

we gain only partial knowledge about $C^{(0)}$ from

$$T_2^{(1)} = \Sigma_0(A^{(0)}) \boxplus \text{Maj}(A^{(0)}, B^{(0)}, C^{(0)}).$$

Only the bits where $A^{(0)} \neq B^{(0)}$ can be obtained systematically. One could argue that, if the security of a device depends on less than 32 secret bits, it can be compromised through brute force, but this would only work if there were not two consecutive series of SCAs that must succeed until one is able to verify the correctness of her assumptions. In our opinion, this is not sufficient since $C^{(0)}$ can not be computed by substitution, and in addition is unnecessary, since we are able to carry out a 9th CPA in the second round of the compression function. Here, we can gain absolute knowledge about $D^{(1)}$ which is $C^{(0)}$.

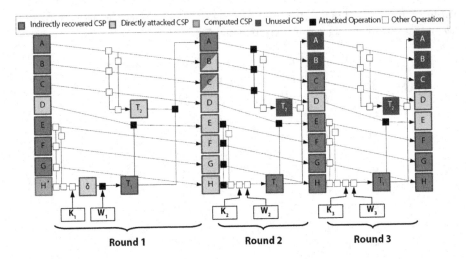

Fig. 2. The schematic analysis path. To recover the hashed key (chaining value), one targets the second hash-block, where the first words $W_1, ..., W_{16}$ correspond to the first 512 bits of the known message M. The red values are directly targeted with an SCA, while the blue values can be obtained indirectly (*i.e.* by substitution) from the recovered parameters. The orange value can be computed in the end from known parameters. (Color figure online)

Belaïd *et al.* [1] have already established that attacks on modular addition converge faster than on logical AND. In addition, as we will discuss later, the inherent structure of the logical AND operations present constraints that make it practically more difficult to analyze. Therefore we have been looking for alternatives in the analysis path for replacing CPAs 4, 5, 6 and 7 (computed on logical AND operations) by CPAs on modular additions. We have found such alternatives for SCAs 6 and 7 by recovering the value D in higher rounds.

We differentiate public parameters that are possibly known, and parameters that the security depends on, which are called CSP. Every value, except from the messages W_i and the constants K_i for $i \in \{1, ..., 64\}$, are CSP in this context. To recover a fixed CSP *via* SCA, it must be involved in an operation that contains a variable public parameter or known CSP. It is noteworthy that there are no more fixed CSP after the 4th round, hence all 8 words of k_1 must be recovered before that round. We give a schematic analysis path in Fig. 2.

Most of the targeted operations are situated in the first two rounds, the alternative path 7b is performed during the 3rd round and 6b during the 4th one. To perform the alternative analyses of CPAs 6 and 7, one must still be able to recover $F^{(1)}$ and $G^{(1)}$. We can not obtain $\delta^{(2)}$ as in the first round, because in the higher rounds this particular value is not a fixed CSP anymore.

2.3   Side-Channel Evaluation of the Internal SHA-2 Operations

Assuming that the Hamming Weight (HW) of a computations outcome can be recognized in the power consumption, we have studied the theoretical leakages of both, modular addition and logical AND operations. For any word $w = w_0, ..., w_{N-1} \in \{0,1\}^N$, the HW is defined as the number of strictly positive bits:

$$HW(w) = \sum_{i=0}^{N-1} w_i.$$

To get more realistic results, we applied a Gaussian distributed random noise $X \sim \mathcal{N}(0, \sigma)$ with mean 0 and standard deviation $\sigma = 5$ to the traces. For an operation $\circ$ and $a, b \in \mathbb{Z}/2^{32}\mathbb{Z}$, we defined the theoretical side-channel activity as:

$$HW(\circ(a,b)) = \sum_{i=0}^{31} \circ(a,b)_i + X.$$

For the analysis, we fixed one of the two entries of each operation, and chose $N \in \mathbb{N}$ random and varying values for the other parameter. Then we recovered the fixed value using a CPA with an evolving HW leakage model, depending on the targeted operation. All words are composed of 32 bits which we have split into *sub-words* of different sizes s of $1, 2, 4$ or 8 bits. For the two operations, we define a learning leakage model which uses the knowledge about the already recovered bits.

In the following, let $k \in \mathbb{Z}/2^{32}\mathbb{Z}$ be the CSP that we want to recover, and $m = m_0, ..., m_{N-1} \in (\mathbb{Z}/2^{32}\mathbb{Z})^N$ the vector of known variable words that the CSP is interacting with. For all sub-parts of the CSP that are indexed with $i \in \{0, ..., \frac{32}{s}\}$, the HW leakage model LM_i takes a hypothesis $k' \in \{0, ..., 2^s - 1\}$ and uses it on the corresponding part of the message, from the LSB to the MSB. This is then compared to the theoretical traces. For $i \in \{0, ..., N\}$ the HW leakage model is computed as:

$$LM_i(k', m) = \begin{bmatrix} HW(\circ(k', m_0 \ggg s \cdot i \bmod 2^s)) \\ \vdots \\ HW(\circ(k', m_{N-1} \ggg s \cdot i \bmod 2^s)) \end{bmatrix}. \tag{2}$$

At every step, the most likely key $\hat{k}_i$ is recovered *via* CPA and is then used to improve the following analyses. The evolutive leakage model ELM_i is defined as:

$$ELM_i(k', m) = LM_i(k', m) + \sum_{j=0}^{i-1} ELM_j(\hat{k}'_j, m). \tag{3}$$

We will show the theoretical results of these analyses for both cases, and discuss some problematic behavior that can be observed, when the same analyses are naively applied on real traces of the HMAC protocol.

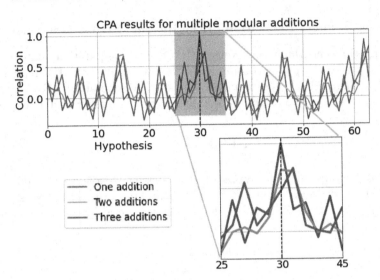

Fig. 3. CPA on a value ($k = 30$) that is involved in multiple additions with zoom on true key hypothesis. In case of only one addition (blue), there is a clear correlation peak at the correct value ($k' = 30$). With two additions (yellow), the peak is less significant, while with three additions (red), the peak does not correspond to the correct value (maximum correlation for $k' = 31$). (Color figure online)

Modular Addition. For an arbitrary but fixed value $w \in \mathbb{Z}/2^{32}\mathbb{Z}$, the addition with w is a bijective function:

$$\mathbb{Z}/2^{32}\mathbb{Z} \longrightarrow \mathbb{Z}/2^{32}\mathbb{Z}$$
$$x \longmapsto w \boxplus x = w + x \pmod{2^{32}} \tag{4}$$

When using the HW leakage model on the modular addition, the CPA on the theoretical traces detects the CSP with $10 - 100$ traces. Using hypotheses on more bits at a time leads always to faster results. The converging success rate for different s can be seen in Fig. 5a. In this ideal setting the result is not surprising, but when a parameter is used in multiple operations in a row, the correlation might be influenced. We stated a problematic behavior of the HW leakage model, when the analyzed value is used in multiple additions. For two additions, the maximum correlation did still occur in the target value, but the correlation peak of other candidates became more significant. For three or more additions, the maximum value did not correspond to the analyzed parameter. The correlations of an experiment with a fixed 6-bit value $k = 30$ is given in Fig. 3. The theoretical traces were generated by adding k to a random vector of known values $m_0, ..., m_N$ and two fixed disturbing factors, $d_0 = 15$ and $d_1 = 59$. For other values we obtained similar results. In some cases the value k could still be recovered, but the example shows that the leakage model does not always lead to the correct key.

Another attribute of this operation is the linearity. In a realistic attack situation, there are many other points that may depend on the known public parameter. Such values when used before or after the targeted operation are going to provoke correlation peaks for arbitrary incorrect hypotheses. In the specific SHA-2 case, the first SCA targeting $\delta^{(0)}$ uses a known variable parameter W_1, which by itself is already used during the ExpansionFunction. Furthermore the following values all can be obtained by the addition with a constant to W_1:

- $T_1^{(1)}$
- $E^{(1)} = F^{(2)} = G^{(3)} = H^{(4)}$
- $A^{(1)} = B^{(2)} = C^{(3)} = D^{(4)}$

These problems do not occur in the theoretical analysis, since we are analyzing only the one temporal instance that happened to be exactly the target operation. But during the computation of SHA-2 many modular additions involve a previously analyzed CSP. Therefore we can not simply choose the highest overall correlation to determine the most likely key. Instead, we must use PoI detection that we describe in Sect. 3.2 in order to find the exploitable points in time.

Logical AND. The logical AND operation is defined bitwise for $a, b \in \{0, 1\}$ as 1 if and only if $a = b = 1$. Otherwise the result is 0.

$$\wedge : \{0, 1\} \times \{0, 1\} \longrightarrow \{0, 1\}$$
$$b_0, b_1 \longmapsto b_0 \wedge b_1 \tag{5}$$

We define the logical AND for a fixed value $w = [w_0, ..., w_{31}] \in \{0, 1\}^{32}$:

$$\wedge_w : \{0, 1\}^{32} \longrightarrow \{0, 1\}^{32}$$
$$x \longmapsto w \wedge x = [(x_i \wedge w_i)_{i \in \{0, ..., 31\}}] \tag{6}$$

This function is not bijective which is the major problem, when it comes to SCA. At any position where the hypothetical CSP is 0, all information about the variable parameter gets erased in the leakage model. In a bitwise analysis, one has basically only one hypothesis ($k = 1$). It can not be evaluated for the 0 hypothesis, since the observed correlation is merely due to random noise. Regardless if it is higher or lower than the correlation of the 1 hypothesis, there is no information gain from such a result. Only the evidence for the 1 hypothesis is used to recover the secret bits. In Fig. 4, we see the theoretical outcome of such an analysis. The entire secret parameter can be recovered by observing the high and low phases of the correlation.

Another issue with logical AND operations is the context where these operations are used. In case of Maj, the expressions in question can be replaced by equivalent computations with different intermediate results (see Example 1).

Example 1. We can express the Maj function in different equivalent ways:

$$\begin{aligned} \text{Maj}(A, B, C) &= (A \wedge B) \oplus (A \wedge C) \oplus (B \wedge C) \\ &= (A \wedge B) \vee (A \wedge C) \vee (B \wedge C) \\ &= (A \vee B) \wedge (A \vee C) \wedge (B \vee C) \\ &= (A \wedge B) \vee (C \wedge (A \oplus B)) \end{aligned}$$

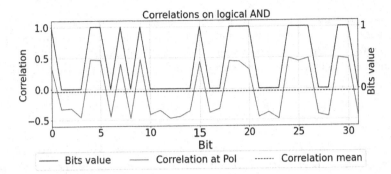

Fig. 4. Example of CPA applied on logical AND, with 1 hypothesis, compared to the real key bits. The real key bits can be guessed by classifying maximum correlation results. Correlations above the defined threshold correspond to the 1 value, while correlations below correspond to 0.

We do not always know which of the expressions have been used in the underlying implementation, and even if we did, some compilers (in case of software) may eventually compute the value with another expression. This makes it desirable to replace the SCA on logical AND operations by analyses on modular additions.

Just as for the modular addition, we tested the logical AND operation in a theoretical context with the HW leakage model but we accepted all 0 values, if no other hypothesis did show any significant correlation. The resulting success rate for different s is given in Fig. 5b. We see that, here, about 10,000 traces are required to guarantee a successful exploitation when using several bits at a time. Interestingly, the bitwise analysis is much faster $(100 - 1,000$ traces), unlike in the case of modular addition, where the bitwise analysis was the least effective.

3   Experimental Results

The approach in [1] describes how a vertical SCA can be performed. However, as there is no specification given about the implementation or the device performing the SHA-2 computation, we assume that the results are of purely theoretical nature. Our theoretical analyses of basic operations in Sect. 2.3 show the same results. This means that the side-channel activity is estimated with the same consumption model that is later used to analyze this operation. To get more realistic results, a Gaussian noise is added to this model. But proceeding in this manner implies that the activity does not contain other points than the PoIs. This section treats a real situation for a HMAC-SHA-256 embedded software, to which the analysis has been applied.

3.1   Experimental Setup

The whole attack path has been tested on a real scenario. Our target is a basic unprotected SHA-2 C implementation used in the HMAC-SHA-2 context. More

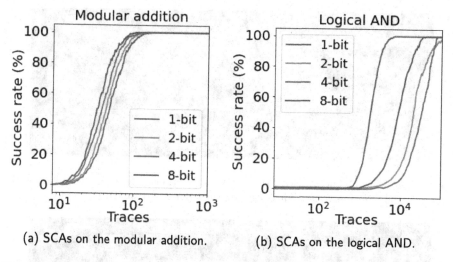

(a) SCAs on the modular addition.   (b) SCAs on the logical AND.

Fig. 5. The success rate of the analysis with the modular addition (left) and logical AND (right), and the HW leakage model for 1 bit, 2 bits, 4 bits, and byte-wise correlation with rising number of traces. All analyses converge eventually. Bigger subkeys lead to faster results, except for the bitwise AND leakage model, which converges much faster.

precisely, the SHA-2 messages were randomly chosen, while the HMAC key was fixed to a specific value. This padded, then hashed, key leads to the following SHA-2 chaining value:

$$A = \text{6699d416}, \quad B = \text{ba662d99}, \quad C = \text{a374edda}, \quad D = \text{a32bbf29},$$
$$E = \text{d5bc7a2e}, \quad F = \text{eeb700b3}, \quad G = \text{732617d2}, \quad H = \text{a4c49928}.$$

The target runs on an embedded CPU, the TOE being an STM32 NUCLEO-F334R8 with a Cortex M3 chip (Fig. 6). 50,000 EM traces of the first ten SHA-2 rounds (over 64) have been acquired using a Langer EMV-Technik RF-U 5-2 probe, and with a sampling rate of 2 GSa/s. Figure 7 is a screenshot of the used oscilloscope during acquisitions. It also shows one of the acquired SHA-2 EM activity traces. Not only can we see the repeating patterns, also a comparison with the later detected leakages (Sect. 3.2) permits us to confirm the delimitation of the different SHA-2 rounds.

3.2   Point of Interest Detection

Because of the linearity of the processed operations, the targeted values are arithmetically close to other processed values. Take for example a fixed key-dependent value x and a known variable y, the sum $x \boxplus y$ is processed, hence it will be recognizable with an appropriate leakage model. But if another value $y \boxplus \hat{x}$ is also processed, we can not know if we have found the correct x or the

Fig. 6. The STM32 Nucleo used for the EM acquisitions. The probe is put over the chip, and the CN7 pin is used for triggering the SHA-2 start and 10th round's end.

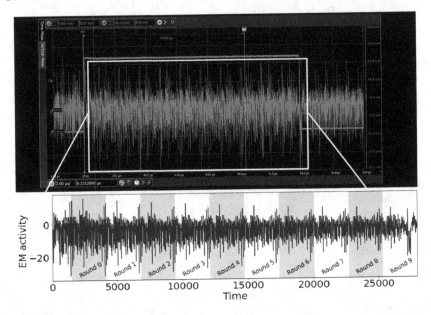

Fig. 7. The SHA-2 traces have been acquired on a Keysight MXR404A oscilloscope. The red signal corresponds to the SHA-2 activity, while the blue signal corresponds to the trigger which delimitates the first ten SHA-2 rounds, zoomed in and highlighted in gray and white. (Color figure online)

false friend $\hat{x}$. In particular when $\hat{x} = 0$, we have $y \boxplus \hat{x} = y$, which is a value that usually appears in the side-channel activity.

For handling this issue, we detect the PoI in advance to determine at which point in time the actual value is used. In this way, we can focus on the correlation peak at the PoI even if it is not the highest peak compared to other time samples. We can detect PoIs using general leakage detection metrics like Welsh's t-test [17], NICV [4] or Pearson-χ^2 [14]. This way, there is no need to evaluate the correlation at every point in time, but only where the CSP really leaks.

Modular Addition. We have used the NICV, which can detect leakages through operations that involve a variable parameter. For each CPA, we perform an NICV on the known parameter (Fig. 8a), and we evaluate the CPA results only at the points where leakages have been detected. Since for every target node multiple PoIs have been detected, the CPA trace was evaluated at multiple samples at once, taking the bit value indicated by the majority of the PoIs. This makes it preferable to perform the analysis bitwise rather than bytewise, even though the theoretical attack was faster when performed on bytes.

Logical AND. It is necessary to conduct PoIs detection of logical AND operations within a White-Box context (or with a particular profiling phase on a controlled clone device). Indeed, in this case, the NICV is not sufficient for detecting the best sample for a leakage evaluation (Fig. 8b and Fig. 9a). Therefore, we propose another way, which basically consists of the CPAs classification according to the real subkey value (0 or 1 within a bitwise analysis). When sorting the CSP, traces with respect to the real CSP, we can see that, at some samples, the correlations of the 1 hypothesis split up according to the true values. This is the PoI where the bitwise analysis can be performed. The area of ± 20 time-samples around the PoI of $B^{(1)}$ is illustrated in Fig. 9c. When the PoIs are detected, one can easily recover any CSP with the strategy described in Sect. 2.2. We show the results of the embedded SHA-2 software analysis in the following section.

3.3 Leakage Evaluation

For the leakage evaluation, we carried out the analysis within a White-Box context. In other words, we have used the knowledge about the secret parameters to find PoIs and evaluated the leakages for the possibility of exploitation. We do not claim that an attacker has these resources to carry out a White-Box profiling then exploiting phases[1]. Therefore, the hereafter presented analyses should rather be seen as a practical application within a security evaluation context.

The evolutive leakage model *ELM* (3) is confirmed in a real SCA experiment with an unprotected software implementation on the STM32 Nucleo. When experimenting on our leakage model in practice, we observe that it can even pass beyond the NICV in some cases, which is unusual. This, we assume is because for later subkeys the accumulated correct information about earlier findings is taken into account, while the NICV evaluates leakages of one subkey only. While a more profound analysis of this observation would be interesting, it is out of the scope of this work.

All target nodes have been analyzed and the CSP could have been recovered with the exception of one bit in the SCA-5, and one bit in the SCA 7*b* (which can be recovered by the SCA 7 anyway). We give SCA-8 with the bytewise analysis of $H^{(1)}$ as an example for a modular addition, and SCA-6 with a bitwise analysis

---

[1] This is also the sense of the Common Criteria quotations, where rating tables are used for estimating the Target Of Evaluation resistance against an attacker who would have particular means. Common Criteria call this the attack potential.

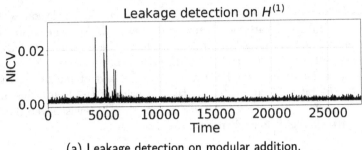

(a) Leakage detection on modular addition.

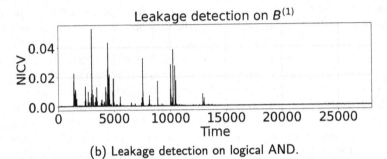

(b) Leakage detection on logical AND.

Fig. 8. PoI detection with NICV, for both a modular addition (here $H^{(1)}$) and a logical AND operation ($B^{(1)}$).

of $B^{(1)}$ as an example for a logical AND. The best round to analyze a CSP is not necessarily the round when it is computed. In case of $H^{(1)}$ every related value is situated in round 1, while $B^{(1)} = C^{(2)} = D^{(3)}$ can be analyzed in one of the subsequent rounds. Since the targeted and other linearly dependent values, are processed in all rounds from round 0 to round 3, one can observe leakages in all these rounds.

When the PoIs are detected, one can easily recover any CSP with the strategy described in Sect. 2.2. We show the results of the embedded SHA-2 software analysis in the following paragraphs.

CPA-8: Bitwise Analysis on Modular Addition. Figure 8a shows the NICV result. There is a peak at the 5792nd time sample and some other related activities in the area between the samples 4500 and 6800. These peaks confirm the presence of values related to the CSP. In Fig. 10, we see how one particular bit of $H^{(1)}$ is recovered at the PoIs. Even though the highest correlation peak often indicated the correct key value, this was not always the case as one can see from the first graph, hence the necessity of using the majority of the PoIs and not only one. Thus, all bits of

$$H^{(1)} = 732617d2$$

which is 01110011001001100001011111010010 when separated into bits, counting from right to left, have been found correctly following this method.

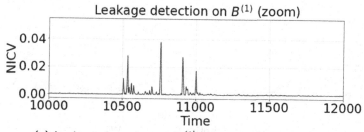

(a) Leakage detection on $B^{(1)}$ using the NICV (zoom).

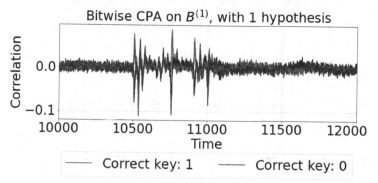

(b) CPA on $B^{(1)}$: the correct key bit can not be distinguished from the wrong one using the maximum CPA value.

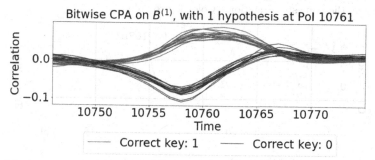

(c) Zoom on the CPA on $B^{(1)}$ at the PoI (10761) with 1 hypothesis: correlation curves can be classified according to the correct key bit.

Fig. 9. The bitwise analysis of a logical AND (targeting $B^{(1)}$). The NICV is used for detecting the PoI, which is then used for classifying the CPA curves according to the correct key bit.

CPA-6: bitwise analysis on logical AND. As expected, the 0 hypothesis can only detect random correlation at low level, whereas the 1 hypothesis leads to correlation peaks all over the area of interest (Fig. 9b). When we look closer at the PoI, we can see that the traces are distributed into high and low correlation

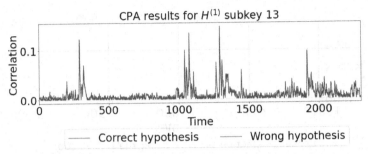

(a) CPA results of subkey 13, which is equal to 0.

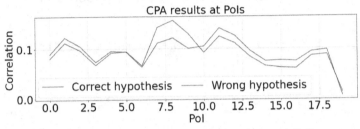

(b) CPA results of subkey 13, at 20 detected PoI.

Fig. 10. The bitwise analysis of $H^{(1)}$ with the use of NICV peaks (see Fig. 8a) as PoIs. One can see that the correct hypothesis (red curves) almost always is a little bit higher than the wrong hypothesis (gray curves). But at some points, the wrong key guess can even provide the highest correlation. (Color figure online)

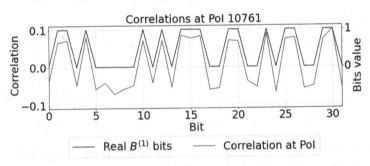

Fig. 11. CPA on logical AND with 1 hypothesis compared to the real key bits.

classes with respect to the CSP bits at the PoI. In Fig. 11 we can see how all bits of $B^{(1)}$ are recovered, as at the PoI the high activity indicates that the bit is 1, and the low activity is 0.

From LSBs to MSBs, we get the bits sequence

$$01100110100110011101010000010110$$

which is

$$B^{(1)} = A^{(0)} = \text{6699d416},$$

the correct CSP. All key parts have been successfully analyzed using the same method as we show in the next section.

4   Comparison

In this section we compare our results with different approaches. First we observe the shortcomings of a naive application of a classical CPA - that is, a CPA which ranks the hypotheses by maximum correlation values without considering the PoIs. After that, we provide quotation results for SCAs, on two HMAC-SHA-2 use-cases (profiled and unprofiled) to highlight the findings of our paper regarding the quotation of the different hurdles we identified in the attack.

Comparison to Naive CPA. Using the approach described in our paper, we can observe a success rate converging to 100% when evaluating the target with a rising number of traces in Fig. 12. We see that the logical AND converges much faster than the modular addition. However, since it was analyzed in a White-Box context one can not really draw the conclusion that these are easier. In comparison a classical CPA does not evolve to a better result for neither operation. In case of the logical AND the best hypothesis is always the one that does not erase any information about the variable parameter - which is ffffffff. Therefore the detected correct bits are all the bits that are 1 and we obtain the HW.

All key parts and alternatives have been analyzed using both methods. The results for all targeted operations compared to the case when adapting a classical CPA are summarized in Fig. 13a). On the one hand, with our method every key value has been recovered, with the exception of 1 bit. Also the alternative modular addition 7b failed to recover 1 bit. However, this has been found on the original logical AND analysis. The classical CPA on the other hand does not recover a single key part entirely with this amount of recorded side-channel activity. The convergence graphs suggest that there is no improvement when adding more traces, which can be explained by linear dependencies, ghost peaks and other structural problems we described in Sect. 2.3. In conclusion, the evaluation at PoIs is a crucial aspect of the analysis and not a mere improvement of an otherwise functional methodology.

Comparison to Template Attacks. Hereafter, we compare the attack potential of an unprofiled attack that is suggested by the works of McEvoy *et al.* [13] and Belaïd *et al.* [1], and that's practicability is discussed and improved throughout this paper to a template attack as proposed by Belenky *et al.* [2]. A direct comparison (Fig. 13b)) of the effectiveness *i.e.* with regard of required traces does not make a lot of sense, since the two approaches handle very different use cases. While our correlation based method is only applicable on software implementations, the template attack handles a parallel hardware design, which

346   L. Vlasak et al.

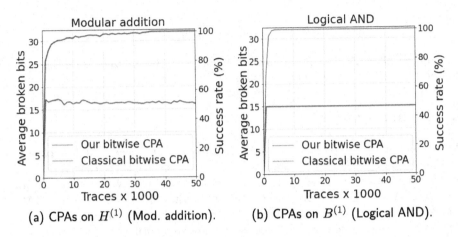

(a) CPAs on $H^{(1)}$ (Mod. addition).   (b) CPAs on $B^{(1)}$ (Logical AND).

Fig. 12. Success rates of CPAs on both modular addition and logical AND operation, on real EM traces. Our approach with PoI detection and *ELM* (3) (red) converges to 100%. The classical CPA with peak correlation using also *ELM* (blue) does not converge. (Color figure online)

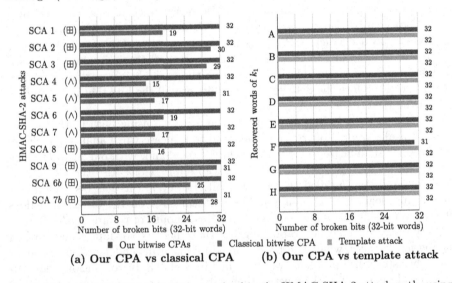

(a) **Our CPA vs classical CPA**   (b) **Our CPA vs template attack**

Fig. 13. Result of the different SCAs involved in the HMAC-SHA-2 attack path, using 50,000 EM traces. They have been tested with our enhanced bitwise CPAs for modular addition and logical AND, and with classical bitwise CPA. The recovered key parts A, B, C, D, E, F, G, H are compared to the ones recoverable through template attacks on a hardware implementation.

might not even be fully exploitable by CPA due to values that are "hidden in the combinational logic" [2]. However, as we have shown the our method requires about $50k$ traces for the PoI detection and exploitation of 99.6% of the CSP bits.

The template approach uses $3M$ traces to build 100% exploitable templates and requires at least $400k$ traces for a complete exploitation.

Therefor, we leverage the rating factors of Common Criteria, as per [8, Tab. 12, page 29]. This table leverages CC v3, because it makes the difference between *identification* and *exploitation* phases. Besides, we opt for the JIL version because it uses a sixth criteria, namely the availability of *open samples*, which is a prerequisite of the template attack [2]. In Table 2 the quotation for the attacks are compared.

We attributed a lower score to the unprofiled attacks in the identification phase of the "elapsed time" category, since PoI detection using NICV or comparable metrics are much easier to perform then data collection, removal of incorrect traces, and building the templates. On the other hand the exploitation of existing templates are almost instantaneous and require a low amount of traces. The "expertise" required for templates is slightly more than for correlation attacks. Both template analyses and correlation analyses are largely described in the open literature. However, we have demonstrated in this paper that unprofiled analyses require an educated choice of leakage models. Both, "Knowledge of TOE" and "Access to TOE" are basic in ether case. The only required knowledge is the knowledge of the algorithm to attack and only one sample (the attacked one) is required for the exploitation phase for both unprofiled and profiled attacks. Equipment is required in both identification (profiling stage) and exploitation (matching/attacking stage) for the profiled attack, whereas there is no identification phase for unprofiled attacks. Considering the rating of "Open samples", as the key is the most precious asset, we rate it as "critical" in the template attack, namely with 9 points. There is no such need for unprofiled attacks. Based on this analysis, we notice that unprofiled attacks constitute an easier attack path (16 points) than profiled attacks (23 points). Notice that the scores in Table 2 are only valid for vanilla implementations of HMAC-SHA-2. As soon as countermeasures are applied, the scores will rise, however wee deem it reasonable to assume, that the efforts increase similarly for unprofiled and profiled attacks.

5   Discussion

As stated before, this methodology is not a new attack but a White-Box evaluation method for vertical unprofiled SCA mitigation. It cannot directly be used to maliciously exploit leakages, as it requires knowledge about the key, to establish the PoI where the logical AND operations are targeted. However, a vertical side-channel attack might still be possible. In an evaluation context it is very useful to understand which are the practical constraints in such an analysis because, generally, a naive implementation of a correlation-based attack does not succeed.

A topic that deserves to be explored more fundamentally is the learning leakage model and the nature of misleading correlation peaks when multiple additions are involved. We have presented both briefly as they motivate a more sophisticated evaluation method. We also have not elaborated on masked or otherwise protected implementations of HMAC-SHA-2. For evaluations of such

Table 2. Quotation of the unprofiled and profiled SCAs on HMAC-SHA-2.

Factor	HMAC-SHA-2			
	Unprofiled analysis (from [1] and this paper)		Profiled analysis (from [2])	
	Identification	Exploitation	Identification	Exploitation
Elapsed time	1	3	3	0
Expertise	5	2	2	2
Knowledge of TOE	0	0	0	0
Access to TOE	0	0	0	0
Equipment	1	4	3	4
Open samples	0	-	9	-
Subtotal	7	9	17	6
Grand total	16		23	
Rating	Basic		Enhanced-Basic	

protections, one can use our method by evaluating his target with all countermeasures deactivated. When the analysis is successful, the masking can be reactivated and one can observe if the analysis becomes more difficult, or even impossible. Analyzing protected implementations and the possibilities to adapt CPA for hardware designs could be subject of future work.

6   Conclusions

We have shown on real EM emanations that vertical SCAs on HMAC-SHA-2 targets are not only theoretical, but can be carried out in practice. The required modifications for improving the theoretical analysis have been identified as PoI detection and evidence-based bitwise leakage models. As our methodology works on unprotected implementations and is easy to apply in a White-Box context, it can be used to evaluate the effectiveness of implemented countermeasures against SCA. To thoroughly apply such a testing method is of particular importance as our discussion about Common Criteria suggests. Generally unprofiled attacks are easier than profiled attacks. They are thus a main threat to consider and must be taken into account when evaluating a HMAC-SHA-2 implementation.

References

1. Belaïd, S., Bettale, L., Dottax, E., Genelle, L., Rondepierre, F.: Differential power analysis of HMAC SHA-1 and HMAC SHA-2 in the hamming weight model. In: Obaidat, M.S., Holzinger, A., Filipe, J. (eds.) ICETE 2014. CCIS, vol. 554, pp. 363–379. Springer, Cham (2014). https://doi.org/10.1007/978-3-319-25915-4_19
2. Belenky, Y., Dushar, I., Teper, V., Chernyshchyk, H., Azriel, L., Kreimer, Y.: First full-fledged side channel attack on HMAC-SHA-2. In: Bhasin, S., Santis, F.D. (eds.) COSADE 2021. LNCS, vol. 12910, pp. 31–52. Springer, Cham (2021). https://doi.org/10.1007/978-3-030-89915-8_2

3. Benoît, O., Peyrin, T.: Side-channel analysis of six SHA-3 candidates. In: Mangard, S., Standaert, F.X. (eds.) CHES 2010. LNCS, vol. 6225, pp. 140–157. Springer, Heidelberg (2010). https://doi.org/10.1007/978-3-642-15031-9_10

4. Bhasin, S., Danger, J.L., Guilley, S., Najm, Z.: NICV: normalized inter-class variance for detection of side-channel leakage. In: International Symposium on Electromagnetic Compatibility (EMC 2014/Tokyo). IEEE (12–16 May 2014), Session OS09: EM Information Leakage. Hitotsubashi Hall (National Center of Sciences), Chiyoda, Tokyo, Japan (2014)

5. Brier, É., Clavier, C., Olivier, F.: Correlation power analysis with a leakage model. In: Joye, M., Quisquater, J.J. (eds.) CHES 2004. LNCS, vol. 3156, pp. 16–29. Springer, Heidelberg (2004). https://doi.org/10.1007/978-3-540-28632-5_2

6. Dang, Q.H., et al.: Secure hash standard (2015). https://doi.org/10.6028/NIST.FIPS.180-4

7. Do, N.T., Hoang, V.P., Pham, C.K.: Low complexity correlation power analysis by combining power trace biasing and correlation distribution techniques. IEEE Access 10, 17578–17589 (2022)

8. Joint Interpretation Library: Application of Attack Potential to Smartcards and Similar Devices, Version 3.1 (2020). https://www.sogis.eu/documents/cc/domains/sc/JIL-Application-of-Attack-Potential-to-Smartcards-v3-1.pdf

9. Jungk, B., Bhasin, S.: Don't fall into a trap: physical side-channel analysis of chacha20-poly1305. In: Design, Automation & Test in Europe Conference & Exhibition (DATE), pp. 1110–1115. IEEE (2017)

10. Kim, H., Kim, T.H., Yoon, J.C., Hong, S.: Practical second-order correlation power analysis on the message blinding method and its novel countermeasure for RSA. ETRI J. 32(1), 102–111 (2010)

11. Kocher, P.C., Jaffe, J., Jun, B.: Differential power analysis. In: Wiener, M. (eds.) CRYPTO 1999. LNCS, vol. 1666, pp. 388–397. Springer, Heidelberg (1999). https://doi.org/10.1007/3-540-48405-1_25. http://dl.acm.org/citation.cfm?id=646764.703989

12. Krawczyk, H., Bellare, M., Canetti, R.: HMAC: keyed-hashing for message authentication. Technical report (1997)

13. McEvoy, R., Tunstall, M., Murphy, C.C., Marnane, W.P.: Differential power analysis of HMAC based on SHA-2, and countermeasures. In: Kim, S., Yung, M., Lee, H.W. (eds.) WISA 2007. LNCS, vol. 4867, pp. 317–332. Springer, Heidelberg (2007). https://doi.org/10.1007/978-3-540-77535-5_23

14. Moradi, A., Richter, B., Schneider, T., Standaert, F.: Leakage Detection with the χ^2-Test. IACR Trans. Cryptogr. Hardw. Embed. Syst. 2018(1), 209–237 (2018). https://doi.org/10.13154/tches.v2018.i1.209-237

15. Oswald, D.: Side-channel attacks on SHA-1-based product authentication ICs. In: Homma, N., Medwed, M. (eds.) CARDIS 2015. LNCS, vol. 9514, pp. 3–14. Springer, Cham (2015). https://doi.org/10.1007/978-3-319-31271-2_1

16. Schuhmacher, F.: Canonical DPA attack on HMAC-SHA1/SHA2. In: Balasch, J., O'Flynn, C. (eds.) COSADE 2022. LNCS, vol. 13211, pp. 193–211. Springer, Cham (2022). https://doi.org/10.1007/978-3-030-99766-3_9

17. Welch, B.: The generalization of "student's" problem when several different population variances are involved. Biometrika 34(1/2), 28 (1947)

18. Zheng, Y., Zhou, Y., Yu, Z., Hu, C., Zhang, H.: How to compare selections of points of interest for side-channel distinguishers in practice? In: Hui, L., Qing, S., Shi, E., Yiu, S. (eds.) ICICS 2014. LNCS, vol. 8958, pp. 200–214. Springer, Cham (2015). https://doi.org/10.1007/978-3-319-21966-0_15

Author Index

Printed in the United States
by Baker & Taylor Publisher Services